Discovering Electronics

An Active-Learning Approach

Roy Edwards
Senior Lecturer in Electronics at East Birmingham College

DP Publications Ltd
Aldine Place
London W12 8AW

1994

Author's Acknowledgements

The completion of this book has only been possible with the support the author has received from family, colleagues and students.

Firstly, I would like to thank Nick for all the help with the diagrams. Emma and Victoria for their word processing skills and my wife Sue for all her patience.

Secondly, I am grateful to John for his technical criticism and help with the practical tasks.

Lastly, but most importantly, my grateful thanks to all the students studying electronics at East Birmingham College for all their help, suggestions and constructive criticism.

Publisher's Acknowledgements

The publishers would like to thank Hameg Ltd for kind permission to reproduce the illustrations of the Hameg 203–6 on pages 7 and 8.

A catalogue record for this book is available from the British Library

ISBN 1 85805 088 X

Typeset by Adam Haynes, Ace Typesetting, Acton Burnell, Shropshire

Printed by Ashford Colour Press, Gosport, Hampshire

Preface

Aim

Discovering Electronics provides an activity-based approach to the study of electronics for students of BTEC National course NII Electronics modules. It meets the NII objectives and, because of its practical approach, will also be of great value to students studying the Part II (Analogue and Digital) CGLI 224 Electronic Servicing Course.

Need

Many institutions now run courses that have one or more of the following features

a) reduced lecturer contact time, and more student self-study.

b) recognition of prior learning and experience

c) multiple entry points to courses throughout the academic year.

This has resulted in the need for a new type of text book. A text book that will allow the students to have more control over their program of study, utilise fully any previous electronic knowledge or experience and allow the lecturer to delegate a proportion of the teaching process to the students themselves. To achieve this the text has to be interesting, motivating and practical.

The text has been written so that it can be used in any one of the following ways:

a) as a support text for a fully taught program of study

b) as a self-learning package with minimum lecturer support

c) as a combination of a) and b) in a negotiated programme of study, that can accommodate accreditation of prior learning.

Approach

The book is project driven, with all the BTEC objectives being met via practical and realistic tasks intended to build up confidence in the use of electronic equipment and components.

It assumes very little prior knowledge of electronics and provides all the background information, practical hints and suggestions (in the Section 2 Information and Skills Bank) to complete the projects.

It also, however, allows for prior learning, and students with previous knowledge and experience need only complete the three main design projects to demonstrate their competence and meet the NII aims and objectives.

The treatment is largely non-mathematical with the design projects being achieved through basic laws and rule of thumb. Wherever possible, an empirical approach has been adopted.

Structure of the book

There are three main parts to the book which deal with each of the main topics of the Electronics NII module:

Part 1: Power Supplies

Part 2: Amplifiers

Part 3: Digital Electronics

Each of these Parts is subdivided into two sections:

Section 1: The Design Project (including practical tasks and investigations)

This section contains all the activities needed to cover the main topic (ie either Power Supplies, Amplifiers or Digital Electronics) in the form of a single design project, which can be tackled as a whole (by students with prior knowledge) or as separate tasks and investigations (by students with little or no prior knowledge). Each task or investigation has a 'helpline' which refers the student to the appropriate unit of Section 2 (see below) to find the information or skills he/she needs to complete the task/investigation. All the tasks are essentially practical, and guidance is given to the correct outcomes where appropriate.

Section 2: Information and Skills Bank

This section contains all the necessary information to enable the practical tasks and investigations of Section 1 to be completed. It therefore covers all the requirements of the Electronics NII. It is organised into the topic units required for answering each of the Section 1 tasks and investigations, and each topic has an introduction and self-assessment questions or practical tasks (as appropriate to the topic) to help students check their understanding and to give further practice. (Answers to self-assessment questions can be found at the end of each part.)

How to use the book

The book is flexible enough to be used in a number of ways:

a) *As a support text for a fully taught course*

   With the aid of the syllabus/topic coverage charts at the beginning of each of the three parts, lecturers can plan to set the tasks and investigations of each of the projects after teaching the appropriate topics in class. Students can refer to the Section 2 Information and Skills Bank to back up their class notes. Tasks and investigations are ordered to give a natural progression, however they can be tackled in any order to suit a particular lecturer's teaching plan.

b) *As a self-learning package with minimum lecturer support*

   Students can work through the tasks and investigations of each project, acquiring the necessary knowledge and skills to complete each task via the references to the Section 2 Information and Skills Bank.

c) *As a combination of a) and b) in a programme of study that can accommodate prior learning accreditation and different entry points to the course*

   Students can be given specific tasks to accomplish on their own using the skills and information in Section 2 if they have not attended class teaching for these topics. Students with prior learning can be set specific projects or tasks to demonstrate their competence.

Part 1 Power Supplies is a prerequisite for Part 2 Amplifiers, but Part 3 Digital Electronics is independent of the other parts and may be studied at any time.

Depending upon your previous knowledge and experience you may be able to complete all the practical tasks and design projects within each part. If, however, you have little previous electronic experience, you will need to refer to Section 2, **The Information and Skills Bank**, before attempting each practical task.

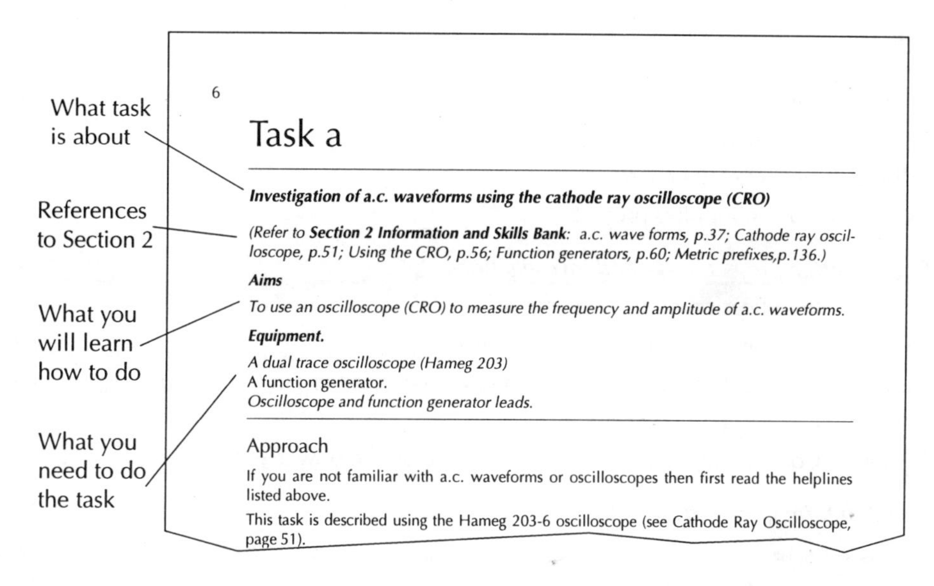

Practical equipment

It is possible to study Discovering Electronics and complete all three design projects with only the minimum of practical support. However, it is strongly recommended that at the beginning of each design project, investigations are made into what equipment is needed, what is available, and what help and support is needed with its use. The practical work is the key to the full understanding of electronics.

The equipment mentioned in this book is the basic standard minimum equipment that can be found in any electronics environment, e.g.

Test equipment:

- Function generator
- Logic pulser and probe
- Variable low voltage power supply
- Digital multimeter x 2
- Dual trace oscilloscope

Components:

- Breadboards and wires
- A range of capacitors
- Diodes, transistors
- A fully enclosed mains transformer (12V secondaries).
- A range of resistors
- Batteries
- FETs and logic chips

Contents

Part 1: Power Supplies

Contents

How to use this part

The book assumes very little previous knowledge or experience in electronics. However, it is assumed that you will have some basic electrical understanding, for example, a knowledge of basic electricity – voltage, current, resistance, power – and an ability to use Ohm's law and manipulate simple formulae related to voltage, current and power.

Students with little prior electrical knowledge

For these readers it is suggested that you look at the first practical task in Section 1 (investigation of a.c. waveforms) and then study in-depth the associated information and skills in Section 2. The text in Section 2 also includes some practical tasks. When you are confident with the self-assessment questions and practical tasks within Section 2, then attempt the first practical task. Continue in this manner until you have completed all the practical tasks in Section 1. You will by then have designed, constructed and tested a power supply!

Students with some prior electrical knowledge and experience

For example, a knowledge of a.c. waveforms, semi-conductor diodes, half-wave and full-wave rectifier circuits, and the use of test instruments.

You need to attempt all the practical tasks, referring to the associated information and skills only when necessary.

Students with considerable previous experience

You may immediately tackle the design project without first attempting the practical tasks.

Syllabus/Topic Coverage Chart

Electronics NII Syllabus topic		Tasks Section 1	Information and Skills bank Section 2
A-1-e	Uses a CRO to investigate power supplies.	*Task a* Investigate a.c. waveforms	a.c waveforms Cathode ray oscilloscope Using the cathode ray oscilloscope Function generators
A-1-a A-1-b A-1-c	P and N semi-conductors PN junction diode Characteristics of diodes	*Task b* Investigation of diode characteristics	Simplified semi-conductor theory Semi-conductor diodes Use of multi-range instruments How to use breadboards Resistors and resistor colour coding
A-1-e	Operation of power supplies	*Task c* *Investigation of half-wave and full-wave rectifier circuits*	Power supply specifications Half-wave and full-wave rectifier circuits Transformers Capacitance and capacitors
A-1-e A-1-f D-4-a D-4-d	Operation of power supplies Tests on power supplies Need for stabilisation Integrated circuit voltage regulator	*Tasks d and e* *Power supply design and construction*	Power supply specifications Half-wave and full-wave rectifier circuits Transformers Capacitance and capacitors Resistors and resistor colour coding Using the cathode ray oscilloscope How to use breadboards

Study Note: The Part 2 Section 2 Information and Skills Bank also needs to be studied to achieve the BTEC syllabus topic D-4-b objectives.

D-4-b	Series voltage regulator using 2 transistors and zener diode.		Regulator employing discrete components Bi-polar transistor The common emitter connection

Outline of Power Supplies project

You are a new employee of a small electronics firm (BIT & BYTE Ltd). Your training supervisor gives you a project to assess your knowledge and practical skills.

You are given the following information about the design project, together with access to a skills and information bank: electronic components, manufacturers' data, catalogues and test equipment.

Design project

An enquiry is received from SAS Alarms Ltd to quote for the supply of 1,000 small power supply units. These units are to provide the power for an alarm panel. Each unit should contain a step-down isolating transformer, a bridge rectifier and a voltage regulator. Your task is to design, construct and test a power supply unit that will meet the specifications given below.

Power supply specification

Range:	8.3V at 300mA
Ripple:	better than 5% at full load
Load regulation:	within 5%
Efficiency:	better than 20%
Line regulation:	A 10% change in the **input to the voltage regulator** should give less than 1% change in the output voltage. Note: normally, the line regulation would involve varying the mains (a.c.) input voltage to determine the effect on the output (d.c.) voltage. Since the measurement and variation of the mains voltage is undesirable from a health and safety viewpoint (only after adequate training should this be attempted), we will determine the line regulation of the actual voltage regulator only.

In order to help you with your project, your training supervisor has broken down the project in to a number of smaller tasks and investigations (see p4).

Part 1: Power Supplies

Section 1
Design project: tasks and investigations

Below is a summary of the practical tasks for this project, together with the associated cross references to the **Section 2 Information and Skills Bank**. The practical tasks follow a logical order of progression but, depending on your previous knowledge and experience (see above), completing all the tasks may not be necessary (unless your teacher instructs you to do so).

Summary of practical tasks

a **Investigation of a.c. waveforms** **page 6**

Use of the cathode ray oscilloscope and signal function generators.

Information and skills
a.c. waveforms
Cathode ray oscilloscope
Using the cathode ray oscilloscope
Function generators

b **Investigation of diode characteristics** **page 10**

Practical determination and plotting of diode characteristics, use of multi-range instruments.

Information and skills
Simplified semi-conductor theory
Semi-conductor diodes
Multi-range test instruments
How to use breadboards
Resistors and resistor colour coding
Metric prefixes

Remember!
If you already have some electronic knowledge and experience, then it may not be necessary for you to complete all the tasks (a) to (e) above. For example, if you are familiar with a.c. waveforms and the use of the CRO, then you can miss out **task a** and move straight on to **task b**. If, however, you have no previous knowledge, you will need to do all the tasks and refer to the **Section 2 Information and Skills Bank** when necessary.

Task a

Investigation of a.c. waveforms using the cathode ray oscilloscope (CRO)

*(Refer to **Section 2 Information and Skills Bank**: a.c. wave forms, p.37; Cathode ray oscilloscope, p.49; Using the CRO, p.56; Function generators, p.60; Metric prefixes,p.136.)*

Aims

To use an oscilloscope (CRO) to measure the frequency and amplitude of a.c. waveforms.

Equipment.

A dual trace oscilloscope (Hameg 203)
A function generator.
Oscilloscope and function generator leads.

Approach

If you are not familiar with a.c. waveforms or oscilloscopes then first read the helplines listed above.

This task is described using the Hameg 203-6 oscilloscope (see Cathode Ray Oscilloscope, page 51).

If you do not have access to a Hameg 203-6 CRO then you will need to read the operating manual of the CRO that you will be using. Other Hameg models do differ slightly and you should refer to their manual before use.

The method of approach (using the CRO) however will be the same as described below.

The Hameg 203-6 has the following main controls.

Main Controls (Hameg 203-6).

Use the key and the diagram of the front panel to identify all these controls on your CRO.

Description	*Key*
Power Switch	(1)
X-Timebase Switch [TIME/DIV]	(12)
Beam Intensity [INTENS	(2)
Focus	(3)
X-shift [X-POS] CH.1	(6)
Y-Amplifier [VOLTS/DIV] CH.1	(26)
CH.2	(31)
Y-shift [Y-POS1]	(21)
[Y-POS2]	(37)
Channel Inputs CH.1	(23)
CH.2	(35)

Obtaining a Trace on the CRO

A trace is a thin line across the screen. With a dual trace CRO you can obtain two lines and vary their position relative to each other with the Y-POS 1 and Y-POS 2 controls.

If all the CRO controls have been maladjusted, then on power up the trace is often missing.

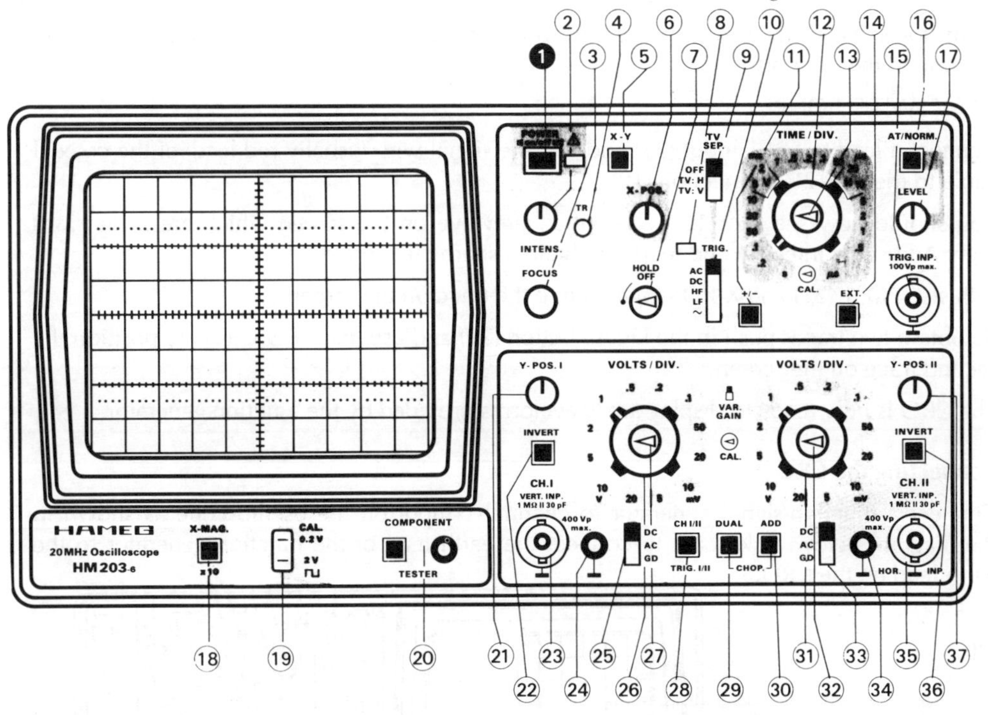

Front Panel of the Hameg 203–6

To Find A Trace

Before switching ON, check that all the following push buttons are in the OUT position.

PushButton	*Key*
Component Tester	(20)
Invert CH.1	(22)
Invert CH.2	(36)
X-Y	(5)
AT/NORM	(16)
EXT	(14)
+/–	(11)
CH.1/11	(28)
DUAL	(29)
ADD	(30)
X10	(18)

Now set the following controls as indicated:

TV SEP to OFF	(9)
TRIG to a.c.	(10)
Input switches to d.c.	(25) & (33)
Y-POS1 to mid position	(21)
Y-POS2 to mid position	(37)
INTENS to mid position	(2)
FOCUS to mid position	(3)
X-POS to mid position	(6)
TIME/DIV to 2ms	(12)
Variable X timebase to the CAL position (fully anti-clockwise)	(13)
VOLTS/DIV to 1volt/div	(26) & (31)
Variable volts controls to the CAL position, (fully anti-clockwise)	(27) & (32)

Note: for some Hameg models the CAL positions are fully clockwise. Look for the CAL symbol on the front display.

NOW SWITCH ON!

Connect a lead to CH.1 input (23) and clip the signal wire (usually red lead) of the coaxial cable to the CAL 2V connection (19).

You should now see 9 or 10 cycles of squarewave on the screen which are 2 divisions high. Adjust the intensity and focus to obtain a sharp clear trace.

Adjust Y-POS1 (21) and X-POS (6) to centre the trace on the screen.

To obtain two traces push in the DUAL button (29) and use the Y-POS2 (37) to position the second trace on the screen.

The CRO is now ready to display a.c. waveforms provided by the function generator.

Connecting the CRO

Connect a function/signal generator to channel-1 input on the oscilloscope as shown in the diagram below. Take care to connect the earth lead of the function generator to the earth lead on the CRO.

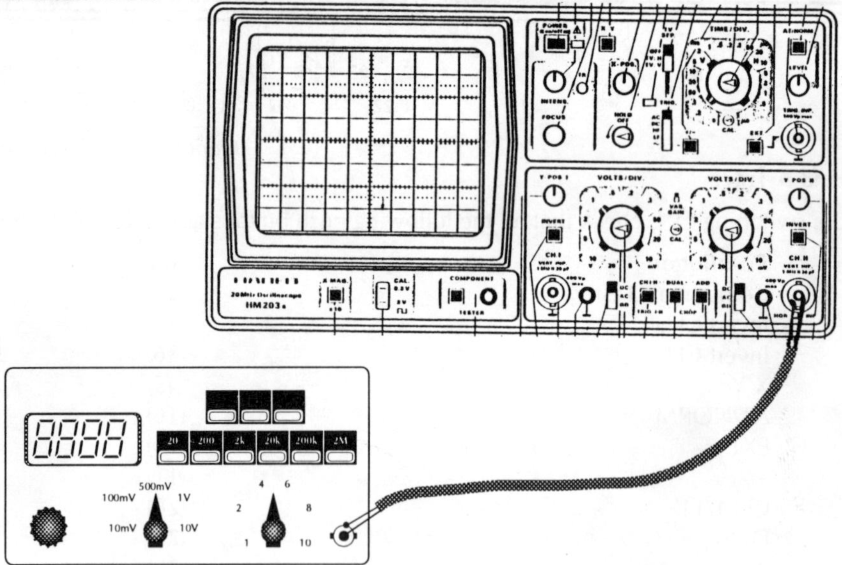

Connecting the Function Generator to the CRO

Displaying a Sinewave on the CRO

1) Adjust the signal/function generator to provide a 1kHz sinewave with an amplitude of about 1V. Adjust the Y-gain control (Volts/div) to obtain a waveform that is 2 to 3 divisions high. Now adjust the X-timebase (Time/div) control to obtain 2 to 3 cycles of the trace across the screen.

2) Check that the Y-gain and X-timebase calibration controls (CAL) are at their calibration settings.

   Ensure that the trace is locked. That is, the Trigger control (TRIG) is set to channel 1 (TRIG 1/11 button is out) and the external/internal trigger switch is set to Internal (EXT button is out).

3) Now use the X-position (X-POS) and Y-position (Y-POS) controls to centre the trace on the screen, e.g. position the trace so that the start of the cycle coincides with a vertical division line and the top peak of the waveform with an horizontal line.
4) Now read off from the vertical axis the number of divisions for the peak to peak amplitude of the sinewave and the number of divisions for one cycle of the sinewave.
5) Observe the Y-gain control setting (Volts/Div) and the X timebase (time/division) settings; then calculate the peak to peak amplitude of the waveform and the frequency

   i.e. use the relationship : $\text{Frequency} = \dfrac{1}{\text{Periodic time}}$

   to determine the frequency.

Further Exercises

(Answers are not supplied for these exercises as results depend on the equipment you use to perform them.)

Measurement of Amplitude

1) Set the signal generator to 1kHz. Set the Y-Gain control on the CRO to its minimum gain setting (e.g. 20V/div) and the output from the signal generator (amplitude) to its maximum setting.
2) Adjust (switch up the gain if necessary) the Y-gain control on the CRO until the amplitude of the displayed waveform is 2 to 3 divisions high.
3) Now determine the Peak to Peak amplitude of the displayed waveform.
4) Switch up the Y-Gain control one position, e.g. from 5V per division to 1V per division. Reduce the output from the signal generator until a waveform 2 to 3 divisions high is obtained.
5) Now determine the Peak to Peak amplitude of the displayed waveform.
6) Repeat methods (4) and (5) until the maximum gain setting is reached on the CRO consistent with the minimum output of the signal generator that gives 2 to 3 divisions of amplitude.

Measurement of Frequency

1) Set the signal generator frequency to 1kHz.
2) Adjust the amplitude control on the generator and the Y-Gain control on the CRO to give a display 2 to 3 divisions high on the screen.
3) Use the CRO to determine the periodic time and frequency.
4) Now adjust the frequency control on the generator to give approximately the following frequencies:

   50Hz, 100Hz, 500Hz,2kHz, 8kHz, 10kHz, 20kHz, 40kHz. 100kHz.
5) Use the CRO to determine the periodic times and frequencies.

Task b

Investigation of silicon diodes

(Refer to ***Section 2 Information and Skills Bank****: Simplified semi-conductor theory, p.65; Semiconductor diodes, p.75; Multirange test instruments, p.89; How to use breadboards, p.96; Resistors and resistor colour coding, p.99; Metric Prefixes, p.136.)*

Aim

To plot the forward and reverse characteristics of a silicon power diode and determine the dynamic forward resistance.
To plot the reverse characteristic of a zener diode.

Equipment

Low voltage power unit providing a variable d.c. output voltage 0-30V, up to 500mA.
A digital multirange meter capable of measuring a few volts d.c.
A digital multirange meter capable of measuring current down to a few microamps.
A breadboard and suitable wires.

Components

1Watt Resistors: 100R, 150R, 180R
1/2 Watt Resistors: 270R, 330R, 470R, 1k, 1k8, 8k2.
A IN4148 silicon diode (or similar type)
An 5.6V zener diode 1.3W (or similar type)

Tools

A pair of wire cutters/strippers and snipe nosed pliers.

Approach

Forward Characteristic

Breadboard the circuit shown below.

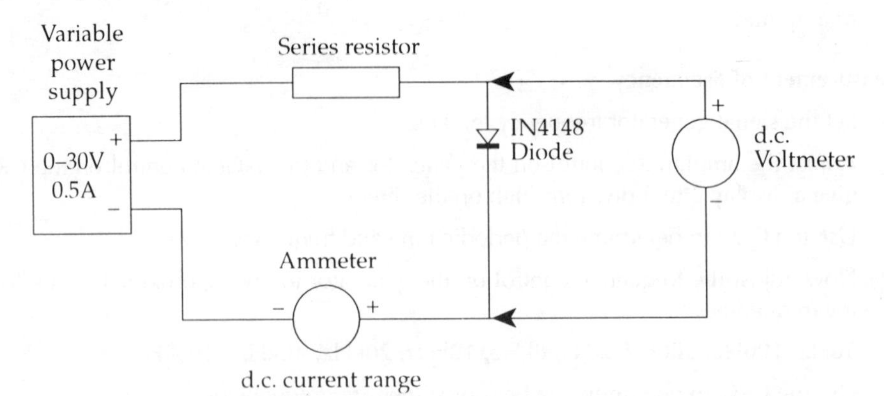

Circuit for determining the forward characteristic

Breadboard Layout

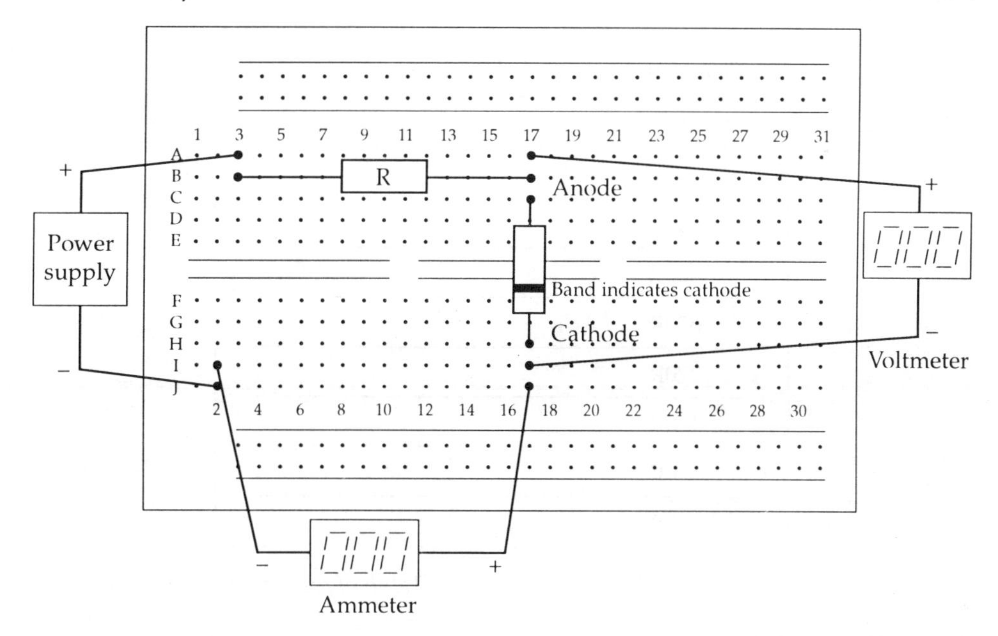

Suggested breadboard layout (diode forward characteristic)

1) Set the Voltmeter to 20V d.c. and the Ammeter to 200mA d.c. or higher ranges depending upon the instruments available.

2) Set the variable voltage supply to 10V and then connect it to the circuit.

   Note: With most laboratory variable voltage supplies there is also a current limit control. This control limits the maximum amount of current that the supply will provide if a short circuit occurs across the supply terminals. If you are using this type of supply, ensure that this control is advanced to provide at least 200mA.

3) Connect each of the series resistors in turn into the circuit and measure the current flowing I_F and the voltage across the diode V_F. Record your results in the table below. Note. You may have to switch ranges on your meters, depending upon the type you have available, to obtain more accurate readings.

Results Table

Series resistor	I_F (mA)	V_F (volts)
8k2		
1k8		
1k		
470R		
330R		
270R		
180R		
150R		
100R		

Table of results (forward characteristic)

Reverse Characteristic

1) Disconnect the power supply; reverse the diode; make the series resistor 100R (this is just to protect your power supply if a short circuit should occur).
2) Set the power supply to 5V d.c. and connect it to the circuit. Switch down the range on the ammeter and record any current flowing.
3) Increase the voltage of the supply in 5V steps, up to a value of 30V. Record the current flowing and complete the table below.

Power supply voltage	Reverse current I_R (μA)	Reverse voltage V_R (volts)
5V		
10V		
15V		
20V		
25V		
30V		

Table of results (reverse characteristic)

4) Now plot a graph of current against voltage for the Forward and Reverse Characteristics on graph paper.

Expected results

Forward Characteristic

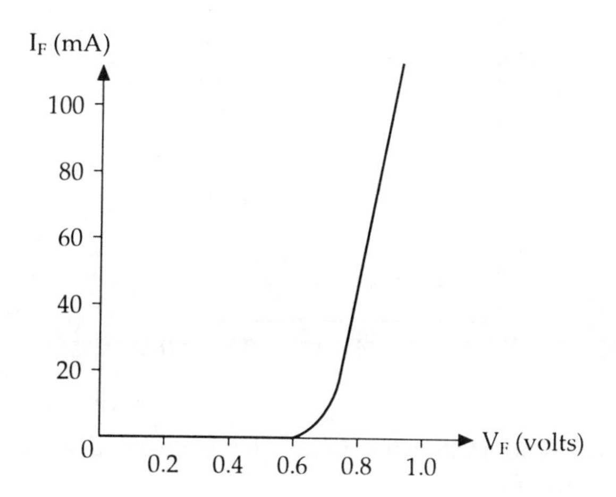

Forward Characteristic IN4148

Reverse Characteristic

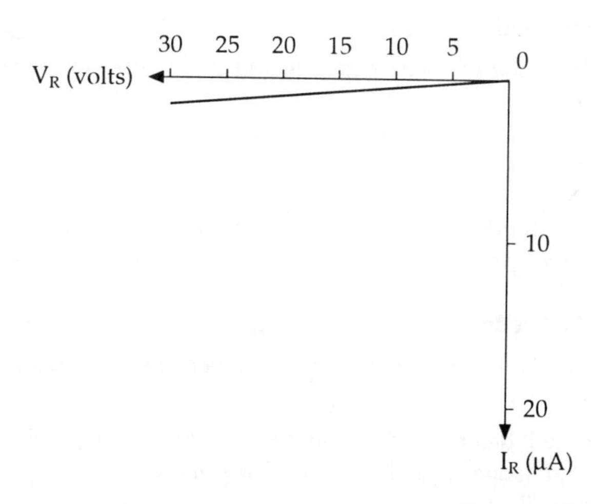

Reverse Characteristic. IN4148

The Forward Volt Drop

The forward characteristic shows that the current flowing through the diode determines the voltage across the diode.

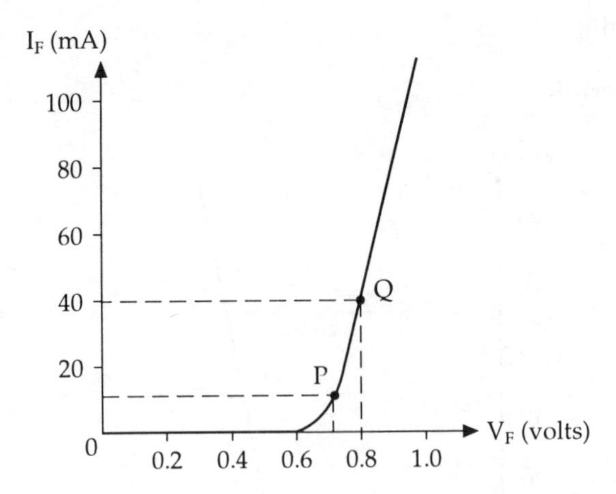

Forward Characteristic IN4148

If the forward current is 40mA then the diode is being operated at point Q and the voltage across the diode is about 0.8V.

If the current through the diode is reduced to about 10mA then the voltage across the diode is about 0.7V (point P).

d.c. Resistance of the Diode

The resistance of the diode will vary depending upon the operating point. At point Q the d.c. resistance will be:

$$R_{dc} = V_F/I_F = 0.8/40\text{mA} = \mathbf{20\ ohms.}$$

At point P the d.c. resistance = 0.7/10mA = **70 ohms**.

If the current through the diode is changing, due to an a.c. sine wave, from say 10mA to 40mA, then we need to work out the a.c. resistance offered by the diode.

a.c. Resistance of the Diode

This resistance is called the dynamic forward resistance and is the resistance offered by the diode to an a.c. signal.

The average current will determine the operating point Q of the graph. The a.c. will cause I_F and therefore V_F to change around this operating point Q.

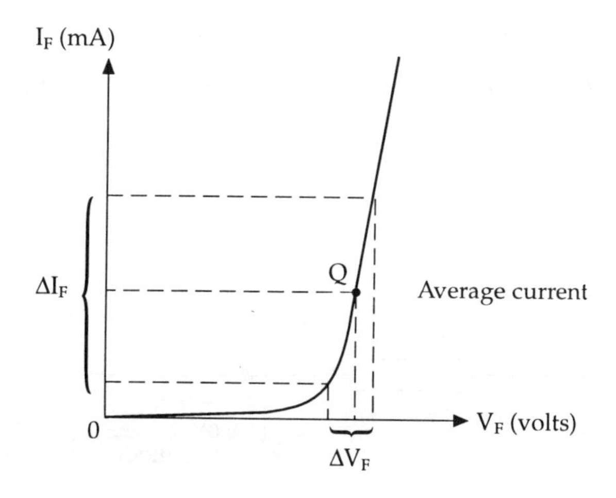

a.c. signal on the forward characteristic

The a.c. resistance is determined by:

$$r_f = \frac{\text{Small change in } V_F}{\text{Small change in } I_F} = \frac{\Delta V_F}{\Delta I_F} \text{ ohms}$$

Note: the symbol Δ means 'a small change in'

r_f is found by drawing a tangent to the characteristic at the operating point Q. A small change in I_F is then chosen and the corresponding change in V_F is read off from the X axis as shown below.

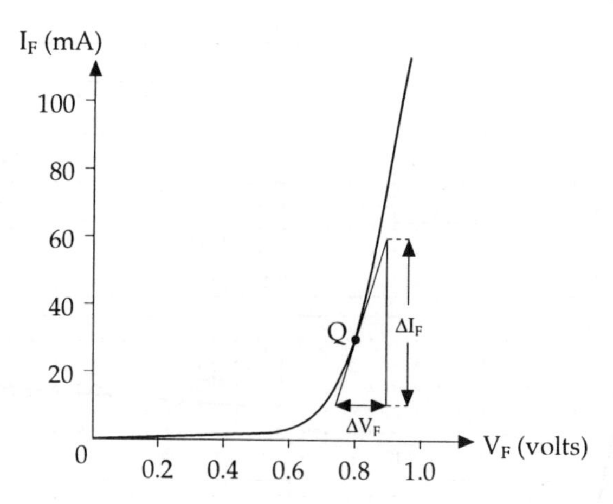

Determination of r_f

From the graph above; I_F varies from 10mA to 60mA and the corresponding change in V_F is from 0.76V to 0.88V.

$$\text{Therefore } r_f = \frac{0.88 - 0.76}{60 - 10} = \frac{0.12\text{V}}{50\text{mA}} = \mathbf{2.4\ ohms}$$

At the operating point (30mA average current) the diode offers 2.4 ohms of resistance to an a.c. signal.

From your graph determine the a.c. resistance when the operating point is 0.86V

The Zener Diode Characteristic

The reverse characteristic of a zener diode may be determined in a similar way.

Approach

Reverse Characteristic

Breadboard the circuit shown below.

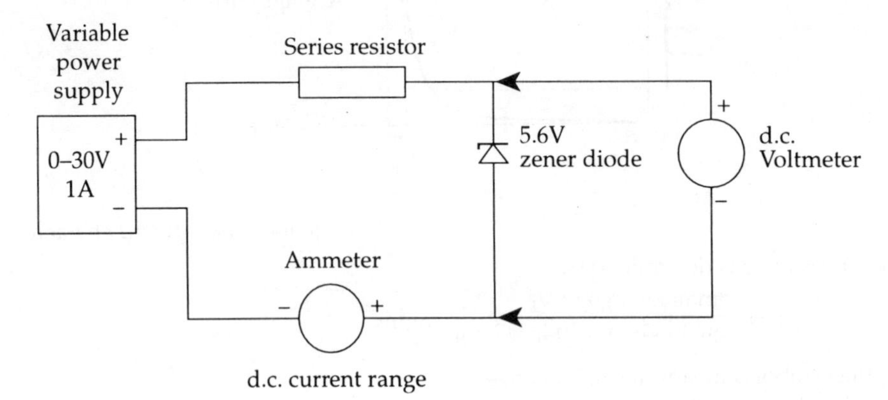

Circuit for determining the Reverse Characteristic

Breadboard Layout

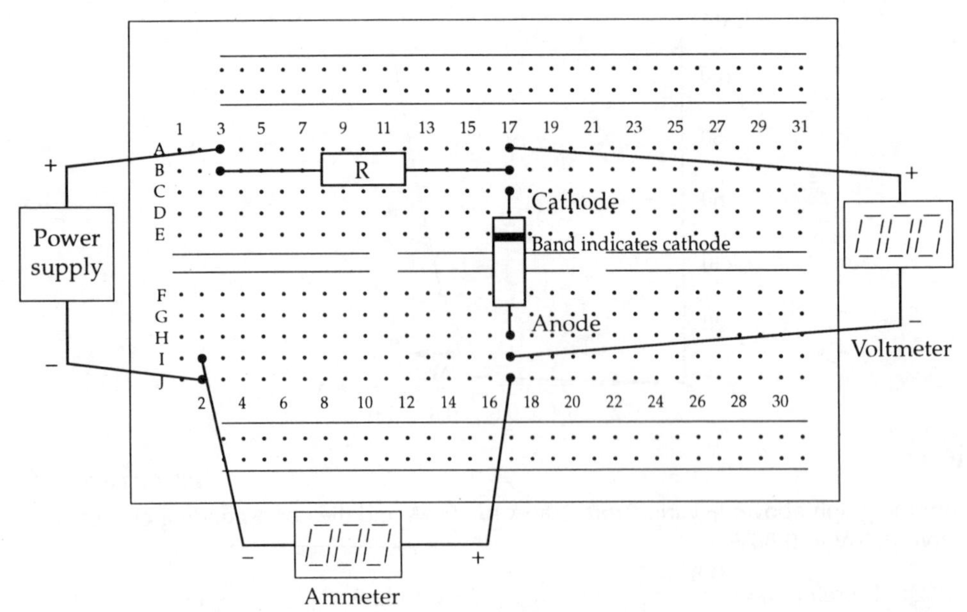

Suggested Breadboard Layout (zener diode reverse characteristic)

1) Breadboard the circuit as shown: connect the diode with the **cathode end** to the series resistor; make the series resistor 330R (this is a suitable value for either a 500mW or a 1.3W zener diode).
2) Set the power supply to 0V d.c. and connect it to the circuit. Switch down the range on the ammeter and record any current flowing.
3) Increase the voltage of the power supply in 5V steps, up to a value of 30V. Record the current flowing and the reverse voltage across the diode at each step. Then complete the table below.

Power supply voltage	*Reverse current I_R (μA)*	*Reverse diode voltage V_R*
5V		
10V		
15V		
20V		
25V		
30V		

Table of results (Reverse Characteristic)

4) Now plot a graph of reverse current against reverse voltage on graph paper.

Expected Graph

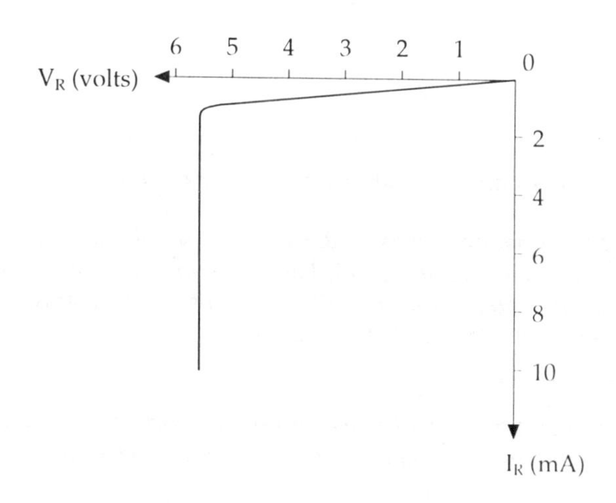

Reverse Characteristic. 5.6V Zener

Self Assessment Assignment

(An answer is not supplied for this assignment as results will vary according to components and equipment used.)

Look at the methods and procedures you have just used to plot the diode characteristics.

In a similar way, devise a procedure to plot the forward characteristics of the zener diode.

Produce a suitable circuit diagram.

Produce a breadboard layout (use the diagram below).

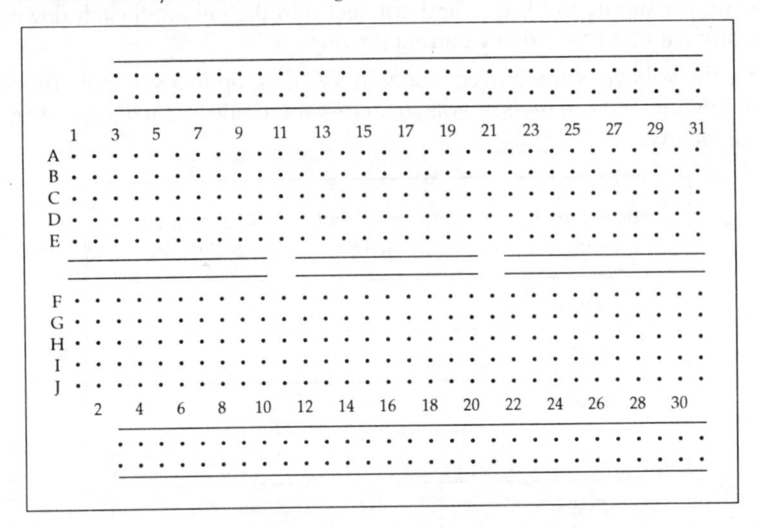

Breadboard Layout (zener diode forward characteristic)

Breadboard the circuit, obtain the results and plot the graph of the forward characteristic.

Compare the forward characteristic of the silicon diode with the forward characteristic of the zener diode. Is there any difference in the forward volt drop?

Task c

Practical investigation of half and full wave rectifier circuits

(Refer to **Section 2 Information and Skills Bank***: Power supply specifications, p.106; Half - wave and full-wave rectifier circuits, p.111; basic principles and operation of transformers, p.121; Capacitance and capacitators, p.130; How to use breadboards, p.96; Resistors and resistor colour coding, p.99; Metric Prefixes, p.136.)*

Aim

To investigate the operation of half wave, full wave and bi-phase rectifier circuits. To observe the effect of the reservoir capacitor on the output waveform.

Equipment

One double trace oscilloscope and suitable test leads: e.g. Hameg 203.
One multirange meter capable of measuring a.c. and d.c. voltages, and test leads.
One breadboard and different lengths of coloured wire suitable for the breadboard.

Components

A completely enclosed step down mains transformer with a centre tapped secondary winding; e.g. 240V/12V/12V. ***It is very important that NO live wire or terminal is accessible*** *(even with a screwdriver) on the mains input side of the transformer. The transformer must be connected to the mains via a suitable 13A mains plug which has a fuse of no more than 5A.*

4 silicon diodes: e.g. type IN 4005.
1 electrolytic capacitor 100µF 30V d.c.
1 1k0 resistor $^1/_2$W.

Tools

A pair of wire cutters/strippers and snipe nosed pliers.

Approach

Breadboard all the test circuits for the six tests in Task c in turn. (See **Section 2 Information and Skills Bank** How to use breadboards p. 96).

A suggested breadboard layout is given for each test. For each test circuit measure the a.c. voltage supplied to the rectifier circuit from the secondary of the transformer (V_s) and the d.c. voltage across the 1k0 load resistor with the multirange voltmeter.(V_L) (see **Section 2 Information and Skills Bank** Multirange test instruments p.89)

For each circuit connect the CRO to the points indicated. Observe and sketch the a.c. input waveform and the output waveform from the CRO. Take care to record the phase relationships correctly for test 4 and note the peak voltage accurately (see **Section 2 Information and Skills Bank** The Cathode ray oscilloscope and Using the cathode ray oscilloscope pp.51–59).

Caution: Remember to have the meter on the correct voltage range and the correct setting; e.g. measure the 12V a.c. on a 20V a.c. range or higher and use at least a 20V d.c. range for the d.c. measurements.

Tip: Always choose a higher range than required and switch down as necessary.

Take care to connect the leads correctly on the CRO. The inner of the coaxial lead is the signal wire and the braid the earth.

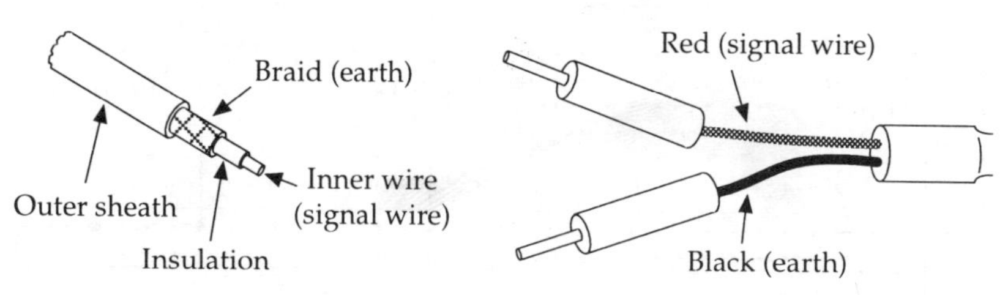

Connections to a coaxial lead.

Most low cost dual trace CRO's have a common earth connection so that if you earth one braid lead on one trace then the other trace is automatically earthed.

Therefore, when using both CRO leads in this task ONLY CONNECT ONE OF THE EARTHS. Leave the other unconnected. This is particularly important for the tests on the bi-phase and full wave circuits. If you use both earth leads then you may earth two parts of the circuit which are at different potentials and cause a short circuit to occur.

Tests 1,2 and 3: Half Wave Circuits

Test 1

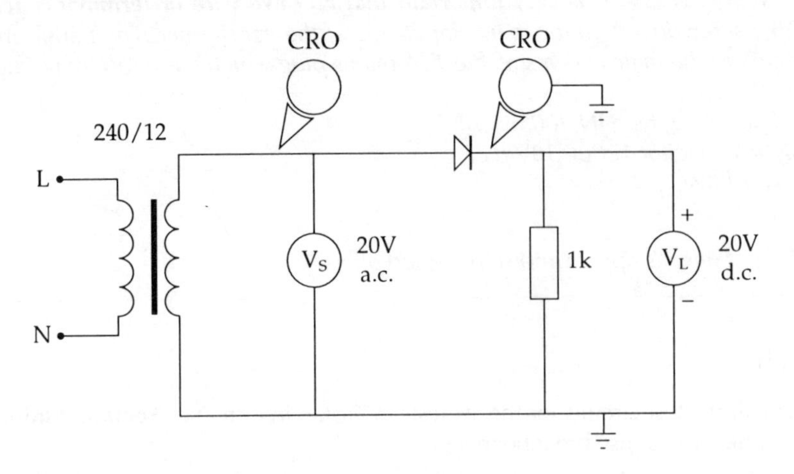

Half wave circuit (negative ground)

Suggested Board Layout for Tests 1,2 and 3

Note:

The 12V a.c. must only be obtained from a completely enclosed mains transformer. The diode has to be reversed for Test 2. The capacitor has to be included only for Test 3.

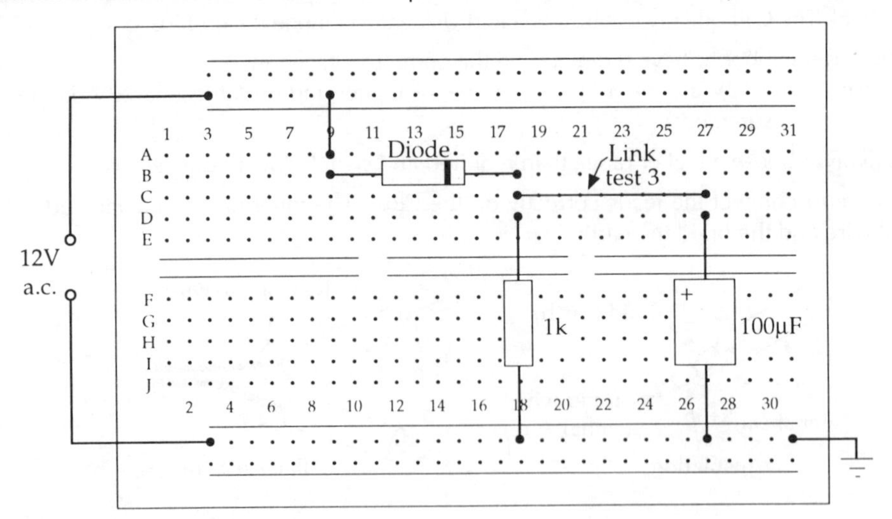

Test 1 Results

From your test circuit!

V_S = ________ r.m.s. V_L = ________ d.c.

Now draw waveforms of V_S and V_L to scale!

Test 2

Use the board layout used in Test 1, and reverse the diode.

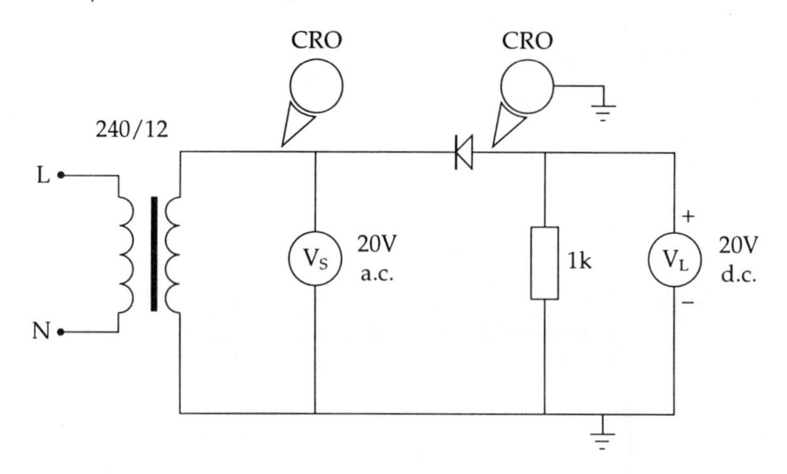

Half wave circuit (positive ground)

Test 2 Results

From your test circuit!

V_S = _______ r.m.s. V_L = _______ d.c.

Now draw waveforms of V_S and V_L to scale!

Test 3 Half Wave circuit: the effect of reservoir capacitor

Use the board layout of Test 1. Reverse the diode again and connect the reservoir capacitor across the load resistor.

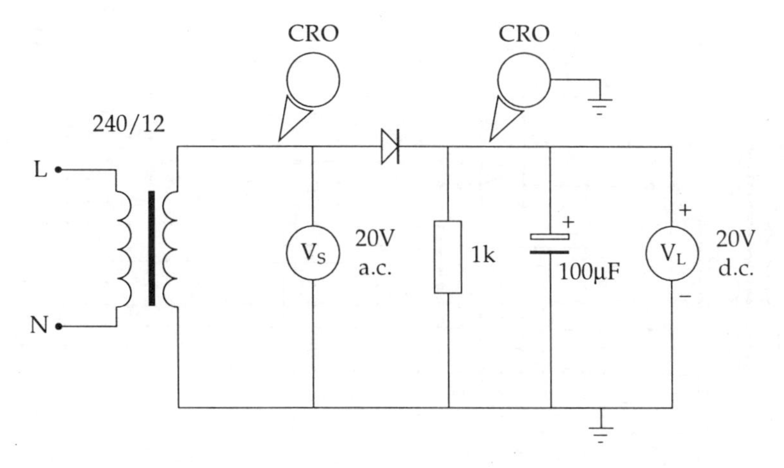

Half wave circuit with reservoir capacitor.

Test 3 results

From your test circuit!

V_S = _______ r.m.s. V_L = _______ d.c.

Now draw the load voltage waveform V_L and waveform V_S.

Test 4. Full wave bi-phase circuit

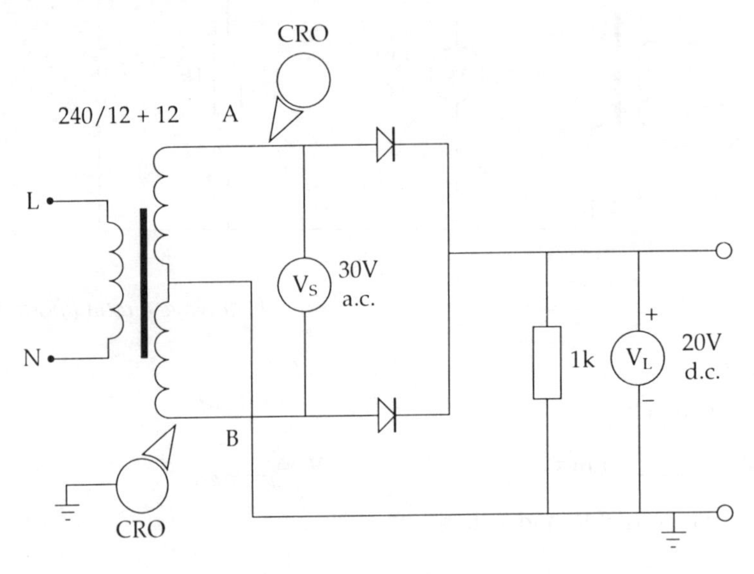

Full Wave bi-phase circuit (negative ground).

Board Layout for Test 4

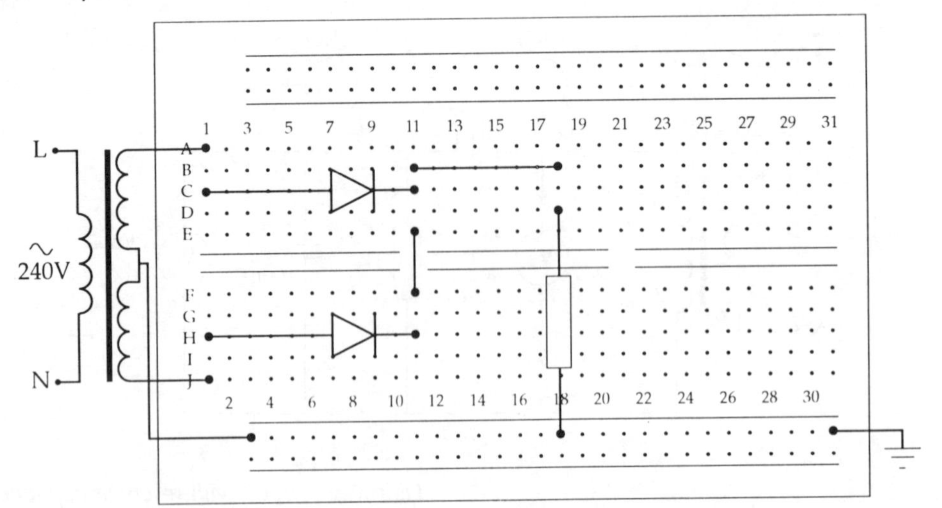

Test 4 results

From your test circuit!

V_S = _______ r.m.s. V_L = _______ d.c.

Now draw waveforms of the voltage at point A, point B and V_L to scale.

Test 5: Full Wave Bridge Circuit and Test 6: Effect of Reservoir Capacitor

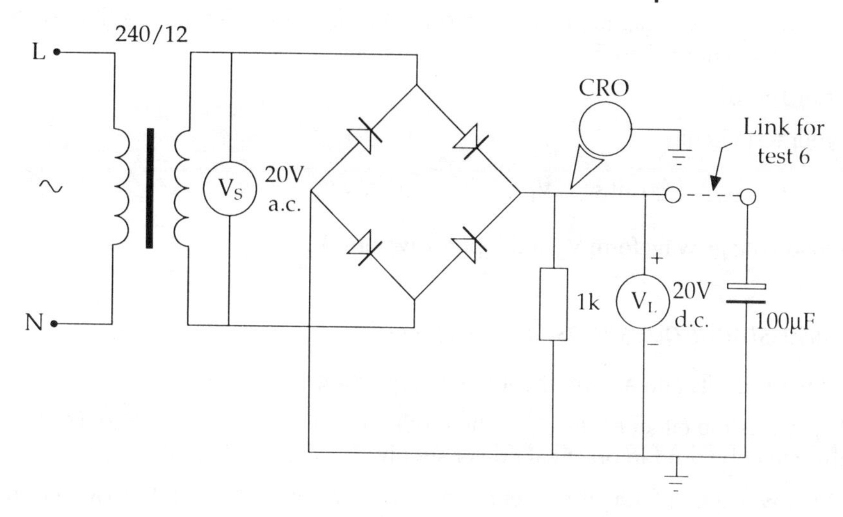

Full Wave Bridge circuit. (negative ground)

Board layout for Tests 5 and 6

For Test 6 the link is inserted as shown to connect the reservoir capacitor into the circuit.

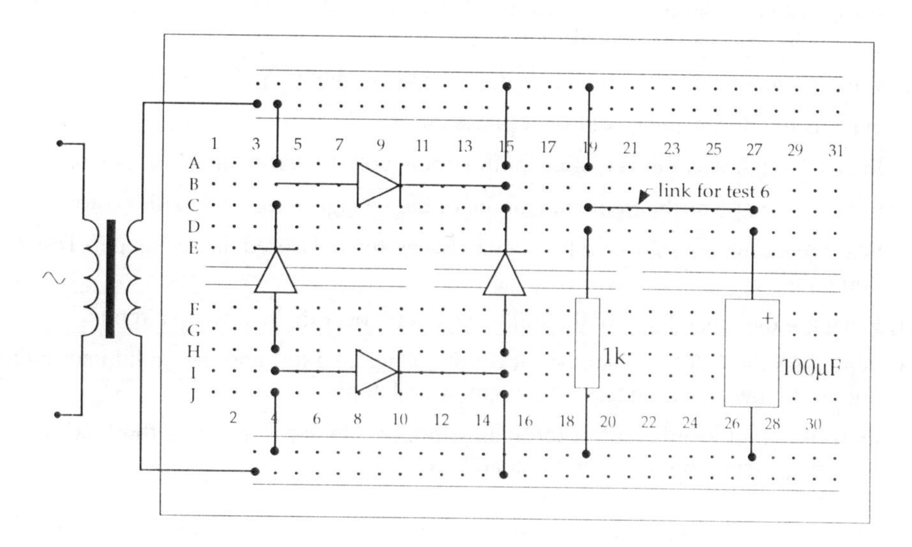

Results of Test 5

From your test circuit!

V_S = ______ r.m.s. V_L = ______ d.c.

Draw load voltage waveform V_L and input waveform V_S.

Test 6: Full wave bridge circuit – effect of reservoir capacitor

Connect the reservoir capacitor across the load resistor by inserting the link as shown in the circuit diagram for Test 5.

Results of Test 6

From your test circuit!

V_S = ______ r.m.s. V_L = ______ d.c.

Draw load voltage waveform V_L and input waveform V_S.

Self-assessment questions (answers page 143)

Look at your results and answer the following questions:

1) What was the effect of reversing the diode in Test 2 compared to Test 1? How would this effect be used in practical power supply?
2) What was the difference between the waveforms of the load pulses (waveform V_L) in Test 1 compared to the load waveforms in Tests 4 and 5.
3) What was the period of the load pulses (waveform V_L) in Tests 4 and 5? How does this compare with the period of one cycle of the mains supply?
4) The bridge circuit uses 4 diodes and the bi-phase only 2; since the volt drop across a silicon diode is about 0.7V, was there any measurable difference between the peak voltage in Test 4 and the peak voltage in Test 5?
5) What is the advantage of the bi-phase circuit over the bridge circuit?
6) What is the disadvantage of the bi-phase circuit?
7) In Test 3 what was the frequency of the ripple voltage across the load resistor?
8) In Test 6 what was the frequency of the ripple voltage across the load resistor?
9) Was there any measurable difference between the measured d.c. voltage in Test 3 and Test 6?
10) What are the advantages of the full wave circuit over the half wave circuit?
11) Why does the CRO indicate one level of voltage in a test and the multimeter indicate another? How are these two measurements related?
12) In Tests 1 and 2 only half of the mains input cycle appears across the load. Is this a way of reducing the power in the load to a half?

Task d

Design a suitable circuit for the power supply to meet the given specifications using manufacturers data

(Refer to ***Section 2 Information and Skills Bank****: Power supply specifications, p.106; Half and full wave rectifier circuits, p.111; Basic principles and operation of transformers, p.121; Capacitance and capacitators p.130; How to use breadboards, p.96; Resistors and resistor colour coding, p.99; Metric Prefixes, p.136.)*

Aim

To design a power supply to the following specification: a d.c. stabilised power supply to provide 8.3V at 300mA. It must incorporate a bridge rectifier, a step down isolating transformer and overload protection in the secondary circuit.

Equipment

Paper, pencil and calculator. Manufacturers' catalogues/data sheets.

Approach

Before you can design the power supply there are a number of things which we need to consider. These are outlined below.

Peak Input Voltage

We need to know what the peak input voltage to the regulator will be. This input voltage is the voltage delivered by the transformer to the rectifier.

R.M.S and Peak Voltage

The voltage specified by the transformer manufacturers is the Root Mean Square (r.m.s.) voltage but the maximum voltage fed into the diode bridge will be the peak of this voltage, i.e. 1.414 x r.m.s. voltage.

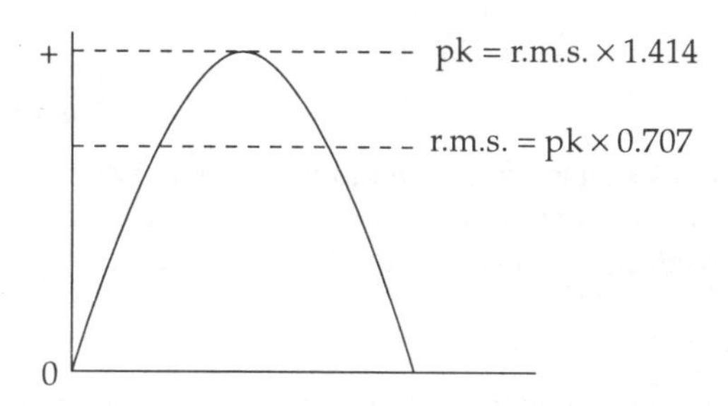

r.m.s and peak voltages of a sine wave.

Diode Bridge Volt Drop

Most diodes produce a forward volt drop of about 0.7V when they are conducting. Because in a full wave bridge circuit there will always be two diodes conducting, this will introduce a volt drop of about 1.4V so the voltage fed to the regulator will be less than the peak value calculated.

Choice of Diodes (bridge rectifier)

The diodes must be capable of carrying the maximum load current and be able to withstand twice the peak mains voltage in the reverse direction; i.e. if the unit is to supply 25V then the peak voltage will be 35.35V and the peak inverse voltage of the diode must be 70.7V. The popular IN4001 diodes are rated at 50V at 1A and would not be suitable in this case.

Minimum Voltage to the Regulator

The minimum input voltage specified by most I.C. regulators is 2.5V higher than the output voltage; e.g. a 5V regulator will require a minimum input of 7.5V. The input voltage to the regulator must not fall below this value.

The required minimum input to the regulator

Ripple Voltage

The difference between the maximum output voltage from the bridge circuit and the minimum input voltage to the regulator is the maximum ripple voltage allowed.

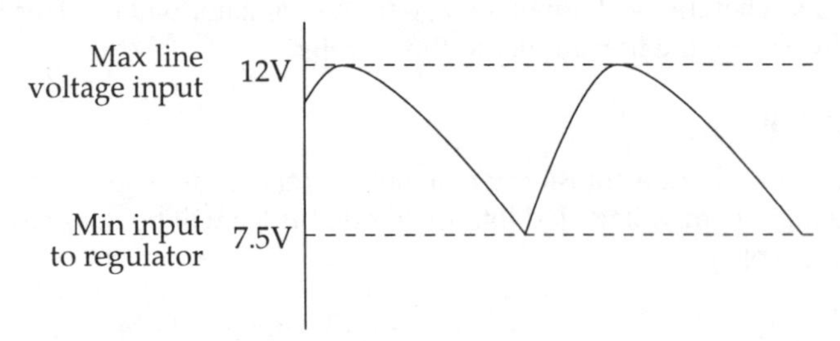

The maximum ripple voltage

In other words, if the peak voltage from the diode bridge is 12V and the minimum input to the regulator is 7.5V then the maximum ripple will be 4.5V. This is the maximum voltage swing that can be allowed on the reservoir or filter capacitor.

Period of the Mains Pulse

The full wave rectified mains will provide pulses at twice the mains frequency (100Hz). The period of each pulse will therefore be 0.01 seconds.

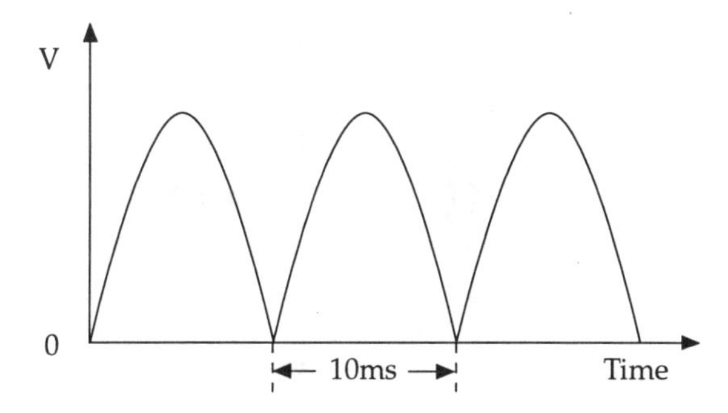

The periodic time of one cycle of the mains supply

Value of Filter Capacitor

A charged capacitor in a circuit which has resistance will take a definite time to discharge the voltage stored on its plates. This time is determined from the time constant of the circuit.

$$t = CR \text{ seconds}$$

Where t = time (secs); C = capacitance (Farads); R = resistance (ohms)

The resistance of a circuit can be determined from Ohms' Law.

$$R = \frac{V}{I} \text{ ohms}$$

If we use this value of resistance in the time constant formula we get:

$$t = C \times \frac{V}{I} \qquad \left(\text{R replaced with } \frac{V}{I}\right)$$

Where: I is the load current; t is the time over which the capacitor is discharged (the time between pulses of mains voltage); V is the voltage lost by the capacitor in time t.

Because the input voltage to the regulator must not fall below a minimum value then V must be equal to V_{RIPPLE}, the maximum allowed ripple voltage.

If we know the maximum load current, the period of the mains pulses and the maximum allowed ripple voltage, then we can calculate the minimum capacitance required:

$$\text{i.e. } C = \frac{t \times I}{V_{RIPPLE}} \text{ farads}$$

Worked Example

Assume we are to design a power supply to provide 7.7V at 400mA.

Choice of Transformer

We will need an input voltage to the regulator in excess of 10V. Therefore we chose a transformer with a 12V secondary (or two 6V windings connected in series).

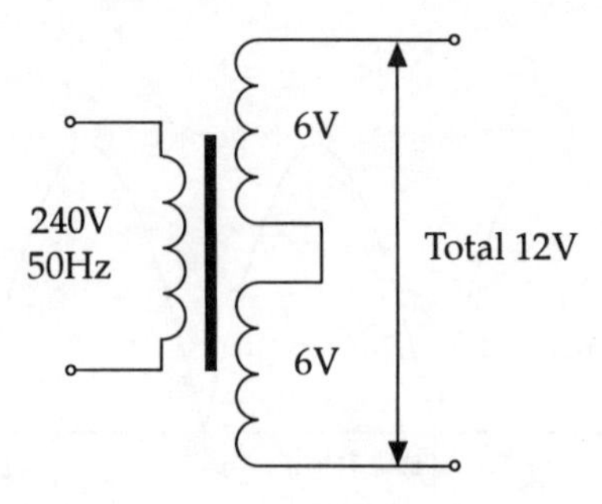

Connecting the secondary windings in series

Power Rating

This can be determined from the simple power formula:

$$P = I \times V$$

Therefore the VA rating of the transformer will be:

P = 400mA × 12V
= 4.8 VA × 1.1 (to account for losses).
= **5.28VA**

Input Voltage to Regulator

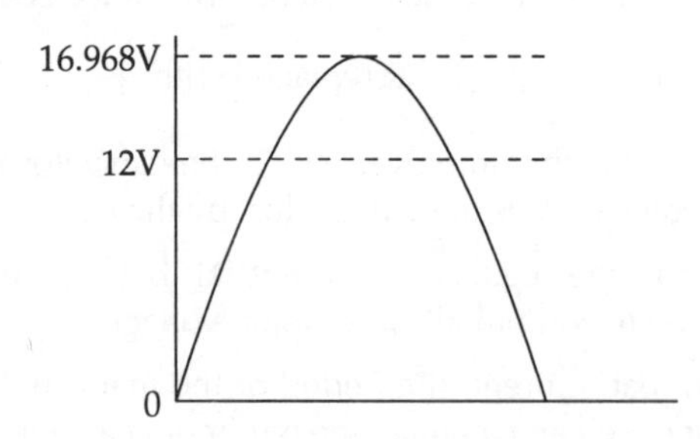

The peak input voltage to the bridge rectifier

The peak input to the bridge will be:

$$12 \times 1.44 = 16.968\text{V}$$

The diode volt drop in the bridge will be 1.4V.
Therefore the peak input voltage to the regulator is

$$16.968\text{V} - 1.4\text{V} = 15.568\text{V}.$$

Ripple Voltage

The regulator will require a minimum voltage of:

$$7.7 + 2.5 = 10.2\text{V}$$

so the maximum that the capacitor can discharge will be:

$$15.568 - 10.2 = 5.368\text{V}$$

This will be our maximum allowed ripple voltage V_{RIPPLE}.

Value of Filter Capacitor

$$C = \frac{tI}{V_{RIPPLE}} = \frac{0.01 \times 400 \times 10^{-3}}{5.368} = 745\mu F$$

This is not a preferred value so the nearest value would be 1000μF.

This is the minimum value required so a larger capacitor may be fitted if the ripple is to be reduced further.

A capacitor of this size would have to be an electrolytic and have a voltage working of at least 25V or more.

Fixed Voltage Regulators

There are a number of 3-terminal regulators available: The 78 series for positive supplies and the 79 series for negative supplies. You will need to check whether a heat sink is required for the level of load current drawn: e.g. the 7812 will only require a heat sink for loads in excess of 500mA up to the maximum of 1A.

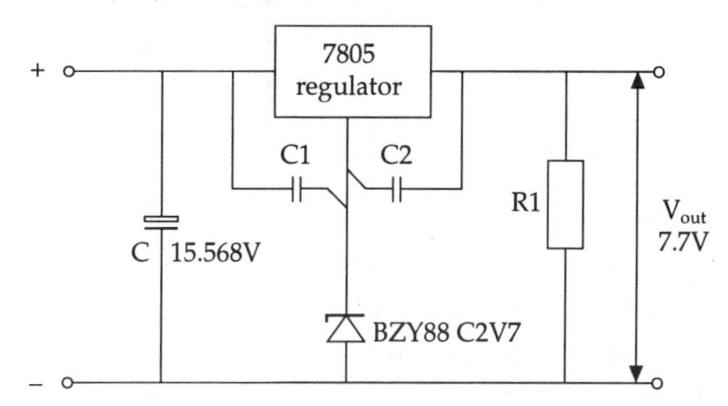

A regulator circuit using a fixed voltage regulator

Voltage Boosted Regulator

The output voltage from the 5V regulator has to be boosted by 2.7V to provide the required 7.7V. This is achieved by inserting a 2.7V zener diode in the common terminal as shown in the previous diagram.

If a suitable variable resistor is fitted in place of the zener diode then the output voltage may be varied: e.g. a 1k ohm variable resistor would give a range of voltages from 5V to about 9.1V

Capacitors C1 and C2

These capacitors are included to prevent the regulator from producing a high frequency oscillation. Their values are specified in the manufacturers' data and are usually around 100nF. **These capacitors must be mounted as close as possible to the pins of the regulator.**

Any non-polarised capacitor with sufficient voltage working will do for these capacitors.

Resistor R1

R1 is included to ensure that the output never operates into an open circuit. It need not be included if the load is permanently connected. R1 will draw some current even if the output is open circuit and prevent damage to the regulator.

Mains Input Switch

The input to the power supply must be fitted with a double pole single-throw switch (DPST). This is in effect two pairs of contacts. One pair is connected in the live and the other pair in the neutral line, which are **ganged** together. That is, when the switch mechanism is operated both pairs of contacts operate at the same time.

Mains Input Arrangement

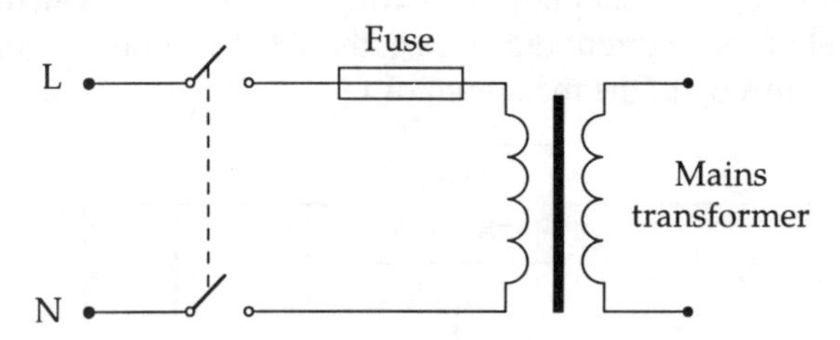

2 pole ganged mains input switch

Fuses

A suitable fuse should be fitted in the live line. This will give protection should a fault occur in the secondary circuit. The fuse should normally be the standard **quick blow** type and be rated at between 1.5 to 2 times the normal working primary current.

For example, if the power supply is to provide 12V at 1A then:

$$\text{secondary power} = 12 \times 1 = 12\text{W}$$

If we assume that the transformer has some losses then the primary power will be about 1.1 times the secondary power. Therefore the primary power will be about 13.2W. The primary current will be approximately:

$$\text{primary current} = \frac{13.2}{240} = 55\text{mA}$$

Therefore a suitable quick blow fuse would be;

$$55 \times 1.5 = 82.5\text{mA}$$

An 80mA or 100mA fuse would be suitable.

In-Rush Current

If the power supply is fitted with a very large reservoir capacitor to reduce the ripple voltage, (e.g. 4,700μF or more) then there will be a large in-rush of current at switch-on to charge this large capacitance. In such cases the fuse may be replaced for a slow-blow fuse with the same nominal current rating.

Overload Protection in the Secondary Circuit

This may also be achieved with a simple fuse fitted between the transformer secondary winding and the bridge rectifier. As the maximum load current is to be 400mA then a suitable fuse would be 500mA. A 500mA fuse is readily available and would allow for a 25% overload before the fuse blows.

Power Supply Design

Now design a power supply to provide 8.3V at 300mA. Draw out the complete circuit of your power supply and include the values of the components on your diagram.

Use the manufacturers' data, or data from suppliers such as RS Components, Farnell, etc., to support your choice of components.

Include where possible suitable extracts from the data sheets in your report on the specification of the power supply.

Task e

Construction and testing of the power supply

*(Refer to **Section 2 Information and Skills Bank**: How to use breadboards, p.96; Resistors and resistor colour coding, p.99; Power supply specifications, p.106; ; Half- and full-wave rectifier circuits, p.111; Capacitance and capacitors, p.130; Metric Prefixes, p.136.)*

Aims

To construct the power supply design on a breadboard and test to see if it meets the specification.

Equipment

Dual trace oscilloscope.
Digital multimeter.
Breadboard and wires.

Components

A fully enclosed and fused mains transformer.
Bridge rectifier (or four suitable power diodes)
A range of capacitors (a large electrolytic for the reservoir).
A 5V voltage regulator and suitable heatsink.
A zener diode.
Load (power!) resistors.
***Note**: The values and ratings of the above components will depend upon your designed values.*

Tools

A pair of cutters/wire strippers.
A pair of snipe nosed pliers.
Small screwdrivers.

Approach

The power supply consists of a number of blocks as shown below.

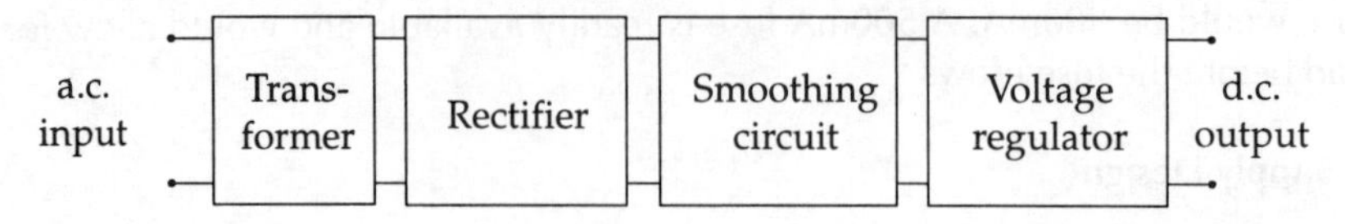

Power supply block diagram

However, for the purpose of constructing and testing, it may be considered as two main blocks.

Block 1) The a.c. input, mains transformer, rectifier and smoothing circuits.

Block 2) The voltage regulator circuit.

It is suggested that these circuits are constructed and tested independently before being connected together to form the complete power supply.

Each block of the circuit may be constructed on two small breadboards of 32 columns (or on each half of a full breadboard 64 columns). See **Section 2 Information and Skills Bank** How to use breadboards, p.96.

Block 1: a.c. Input, Mains Transformer, Rectifier and Smoothing Circuits

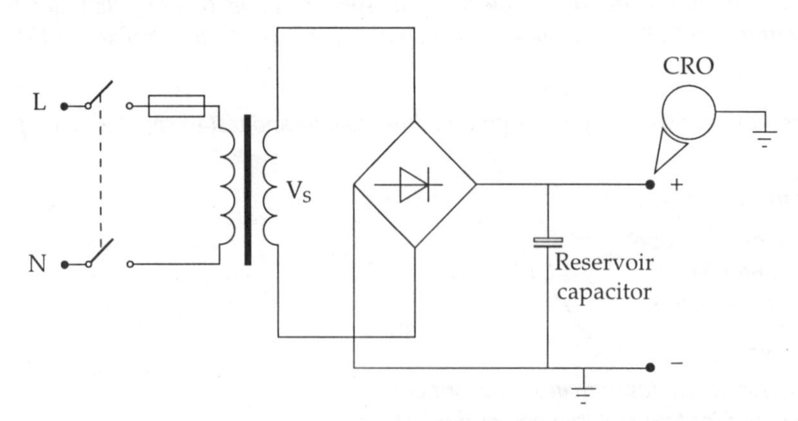

Full Wave Bridge circuit (negative ground)

Suggested Board Layout

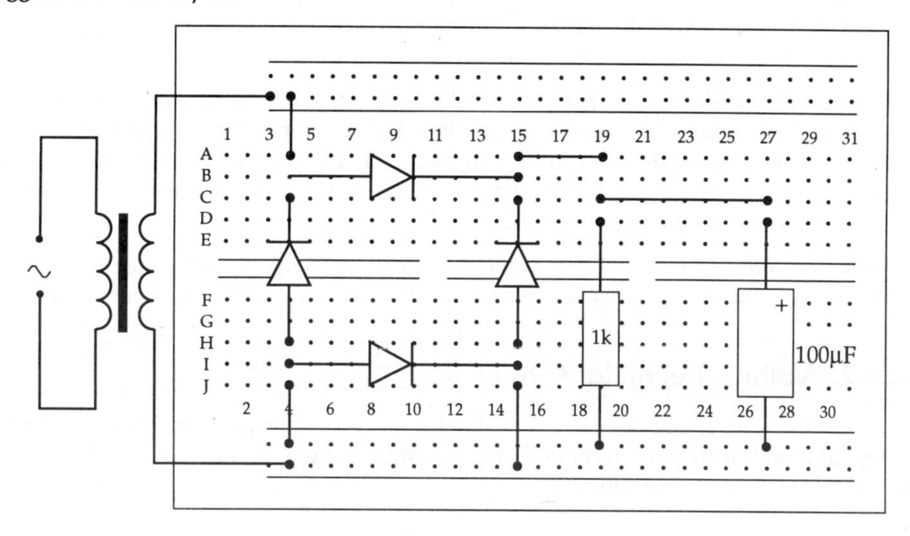

Suggested board layout for rectifier and smoothing circuit

Safety Precaution!

Although your complete circuit diagram will have the input fuse and mains input switch arrangement, because of the danger from electric shock **do not breadboard these components. Only use a completely enclosed mains transformer with no exposed live terminals** to obtain your low voltage a.c. supply for your rectifier circuit.

Testing

Breadboard the circuit using your component values. Apply the required low voltage a.c. supply to the rectifier circuit input and test to see if you have the correct d.c. voltage across the reservoir capacitor. Remember that this part of the circuit is unstabilised so that the output voltage off load will be higher than when the load is connected.

Connect a 1k load resistor and check the level of the ripple voltage across this load with the CRO (see Task c on p.18 for the circuit connections).

Faultfinding on the Power Supply

The first and most important step when faultfinding is to observe the symptoms; i.e. what is the circuit under test doing or not doing as the case may be?

Symptom/s

No output voltage or low output voltage.

Action

a) The most obvious cause for this will be incorrect breadboarding. If the diodes are not connected correctly then you may blow a fuse. Switch off! Check your circuit again, start from the a.c. input and work towards the d.c. output tracing the circuit out.

b) If the circuit is correct, disconnect the mains transformer from the breadboard and check that you have a low voltage a.c. output from the mains transformer.

If there is no voltage then check for a probable blown fuse in the transformer unit. Replace the fuse (with the correct rating of fuse), re-connect to the breadboard and test the complete circuit again.

c) If the breadboarded circuit is correct and you are still not getting any output or incorrect level of output then you have possible faulty component/s e.g. diodes open or short circuit (see **Section 2 Information and Skills Bank** Semiconductor diodes, p.75) or an open or short circuit capacitor (see **Section 2 Information and Skills Bank** Capacitance and capacitors, p.130).

Note: It is very unlikely that new components will be faulty, but if they are not new and have been used for experimental work before, they may have been inadvertently damaged.

Block 2: Voltage Regulator Circuit

Breadboard the voltage regulator circuit using the values of components from your design. The regulator will need to be mounted on a heatsink and wires soldered onto the leads to make connections to the breadboard.

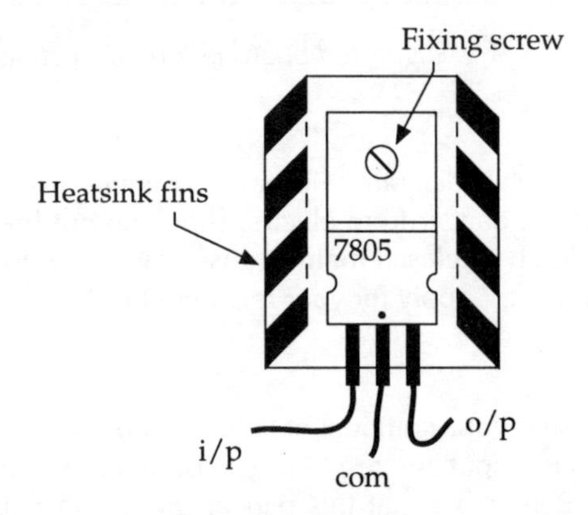

Regulator with soldered connections mounted on heatsink

Remember that the reservoir capacitor has already been included on the rectifier circuit board, and does not need to be included again.

R1 needs to be about 1k ohms. This resistor is included to provide a small load so that the voltage regulator is not operating into an open circuit.

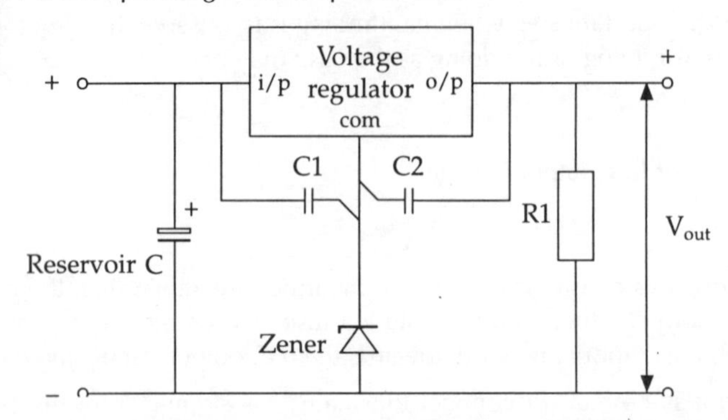

The regulator circuit using a fixed voltage regulator

Suggested Breadboard Layout

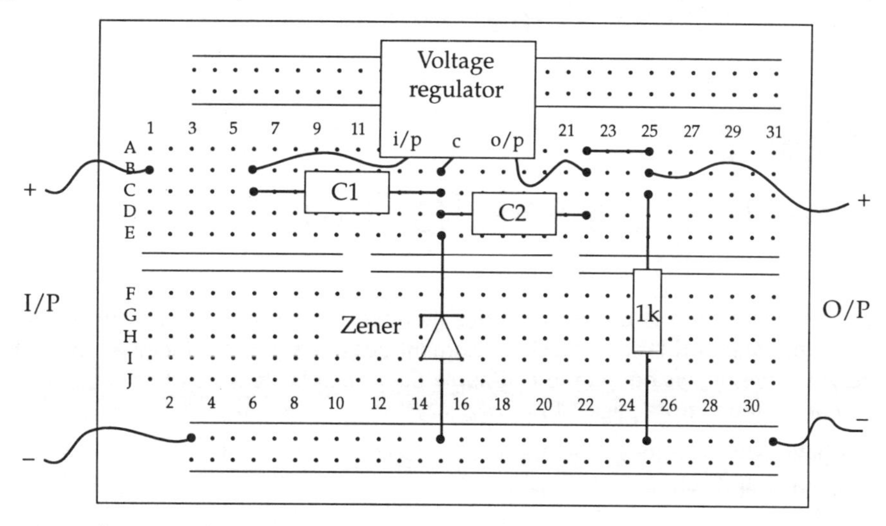

Note: C1 and C2 must be as short as possible and be mounted physically close to the regulator as shown on the breadboard.

Testing

If your rectifier circuit is working correctly then connect it to the regulator circuit and check to see if you obtain the correct output voltage from the regulator. If you wish to test the regulator without using the rectifier circuit then connect a d.c. voltage of about 15V to the regulator input terminals. Again check for the correct output voltage.

Faultfinding (observe symptoms first!)

Again the most probable cause of an incorrect output or no output will be incorrect breadboarding. Check your breadboarded circuit! If the circuit is correct then it can only be faulty components. Substitute each component in turn with another and test the output.

Finally...

Test the performance of the power supply against the power supply specification: i.e. efficiency, regulation ripple voltage etc. Use the information given in **Section 2 Information and Skills Bank** Power supply specifications, p.106; for the definitions of these terms.

Caution!

Note that it will not be possible to check the line regulation unless a suitable well-insulated auto transformer for altering the mains input voltage is available.

If your power supply does not meet the specification then you will have to modify your design.

Part 1: Power Supplies

Section 2
Information and Skills Bank

This section contains all the theory and skills required to complete the design project in Section 1. Some practical tasks and self-assessment questions are also included in the text to ensure you have understood it and can apply the principles (answers to self-assessment questions can be found from page 143).

Once you have worked through the appropriate topic for a particular Section 1 task, turn back to Section 1 and complete that task.

Contents

a.c. waveforms

Introduction

With direct current (d.c.) the current flows around the circuit in one direction only.

With alternating current (a.c.) the current periodically reverses its direction around the circuit. That is, the voltage and current vary with time. The a.c. current or voltage therefore has different properties to d.c.

Fundamentals of Alternating Currents

A direct current is obtained by using a source of direct Electromotive Force (e.m.f.), typically a battery.

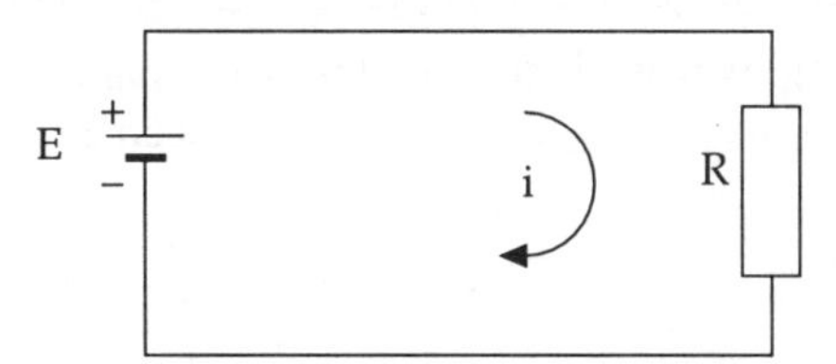

An alternating current is obtained by using an alternating source of e.m.f., typically an alternator, signal generator or function generator.

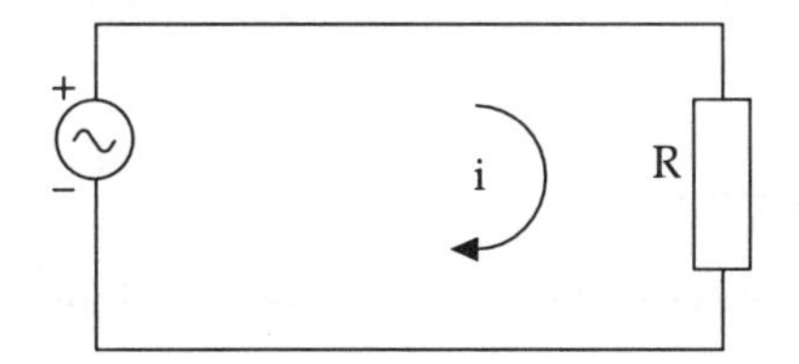

One half cycle

For a short time, referred to as the **positive half cycle**, the e.m.f. and the current are in one direction. During the following time, referred to as the **negative half cycle**, the voltage and current reverse direction.

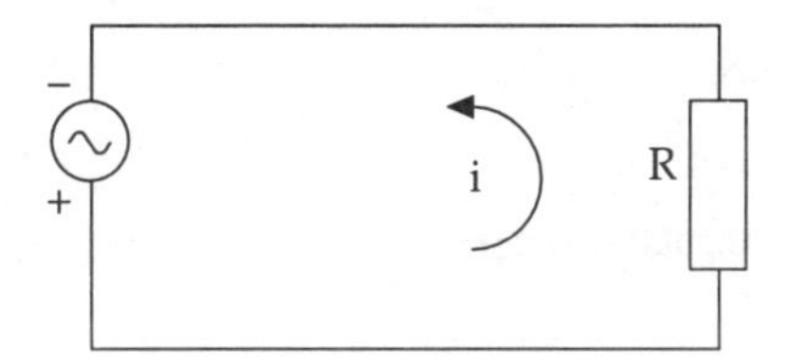

The following half cycle

Graphs and Waveforms

A voltage or current may vary with time. If the variation of a voltage is plotted on a vertical scale (this is called the Y axis) and the time for which it varies on the **horizontal scale** (the X axis) then a **graph** of the variation of voltage against time has been plotted.

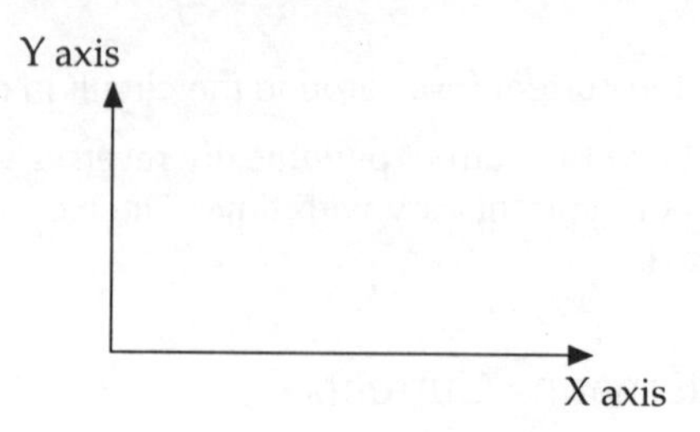

X and Y axes

The shape of the graph of voltage or current plotted against time is called its **waveform**.

Many naturally occurring waves, including the a.c. mains supply, are **sinusoidal** or **sinewave**. The sinusoidal waveform of the mains as seen on an **oscilloscope** is shown below.

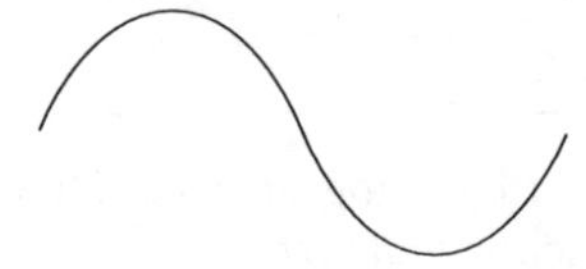

Sine waveform

Common a.c. Waveforms

An alternating current or voltage can have any waveform. Some commonly met waveforms are as shown below.

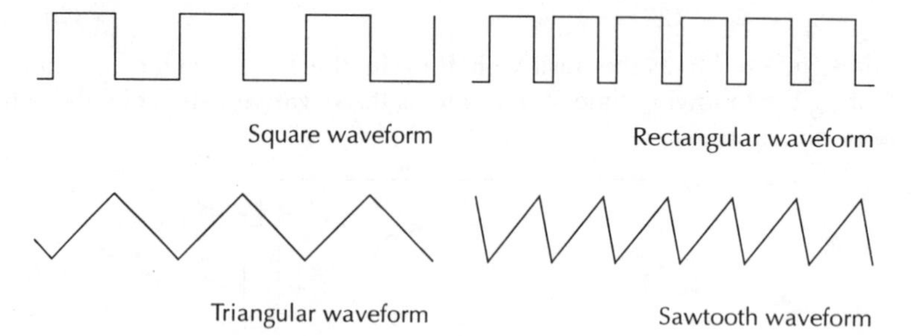

Square waveform

Rectangular waveform

Triangular waveform

Sawtooth waveform

Properties of a.c. Waveforms

Cycle

A complete set of positive and negative values.

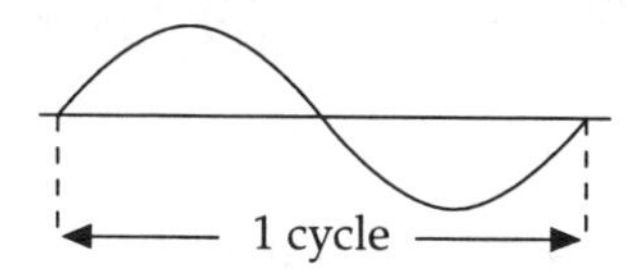

One cycle of a waveform

Frequency

The number of cycles that occur in one second. Units Hertz (Hz).

Period

If there are 50 cycles in one second then each cycle must take $^1/_{50}$th of a second. Therefore the period of a wave is the time for one cycle. One cycle of the mains supply is $^1/_{50}$th of a second.

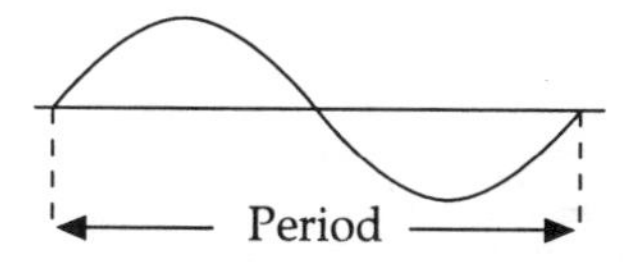

The periodic time of one cycle of the waveform

Amplitude (or Size) of Waveform

There are several ways to describe the amplitude of a waveform:

Peak Value

This is the maximum value of voltage (or current) attained by the wave on the positive or negative half cycles of the waveform.

Peak to Peak Value

This is twice the peak value.

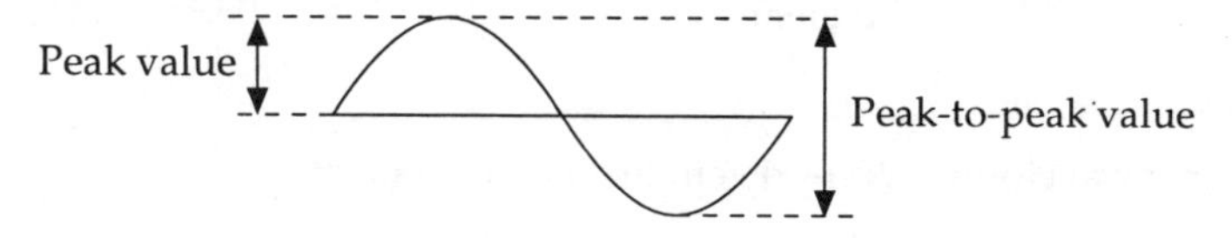

Peak, and peak-to-peak values

Root Mean Square Value

This is the effective value of the a.c. and will produce the same heating effect as a similar value of d.c.

That is, 240V a.c. (r.m.s.) will produce the same heat in a 1kW electric fire as will 240V d.c. Similarly, 10A a.c. (r.m.s) will produce the same heat in a resistance as will 10A d.c.

The r.m.s. value may be determined from the peak value:

$$\text{r.m.s.} = \frac{\text{peak value}}{\sqrt{2}} = \frac{\text{peak}}{1.414} = \text{peak} \times \frac{1}{1.414}$$

From which:

$$\boxed{\text{r.m.s} = \text{peak value} \times 0.707}$$

Example 1

If the r.m.s value is 240V determine the peak.

$$\text{Peak} = \frac{\text{r.m.s.}}{0.707}$$

$$= \frac{240}{0.707} = \mathbf{339.46V}$$

Therefore the peak to peak value will be:

2 x pk = **678.925V**!

Example 2

If the peak value is 10V determine the r.m.s.

$$\text{r.m.s} = \text{pk} \times 0.707 = 10 \times 0.707$$

$$= \mathbf{7.07V}$$

Measuring Instruments

Most analogue and digital multirange instruments measure the r.m.s. value of a sinewave.

Period and Frequency

If the frequency of a waveform is 10 Hz (10 cycles per second) then one cycle would have a period of $^1/_{10}$th of a second. Similarly, if f=1000Hz (1kHz) then the period of one cycle would be $^1/_{1000}$th of a second (1ms).

The relationship between the period and frequency of a waveform is therefore:

$$\text{Period} = \frac{1}{\text{Frequency}} \quad \text{or} \quad \text{Frequency} = \frac{1}{\text{Period}}$$

Example 1

If the period of a waveform is 20ms determine the frequency.

$$\text{Frequency} = \frac{1}{\text{Period}} = \frac{1}{20 \times 10^{-3}} = \frac{10^3}{20} = \frac{100}{2} = 50\text{Hz}$$

Example 2

If the frequency of a waveform is 20,000Hz (20kHz) determine the period.

$$\text{Period} = \frac{1}{20{,}000} \text{ seconds}$$

Now in 1 second there will be 1,000,000 microseconds (μs). Therefore to determine the period in microseconds we simply divide the 20,000 into 1 millon.

$$\text{Period} = \frac{1{,}000{,}000}{20{,}000} \ \mu s = \mathbf{50\mu s}$$

Remember: To get the period in milliseconds (ms), divide the frequency into 1000. If it is wanted in nanoseconds (ns) then we divide the frequency into 1,000,000,000.

Summary of a.c. Waveform

- ❐ Alternating current is where the current periodically reverses its polarity.
- ❐ Common a.c. waveforms are sinewaves, squarewaves, triangular waves, sawtooth waves and rectangular waves.
- ❐ A cycle is a complete set of positive and negative events.
- ❐ A cycle will occur in a certain time known as the periodic time (or periodic time, P.T.).
- ❐ Frequency is the number of cycles per second.
- ❐ The frequency may be obtained from 1/periodic time.
- ❐ The periodic time may be obtained from 1/frequency.
- ❐ The peak amplitude is the maximum value the waveform reaches in the positive or negative direction.
- ❐ The peak to peak amplitude is twice the peak.
- ❐ The r.m.s. value is the value of a.c. that will produce the same heating effect as a similar value of d.c. in a heating element.
- ❐ The r.m.s value = peak value × 0.707.
- ❐ The peak value = r.m.s. value × 1.414. or $^{\text{r.m.s.}}/_{0.707}$

Self Assessment Exercises (answers page 143)

1) Determine the periodic time and calculate the frequency (f) of each of the waveforms shown in A, B and C below.

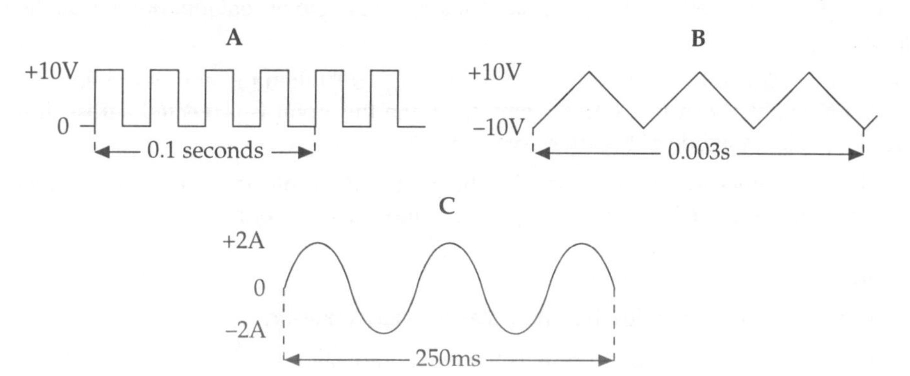

2) Determine the Period, frequency (f), peak value and r.m.s value of the waveform shown.

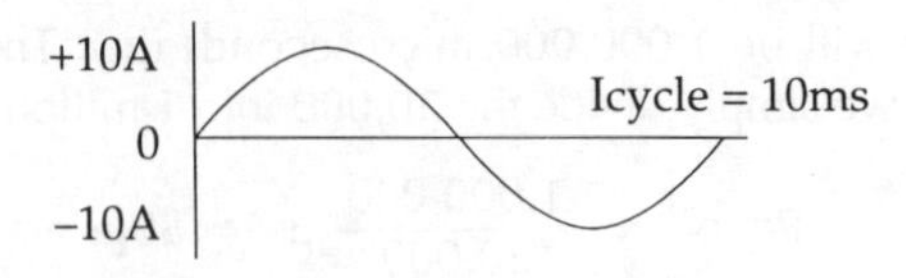

3) Convert the following periodic times to frequency:

   a) 10ms b) 100ns c) 20μs

4) Convert the following frequencies to period times:

   a) 0.5MHz b) 300kHz c) 20kHz

5) Convert the following r.m.s values to peak:

   a) 10V b) 30mA c) 200V

6) Convert the following peak values to r.m.s:

   a) 100mV b) 3A c) 50V

Production of a Sinewave

We can produce one cycle of a sine wave by rotating a line through 360°.

If a line OA is rotated *anti-clockwise* and the height of the end of the line, above the X axis, is plotted against the angle of rotation, then the result is a sinewave.

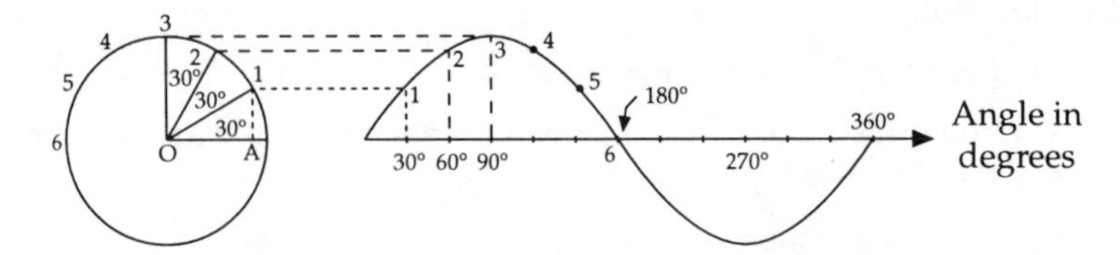

Production of a sinewave by rotating a line O-A

In the diagram above, when the line O-A has rotated through 30°, the vertical height of the line above the X axis is indicated at position 1 on the circle. This height is projected across and plotted against the angle of rotation (30°). (Point 1 on the sine curve.)

If the line is now rotated through another 30° the the new height is shown at point 2. This point is again projected across and plotted against the angle of rotation (point 2 on the sine curve).

After a further 30° (the line has now rotated through 90°) then the end of the line is at the maximum height above the X axis (point 3). When this point is projected across, it represents the peak amplitude of the sinewave.

If the line continues to rotate and the heights of the line projected across at 30° intervals, then when all the projected points are joined a sine wave is plotted.

Phasors

- ❒ The line which generates the sine wave is called a phasor.
- ❒ The length of the phasor represents the maximum value (peak value) of the sine wave.

- If the phasor rotates at f revolutions per second, it will generate a sine wave having f cycles in one second, so the frequency will be f Hertz.
- One cycle of the sinewave is completed by 360° of rotation.

Phase Relationship of Two Sinewaves

When two sinewaves have the same frequency and their peaks occur at the same time, they are said to be **in phase**.

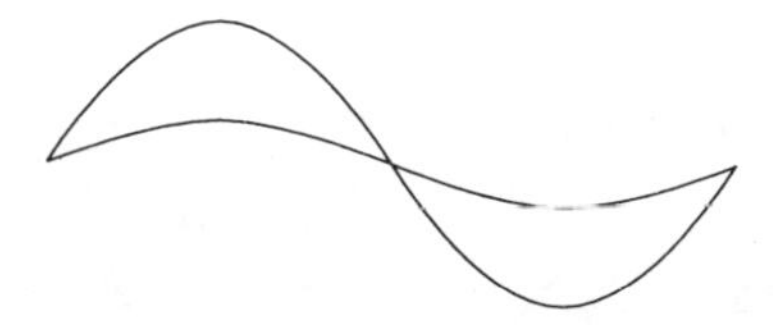

Two Sinewaves Which are In Phase.

When two sinewaves have the same frequency, but their peaks do not occur at the same time, they are said to be **out of phase**.

If we think of the two sine waves being generated by two rotating lines, then both lines are rotating at the same speed *anti-clockwise*, but one line started rotating earlier than the other, and so there is a phase angle of so many degrees between them.

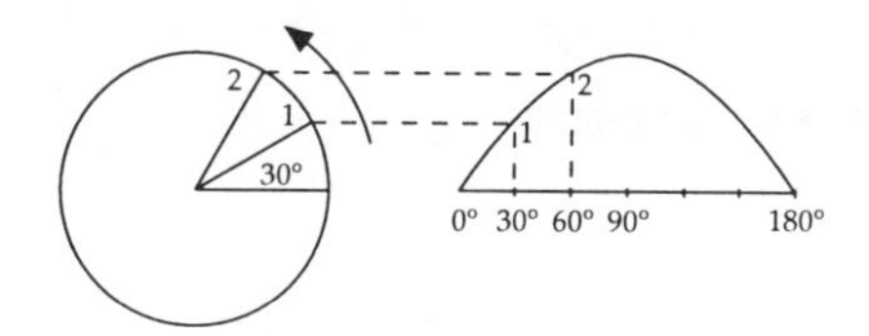

Two lines being rotated at the same velocity but starting to rotate at different moments in time.

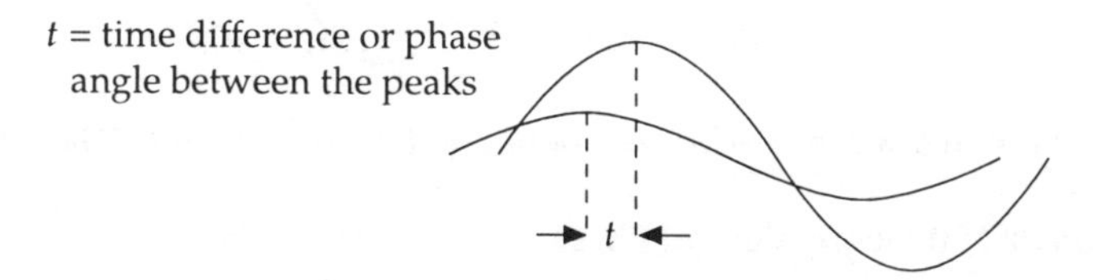

Two sinewaves which are Out of Phase.

Although phase is a difference between the two peaks, **phase difference** can best be shown as the difference in angle between two phasors (rotating lines) which generate the sinewaves.

Phasors

The two sinewaves produced may more easily be represented by two lines. The length of each line represents the maximum value (peak value) that the current or voltage can have and the angle between the lines represents the phase difference.

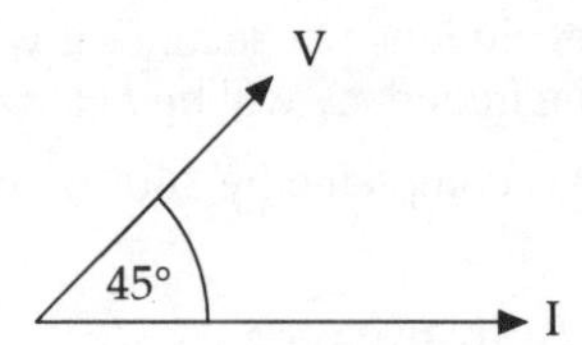

Phasor diagram representing two sinewaves. The peak value of the voltage and current is represented by the length of the lines. The angle between the phasors represents the phase difference between the two sinewaves.

Two Sinewaves 60 degrees out of Phase

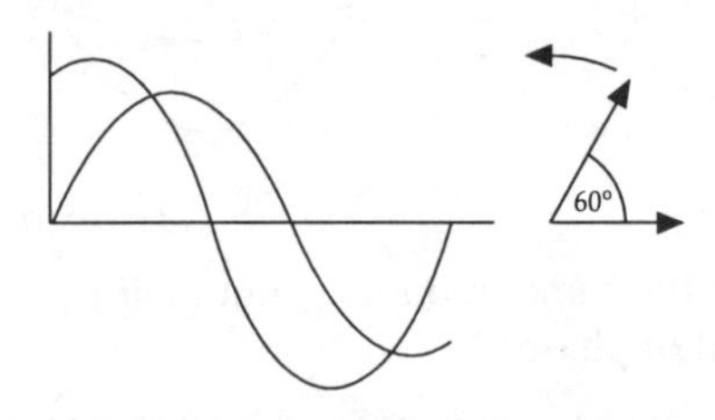

These two waves are 60° out of phase. Their peaks are $^1/_6$th of a cycle apart (a cycle is generated by 360° rotation, and $^1/_6$th of 360° is 60°)

Two Sinewaves 90 Degrees out of Phase

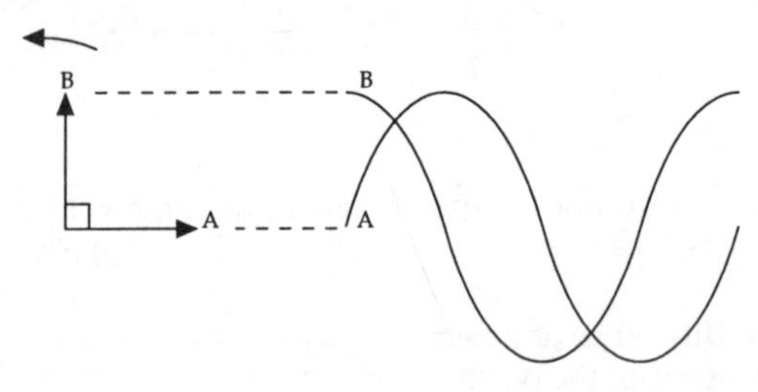

These two waves are 90° out of phase. B leads *A by 90° (OR A* lags *B by 90°)*

Phase Inversion or 180 Degrees out of Phase

The Cathode Ray Oscilloscope is an instrument which enables us to see electrical waveforms. It is often used to observe and compare the input and output waveforms of a voltage amplifier. Any difference in the phase or waveshape can easily be observed.

With single stage (one transistor) amplifiers the output waveform is inverted compared to the input waveform. The same effect will occur if the two waveforms are 180 degrees out of phase.

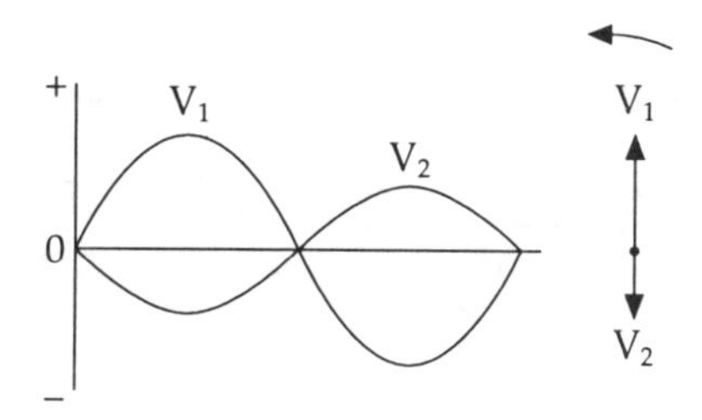

Two waveforms 180 degrees out of phase

Phase Relationship Between Voltage and Current

Capacitors and inductors take currents which are out of phase with the voltage applied to them.

Capacitors

If a capacitor is connected to an a.c. supply, it will take an alternating current which charges, then discharges and re-charges it with the opposite polarity as the e.m.f. of the supply alternates.

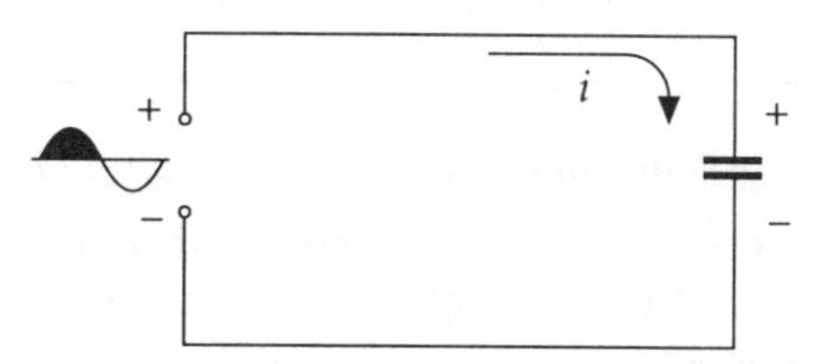

Positive half cycle of the a.c. supply

On one half cycle of the a.c. supply the capacitor is charged with one polarity.

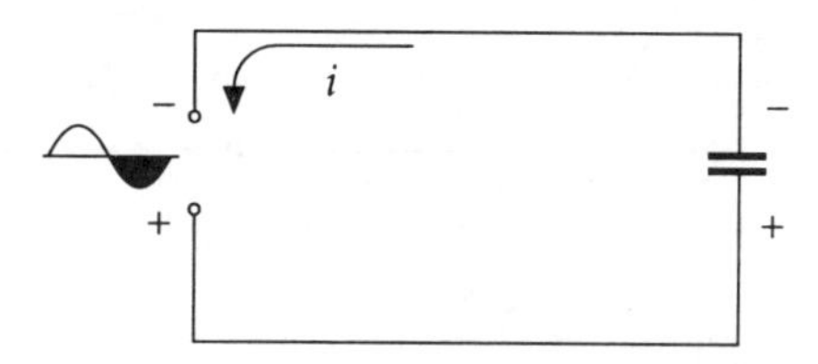

Negative half cycle of the a.c. supply

On the other half cycle of the a.c supply, the capacitor is charged with the opposite polarity.

The alternating charging current leads the applied voltage by one quarter of a cycle (90°) because the **current must flow to charge the capacitor before any voltage appears across it.**

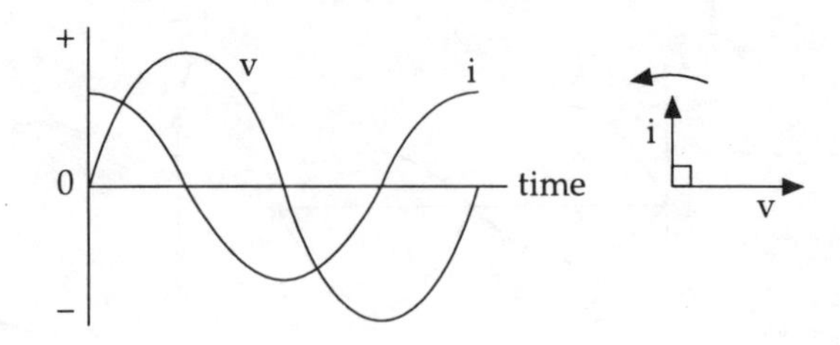

Waveforms of voltage and current in a capacitor. In a capacitor, current leads the voltage by 90 degrees.

Inductors (coils)

If an inductor is connected to an a.c. supply, it will take current which alternates as the supply polarity alternates.

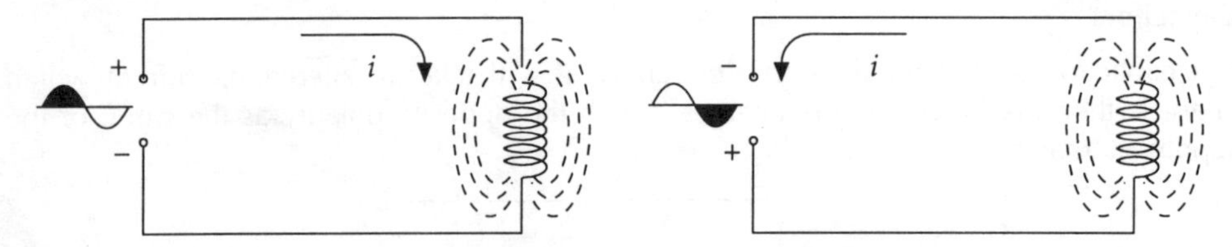

Current flow on positive half cycle *Current flow on negative half cycle*

The alternating supply current causes the magnetic field of the coil to build up and collapse. This changing magnetic field induces an e.m.f. in the coil which opposes the rise and fall of the supply current, and this causes the current to lag behind the voltage by one quarter of a cycle (90°).

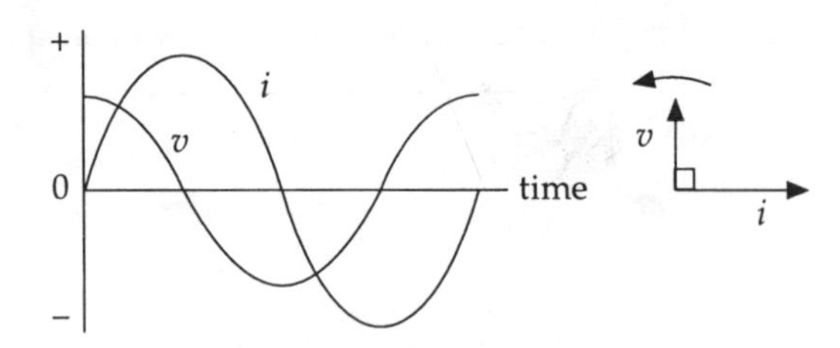

Waveform of voltage and current in an inductor. In an inductor current lags behind the voltage by 90°

Summary of Phase

- A sinewave can be produced by rotating a line anti-clockwise. One cycle of a sinewave is completed by 360° of rotation.
- The height of the line above the X axis gives the amplitude of the sinewave at that moment in time.
- The sinewave has a maximum positive value at 90° of rotation and a maximum negative value at 270° of rotation. It has a value of zero at 180° and 360°.

- ❐ If the line revolves at f revolutions per second then it will generate a sine wave having f cycles per second (hertz).
- ❐ The rotating line is called a phasor: its length represents the peak amplitude of the sinewave.
- ❐ Two sinewaves out of phase; may be represented by two phasors (lines) the angle between the two lines represents the phase difference in degrees.
- ❐ When two sinewaves are in phase their peaks coincide.
- ❐ When two sinewaves are 180° out of phase, if the one sinewave is peak positive the other will be peak negative and vice versa.
- ❐ In a pure capacitor the current leads the voltage by 90°.
- ❐ In a pure inductor the current lags behind the voltage by 90°.

Self-assessment questions *(answers page 143)*

Decide whether the following statements are true or false.

1) The voltage across a capacitor leads the current by 90 degrees.

   True or False: _______

2) The current in a capacitor leads the voltage by 90 degrees.

   True or False: _______

3) The voltage across an inductor leads the current by 90 degrees.

   True or False: _______

4) The angle between the phasors represents the phase difference, whereas the length of the phasors represents the peak value that the voltage or current can have.

   True or False: _______

5) By convention phasors rotate in a clockwise direction.

   True or False: _______

6) In the diagram the current leads the voltage by 90 degrees.

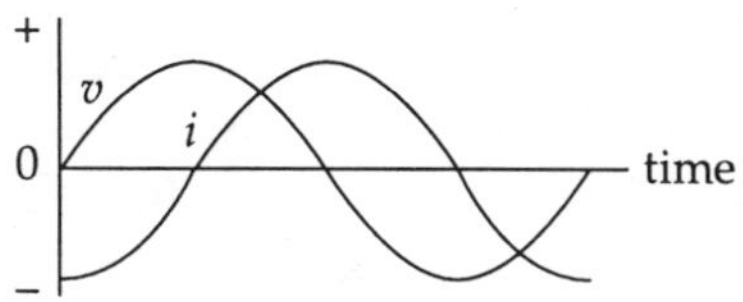

True or False: _______

7) In the diagram the voltage leads the current by 90 degrees.

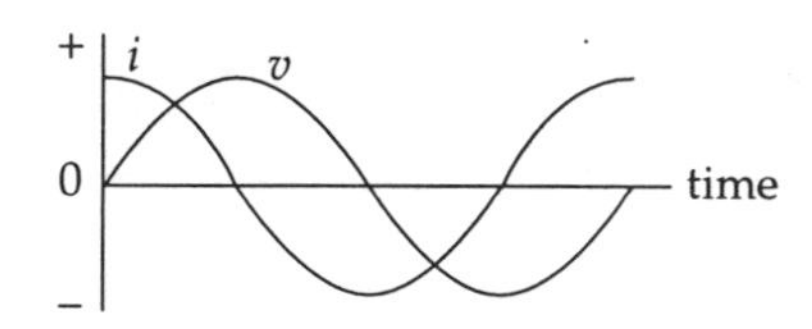

True or False: _______

8) A voltage will appear across an inductor before any current will be seen to flow through it.

   True or False: _______

9) A voltage will appear across a capacitor before any current will flow to charge it.

   True or False: _______

10) Capacitors and inductors take a.c. currents which are out of phase with the voltage applied to them, but in pure resistors the a.c. current and voltage are always in phase.

    True or False: _______

11) Two waveforms are always out of phase if their frequencies are different.

    True or False: _______

12) Two waveforms are always in phase in their frequencies are the same but their peaks do not coincide.

    True or False: _______

13) Two waveforms are always in phase if their peaks coincide and their frequencies are the same.

    True or False: _______

14) Two waveforms are always out of phase if their frequencies are the same and their peaks do not coincide.

    True or False: _______

15) Two sinewaves are in phase; if one of the waveforms is inverted then it becomes out of phase by 180°.

    True or False: _______

Cathode ray oscilloscope

Introduction

The Cathode Ray Oscilloscope (CRO) is an instrument which enables us to 'see' electrical waveforms. The shape of a waveform can be displayed on the screen, and its frequency and voltage can be measured. The main component in a CRO is the Cathode Ray Tube.

The Cathode Ray Tube

The Cathode Ray Tube is a glass tube with a flare at one end, the front of which forms the viewing screen. Air is evacuated from the tube and the tube is sealed.

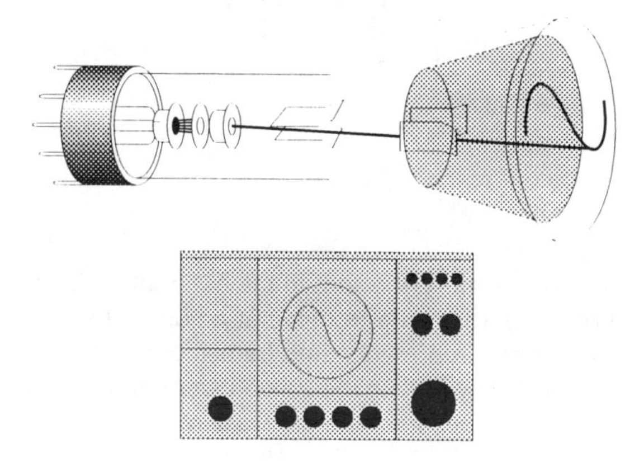

Within the tube is an electron gun which produces a finely focused beam of electrons. The electrons are accelerated towards the screen at the front of the tube by a large positive voltage (several thousands of volts).

The electron beam produces a spot of light where it strikes the fluorescent coated screen. The spot can be moved around the screen by applying voltages to two sets of deflector plates. As the spot moves it draws a visible trace, which is the graph of the waveform being examined.

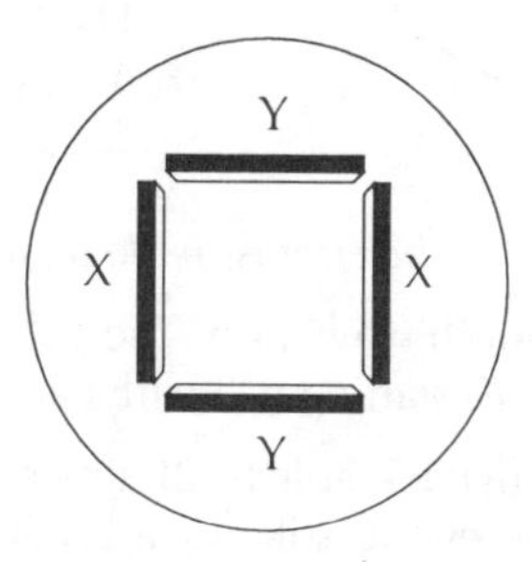

View of the deflector plates looking from the screen face

The two sets of deflector plates are arranged as shown. The X plates deflect the electron beam horizontally and the Y plates vertically.

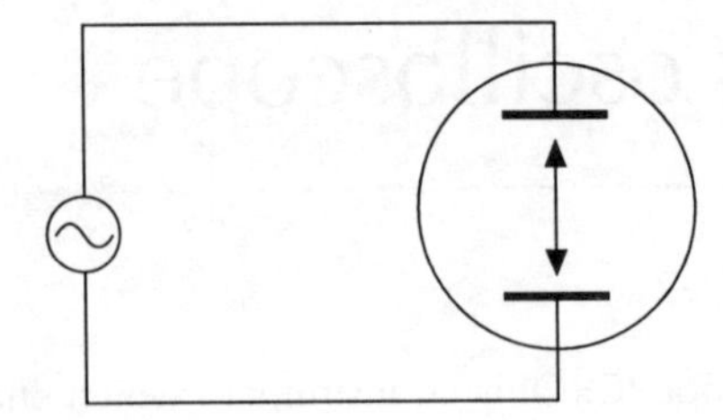

The Y deflection plates

A voltage across the Y plates moves the spot vertically (in the same direction as the Y axis of a graph).

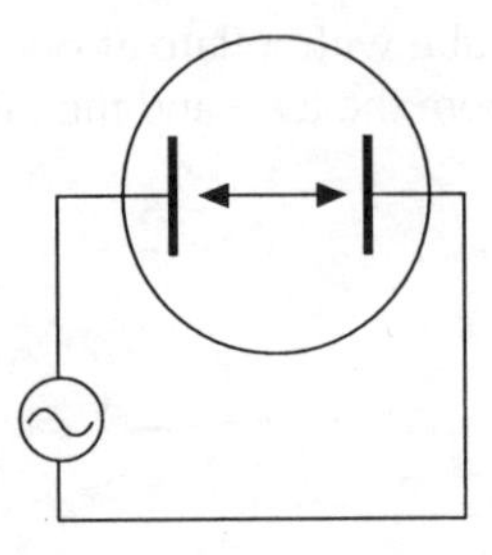

The X deflection plates

Voltages applied to the X plates move the spot in the horizontal direction (along the X axis of the graph). If the spot of light is made to move at a constant velocity along the X axis, then the distance along this axis represents time (remember that velocity is distance covered in a given time). It therefore will take the spot a certain time to travel from left to right across the screen, and the X axis becomes the **timebase** of the graph.

Timebase Generator

The spot is made to move across the screen (sweep) and then **flyback** again to the left hand side.

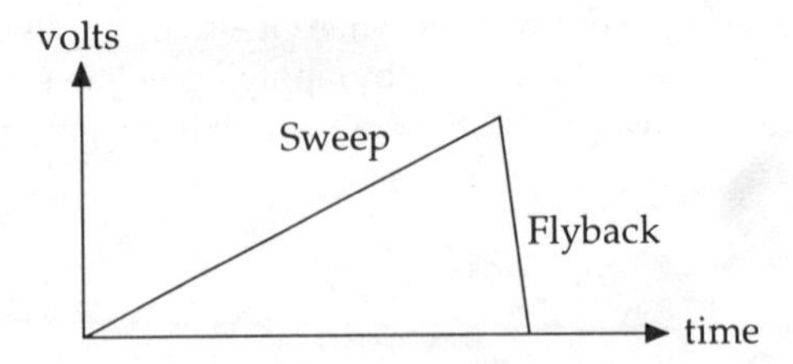

The spot is made to sweep the screen and then flyback

The timebase waveform is a **sawtooth** waveform. The voltage rises at a steady rate (sweep), and then returns at a rapid rate to its starting value (flyback).

If the spot is made to sweep and flyback at least 20 times a second, then due to the persistence of vision, (the ability of the eye to still see a bright light for a fraction of a second after it has disappeared) and the persistence of the cathode ray tube (afterglow of the fluorescent coating), instead of seeing a moving spot we see a line of light called a **trace** on the screen.

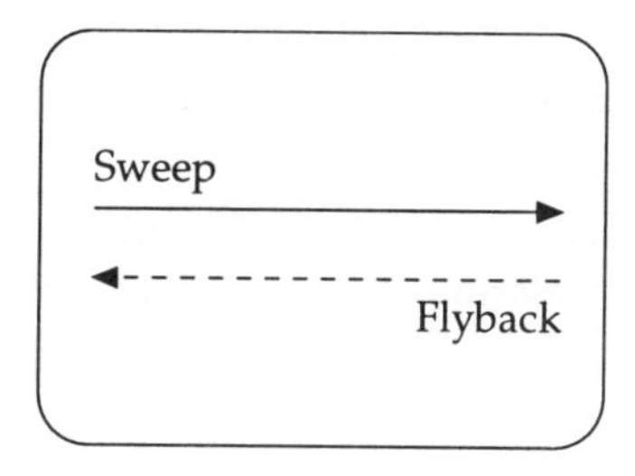

The sweep generates a trace on the screen

The purpose of the timebase generator is to provide these sawtooth waveforms. The frequency of the sawtooth must be variable so that the spot can be made to move across the screen at different rates. The sawtooth waveform is calibrated so that the spot moves one division (e.g.1cm) on the screen, in so many milliseconds or microseconds.

Displaying a Waveform

The signal that we wish to observe is now applied to the Y plates, whilst a suitable sawtooth waveform is being applied to the X plates.

The diagram below shows the effect of these two waveforms being applied to the deflection plates. These deflection waveforms applied to the X and Y plates cause the electron beam to **sweep out** a faithful replica of the Y-input waveform.

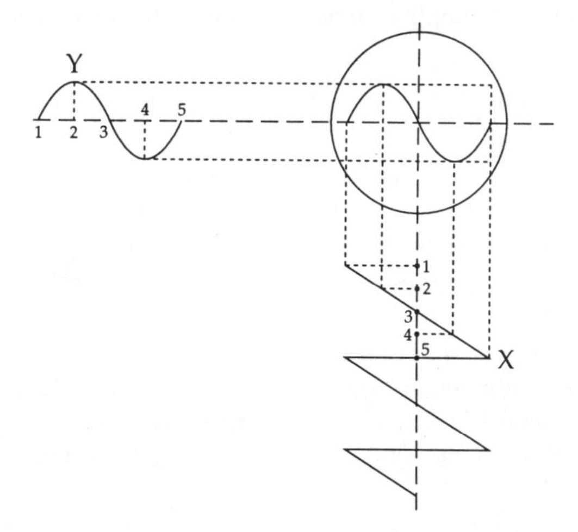

Producing the Y-input waveform. When the sawtooth waveform is applied to the X plates and the waveform to be observed is applied to the Y plates, then the spot traces out the waveform applied to the Y plates

When both deflection waveforms are at position 1 there is no X or Y deflection and the spot on the screen is at the start of the cycle.

When both deflection waveforms are at position 2, then the Y deflection is a maximum positive value, whereas there is only a small amount of X deflection in this time. The spot has therefore moved to indicate the peak value of the Y input signal at 90 degrees through the cycle.

By plotting points 3,4 and 5 in a similar way the spot of light has moved through one complete cycle. If the cycle is repeated many times a second then a complete cycle of the waveform is observed on the screen.

If the timebase generator frequency is exactly equal to the sinewave Y-input frequency, then exactly one cycle of the waveform will be traced out on the screen.

The Y Amplifier

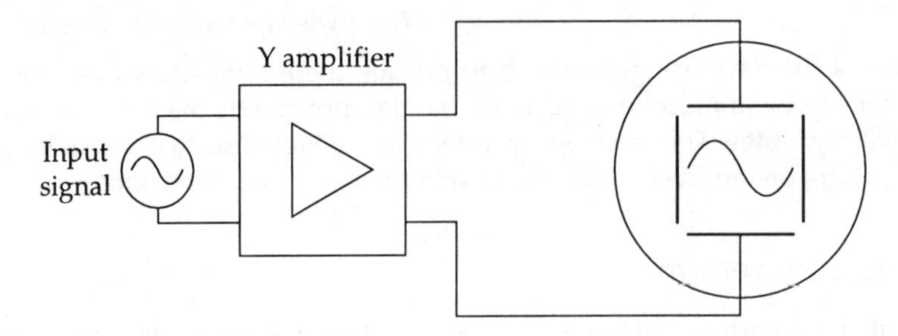

The Y amplifier amplifies the input signal

The cathode ray tube is not very sensitive and requires some tens of volts to drive the spot across, or up and down the screen. So, to produce sufficient deflection from a few millivolts of Y-input signal, a Y-amplifier is used to increase the size of the Y-input signal. Usually only about 10mV at the Y-amplifier input is required to move the spot vertically on the screen by 1cm.

The amplifier is designed to be this sensitive so that small voltages may be investigated easily.

A large input signal will now overload the amplifier causing distortion. To prevent distortion, an attenuator (a device for cutting down the size of the signal, e.g. a voltage divider), can be switched in before the Y-amplifier to reduce large signals.

Switched Gain Y-amplifier

The Y-amplifier gain (amplification) is therefore switched to provide several levels of amplification. It is also calibrated so that a particular level of voltage at the Y-amplifier input terminals will produce a known amount of deflection on the screen face: e.g. 10mV produces 1cm of deflection on the 10mV/division Y-Amplifier setting.

Synchronizing

It is very important that the start of the Y-deflection waveform coincides with the start of the X deflection sawtooth waveform. If this is not the case for every cycle of the 50 cycles per second of the mains frequency, for example, then the displayed waveform tends to 'run' across the screen and makes observation of the waveform difficult.

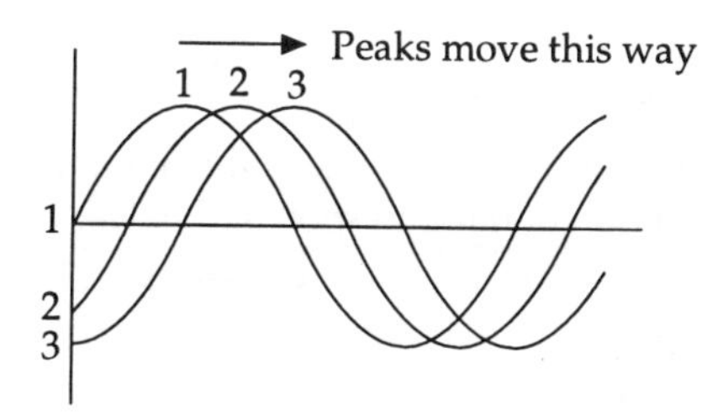

Unsynchronized waveforms appear to 'run' across the screen

In the diagram above three sweeps are shown, each sweep starting at a different place on the Y-input signal. If the X-timebase speed is reasonably fast then the three individual sweeps are not seen. Only one trace is observed which appears to 'run' either to the left or to the right of the screen. To avoid this, the X and Y deflection waveforms must be synchronized to start together. This is achieved by using the amplified Y-input signal to *trigger* the X-timebase at the start of each new sweep.

The Y-input signal is fed to a synchronizing circuit, usually called a *triggering circuit,* to synchronize the X-timebase generator. This ensures that the sweep across the screen face starts at the same point on the Y-input waveform each time and produces a *locked trace.*

The Main Components of a Cathode Ray Oscilloscope

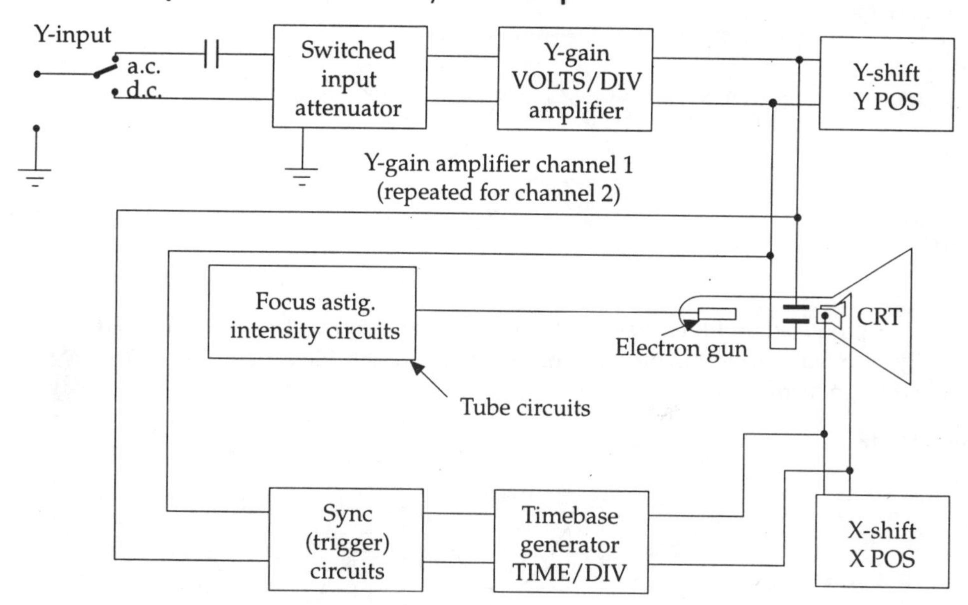

Cathode ray oscilloscope block diagram

Refer to the CRO block diagram above.

There are three main sections to any oscilloscope.

1) *The tube (C.R.T.) circuits.* This usually contains the brightness: (intensity), focus, astigmatism (control to produce a good round spot) and power controls.
2) *The Timebase circuits.* This section contains the switchable timebase generator, the trigger (synchronizing) circuits, the horizontal shift (horizontal position) control and an external/internal trigger switch.
3) *The Y-Amplifier circuits.* This section contains:
   2 sets of Y-Amplifier controls for Channel 1 and Channel 2.
   Two switchable gain Y-Amp controls: (volts/division): CH.1 and CH.2.
   Two vertical shift controls (Y-position).
   Two a.c/d.c.and ground switches. These enable the input to the Y-Amplifiers (CH.1, CH.2) to be connected to either a.c. or d.c. signals or grounded to earth.
   A channel switch. This allows the selection of one trace, CH.1, or two traces, CH.1 and CH.2, to be displayed on the screen.
   Trigger switch. This determines which input, CH.1 or CH.2, is to be used to trigger (synchronize) the X-timebase.
   A CHOP and ALTERNATE switch. This is used to switch between the two traces being produced alternately.

Alternate Mode

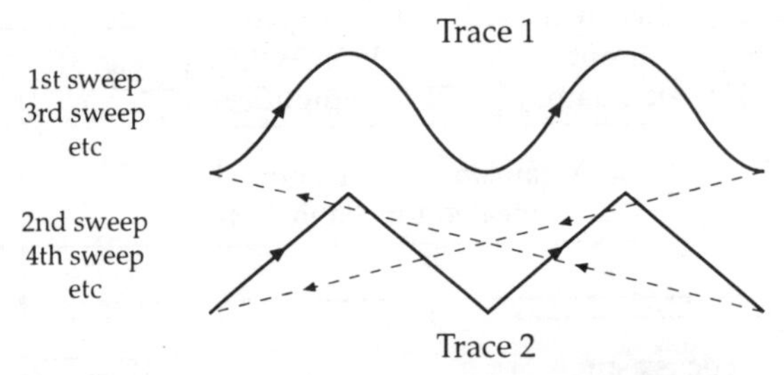

Trace 1 is produced and then trace 2, then trace 1 and so on. This mode is used for the majority of input waveforms on the input channels which have frequencies from a few hundred Hz to many MHz.

Chop Mode

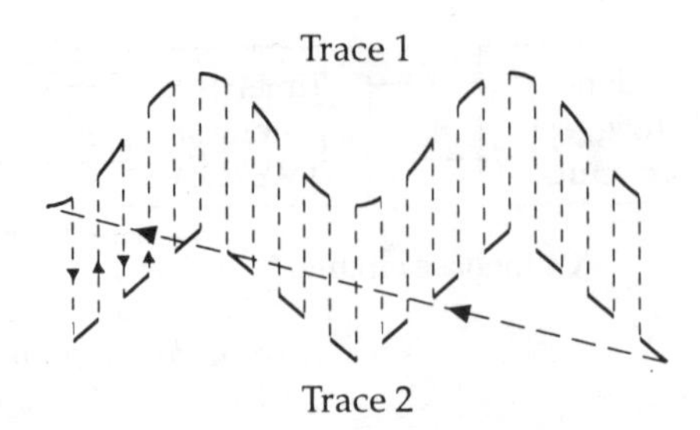

A small part of trace 1 then a part of trace 2 is produced, i.e. the horizontal timebase is 'chopped' between trace 1 and 2 as it produces the sweep across the tube face. Used when observing low frequencies (e.g. 50Hz power frequencies) to produce a flicker free trace.

Summary of Cathode Ray Oscilloscope

- The CRO is an instrument that allows us to see electrical waveforms.
- A spot of light is produced by an electron beam hitting a phosphor coating on the inside of the screen face at high velocity. The colour of the light depends upon the phosphor material.
- The spot is deflected up and down the screen by a voltage on the Y deflection plates and across the screen by a voltage on the X deflection plates.
- To produce a trace (line across the screen) a sawtooth waveform is applied to the X plates from the X timebase generator. The frequency of this sawtooth is variable.
- The waveform to be displayed is amplified by the Y gain amplifier and fed to the Y plates.
- The frequency of the sawtooth waveform from the X timebase generator should be adjusted to be about the same frequency as the Y input signal if only one to two cycles of the waveform are to be displayed.
- To obtain a steady (locked) trace the signal from the Y amplifier is also fed via a trigger circuit to the X timebase. This is called synchronizing.
- For the majority of input signals the alternate mode of display is used, but for low frequencies (e.g. 50Hz) then the chop mode of display is used.

Now study the **Section 2 Information and Skills Bank** Using the Cathode Ray Oscilloscope (page 56) and do the self assessment questions before attempting Task a.

Using the cathode ray oscilloscope

Introduction

The oscilloscope enables us 'to see' and measure waveforms in electronic circuits. For example, the **amplitude** (size) of the waveform can be measured; the **frequency** (number of cycles per second) of the waveform can be determined and the **shape** (sine, square, rectangular, etc) of the waveform can be observed.

Measurement of Amplitude (voltage)

Using a signal/function generator as a signal source, the CRO can be used to measure the peak, or peak to peak, amplitude of the waveform.

Remember that other multirange analogue or digital instruments will measure the r.m.s. value of a sinewave.

The amplitude of the waveform is determined from size of the vertical deflection on the screen. The graticule (square grid pattern on the face of the screen) is often in 1cm squares. This grid is used to determine the height of the waveform.

Once the number of divisions or centimetres is known then this figure is multiplied by the setting on the Y-gain (volts/division) control. This then gives the peak to peak value. The peak value is half the peak to peak.

Example (Measurement of Voltage)

A sine wave displayed on the screen of an oscilloscope is as shown.

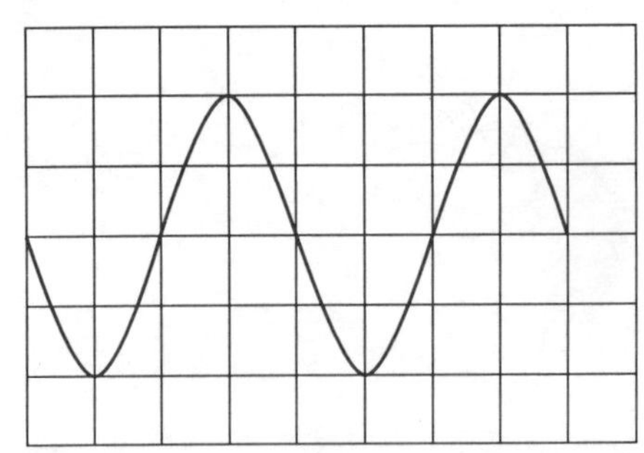

Sinewave displayed on the CRO. The peak-to-peak height is 4 divisions.

We now look at the Y-gain setting to determine the number of volts/division.

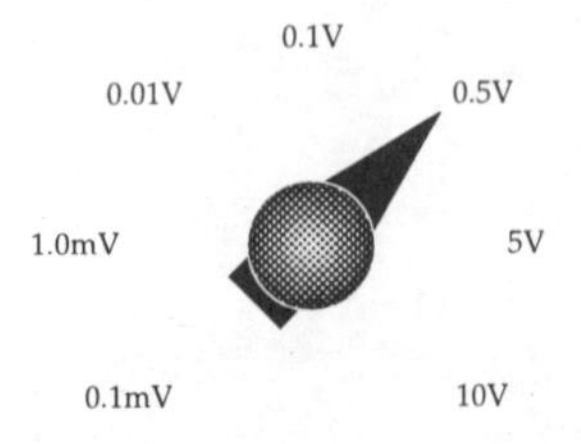

The Y-gain control (Volts per division)

In the example shown the setting of the Y-gain is 0.5V. Therefore the peak to peak amplitude of the waveform is:

$$4 \times 0.5V = 2V \text{ Peak to Peak.}$$

and: the peak amplitude is:

$$\frac{\text{Peak to peak}}{2} = 1V$$

the r.m.s. value is:

$$\text{Peak} \times 0.707 = 0.707V$$

Summary (Measurement of Amplitude)

The Peak to Peak amplitude is:

Height of display in divisions × the Volts/Division setting.

Self Assessment Questions *(answers page 144)*

Determine the Peak to Peak, Peak, and r.m.s. values of the following displays of sinewaves:

	Total height of display on CRO	*Y-Gain setting Volts/division*
a)	3.5 divisions	5V
b)	1.75 divisions	0.1V
c)	2.25 divisions	1mV
d)	3.75 divisions	0.01V

Measurement of Frequency

The CRO can also be used to measure frequency. The spot forming the trace is made to move across the screen in a certain time. If this time is accurately known then the time for one cycle of a waveform can be determined.

The spot is made to sweep across the screen at different speeds by the X-timebase setting. The rate at which the spot moves is accurately calibrated so that the spot moves one division in a given time, e.g. one division per millisecond.

Example (Measurement of Frequency)

A sine wave displayed on the screen of an oscilloscope is as shown.

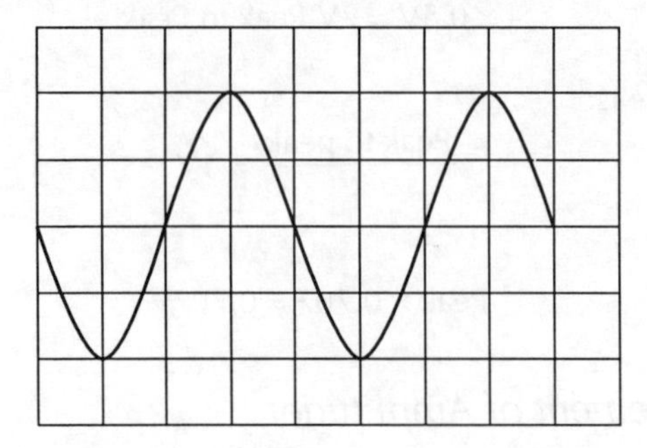

Sinewave displayed on the CRO

The number of divisions for one cycle of the waveform is four divisions.

We now look at the X-gain setting to determine the time/division.

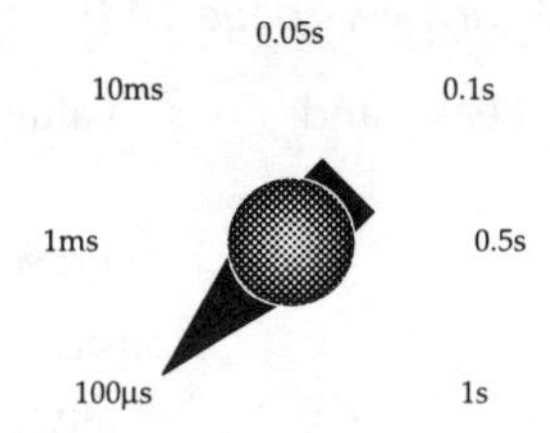

The X-gain control (time per division)

In the example shown the setting of the X-gain is 100μs/division.

The period of one cycle of the waveform is therefore:

$$4 \times 100\mu s = 400\mu s.$$

The frequency is determined from $\dfrac{1}{\text{Period of one cycle}}$

Therefore the frequency is: $\dfrac{1}{400 \times 10^{-6}} = 2{,}500\text{Hz}$

Summary (Measurement of Frequency)

The periodic time is:

number of divisions per cycle × the time/division setting.

The frequency is then determined by:

$$\text{Frequency} = \frac{1}{\text{Periodic time}}\ \text{Hz}$$

Self-assessment questions *(answers page 144)*

Determine the periodic time for one cycle and calculate the frequency from the following displays of sinewaves:

	Number of divisions per cycle	*X-timbase setting*
a)	3.5 divisions	5ms/div
b)	1.75 divisions	100μs/div
c)	2.25 divisions	10ms/div
d)	3.75 divisions	1ms/div

Function generators

Introduction

Function or signal generators are used to provide test signals in order to test the operation or performance of electronic circuits. They usually provide sine and square waves over a wide range of frequency at output levels between a few mV to a few volts r.m.s. You are therefore able to select the type (shape), the frequency and the amplitude of the test signal.

When using a function generator for the first time you need to look for the following main controls.

Selection of Signal

The types of test signals required are: sine waves, square waves, and triangular waves, (see p.37, a.c. waveforms). These are usually selected by push button switches.

Selection of Frequency

The function generator must provide these test signals over a wide range of frequencies. Typically 0.2Hz to 2MHz divided up into several frequency ranges. Each frequency range is selected by a push button switch and the frequencies within a particular range are selected by a variable frequency control.

Many of the latest function generators provide a digital readout of the frequency setting. To do this they have a built in frequency counter and on some models the function generator may also be used as a frequency counter: i.e. the frequency of an external signal may be determined.

Selection of Amplitude

In order to test amplifiers, for example, a test signal of a few volts down to a signal of a few mV peak to peak may be required. Therefore a switched or variable amplitude control is necessary.

The function generator is therefore designed to provide a maximum output of between 10 to 20V peak to peak, which may then be reduced by a switched or variable attenuator down to a few mV.

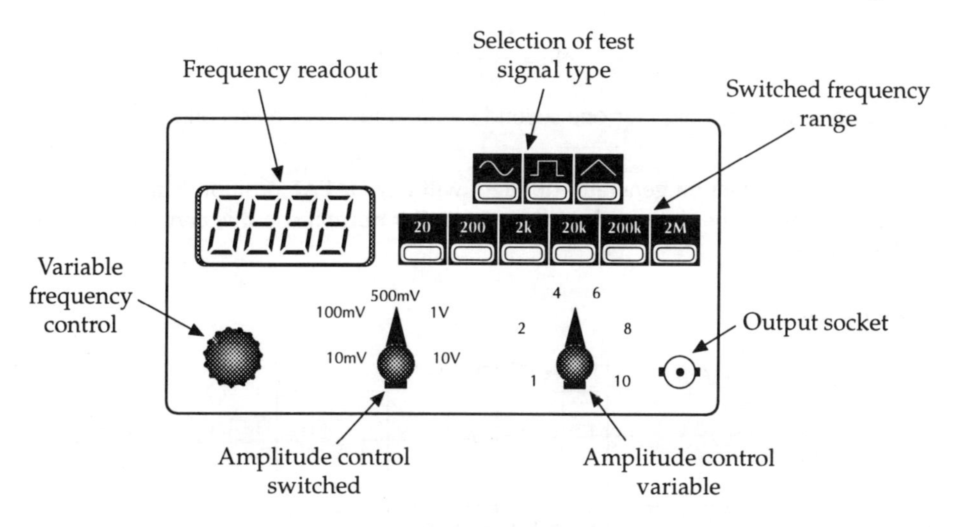

The front panel of a typical Function Generator

Additional Facilities

Most function generators provide extra facilities to those described above.

d.c. Offset Control

This allows the d.c. level of the output waveform to be set as desired. The output waveform can consist of an a.c. signal superimposed upon a d.c. level. This control allows the setting of the d.c. level. For most applications the d.c. level should be set to a minimum.

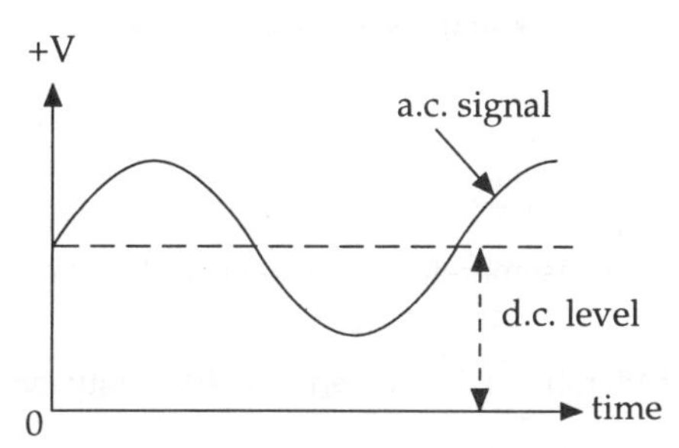

An a.c. signal superimposed upon a d.c. level.

Duty Control

This control is used with square waveforms.

The control allows the time period of one half cycle of the waveform to be changed while the other half remains fixed at any given set frequency. This alters the **duty cycle** of the waveform.

Duty Cycle

The duty cycle of a waveform is the ratio of the **on** time to the **on + off** time and is often expressed as a percentage.

$$\text{Duty cycle} = \frac{t_{on}}{t_{off} + t_{on}} \times 100\%$$

e.g. a waveform which is high for one second (on) and low for one second (off) has a duty cycle of 50%

This control on the function generator (if fitted) will have a **CAL** position (calibration position) and at this point the duty cycle will be 50%. **For normal use this control must be in the CAL position!**

50% Duty Cycle

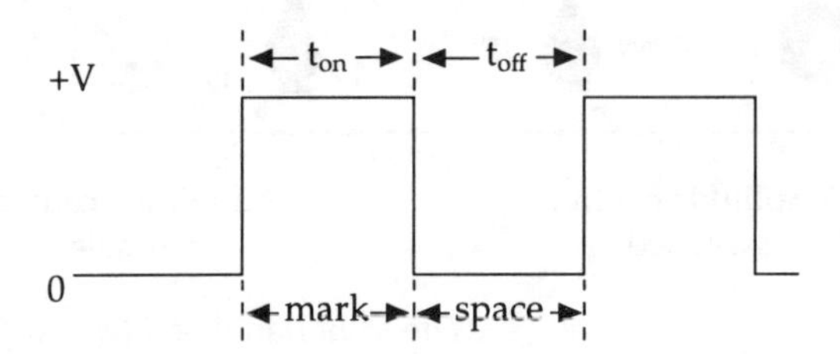

A duty cycle of 50% (mark space ratio 1:1)

40% Duty Cycle

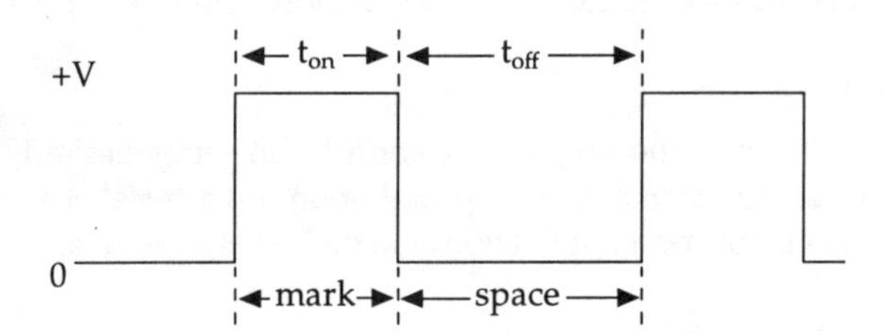

A duty cycle of 40% (mark space ratio 1:1.5)

Mark Space Ratio

Another way of expressing this asymmetry is the ratio of the mark (on time) to the space (off time).

The 50% duty cycle has a M/S ratio of 1:1, whereas a 40% duty cycle has a M/S of 1:1.5.

Digital Logic Facilities

The square and rectangular waveforms may be used to provide test signals for digital logic systems. If this is the case then additional output sockets are provided.

TTL Outputs

For digital logic circuits involving TTL devices a square or rectangular waveform is required that switches between logic 0 (0V) and logic 1 (+5V). If a TTL output socket is provided then these are the voltage levels that will be present at that output at any desired frequency and the normal amplitude controls will be ineffective.

CMOS Outputs

For digital logic circuits involving CMOS devices then a square or rectangular wave with a logic 1 level anywhere between 3V and 15V is required (logic 0 = 0V). For this reason a CMOS level control will be fitted to allow the output to be varied, (usually between +5V and +15V).

Finally...

Many of these additional facilities are for more advanced applications and for first-time operation all you need are the main controls for adjusting the frequency, amplitude and the correct output socket. It is recommended that you read the instrument manual when you need to find out how to use these extra facilities.

Summary of Function Generator

- The function/signal generator provides sine and square waves over a wide range of frequencies at output levels between a few mV to a few volts r.m.s.
- The frequency, amplitude and the shape of the signal must be selected/set by the user.
- Many function generators have extra facilities: d.c. offset control to adjust the d.c. level of the output waveform; the duty control, to vary the ratio of the ON time to the ON + OFF time (mark space ratio) and TTL or CMOS outputs.

Self-assessment questions *(answers page 144)*

1) Which control on the function generator alters the peak-to-peak amplitude of the test waveform by a predetermined amount?
   a) The variable frequency control
   b) The variable amplitude control
   c) The switched amplitude control
   d) The switched frequency control

2) Which control on the function generator continuously varies the peak-to-peak amplitude of the test waveform?
   a) The variable frequency control
   b) The variable amplitude control
   c) The switched amplitude control
   d) The switched frequency control

3) Which control on the function generator varies the frequency range of the test waveform?
   a) The variable frequency control
   b) The variable amplitude control
   c) The switched amplitude control
   d) The switched frequency control

4) Which control on the function generator continuously varies the frequency of the test waveform?
   a) The variable frequency control
   b) The variable amplitude control
   c) The switched amplitude control
   d) The switched frequency control

5) A squarewave has an ON time of 1 second and an OFF time of 3 seconds. The duty cycle is therefore:
   a) 25%   b) 33%   c) 50%   d) 100%

6) The mark-to-space ratio for the waveform in question 5 would be:
   a) 1:3   b) 1:4   c) 3:1   d) 4:1

7) A squarewave has as ON time of 2 seconds and an OFF time of 4 seconds. The duty cycle is therefore:
   a) 25%   b) 33%   c) 50%   d) 100%

8) The mark-to-space ratio for the waveform in question 7 would be:
   a) 1:2   b) 1:3   c) 4:2   d) 2:1

9) The TTL output socket on a function generator provides a waveform with an output that:
   a) is continuously variable
   b) switches between 0V and 5V
   c) switches between 0V and 15V
   d) can be varied to provide a maximum output from 5V to 15V.

10) The CMOS output socket on a function generator provides a waveform with an output that:
   a) is continuously variable
   b) switches between 0V and 5V
   c) switches between 0V and 15V
   d) can be varied to provide a maximum output from 5V to 15V.

Now do Task a on p.6 (investigation of a.c. waveforms using the CRO and function generator.)

Simplified semi-conductor theory

Introduction

In order to understand semiconductor material we need to look at the structure of how materials are formed. A substance or material is called **matter**.

Matter

Matter is anything which occupies space, either as a liquid, a solid or a gas. All matter is made up of **elements**.

Element

An element is a substance which cannot be broken down into simpler substances. For example, copper, gold, silver, hydrogen, oxygen, carbon are all examples of elements. Substances that have more than one element are called **compounds**. All elements contain **atoms**. There are about 100 different kinds of atoms.

Atoms

An atom is the smallest part of an element which can exist and take part in chemical action. All the atoms within the element will be the same and have the same average mass. For example, a piece of gold, which is a basic element, is made up from gold atoms and if the gold is pure no other atoms will be present in the material.

Nucleus

An atom consists of a central nucleus, which contains most of the mass of the atom.

Protons and Neutrons

The nucleus is made up of **protons** which have a small positive charge and **neutrons** which have mass but no charge. They are electrically neutral.

Electrons

Around this nucleus **electrons** orbit. Electrons have very little mass, and are practically weightless, but are negatively charged.

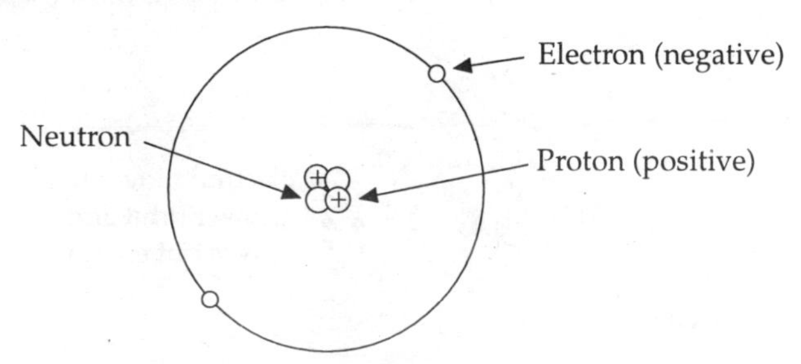

The structure of the atom. The neutrons and protrons form the nucleus of the atom

Electronics

An electric current is the movement of electrons and so the study of the behaviour and control of electrons in materials is called **electronics**.

Properties of the Atom

	Mass	*Charge*
Proton	1.7725×10^{-27}	+e
Electron	9.109×10^{-31}	−e
Neutron	1.63×10^{-27}	No charge

where $e = 1.062 \times 10^{-19}$ coulombs

Protons and neutrons are very difficult to remove from the nucleus and can therefore be considered to be fixed permanently to their parent atoms. They are therefore of little significance to Electronics.

Atomic Structure

Electrons revolve or orbit around the nucleus at fixed distances called shells. These shells are at discrete distances from the nucleus and represent discrete amounts of energy.

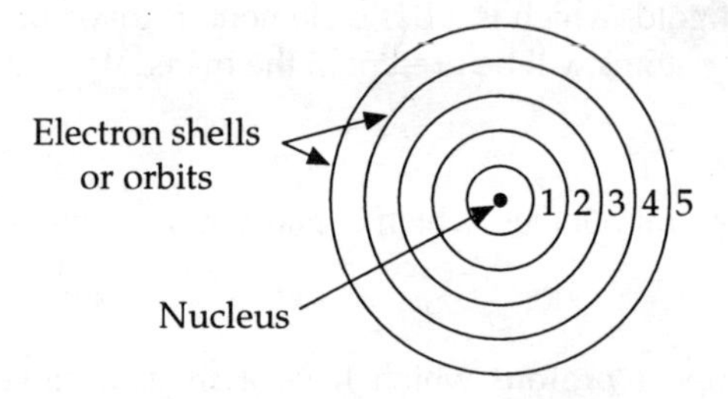

Electrons exist in energy shells around the nucleus of the atom

The electrons can only exist in these shells. If the atom is given energy then an electron may jump to a higher energy level, i.e. move to another orbit further away from the nucleus. When the atom loses energy then the electron jumps to a lower energy level and moves closer to the nucleus. In doing this the electron must give out a small amount of energy.

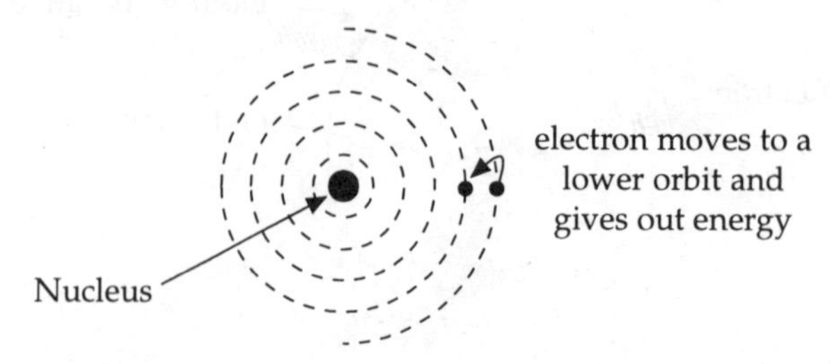

When the atom cools the electron jumps to a lower energy level and in doing so gives out energy.

Positive Ions

Normally an atom has the same number of electrons as protons, and the total positive charges equals the total negative charges. The atom is therefore electrically neutral.

If an atom loses an electron, the protons outnumber the electrons and the overall charge is positive by an amount +e. The atom is then called a positive ion.

Negative Ions

If an atom gains an extra electron then the electrons out number the protons and the overall charge of the atom is negative. Such atoms are called negative ions.

Hydrogen Atom

The hydrogen atom is the simplest atom. It contains a nucleus of one proton and has one electron in orbit around it.

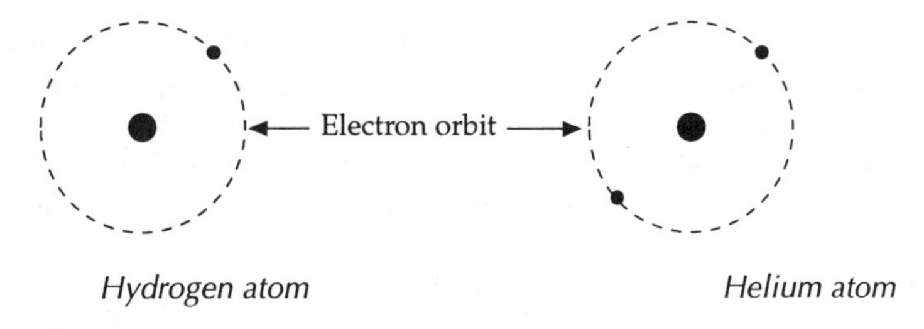

Hydrogen atom *Helium atom*

Periodic Table Of Elements

The number and combination of these particles (electrons, protons, and neutrons) in any element determines the properties of that element.

High order atoms which have large numbers of electrons and protons, have many shells to contain all the electrons.

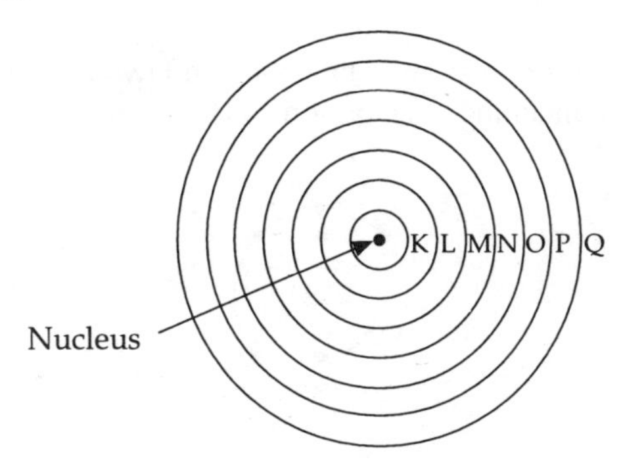

Shells of the atom

Valence Electrons

The Shell which is of most importance in electronics is the outermost shell. The electrons in this shell are called **valence** electrons and determine the chemical properties of the element.

The shells are labelled K,L,M,O,P,Q from the innermost shell out. The electrons contained in these shells for group III,IV, and V elements are given in the table below.

		Number of electrons in shells K L M N O P Q
Arsenic	Donors	2 8 18 **5**
Phosphorous	Group V	2 8 **5**
Antimony		2 8 18 18 **5**
Germanium	Group IV	2 8 18 **4**
Silicon		2 8 **4**
Gallium	Acceptors	2 8 18 **3**
Aluminium	Group III	2 8 **3**
Boron		2 **3**
Indium		2 8 18 18 **3**

Properties of elements in groups III, IV, and V

The valence electrons are shown in bold in the table above. The elements are grouped in terms of their valence electrons. As can be seen from the Table all elements in group III have 3 valence electrons and those in group IV, 4 valence electrons. The Table only shows groups III, IV, and V because these are the most important groups for semiconductors.

Molecules

Atoms of different elements will tend to group together to form **molecules**.

A molecule is the smallest particle of a substance that can exist while still retaining the properties of the substance. For example, one molecule of water contains one atom of oxygen and two of hydrogen. If the water molecule is further divided then the result is oxygen and hydrogen and not water.

The Water Molecule

The water molecule contains one atom of hydrogen and two of oxygen (H2O). The oxygen atoms have a nucleus containing one proton (+1), whereas the hydrogen atom has a nucleus of eight protons (+8).

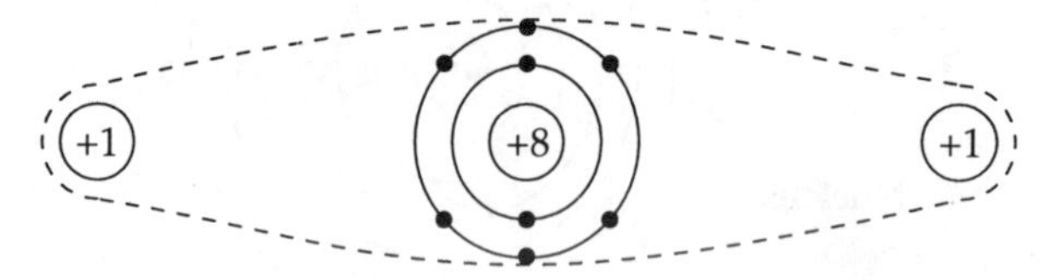

The water molecule

The Silicon Atom

For convenience, we draw a simplified diagram of the atom, since we are concerned only with the outer valence electrons.

In the simplified diagram below, only the four negative valence electrons are shown, together with four positive protons in the nucleus (+4)

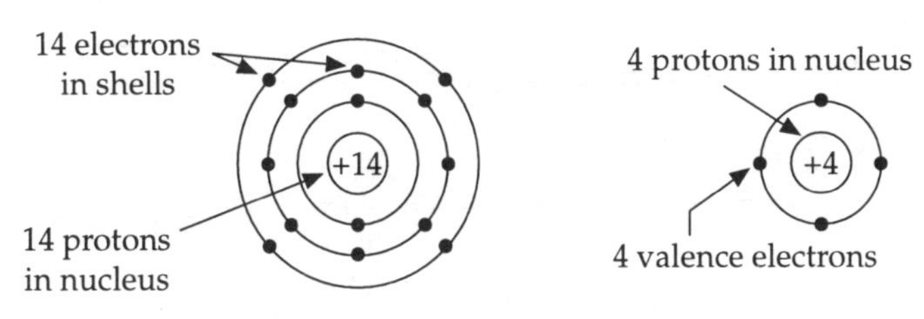

The silicon atom *Simplified silicon atom*

Conductors and Insulators

In some materials the electrons are loosely bound to their parent nucleus and these materials make good conductors (materials with low resistivity). In other materials the bonding is very strong and these materials make good insulators (high resistivity).

Semiconductors

These materials have a resistivity between that of an insulator and a conductor and therefore are neither a good conductor nor a good insulator.

Crystal Lattice

Silicon, Germanium and Carbon will all form a crystal lattice. In this structure, every atom is linked to four neighbour atoms by a pair of shared electrons which form **co-valent bonds**.

Co-valent Bonds

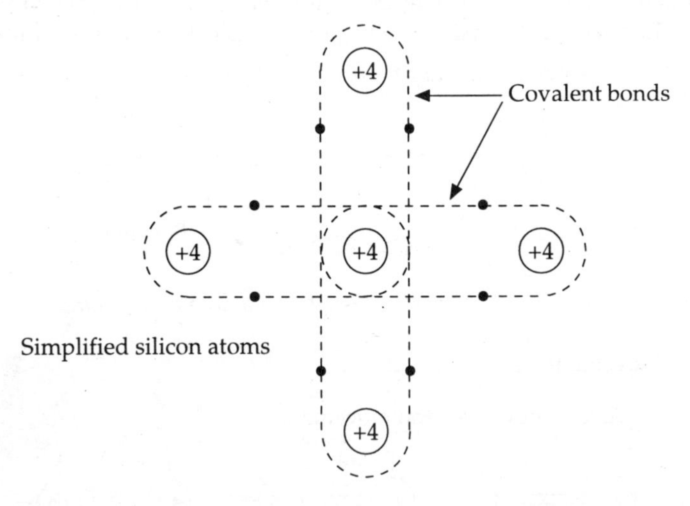

Co-valent bonding between simplified silicon atoms

Every atom is orbited by eight valence electrons, and therefore appears to have a full outer shell. This bonding makes the crystal a very stable structure, both chemically and physically, and gives it its characteristic surface appearance.

Simplified Diagrams to Represent the Crystal Lattice

To further simplify the diagram, the valence electrons are shown by lines, each line representing an **electron shared** with another atom.

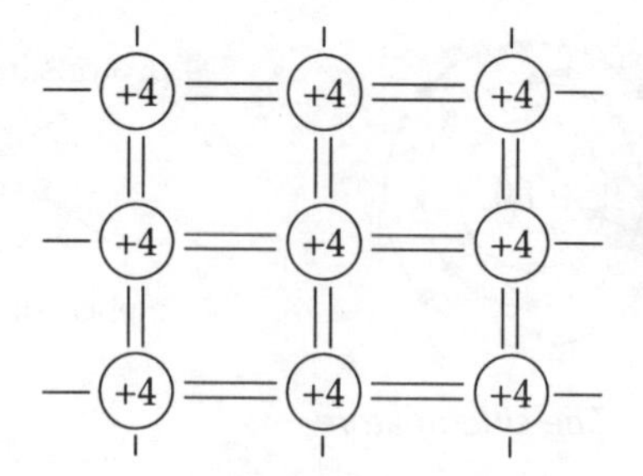

Simplified diagram representing co-valent bonding

The simplified diagram of co-valent bonding shows a two dimensional diagram of a piece of crystal with the co-valent bonds. Remember that this should really be a three dimensional structure.

Holes and Electrons

The perfect bonding shown in the crystal lattice diagram can only exist at absolute zero temperature (–273 degrees celcius or zero degrees Kelvin). There is then no free electrons, no current, and the material is a perfect insulator.

At room temperatures (293 degrees Kelvin) the electrons absorb sufficient heat energy to break free from their bonds. If we apply an e.m.f across the crystal then these electrons can be moved to form an electric current through the crystal.

Whenever an electron breaks out of its bond, it leaves behind a vacancy in the crystal lattice structure. This vacancy is positively charged, and is called a positive **hole**.

These holes will move through the crystal structure the opposite way to the electrons.

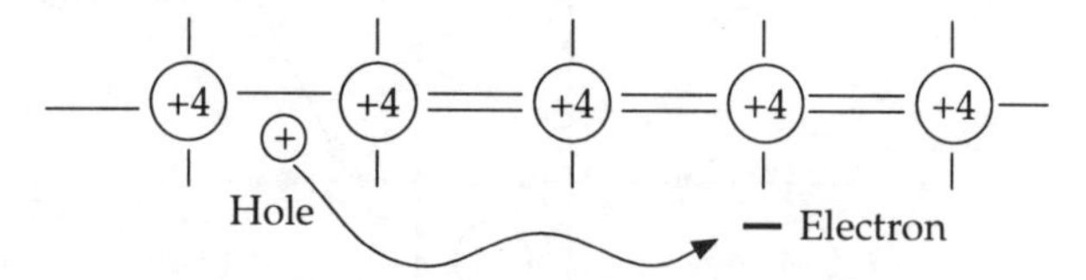

Whenever thermal energy breaks a bond a hole and electron pair are generated

Hole Electron Movement

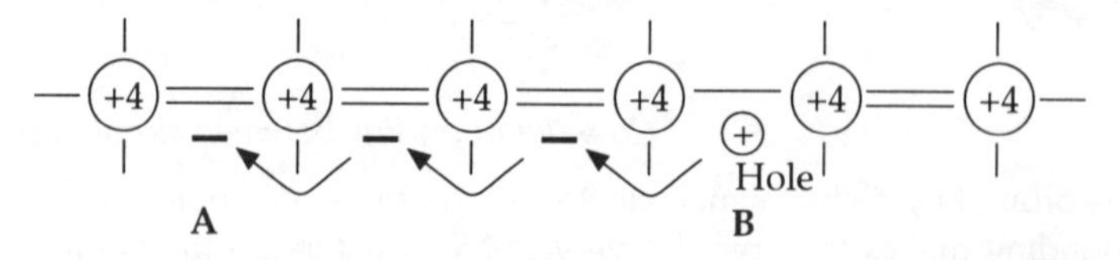

A series of electron movements from atom to atom will effectively move the hole from A to B.

Summary

- **Semiconductor**: Semiconductor material has a resistance between that of a conductor and an insulator. The resistance will **decrease** as the temperature **increases**. Common semiconductor materials are silicon and germanium.
- **Pure semiconductor**: Co-valent bonding produces crystal lattice. The number of electrons is equal to the number of holes at room temperature. Heat **increases** the generation of hole-electron pairs of charges.

Impurities, P and N Type Semiconductor Material

In order to improve the conductivity of the semiconductor material, impurities are added by a process called doping.

Doping

Silicon is produced in very pure form by **zone refining**. This reduces the impurity content to less than one part in 10,000,000,000. To give the semiconductor the correct characteristics it is then doped with controlled amounts of impurities to produce an impurity level of one part in 100,000,000.

Impurities are added from either Group III elements to create **P-type material**, or from Group V elements to create **N-Type material**.

P-Type Material

Boron, Indium and Aluminium are Group III elements and therefore only have three valence electrons in their outer shell.

A Group III impurity atom can only complete three of the co-valent bonds with neighbouring silicon atoms, and at the site of the broken bond a hole is introduced into the lattice.

The positive hole can accept a free electron and increase the flow of current through the material, i.e. when an electron breaks free from the other bonds to fill the vacant hole it leaves behind a hole. The majority of free charges are holes in P-type material.

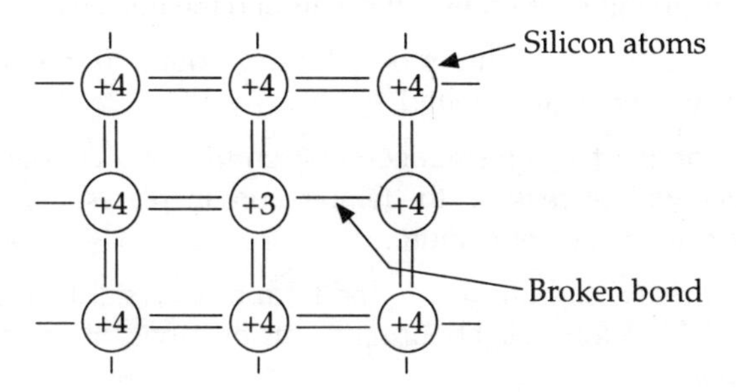

P-type Impurity Atoms (the impurity atom bonds with only three of the four surrounding silicon atoms)

N-Type Material

N type semiconductor material has impurity atoms with five valence electrons added. Phosphorus, arsenic or antimony are commonly used materials.

These materials all have five electrons in their outer shells. Four of these electrons will complete the bonding with silicon atoms but the fifth electron cannot be bonded and so is available to help form an electric current.

In N-type the majority of free charges are electrons.

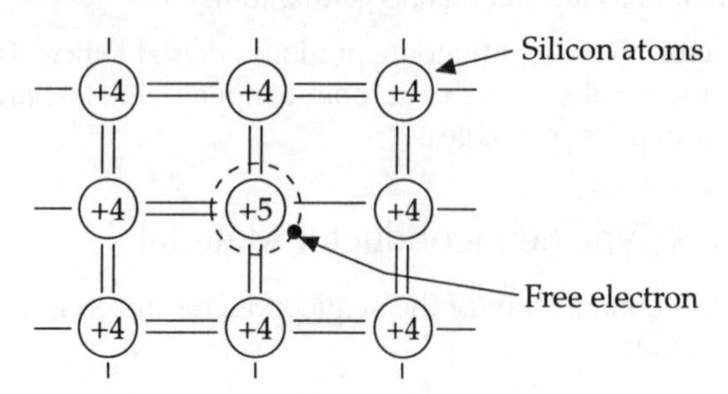

N-type impurity atoms (the impurity atom bonds with all four surrounding silicon atoms, but a fifth electron cannot be bonded).

Summary of Simplified Semi-conductor Theory

- Anything which occupies space contains matter.
- All matter is made up of elements.
- An element is a substance that cannot be broken down into a simpler substance. A molecule of a material contains two or more different elements.
- All elements contain atoms.
- An atom is the smallest part of an element that can exist and take part in chemical action.
- An atom has a nucleus, made up from protons and neutrons, at its centre, which is orbited by electrons. Electrons have very little mass.
- The number of electrons in any element determines the properties of that element. The electrons exist in different shells (orbits) around the nucleus of the atom.
- A conductor has electrons which are only loosely bound to the nucleus, whereas in an insulator they are very tightly bound.
- Semiconductor material has a resistance between that of a conductor and an insulator. The resistance will decrease as the temperature increases. Common semiconductor materials are silicon and germanium.
- Silicon, germanium and carbon exist in the form of a crystal lattice. In this structure, every atom is linked to four neighbouring atoms by a pair of shared electrons which form co-valent bonds.
- In pure semiconductor the number of electrons is equal to the number of holes at room temperature (20°C). Heat increases the generation of hole-electron pairs of charges; i.e. electrons may break their bonding and leave a positive hole.
- *P-Type Material.* This material is doped with tri-valent impurity atoms which cannot complete the bonding in the lattice and a positive hole exists which can accept a

negative electron. They are therefore called acceptor atoms. The majority of free charges are holes.

- *N-Type Material.* This material is doped with pentavalent impurity atoms which donate a free electron. All four bonds are completed leaving a free electron available for conduction. For this reason they are called donor atoms. The majority of free charges are electrons.

Self-assessment questions *(answers page 144)*

1) Which of the following are properties of semiconductors?
   a) electron emission.
   b) resistance decreases as temperature rises.
   c) has a crystalline structure.
   d) will only conduct in one direction.

2) Which of the following are semi-conductor devices?
   a) reed relay.
   b) silicon diode.
   c) thermionic diode.
   d) germanium signal diode.

3) The number of holes is equal to the number of electrons and electron-hole pair generation increases greatly as the temperature rises in:
   a) P-type material.
   b) N-type material.
   c) pure semiconductor material.
   d) insulators.
   e) conductors.

4) Positive holes are the majority charge carriers in:
   a) P-type material.
   b) N-type material.
   c) pure semiconductor material.
   d) insulators.
   e) conductors.

5) Negative electrons are the majority charge carriers in:
   a) P-type material.
   b) N-type material.
   c) pure semiconductor material.
   d) a perfect insulator.

6) In P-type material at room temperature:
   a) there are free electrons.
   b) there are no free electrons.
   c) There are no mobile positive holes.

7) In N-type material at room temperature:
   a) there are mobile positive charges.
   b) there are no free electrons.
   c) There are no mobile positive holes.

8) A co-valent bond is a bond formed by:
   a) two atoms uniting.
   b) four atoms uniting.
   c) two atoms sharing one valence electron.
   d) two atoms sharing two valence electrons.

9) Doping a semiconductor with three valent impurities produces:
   a) N-type material.
   b) P-type material.
   c) an insulator.
   d) a good conductor.

10) Doping a semiconductor with five valent impurities produces:
   a) N-type material.
   b) P-type material.
   c) an insulator.
   d) a good conductor.

Semi-conductor diodes

Introduction

One of the main uses of a diode is to convert a.c. voltage into d.c. To do this it must not allow current to flow in both directions around the circuit. The current must only flow in one direction.

The diode is therefore a device that has a low resistance in one direction and a much higher resistance in the reverse direction.

Properties of the Diode

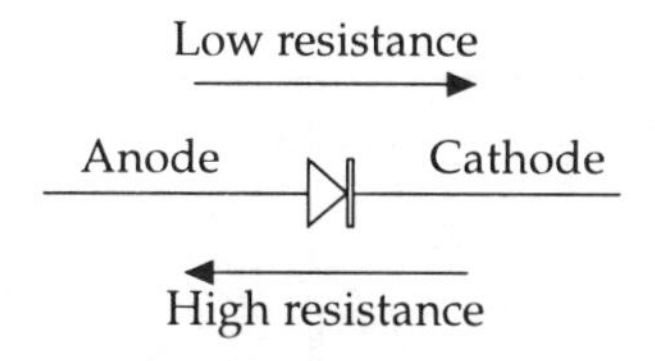

The diode symbol: the arrow on the diode symbol indicates the direction of low resistance.

If a d.c. voltage is applied to the diode, the diode will either conduct or block the current depending upon the polarity of the voltage.

Diode Conducting

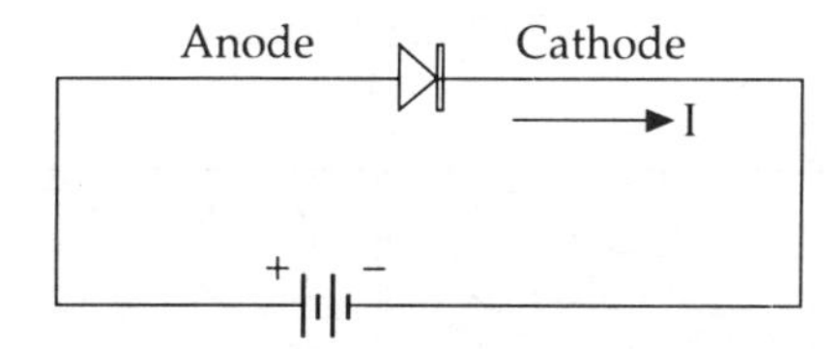

When the anode is positive, the diode conducts and a large current flows producing a small forward volt drop.

Diode Blocking

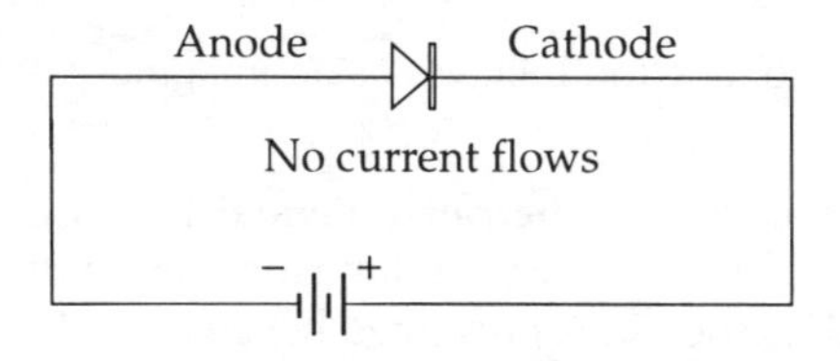

When the anode is negative, the diode blocks current and only a small leakage current flows.

Semi-conductor Diode

The diode can be formed from P–and–N type semi-conductor material.

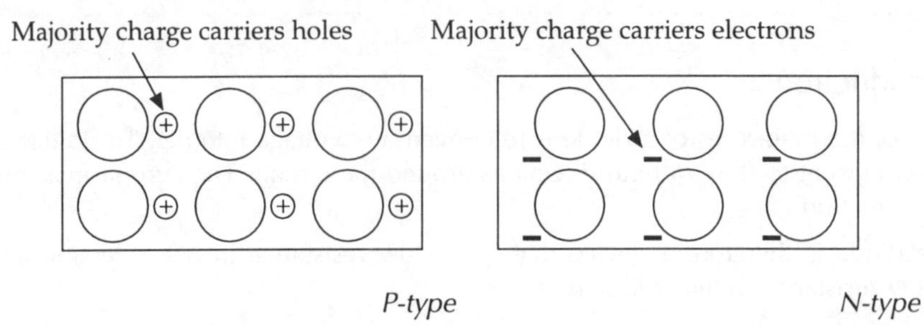

If we consider a bar of silicon doped with P–type material in the one half and N–type in the other half, then the electrons and holes near the junction of the two halves move across and combine with each other and effectively disappear.

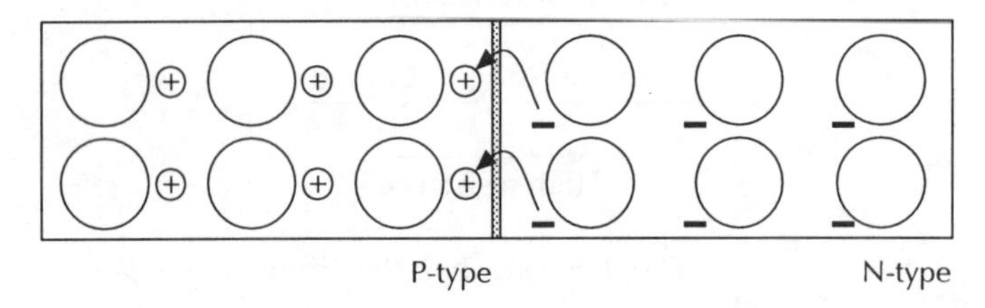

The P–type gains electrons and the N–type loses electrons

That is, electrons leave the N–type and cross over into the P–type in the region of the junction; the P–type acquires a small negative charge (gained electrons) and the N–type a small positive charge (lost electrons).

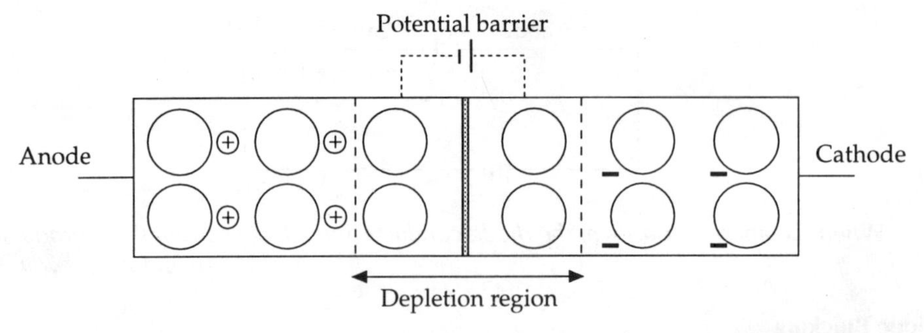

Electrons combine with holes creating a potential barrier and a depletion area

This small charge across the junction behaves as an imaginary battery and typically for silicon is about 0.7V.

This imaginary battery is called the **barrier potential** because it is a 'barrier' to further movement of holes and electrons across the junction. Any further movement of electrons from the N–type across the junction is prevented because the P–type now has a small negative charge. (*Remember*: like charges repel each other).

Since the region around the junction is now devoid of free holes and electrons it is called the **depletion region** and it acts like an insulator.

Note that it is not possible to produce a P-N junction simply by 'sticking' pieces of P-type and N-type material together. To create a diode a piece of intrinsic silicon is taken and each half is doped to produce the P–and N–type material.

We have now created a semi-conductor diode which can either **block** or **conduct** current depending upon which way round the voltage is applied.

Blocking Mode

In the blocking mode the voltage is applied across the diode to assist the potential barrier (negative to the P-type or anode).

The reinforced barrier potential will stop any majority carriers (electrons in N-type and holes in P-type) from crossing the junction. However, electron-hole pairs are generated on or near the depletion region, due to the effect of temperature, and give rise to a small reverse current.

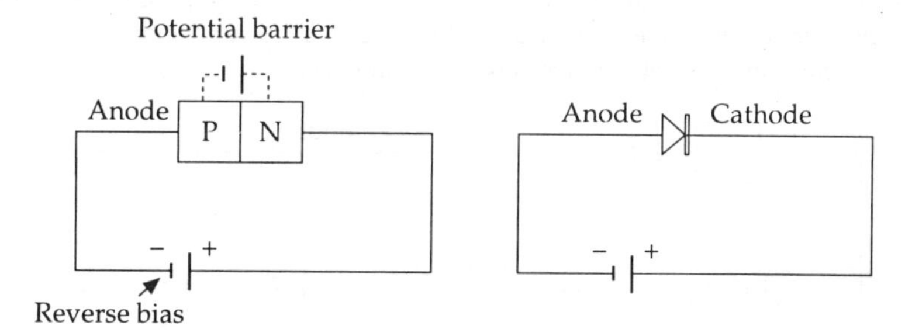

The reverse bias voltage assists the potential barrier and prevents movement of charge carriers.

Only a very small current flows which is called the **reverse current** or **leakage current**.

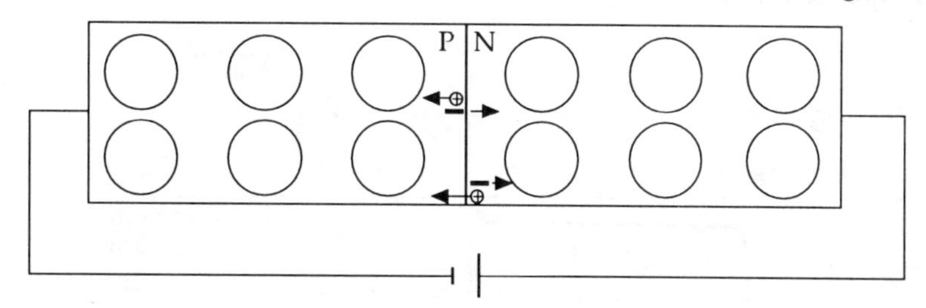

Electron-hole pairs are generated near the junction due to temperature.

The barrier potential across the junction is such that minority carriers (holes on N–side, electrons on P–side) are **attracted across the junction**. This becomes the reverse current.

This small current is in the order of microamps but will increase as the temperature of the semi-conductor material increases. However the diode resistance in this direction is very large, several million ohms, so the diode blocks current.

If the reverse voltage is increased, then the negative terminal of the supply attracts the holes in the P-type away from the junction. Similarly the positive terminal of the supply attracts the electrons from the N-type material. The majority current carriers in both the P–and N–type material are therefore attracted away from the junction. This depletion in the number of current carriers has the effect of increasing the width of the depletion region.

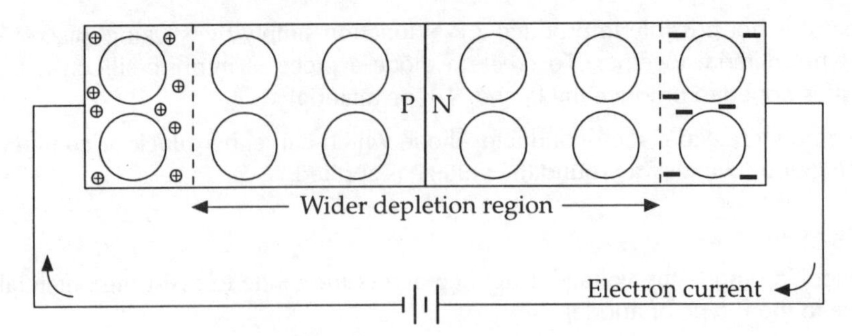

Increasing the reverse bias voltage widens the depletion region

Conducting Mode

If the supply is connected in such a way that it **breaks down** the potential barrier (positive to the P-type anode, negative to cathode) then there is no longer any opposition to the movement of majority current carriers across the junction.

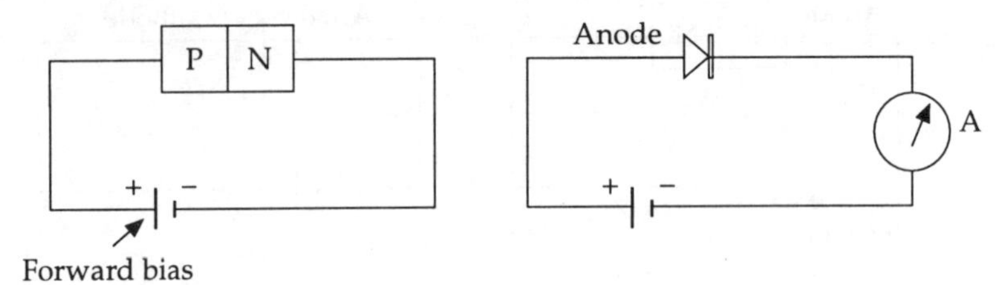

Forward bias voltage breaks down the potential barrier and a large current flows

The Diode Characteristic

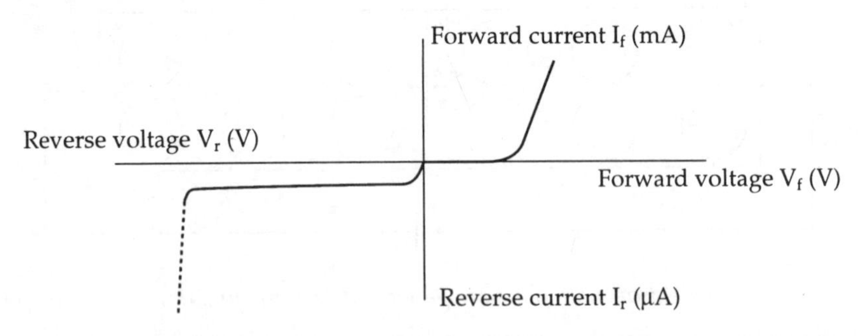

A typical Forward and Reverse Characteristic for a semi-conductor diode

The Forward Characteristic

As the applied forward voltage (V_f) is increased from zero, no current is measured until a value of voltage known as the turn-on voltage is reached (about 0.7V for silicon). When this happens, forward current I_f begins to flow.

Any further increase in supply voltage causes I_f to increase very rapidly, while the value of voltage across the diode, V_f, remains at a fairly constant value around 0.7V for silicon.

The forward volt drop (V_f) and the reverse leakage current (I_r) depend upon the construction of the diode, and, principally upon the material used.

The Reverse Characteristic

A graph of reverse voltage V_r plotted against reverse current I_r at a certain temperature is called the reverse characteristic of the P–N junction diode.

Only a very small leakage current flows in the reverse direction and increasing the reverse voltage V_r does not alter the reverse current I_r very much. (However too large a reverse voltage will cause the diode junction to break down; allowing too much current to flow and possibly destroying the junction!) An increase in temperature has a greater effect!

The reverse current is not very dependent on the reverse voltage, as this current is due to electron-hole pairs being generated by thermal energy.

The temperature of the semi-conductor therefore has more effect on the reverse current since more electron-hole pairs are produced as the temperature increases.

Effect of Temperature on the Reverse Characteristic

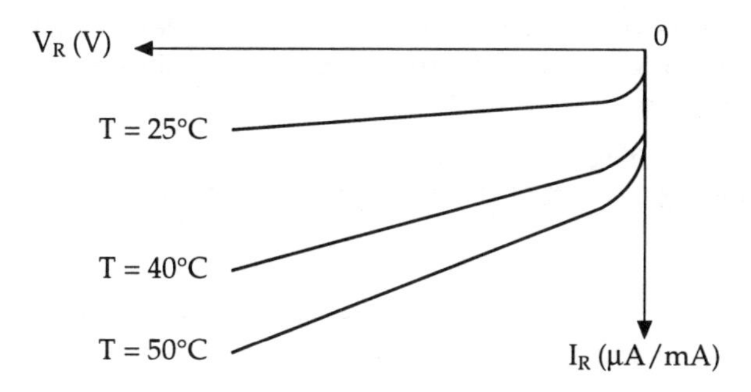

Reverse current increases as the temperature increases

Typical Values

	Forward voltage V_f	*Reverse current* I_r
Silicon	0.6V to 0.8V	a few nA
Germanium	0.2 to 0.3V	a few μA

Types of Diode

Diodes are often classified into **signal** or **power** types.

Signal diodes are used to process a.c. signals and require a low forward volt drop. Germanium diodes are therefore used for processes like demodulation.

Power diodes are used for rectification and need to cope with large forward currents and large reverse voltages.

When choosing power diodes the following factors need to be considered.

Average Forward Current I_{Fav}.

This is the average forward rectified current and is quoted for a range of temperatures. If this temperature range is exceeded then this figure must be derated: e.g. at a temperature $T_{ambient}$ = 60°C it has a rating of 1A and at T_{amb} = 100°C a rating of 0.75A.

Repetitive Peak Reverse Voltage V_{RRM}.

This is the maximum repetitive peak voltage which the diode can withstand in the reverse direction, e.g. 100V.

Forward Voltage V_F.

This is the forward volt drop, usually quoted for a particular value of forward current and junction temperature of the diode: e.g.

$$V_F = 1.3V \text{ at } I_F = 2A,\ T_j = 25°C.$$

Diode Faults

As already indicated above, the main causes of failure in diodes are due to excessive repetitive reverse voltages or excessive forward current which causes the power rating (P_{tot}) of the diode to be exceeded.

The result is usually a catastrophic failure with the diode either going short circuit (very low resistance) blowing fuses, or open circuit (high resistance).

Testing Diodes

Use an Ohmmeter and measure the forward resistance (a few ohms) and the reverse resistance (hundreds of thousands of ohms).

Note: if the diode is still in the circuit under test then the measured reverse resistance may be altered by the shunting effect of other components, transformer windings, resistors etc,. If in doubt unsolder one end of the diode and repeat the test.

Summary (P–N Junction Diodes)

- P–and N–type regions in a slice of crystal form a P–N junction diode

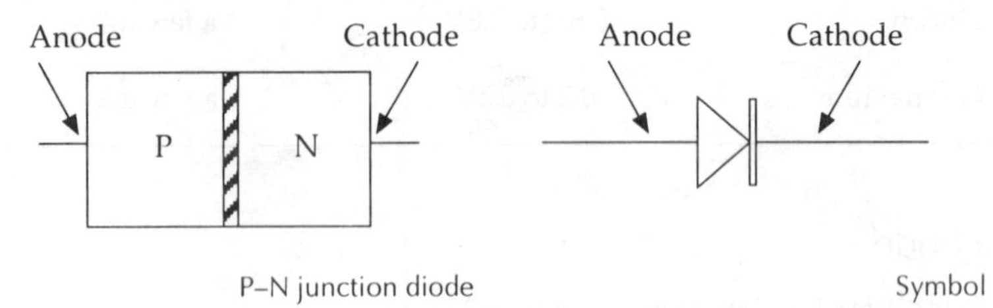

P–N junction diode   Symbol

- This device conducts more easily in one direction than the other, so it will rectify a.c.
- At room temperature, mobile charges have enough energy to cross between the P and N junctions.

- The result is that there is an excess of positive charge on the N side, and an excess of negative charge on the P side of the junction.
- This builds up a **potential barrier** of a few tenths of a volt which repels further charge movement.

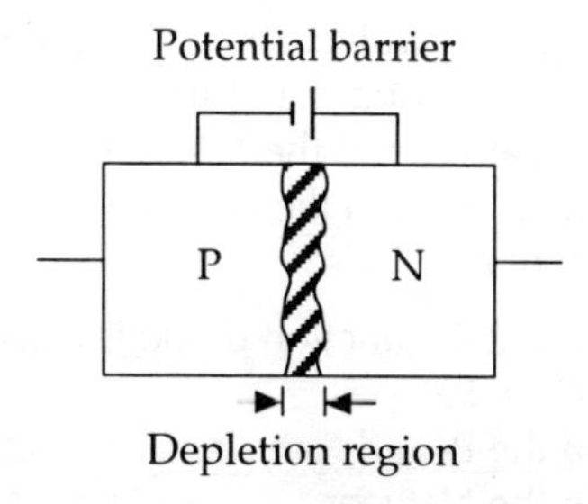

A potential barrier is formed at the junction and a depletion region exists across it.

- When the junction is first formed, charge movement leaves a narrow strip between P and N regions depleted of mobile charges. It is across this **depletion layer** that the potential barrier is formed.
- *Forward Bias.* The externally applied e.m.f. **reduces** the potential barrier and allows charges to flow across the junction. **The diode offers a low resistance** (few ohms).

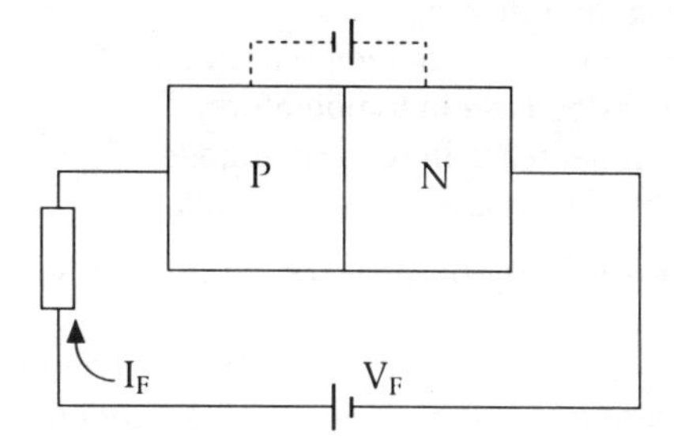

Forward Bias. Conducting mode.

- Reverse Bias. The externally applied e.m.f. **increases** the potential barrier. Only a very small leakage current, due to minority carriers, flows in the reverse direction. **The diode offers a very high resistance** (M ohms).

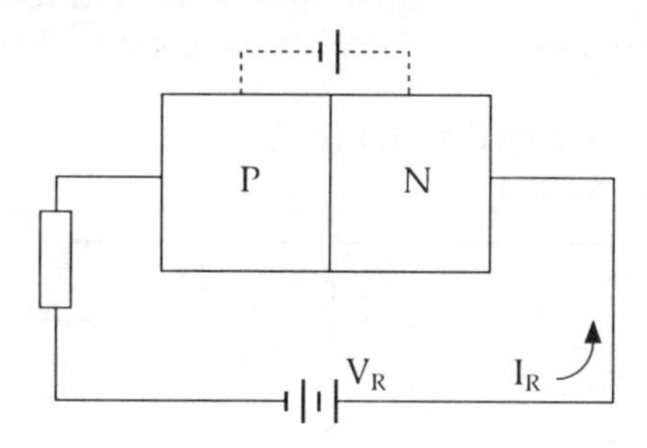

Reverse Bias. Blocking mode.

Self-assessment questions (answers page 145)

1) A P-N junction forms:
   a) a diode
   b) a short circuit
   c) a bi-directional switch
   d) a transistor.

2) Forward biasing a P-N junction involves the connection of an external e.m.f which:
   a) makes the P side positive, and the N side negative.
   b) makes the P side negative, and the N side positive.
   c) makes the anode negative and the cathode positive.
   ċ) assists the potential barrier.

3) Reverse leakage current in a P–N junction diode is caused by the presence of:
   a) majority carriers in the P–type.
   b) majority carriers in the P–and N–type.
   c) minority carriers in the N–type.
   d) minority carriers in the N–and P–type.

4) Reverse biasing a P-N junction involves the connection of an external e.m.f which:
   a) makes the P side positive, and the N side negative.
   b) makes the P side negative, and the N side positive.
   c) makes the anode positive and the cathode negative.
   d) breaks down the potential barrier.

5) Reverse leakage current in a diode:
   a) increases with an increase of temperature.
   b) increases with a decrease in temperature.
   c) increases with a decrease in reverse voltage.
   d) decreases with an increase in reverse voltage.

6) A diode having a forward volt-drop of 0.25v and a reverse leakage current of 8μA would be:
   a) a germanium diode
   b) a silicon diode
   c) a thermionic diode
   d) a light emitting diode.

7) A diode having a forward volt-drop of 0.7v and a reverse leakage current of 10nA would be:
   a) a germanium diode
   b) a silicon diode.
   c) a thermionic diode
   d) a light emitting diode.

8) The two diodes shown in the diagram have a forward resistance of 2 ohms and an infinitely large reverse resistance. Calculate the currents flowing in each branch of the circuit when:
   a) the battery is connected as shown.
   b) the battery is reversed.

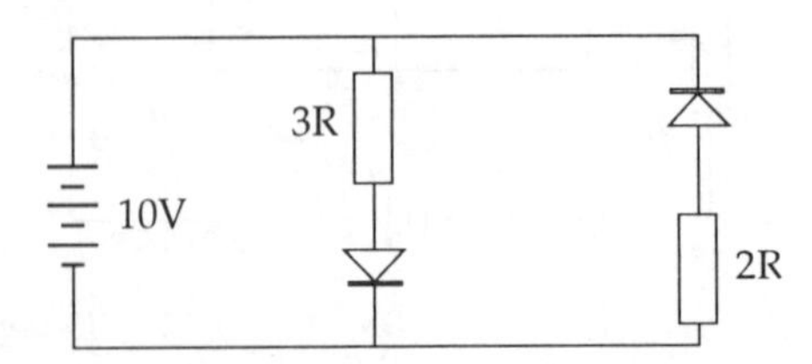

Remember that 2R means 2 ohms!

9) A silicon diode has a leakage current of 10nA at 20°C. If the leakage current doubles for every 5°C, what will be the leakage current at 100°C?

10) Consider that the diagrams below have diodes with a forward resistance of 4 ohms and an infinite reverse resistance. Calculate the readings on the ammeters A1, A2 and A3.

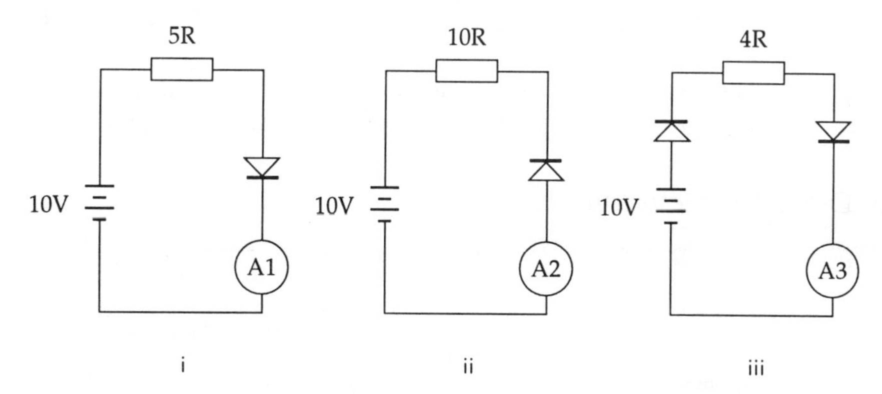

Zener (Voltage Reference) Diodes

Zener diodes are P-N junction diodes that are designed to break down in the reverse direction when the reverse voltage exceeds a certain value.

With the temperature constant, the reverse current I_r is fairly constant. If the breakdown voltage is exceeded then the reverse current suddenly increases rapidly.

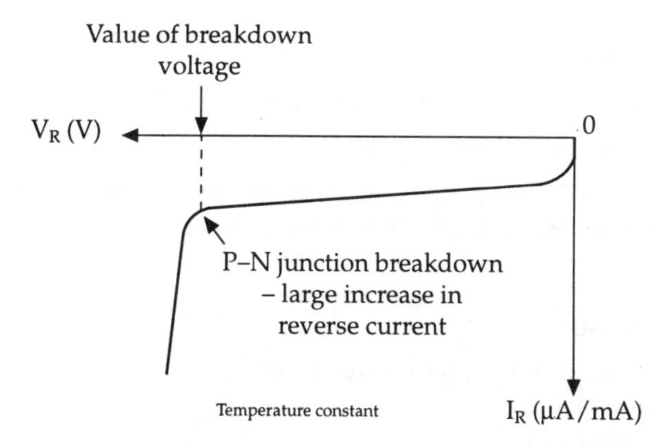

The diode breaks down with reverse voltage; the voltage across the diode is constant and a large reverse current flows

Reverse Breakdown

Zener diodes are made to operate in this reverse breakdown region with breakdown voltages in the E12 and E24 series ranging from 2.7V to around 68V. (The E12 and E24 series are also used with the preferred ranges of resistor values.)

To prevent destruction of the diode a current limiting resistor in series with the diode is needed. This will limit the power dissipated by the diode and prevent its destruction.

That is, the zener voltage × reverse current must not exceed the maximum power rating P_{tot} of the diode.

Uses of Zener Diodes

When the diode has broken down the voltage across the diode is constant (as long as the temperature remains constant) and the zener diode can be used to provide a reference voltage.

Also, when the diode has broken down the voltage across the diode is independent of the reverse current and the zener diode may be used to provide a stabilised supply.

Zener Diode Characteristic

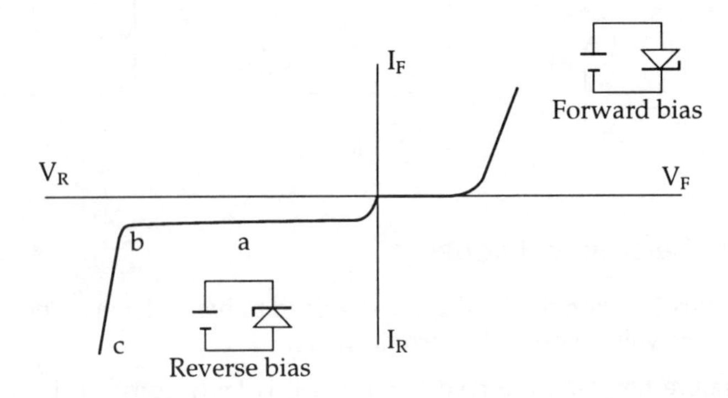

(a) Small I_r (b) breakdown (c) large I_r

The Reverse Characteristic

At (a) the diode has reverse bias, and passes only a very small leakage current.

At (b) the diode is on the point of breaking down.

At (c) the diode is in reverse breakdown. Reverse current can increase rapidly with little further increase in the reverse voltage. The reverse current is independent of the reverse voltage.

The Forward Characteristic

This is similar to a normal diode and the zener will have low forward resistance but a slightly higher forward volt drop.

Range of Zener Diodes

Series	*Power rating*	*Zener Voltage*
BZY88	500mW	2.7V to 15V
BZX85	1.3W	2.7V to 6.8V
BZX61	1.3W	7.5V to 72V
BZY93	20W	9.1V to 75V
IN5333	5W	3.3V to 24V

When selecting zener diodes you must consider the power rating as well as the zener voltage!

Case Outlines

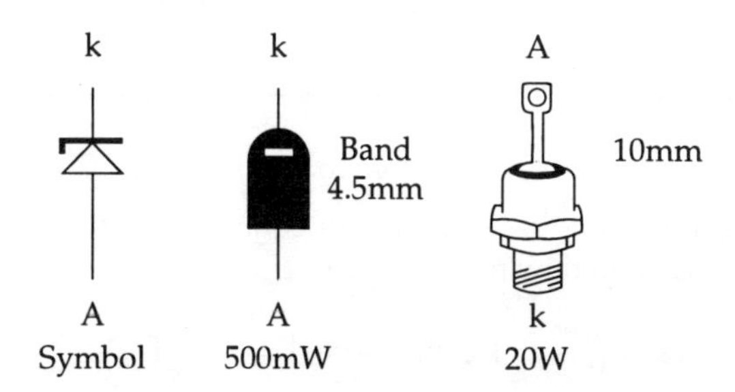

Typical case outlines for zener diodes

Note: The zener diode is normally operated with the cathode connected to the positive rail of the supply!

Testing Zener Diodes

The zener may be tested with an ohmmeter in the same way as a normal diode. It should have a low forward resistance and a high reverse resistance (providing the battery voltage in the ohmmeter does not exceed the breakdown voltage of the zener).

When checking the zener in the circuit under test, if the correct zener voltage is not obtained, then either there is insufficient reverse voltage to break the zener down or the zener diode is faulty.

Using the Zener Diode to Stabilise a d.c. Supply

The use of the zener diode to stabilise a supply is best shown by an example. Consider the circuit shown below.

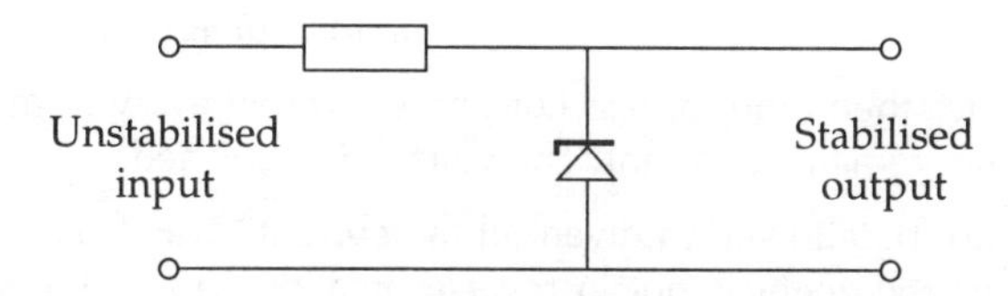

The zener diode and series resistor are connected as shown to provide a d.c. stabilised supply.

Current Limiting

To prevent damage to the zener diode then the maximum power dissipated by the diode must be limited by a series resistor. If we assume that a 5V, 5W zener is to be used, then an unstabilised supply of say 10V will be needed to allow a volt drop across the series resistor.

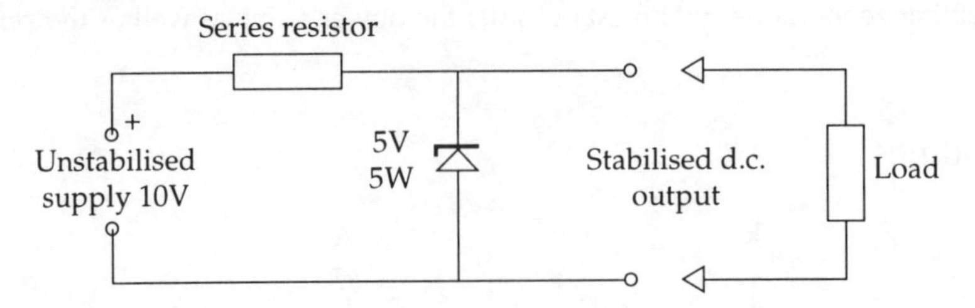

No load; the zener dissipates maximum power

From the above Figure, if the load resistor was not connected and the zener diode was dissipating its 5W then the current flowing would be:

$$\frac{\text{Power}}{V} = I \ : \ \text{therefore, } I = \frac{5}{5} = 1\text{A}.$$

The supply voltage is 10V therefore 5V must be dropped across the series resistor.

The value of the series resistor must be:

$$R = \frac{V}{I} = \frac{5\text{V}}{1\text{A}} = 5 \text{ ohms}$$

Note that without a load being connected the zener is dissipating maximum power.

If now the load resistor is connected then current will be diverted away from the zener and less power will be dissipated in the zener: i.e. if the load resistor dissipates 3W then only 2W will be dissipated in the zener.

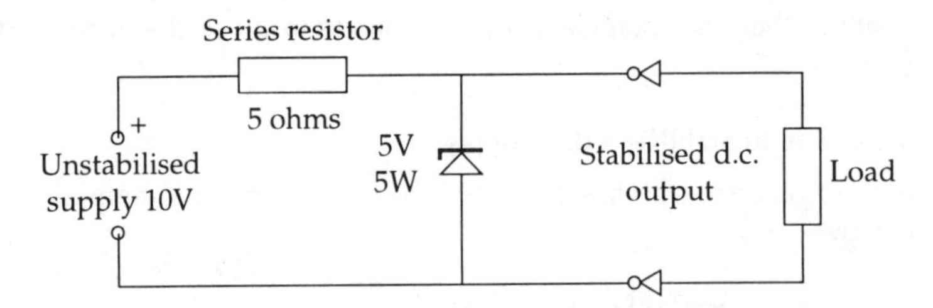

The load diverts current away from zener

If the load current varies then more or less current is diverted away from the zener but the voltage across the zener remains constant. The voltage is stabilised.

Note: the load must not be allowed to divert all the current away from the zener, because under these conditions the zener is not in reverse breakdown and the voltage across the diode will no longer remain constant.

The stabilising circuit above may be connected across the output of a d.c. power supply to provide a constant output voltage.

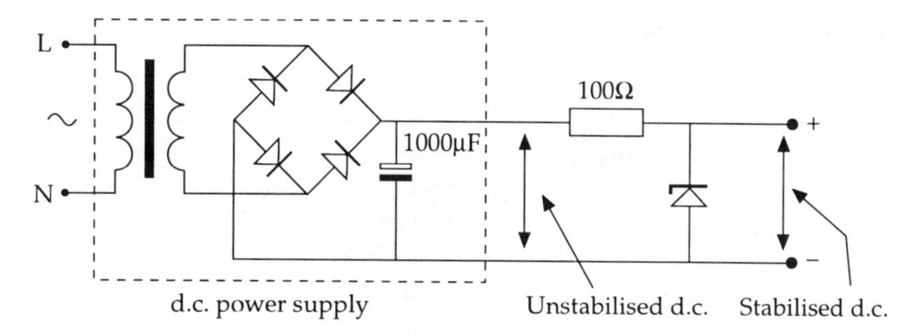

Obtaining a stabilised output from a d.c. power supply

Summary (Zener Diodes)

- Zener diodes are designed to break down in the reverse direction without damage, as long as the power rating of the diode is not exceeded.
- When the diode breaks down a large current flows in the reverse direction and the voltage across the diode remains constant.
- To limit the reverse current the zener diode is always operated with a series resistor.
- The zener diode may be used as a voltage reference or to provide a stabilised supply.

Self-assessment questions *(answers page 145)*

1) The zener diode in normal use has its:
   a) anode connected to the positive rail.
   b) cathode connected to the positive rail.
   c) cathode connected to the negative rail.

2) A resistor is placed in series with the zener diode to:
   a) reduce the supply voltage.
   b) isolate the load from the unstabilised power supply.
   c) limit the current taken by the load.
   d) limit the current taken by the zener.

3) When the zener diode breaks down then the voltage across it is:
   a) constant.
   b) constant if the temperature remains constant.
   c) dependent upon the reverse current that is flowing.
   d) independent of temperature and reverse current.

Questions 4 to 8 refer to the circuit below

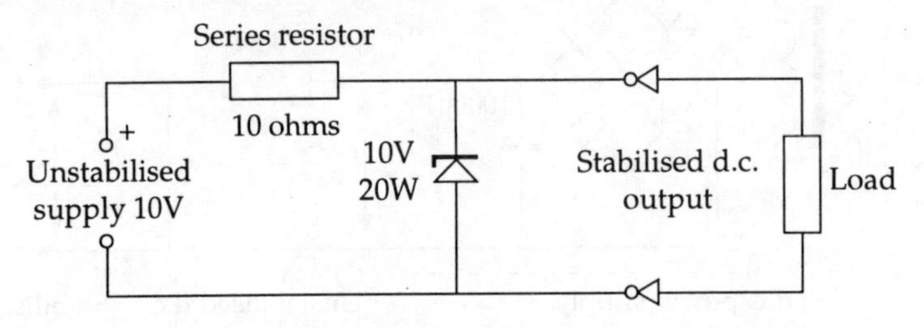

4) What current flows through the zener diode when the unstabilised input voltage is 25V and the load is disconnected from the output?

   a) 1A b) 1.5A c) 2.0A d) 2.5A

5) What current flows through a load of 10R connected to the stabilised output of the circuit?

   a) 0.5A b) 1A c) 1.25A d) 2A

6) A load is connected which takes a current of 0.5A from the circuit. If the unstabilised d.c. input voltage is 25V, what current flows through the zener diode?

   a) 0.5V b) 1A c) 2A d) 2.5A

7) A load of 10 ohms is connected to the stabilised d.c. output of the circuit. If the unstabilised input voltage is 30V, what current flows in the load?

   a) 0.5V b) 1A c) 2A d) 3A

8) The unstabilised d.c. input voltage to the circuit remains constant at 25V, but the load resistance connected to the stabilised output terminals is reduced. Which of the following then happens?

   a) the load current is reduced.
   b) the load voltage increases.
   c) the volt drop across the 10R resistor is reduced.
   d) the current through the zener is reduced.

Multirange test instruments

Introduction

When measuring, testing and faultfinding we often need to measure a large range of a.c.and d.c. voltages and currents, as well as resistance. To do this we need a **general purpose multirange instrument**.

Both analogue (pointer type of display) and digital (7 segment display) multirange testmeters are available but it is recommended that a digital instrument is used as these are much easier to read. Also, if you connect it the wrong way around (positive lead where the negative should be) you won't do any damage to the instrument.

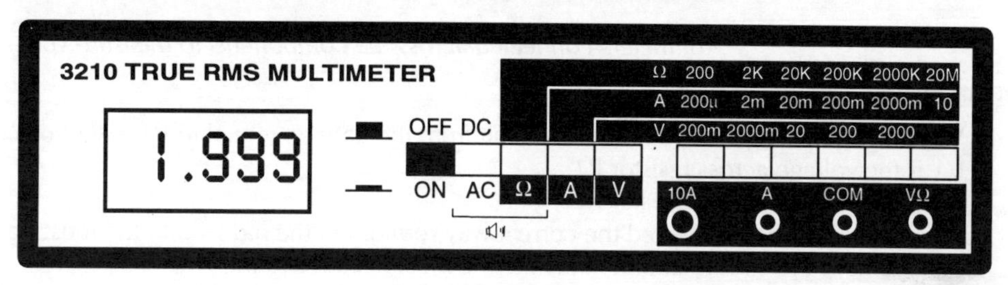

A Typical Digital Multirange Meter

Using Multirange Testmeters

Precautions

1) When measuring voltages, always set the scale range to a suitable scale *before* connecting the meter to the circuit.

   If the size of the voltage being measured is not known set the meter to the *highest voltage range* and reduce the range in steps after connecting the meter.

2) Never change to a current or resistance range when you are connected to a circuit and measuring a voltage.

3) When measuring current, switch the power supply to the circuit OFF, connect the meter, set the current range to the highest range possible, switch ON, and then switch down the ranges, if necessary, to obtain the most suitable reading for the current flowing.

Note on a.c. Supplies

1) On a.c. the polarity of the connections is not important; the meter can be connected either way round.

2) A meter set for a d.c. range will read zero if it is connected to an a.c. supply.

3) The voltage indicated on a.c. will be the r.m.s. value.

4) The meter reading will only be accurate on a.c. supplies for a true sine wave.

5) A meter set for an a.c. range will indicate a reading when connected to a d.c. supply but the reading will be inaccurate.

Measurement of Voltage

To measure voltage the meter must be placed across the component as shown below.

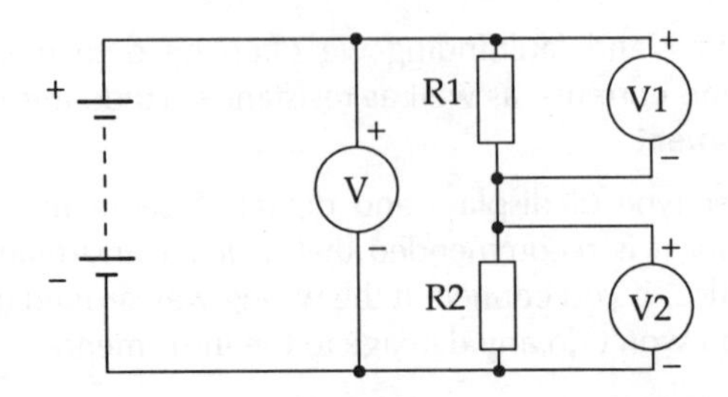

Voltmeters connected across the components to measure voltage

In this Figure:

V is the total voltage supplied to the circuit from the battery, (equal to V1 + V2 volts).
V1 is the voltage across resistor R1.
V2 is the voltage across resistor R2.
The meter must be connected the correct way round: i.e. the red lead to the most positive point in the circuit, and the black to the most negative point.
For d.c. supplies the meter must be set to d.c. ranges.
For a.c. supplies the meter must be set to a.c. ranges.

Measurement of Current

When measuring current the meter must be placed in series with the component as shown in the Figure below.

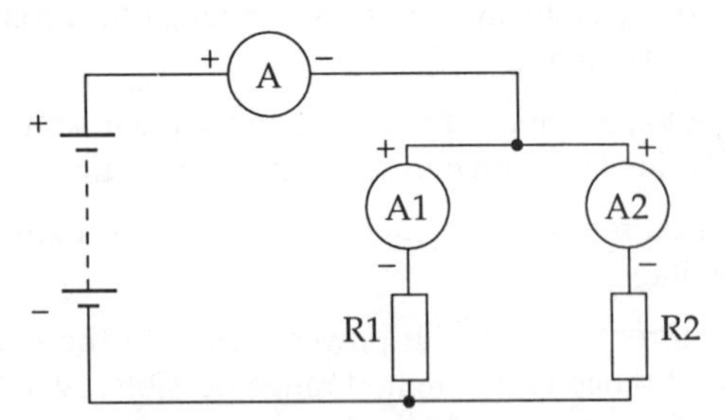

The meters are connected in series with the components to measure current

Once again the meter must be connected to observe the correct polarity.

To Measure Current:

Switch off the supply.
Connect the meter in series with the component.
Select a.c. or d.c. range as appropriate.
Note: With most digital instruments there is a separate input socket for current measurement. Re-connect the meter if necessary to the current socket.
Switch to the highest current range available.

Switch on the supply.
Switch down the current ranges until a suitable reading is obtained.

In the Figure above:
Current A is the total current supplied to the circuit (Equal to A1 + A2 amps).
Current A1 is the current flowing in resistor R1.
Current A2 is the current flowing in resistor R2.

Measurement of Resistance

To measure the resistance of a component in a circuit first **switch off** the **supply** to the circuit.

Then select **Resistance** and switch to the **appropriate range** on the multirange meter.

Connect the meter leads **across the component** and measure the resistance, switching up or down ranges to obtain the best reading.

If a lower than expected resistance is obtained check to see if another smaller resistance is in parallel and is shunting the resistor you are testing. If this is the case you will have to remove one end of the component under test from the circuit.

Continuity (Short Circuit) Test

Many of the digital instruments have a very low resistance or continuity test setting. This is very useful for checking if wires or tracks on printed circuit boards are continuous (not broken). Many of these continuity testers also have a buzzer, so that if the test leads are shorted together whilst on this range, an audible buzz is heard.

Diode Test

Another facility that is provided with the digital instrument is a diode test. To test a silicon diode in the forward direction, a voltage of at least 0.7V is required, otherwise the diode will not be forward biased and a high resistance will be indicated even in the forward direction!

On the diode test setting sufficient voltage is available to forward bias the diode!

Practical Exercises

Tips:

1) Use a breadboard (see How to use breadboards, p.96) and **lay out the components** on the board **as they appear on the circuit diagram**.
2) Always breadboard the circuit before connecting the battery or power supply.
3) If you are using a variable voltage power supply, always switch on and set the output voltage **before connecting it to the circuit**.
4) When making any alterations to your circuit always switch off the power supply first, make the change, then switch on.

Practical Exercise 1: Measurement of Voltage

a) Using the circuit shown at the top of page 90, connect two 100 ohm 1W resistors in series across a 9V supply.

Measure and record the voltages V, V1 and V2.

b) Repeat exercise (a) with two 1k (1000) ohm resistors.

Measure and record the voltages V, V1 and V2.

c) Change R1 for a 6k8 ohm (6.8k) resistor, and R2 for a 2k2 ohm resistor. Measure and record the voltages V, V1 and V2

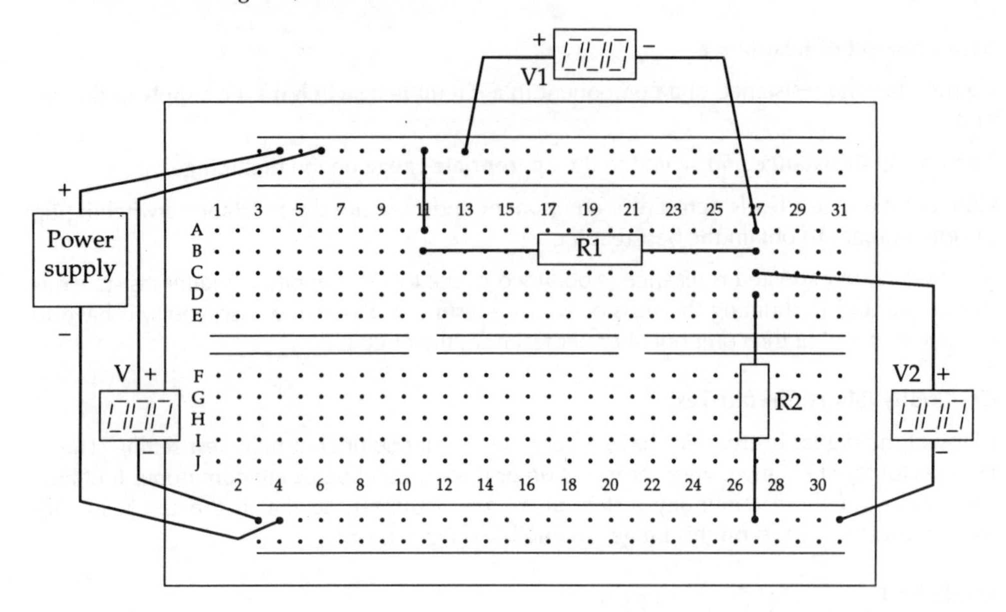

Suggested board layout for exercise 1

Note: If you only have one voltmeter then connect it across the supply, then across R1 and then R2 in turn to measure the voltages.

Practical Exercise 2

a) Using the circuit at the bottom of page 90, connect two 100 ohm 1W resistors in parallel across a 9V supply. (Remember to reconnect the meter to read current if necessary, and set it to a high current range). Measure and record the currents A, A1 and A2.

b) Repeat the exercise using two 1k ohm resistors.

c) Make R1 a 6k8 ohm resistor and R2 a 2k2 ohm resistor and measure and record the currents A1 and A2.

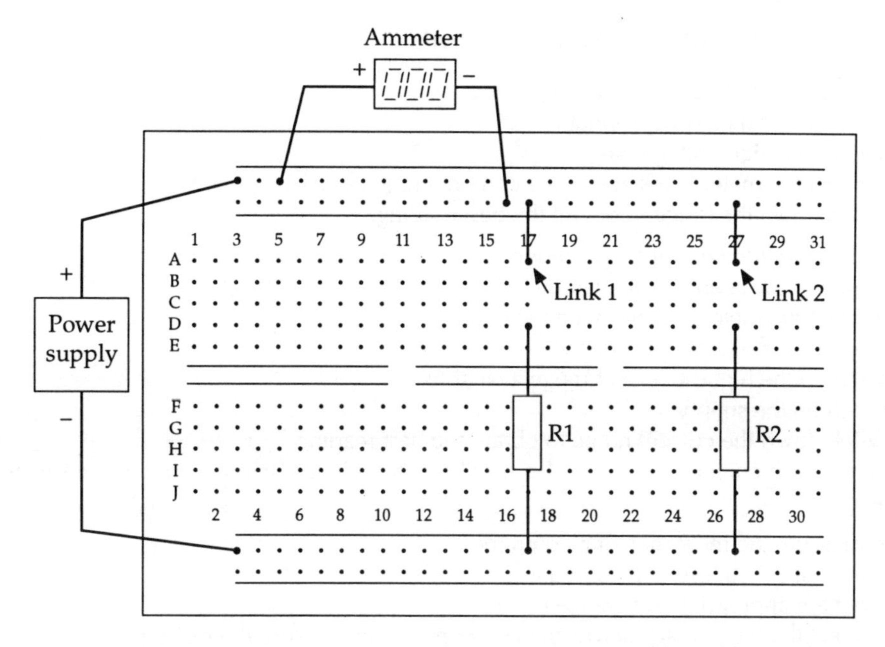

Suggested board layout for Exercise 2

Note: If you only have one ammeter available then connect up the breadboard as shown with link 1 and link 2 inserted.

Measure and record the supply current A.

Now remove the ammeter and complete the circuit using a wire link. Remove link 1 and insert the ammeter in its place.

Measure and record the current A1.

Replace link 1, remove link 2 and insert the ammeter in its place. Measure and record current A2.

Practical Exercise 3 measurement of resistance

Disconnect the power supply from your breadboard!

a) Use the ohms range on your instrument to measure the resistance of two 100 ohm resistors and then the combined resistance of the two resistors in series.

b) Use the ohms range on your instrument to measure the resistance of two 1000 ohm resistors and then the combined resistance of the two resistors in series and in parallel.

c) Use the ohms range on your instrument to measure the resistance of two 6k8 ohm resistors and then the combined resistance of the two resistors in series and in parallel.

Summary of Multirange Test Instruments

- *To measure voltage across a component:*
  Select a.c. or d.c. as appropriate.
  Switch to a high voltage range.
  Connect the meter leads across the component (observe the polarity).
  Switch down the range to obtain the best reading.
- *To measure the current in a component.*
  Switch off the supply to the circuit.
  Connect the meter in series with the component.
  Select a.c. or d.c. as appropriate.
  Switch to the highest current range available.
  Switch on the supply.
  Switch down the current range to obtain the best reading.
  After measurement, switch off the supply to the circuit before removing the instrument.
- To measure the resistance of a component.
  Switch off the supply to the circuit.
  Select the appropriate resistance range.
  Connect the meter leads across the component and measure the resistance.
  If a lower than expected resistance is obtained check to see if another smaller resistance is in parallel and is shunting the resistor you are testing. If this is the case you will have to remove one end of the component under test from the circuit.
  Remember to switch the instrument from the ohms range after use and turn the instrument OFF.

Self-assessment questions *(answers page 145)*

These should be attempted after doing the practical exercises above.

1) When two resistors in series are connected across a power supply then:
   a) V1 – V2 = V.
   b) V1 + V2 = V.
   c) V1 + V2 is less than V.
   d) V1 + V2 is greater than V.
2) When two equal value resistors are connected in series across a power supply then:
   a) The combined voltage drop across them will equal the supply voltage but V1 and V2 will differ.
   b) The combined voltage drop will not be equal to the supply but V1 will equal V2.
   c) The combined voltage drop will equal the supply voltage and V1 will equal V2.
   d) The combined voltage drop will not be equal to the supply voltage and V1 and V2 will be different.
3) When two unequal value resistors are connected across a power supply then:
   a) The combined voltage drop across them will equal the supply voltage but V1 and V2 will differ.

   b) The combined voltage drop will not be equal to the supply but V1 will equal V2.
   c) The combined voltage drop will equal the supply voltage and V1 will equal V2.
   d) The combined voltage drop will not be equal to the supply voltage and V1 and V2 will different.

4) When two unequal value resistors are connected in series across a power supply then to calculate the volt drop across each resistor we only need to know:
   a) The supply voltage.
   b) The supply current.
   c) The supply voltage and the values of the two resistors.
   d) The supply voltage and the difference in value between the two resistors.

5) When two resistors are connected in parallel across a power supply then:
   a) A1 – A2 = A
   b) A1 + A2 = A
   c) A1 + A2 is less than A
   d) A1 + A2 is greater than A

6) When two equal value resistors are connected in parallel across a power supply then:
   a) The combined currents in them will equal the supply current but A1 and A2 will differ.
   b) The combined currents in them will not be equal to the supply current but A1 will equal A2.
   c) The combined currents in them will equal the supply current and A1 will equal A2.
   d) The combined currents will not be equal to the supply current and A1 and A2 will differ.

7) When two unequal value resistors are connected in parallel across a power supply:
   a) The combined currents in them will equal the supply current but A1 and A2 will differ.
   b) The combined currents in them will not be equal to the supply current but A1 will equal A2.
   c) The combined currents will equal the supply current and A1 will equal A2.
   d) The combined currents will not be equal to the supply current and A1 and A2 will differ.

8) When two unequal value resistors are connected in parallel across a power supply then to calculate the current flowing through each resistor we only need to know:
   a) The supply voltage.
   b) The supply current.
   c) The supply voltage and the ratio of the two resistors.
   d) The supply voltage and the value of each resistor.

How to use breadboards

Introduction

A breadboard consists of a board with a large number of holes, into which a wire may be inserted. The wire will be gripped and held firm and make a good electrical connection. This makes the device ideal for constructing and testing temporary circuits.

Below is a diagram of a typical breadboard.

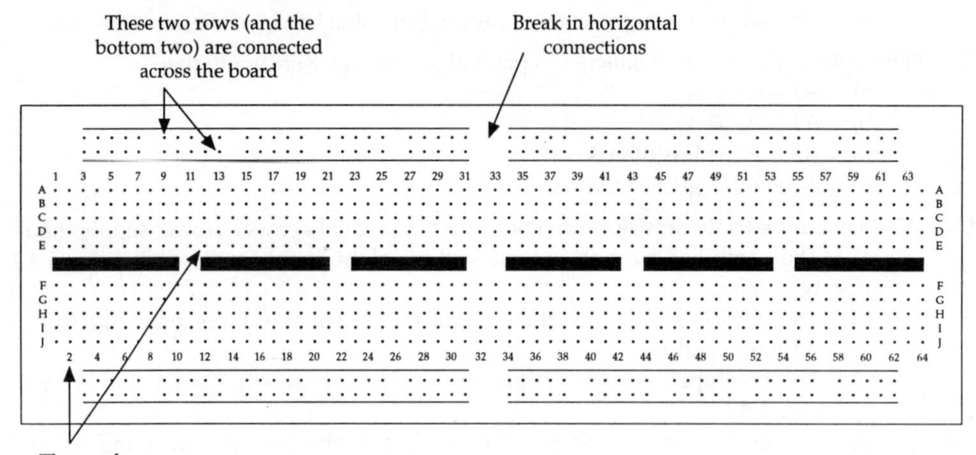

The central part of the board has rows of holes labelled A,B,C,D, etc, and columns of holes numbered by odd numbers on the top (1 to 63) and even numbers on the bottom (2 to 64).

If we consider **column 1** then **A,B,C,D**, and **E** are all **connected** together underneath the board. Similarly **F,G,H,I** and **J** are all **connected** together. The same is true for all the other columns.

Across the top of the board are two sets of holes that may be used as power supply rails to the breadboarded circuit. The top row of holes are all connected together underneath the board, similarly with the second row of holes. However, as shown above, there is a break in the connections in the centre (column 33) and if both halves of the board are used a bridge wire is needed across this gap. There are similar holes at the bottom of the board.

Breadboarding is quite simple but does require a little care and attention if you are to be successful. The following points will help to make your breadboarded circuit more successful and will also help to preserve the life of the breadboard.

Using Breadboards

1) Use different coloured wires: i.e. red for positive supplies, black for negative supplies, yellow or green for inputs, blue for outputs etc. This will help you to identify parts on the circuit more easily.
2) Do not use stranded wire, as strands break off and remain lodged in the breadboard.

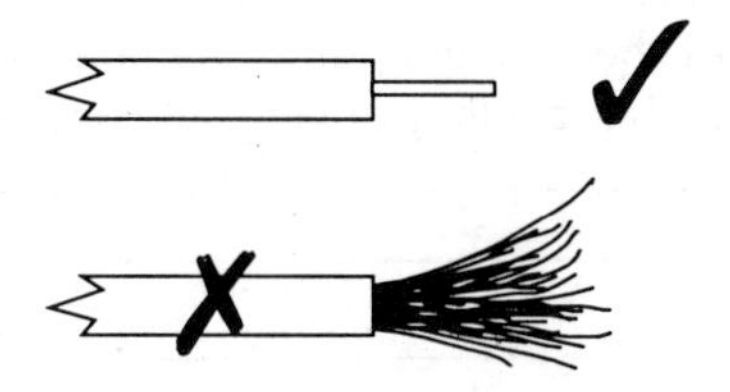

Do not use stranded wire

3) Wherever possible use a flat layout. Lay all components flat against the board, cut components to the correct length.

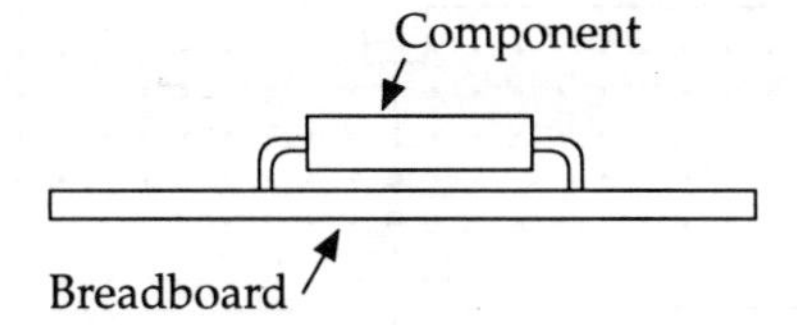

Use a flat layout

4) However if you are short of space then it is quite in order to stand large components on end rather than flat against the board.

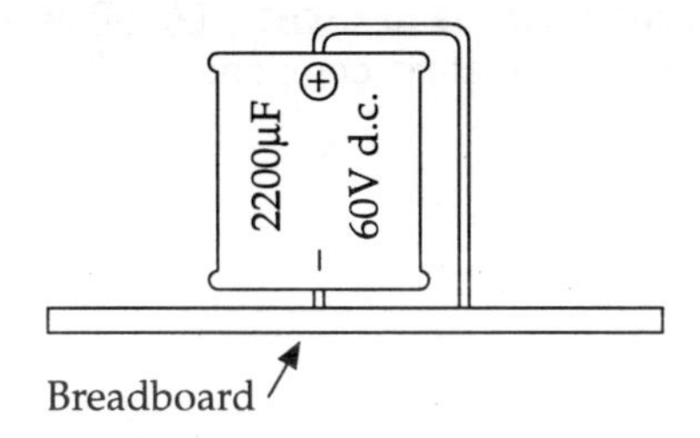

Large components such as electrolytic capacitors may be stood on end

5) Arrange the circuit on the board to correspond with the circuit diagram as much as possible: e.g. inputs on the left, outputs on the right hand side.

6) Only one component/wire per hole.

7) Use snipe-nosed pliers to insert and withdraw wires/components. Grip the component/wire about 3mm (1/8th inch) from the end. Do not use too much force however and insert/withdraw the wire vertically, never at an angle.

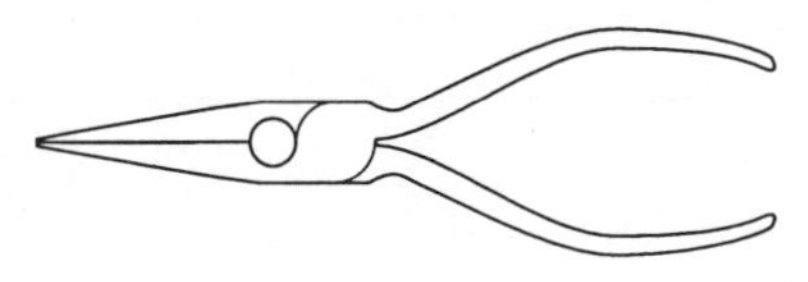

Use snipe-nosed pliers to insert wire/components

8) Use the whole of the board. If you are using the same breadboard for a lot of exercises then use different parts of the board for each exercise. This will reduce the wear and tear on the board.

9) Do not have wires crossing over each other and particularly over components (see diagram below). It makes the changing of a component very difficult. For neatness arrange components either horizontally or vertically on the board.

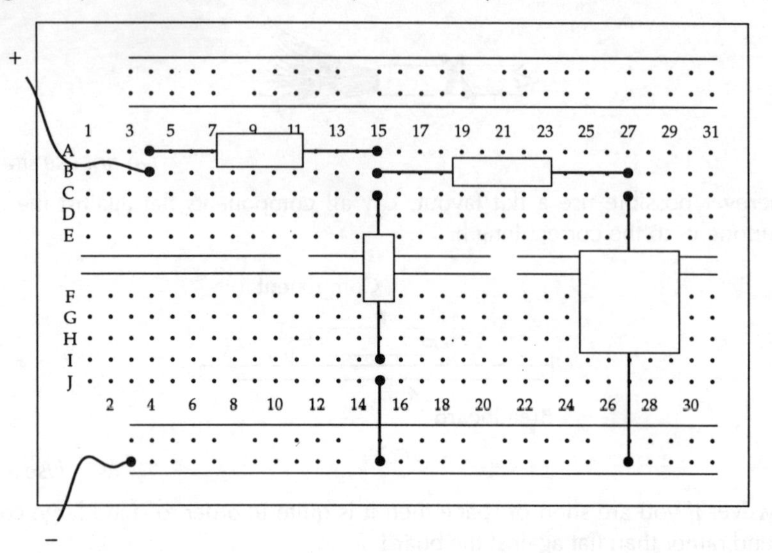

10) No circuit current on the board is to exceed 1A. Power resistor leads may be too large for the holes. Heat generating components should be mounted clear of the board.

Resistors and resistor colour coding

Introduction

Resistors are either used to control the current in electronic circuits, to divide up the voltage (potential divider), or to convert a changing current in the output of a system into a corresponding change in voltage.

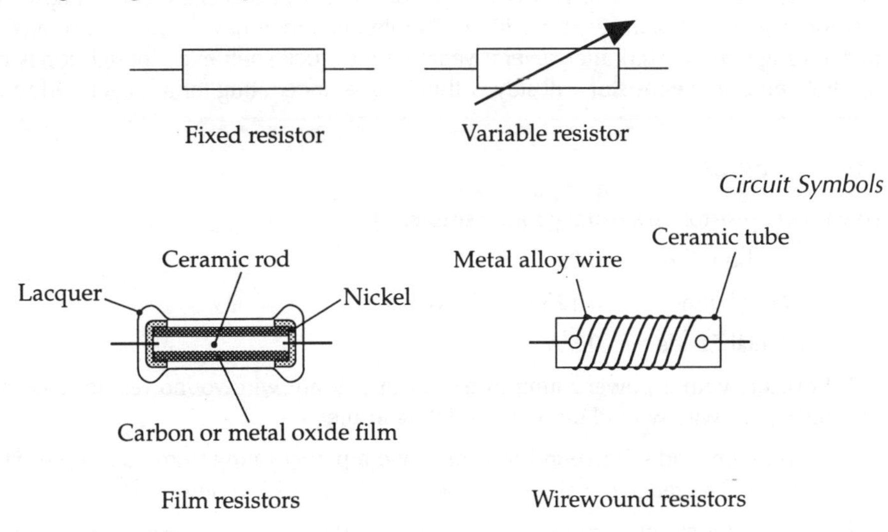

Circuit Symbols

Typical Resistor construction

The Main Properties of Resistors

Resistance value

Expressed in ohms (Ω kΩ or MΩ).

Power Rating

The maximum power that can be dissipated by the resistor determines its power rating. The power dissipated produces heat and if the rating is exceeded then the resistor overheats and burns out.

Power dissipated can most easily be determined from measuring the voltage across the resistor and then calculating the power:

$$\text{e.g.: Power} = V^2/R$$

Tolerance

It is very difficult to manufacture a large number of resistors to have a value of say 100 ohms exactly. Some of the resistors will be below and others above this value. Resistors are therefore produced and all those within 5% of the nominal (stated) value will be given a

5% tolerance rating. That is, all resistors that have an actual value of between 95 ohms to 105 ohms for a 100 ohm resistor will be within the tolerance rating.

Noise

When a current flows through a resistor it generates electrical noise. This is important in high gain amplifiers.

Stability

The value of the resistor will vary over a period of time due to chemical changes within the resistor: e.g. a 100 ohm resistor with a 5% tolerance may have values outside of the permitted range (95 to 105) after several years on the stock shelf even though it has not been used! A more stable resistor will stay within its tolerance rating for a longer period of time.

Resistor types

Low power resistors are usually *film* resistors:

Carbon film	0.25W to 2W
Metal film	0.125W to 0.5W
Metal Oxide	0.5W

- Resistors with a power rating in excess of 2W are wirewound resistors: i.e. they use resistance wire wound onto an insulating former.
- Aluminium clad wirewound resistors have a power rating from 25W up to 50W with heatsink. The range of values are from 0.1 ohms to 1000 ohms (1kΩ).
- Ceramic wirewound resistors: power rating 4W, 7W, 11W, and 17W. Range of values: 0.47 ohms to 22k ohms.
- Silicon and vitreous enamel wirewound resistors: power rating: 25W. Range of values: 0.1 ohms to 22k ohms.

Preferred Values of Resistors

A 100 ohm resistor with a 10% tolerance will have values between 90 and 110 ohms. There is little point therefore in manufacturing a resistor of 95 ohms with 10% tolerance. Because of the overlap a *preferred* range of resistor values are manufactured.

The 10% (E12 series) range of resistor values is:

1.0, 1.2, 1.5, 1.8, 2.2, 2.7, 3.3, 3.9, 4.7, 5.6, 6.8, 8.2.

i.e. 1k, 1k2, 1k5, 1k8, 2k2 etc.

10k, 12k, 15k, 18k, 22k etc.

100k, 120k, 150k, 180k, 220k etc.

The 5% (E24 series) range is:

1.0, 1.1, 1.2, 1.3, 1.5, 1.6, 1.8, 2.0, 2.2, 2.4, 2.7, 3.0, 3.3, 3.6, 3.9, 4.3, 4.7, 5.1, 5.6, 6.2, 6.8, 7.5, 8.2, 9.1,

Resistor Colour Codes

Small resistors up to 1W rating are colour coded using four bands of colour painted around the body of the resistor towards one end.

The colour bands indicate the nominal (stated) value of the resistance and the tolerance relative to the stated value.

Four Band Resistor Colour Code

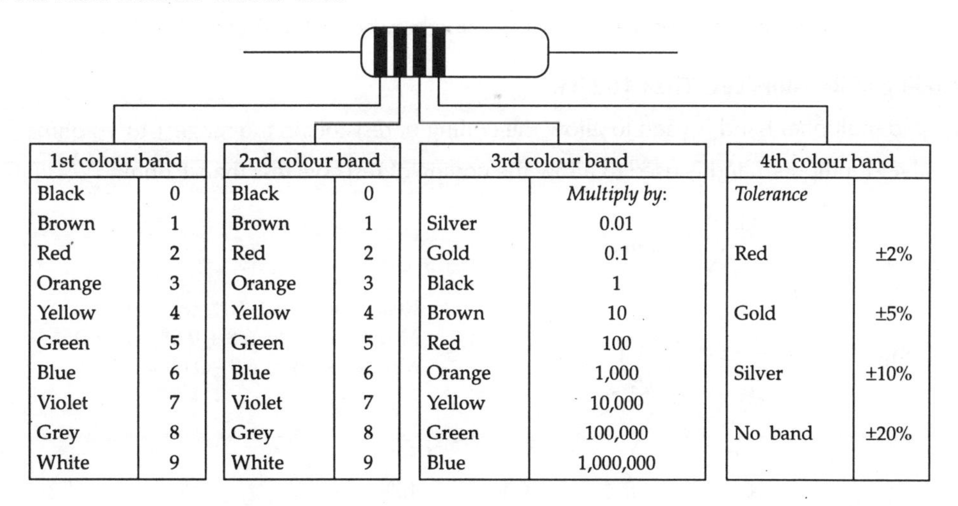

1st colour band		2nd colour band		3rd colour band		4th colour band	
Black	0	Black	0		*Multiply by:*	*Tolerance*	
Brown	1	Brown	1	Silver	0.01		
Red	2	Red	2	Gold	0.1	Red	±2%
Orange	3	Orange	3	Black	1		
Yellow	4	Yellow	4	Brown	10	Gold	±5%
Green	5	Green	5	Red	100		
Blue	6	Blue	6	Orange	1,000	Silver	±10%
Violet	7	Violet	7	Yellow	10,000		
Grey	8	Grey	8	Green	100,000	No band	±20%
White	9	White	9	Blue	1,000,000		

Colour Code

Example of Resistor Colour Code Banding.

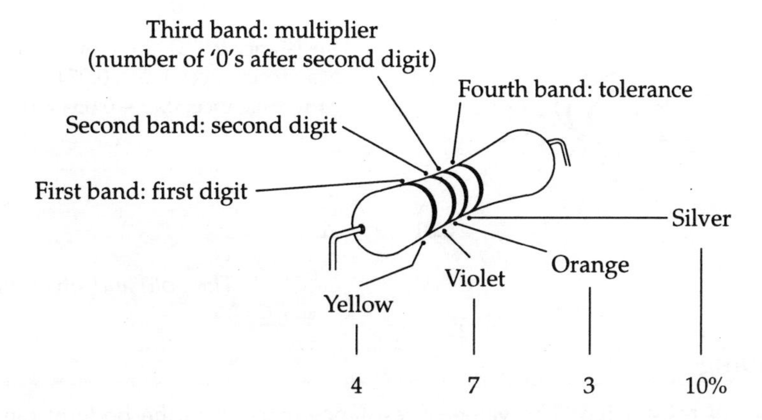

- ❒ The first two digits are 47.
- ❒ There are three 0's following these digits.
- ❒ The tolerance is 10%.

The value of the resistor is therefore 47,000 ohms with a tolerance of 10%. The symbol Ω (Greek letter omega) is used to represent ohms.

Maximum and Minimum Values

10% of 47,000 = 4700

The maximum resistance is

47,000 + 4700 = 51,700 ohms.

The minimum resistance is

47,000 – 4700 = 42,300 ohms.

Therefore a 47k ohm resistor may have any value between 42,300 and 51,700 ohms.

Coding of Resistors Less Than 10 ohms

A gold multiplier band is used to allow the coding of resistors in the range 1 to 10 ohms.

A silver multiplier band is used to allow the coding of resistors less than 1 ohm.

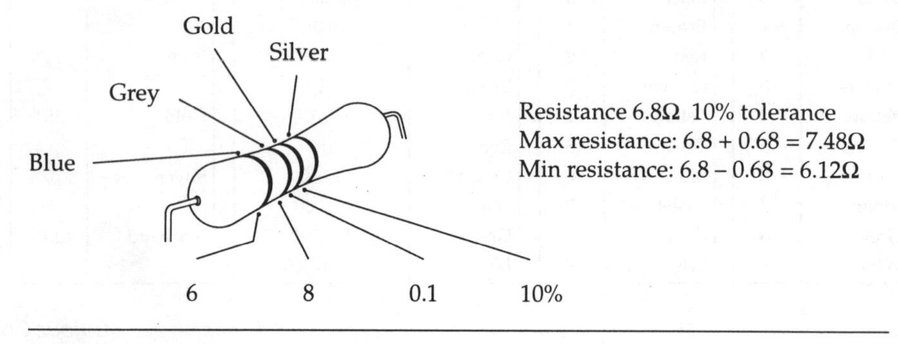

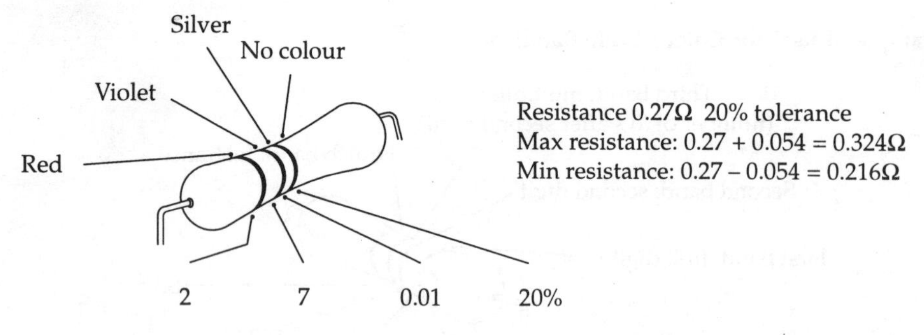

The gold and silver multipliers

Letter Coding

Certain types of resistor have the value of resistance marked on the body of the resistor. The multiplier is a letter marking the position of the decimal point.

M is used to signify millions (1,000,000)
k is used to signify thousands (1000)
R is used to signify units (1)

Examples

10M = 10,000,000 ohms	2M7 = 2,700,000 ohms
5R6 = 5.6 ohms	120R = 120 ohms

Tolerance letter

A further letter may be added to indicate the tolerance:

F = 1%, G = 2%, J = 5%, K = 10%, M = 20%

Examples

R22M = 0.22 ohms	20% tolerance
4R7K = 4.7 ohms	10%
68RJ = 68 ohms	5%
1MOF = 1 M ohm	1%
3M3M = 3.3M ohm	20%

Resistor Faults

Fixed value resistors are generally very reliable. They will however fail due to: chemical ageing, mechanical stress, vibration, heat, applied voltage and humidity.

When they fail most resistors, simply burn out and become open circuit. Carbon composition resistors may go high in value: e.g. a 1kΩ coded resistor has a resistance in excess of 100kΩ.

The only other problem that can be experienced is that they become noisier. This can be a problem in high gain voltage amplifiers.

Variable resistors on the other hand are not very reliable. The wiper contact causes wear on the track, and dust and dirt cause poor electrical contact. The variable resistor becomes noisy or open circuit.

Testing: measure the resistance with a good ohmmeter.

Summary of Resistors and Resistor Colour Coding

- When choosing a resistor the following properties must be considered; resistance value, power rating, tolerance, stability and in certain circumstances noise.
- These properties are governed by the type of resistor; e.g. carbon film, metal film or metal oxide.
- Resistors are manufactured in preferred ranges. The most popular are the 10% (E12) and the 5% (E24) series.
- The nominal value of the resistor and its tolerance are indicated on the component in the form of a colour code.
- Fixed resistors age chemically and may be out of tolerance after a number of years even if they have never been connected in a circuit. They have a certain shelf life.

Self-assessment questions (answers page 146)

1) Identify the nominal value and the resistance of the following resistors.

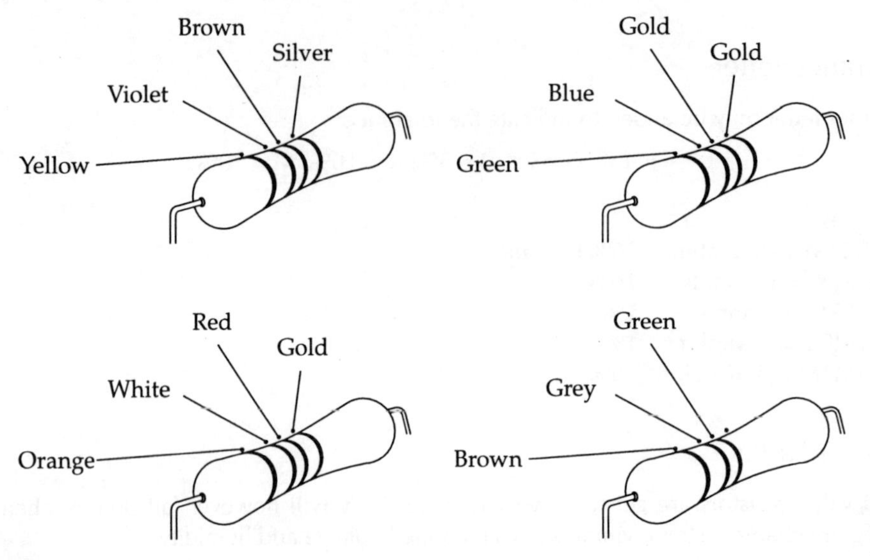

2) Calculate the maximum and minimum allowable values of the resistors below if they are within the stated tolerance.

Nominal resistance	*Tolerance*	*Resistance*	
		Minimum	*Maximum*
100k	10%		
3.3k	5%		
150R	10%		
2R7	1%		

3) State the nominal value and tolerance of the following resistors.
a) R33M b) 4k7F c) 6M8M d) 22KK

4) Give the colour coding for the following resistors:
a) 4R7 10% b) 1R2 2% c) R47 5%

Practical Assignment

Obtain 10 different resistors: i.e. with different colour codes. Determine their nominal resistance value and tolerance from the colour coding.

Enter these values in the table on the next page.

Using the resistance range of a multi-range meter, measure the resistance of each resistor in turn, and enter the value in the table.

Are any of the resistors outside the specified tolerance?

Resistor	*Band colour*				*Nominal resistance*	*Tolerance*	*Resistance*		*Measured resistance*
	1st	*2nd*	*3rd*	*4th*			*Min*	*Max*	
R1									
R2									
R3									
R4									
R5									
R6									
R7									
R8									
R9									
R10									

Table 1 Resistor values

Power supply specifications

Introduction

Most of the present electronic circuits employ semiconductor devices such as transistors and integrated circuits which require a low voltage d.c. supply to operate them.

Alternating current is distributed over the National Grid Network because it is easier to generate and distribute than d.c. Using transformers the a.c. voltages can be stepped-up, for distribution, to a very high voltage. This has the advantage of reducing the power lost in distribution.

For example, 1kW of electricity can be supplied either as 1000V at 1A (the small current will only give rise to a small volt drop and power loss), or 100A at 10V (the larger current would create a larger volt drop with a greater power loss unless the cable size was increased in diameter).

Power Supplies

A power supply is a unit that converts the a.c. mains into d.c power. Usually the d.c. power is at a low voltage, typically 5V, 9V, 12V or 15V, to operate the semiconductor devices.

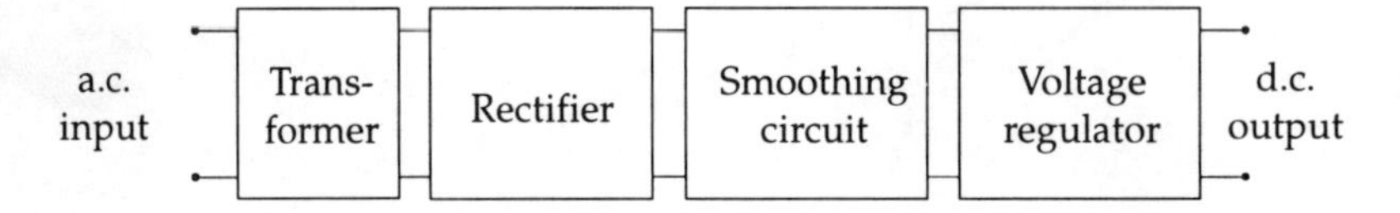

Block Diagram of a Power Supply

Transformer

The transformer isolates the unit from the a.c. mains and usually **steps down** the 240V to a much lower value: e.g. 6V, 9V, 12V or 15V r.m.s.

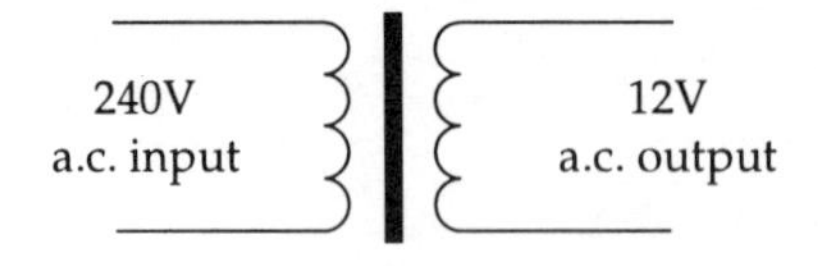

A step down transformer

Rectifier

The rectifier converts the bi-directional a.c. waveform into uni-directional pulses.

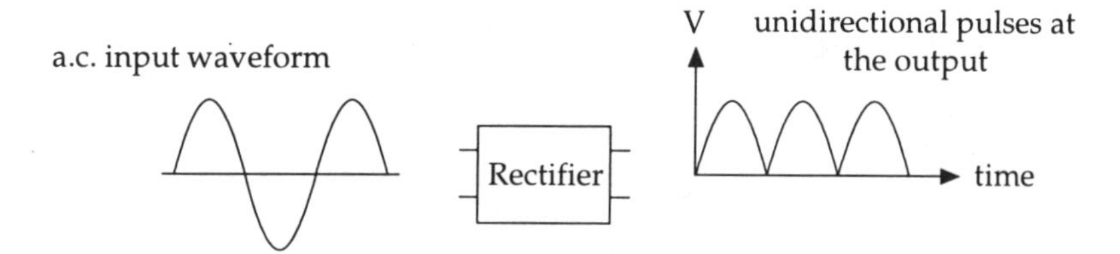

The action of the rectifier

Smoothing Circuit

The smoothing circuit smooths out the unidirectional pulses to give d.c.

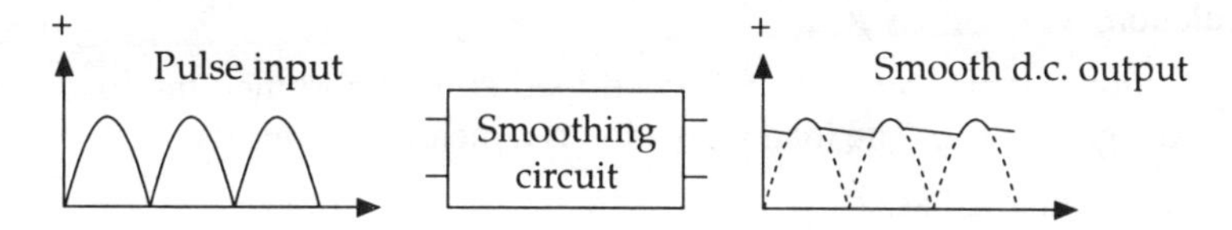

The action of the smoothing circuit

Note that the d.c. output may still contain a small fluctuation called a **ripple**, when the load current is drawn from the supply, as shown in the Figure immediately above.

The Voltage Regulator

The voltage regulator ensures that, as current is drawn from the supply unit, the output voltage (V_{out}) remains constant.

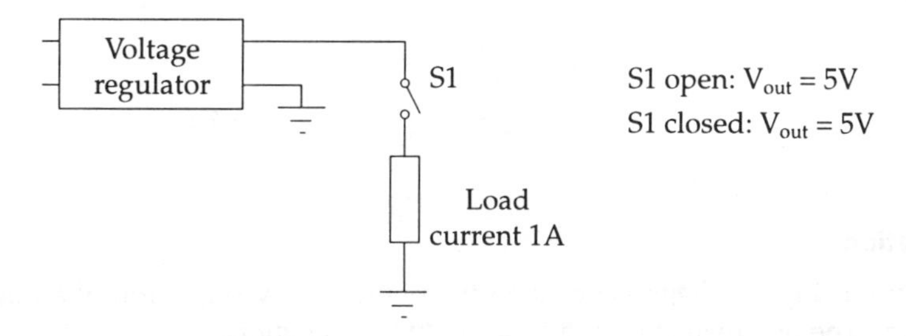

Action of the voltage regulator

Power Supply Specification

There are many different types of power supply which all have different properties and specifications. With all power supplies however we are interested in the following:

Voltage and Current

The input voltage, output voltage and load current.

Efficiency

Not all the power drawn from the a.c. mains is available as output power to the load. Some power is lost within the power supply unit itself.

The efficiency of a power supply is given by:

$$\text{Efficiency} = \frac{\text{d.c. power out}}{\text{a.c. power in}} \times 100\%$$

Example

240V at 200mA — Power supply — 24V at 1.2A

$P_{in} = 48W$ $\qquad$ $P_{out} = 28.8W$

$$\text{Efficiency} = \frac{28.8}{48} \times 100\% = 60\%$$

Load Regulation

If the power supply is delivering its full rated load current and then the load is removed, the output voltage will rise. The load regulation is a measure of this rise.

Load regulation is defined as:

$$\frac{V_{out}\text{ (no load)} - V_{out}\text{ (full load)}}{V_{out}\text{ (no load)}} \times 100\%$$

Example

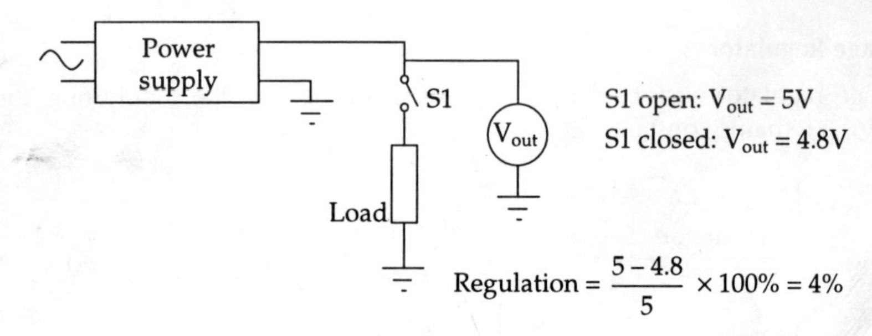

$$\text{Regulation} = \frac{5 - 4.8}{5} \times 100\% = 4\%$$

Line Regulation

If the a.c. mains input voltage varies then the output d.c. voltage from the power supply will also vary. The line regulation is a measure of this variation.

It is often quoted as a percentage ratio: i.e. 10% mains change to 0.01% change in output voltage.

Example

A change of 6V in the 240V mains supply gave rise to a change of 0.0012V in the 12V secondary d.c. voltage.

Therefore a 2.5% change in line voltage caused a change of 0.01% in output voltage.

Ripple

This is the peak to peak or r.m.s. value of any ripple superimposed upon the d.c. output voltage. The value of the ripple is usually quoted at full load: e.g. ripple: less than 10mV peak to peak at full load.

Output Impedance

For a good power supply the output impedance should be as low as possible, typically less than 0.1 ohms.

$$\text{Output impedance} = \frac{V_{out}\text{ (no load)} - V_{out}\text{ (full load)}}{\text{Full load current}} \times 100\%$$

Example

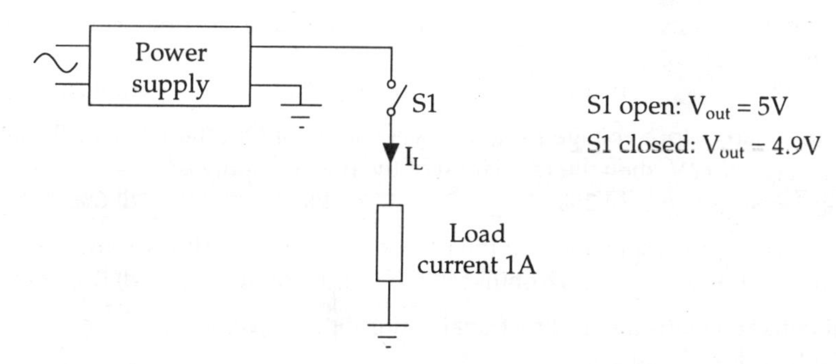

If the output changes from 5V to 4.9V and the full load current is 1A then the output impedance will be:

$$Z_{out} = \frac{5V - 4.9V}{1A} = 0.1\Omega$$

Summary of Power Supply Specifications

- A power supply is used to convert the 240V a.c. mains to a low voltage d.c. supply. It also isolates equipment from the a.c. mains.
- Most power supplies contain a transformer, a rectifier, a smoothing circuit and a voltage regulator.
- The transformer steps down the a.c. mains and isolates the equipment from the a.c. mains.
- The rectifier converts the bi-directional mains supply into uni-directional pulses of voltage.
- The smoothing circuit smooths out the unidirectional pulses to provide a d.c. voltage which contains only a small a.c. ripple.
- The voltage regulator holds the d.c. output voltage constant when the output (load) current changes.
- A power supply is specified in terms of its voltage and current (power out), efficiency, load regulation, line regulation, ripple and output impedance (resistance).
- Efficiency is the d.c. power out divided by the a.c. power in × 100%.
- The load regulation is a measure of the change in the d.c. output voltage between the power supply providing its full rated load current and no load current.
- Line regulation is a measure of the change in the d.c. output voltage for a change in the a.c. input mains voltage.

- Ripple is the peak to peak or r.m.s. value of any remaining a.c., after smoothing, that is present in the d.c. output voltage.
- Output impedance is the internal resistance of the power supply as seen at the d.c. output terminals. For a good supply this should be as low as possible.

Self-assessment questions *(answers page 146)*

1) If the input to a power supply is 240 V at 83.3 mA and the output is 12V at 1A then the efficiency is:
   a) 20%   b) 40%   c) 60%   d) 80%

2) If the full load output voltage for the power supply of Question 1 is 11.8V and the no load voltage is 12V, then the regulation of the power supply is;
   a) 1.6%   b) 2%   c) 4%   d) 6%

3) What is the output impedance for the power supply of Questions 1 and 2 ?
   a) 0.01 ohms   b) 0.02 ohms   c) 0.1 ohm   d) 0.2 ohms

4) The rectifier converts the bi-directional mains input waveform to:
   a) Uni-directional pulses.
   b) Bi-directional pulses.
   c) Pure d.c.
   d) The r.m.s. value.

5) The smoothing circuit smooths out the:
   a) bi-directional pulses to give pure d.c.
   b) unidirectional pulses to give pure d.c.
   c) unidirectional pulses to give pure d.c. with an a.c. ripple.
   d) unidirectional pulses to give d.c. with a small a.c. ripple.

6) The main purpose of the voltage regulator is to:
   a) smooth out the a.c. ripple.
   b) stabilise the output voltage if the mains input voltage changes.
   c) Stabilise the output current if the mains input voltage changes.
   d) stabilise the output voltage if the output load current changes.

7) A measure of the line regulation is where:
   a) The output current is measured against changes in mains voltage.
   b) the output voltage is measured against changes in mains current.
   c) the output voltage is measured against changes in mains voltage.
   d) the output voltage is measured against changes in the output current.

8) For most electronic devices fitted with a power supply the purpose of a power supply is to convert the 240V a.c. mains to:
   a) 240V d.c.
   b) A voltage higher than 240V.
   c) A voltage lower than 240V.
   d) A d.c. voltage between 5 and 24V.

Half-wave and full-wave rectifier circuits

Introduction

The purpose of the rectifier is to convert the a.c. mains supply into a d.c. supply, usually at a low voltage.

There are two main types of rectifier circuit: the half wave, which only employs one half of one cycle of the mains, and the full wave which utilises both half cycles.

The Half Wave Rectifier

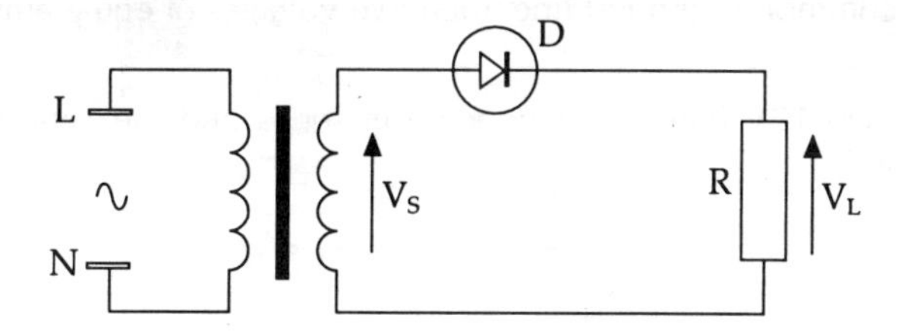

The half wave rectifier circuit

The diode D conducts only when its anode is more positive than its cathode. This results in current flowing through the load only on the positive half cycle of supply voltage, and the voltage across the load resistor is a series of pulses.

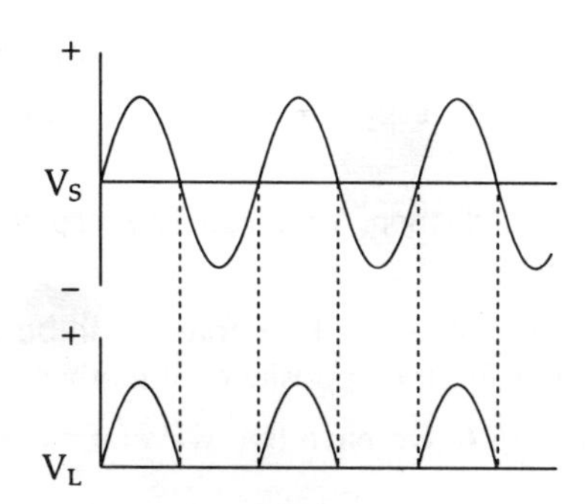

Waveforms of the supply voltage and the voltage across the load resistor R

The average voltage across the load resistor

$$V_L = 0.318 \times V_{peak}$$

where V_{peak} is the maximum or peak voltage of the supply voltage V_S.

Reversing the diode causes it to conduct on the negative half cycles with the result that the polarity of the d.c. output voltage is reversed.

With the half wave circuit, power is only delivered to the load for half of each cycle of the mains supply. If we can provide power to the load on both half cycles then the average d.c. voltage produced will be higher.

Full Wave Bi-phase Circuit

This circuit employs a transformer with two equal secondary windings as shown. It will produce two equal secondary voltages.

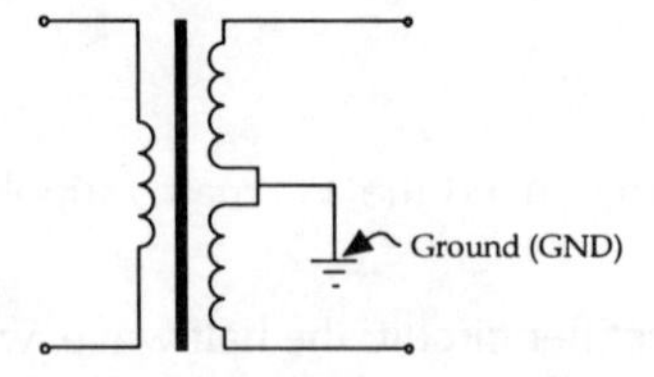

Transformer with two equal secondaries

If the two windings are connected together in the middle as shown and this connection is connected to the common or ground line, then two voltages of equal amplitude and opposite phase may be obtained.

This is similar to two 12V batteries connected in series and the centre connection connected to the ground.

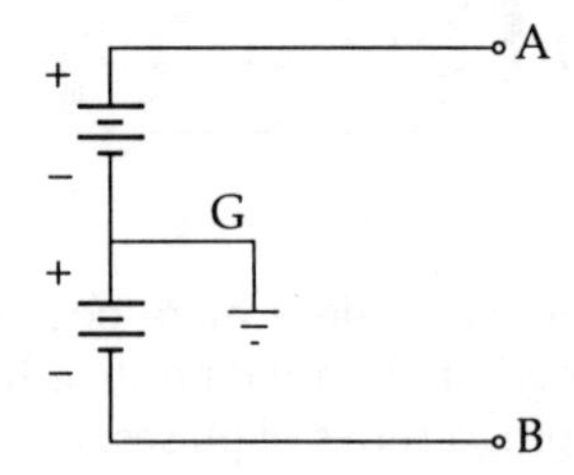

Series connection of two batteries

Point A will be 12V positive with respect to ground G. Point B will be 12V negative with respect to ground G.

If the batteries are now reversed, then A will be 12V negative, and B 12V positive with respect to G.

The a.c. supply reverses each half cycle. Therefore A will be positive and B negative with respect to G on one half cycle and the opposite on the other half cycle.

This means that with respect to G we have two waveforms of equal magnitude but opposite phase at A and B.

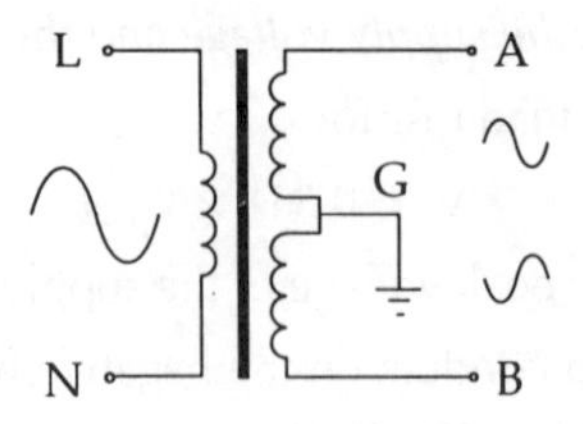

The secondary connection produces two waveforms of equal magnitude but opposite phase

The Bi-phase Circuit

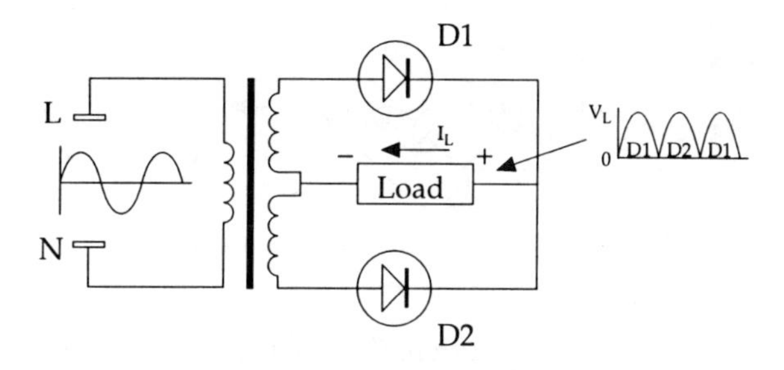

Full wave bi-phase circuit

Because the secondary connection produces two waveforms opposite in phase, when D1 anode is positive, D2 anode will be negative.

Therefore: On one half cycle, D1 conducts and D2 blocks current. On the second half cycle D2 conducts and D1 blocks current.

This causes current to flow in the same direction through the load on both half cycles, giving the load voltage waveform across the load resistor (V_L) as shown:

$$V_L = 0.637\ V_{peak}$$

Advantage: Higher average d.c. output voltage compared to half wave; both half cycles used.

Disadvantage: Need for a transformer with two equal secondaries (or a single secondary winding with a centre tap).

Full Wave Bridge Circuit

If a transformer with only one secondary winding is available then extra diodes can be employed to make a full wave bridge circuit.

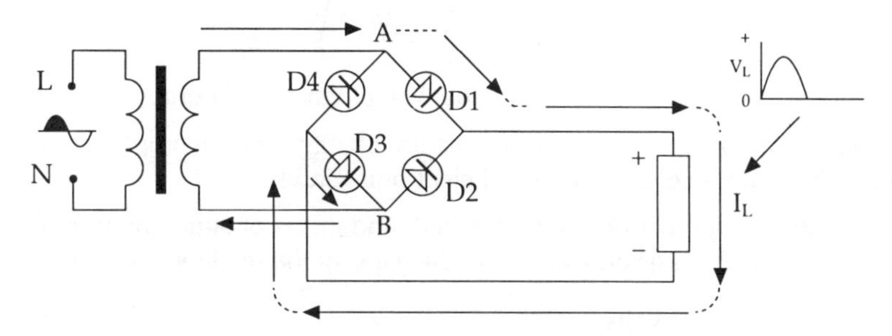

Current flow in a bridge circuit on one half cycle

On one half cycle Point A will be positive with respect to B. Diodes D1, D3 conduct, while D2, D4 block current.

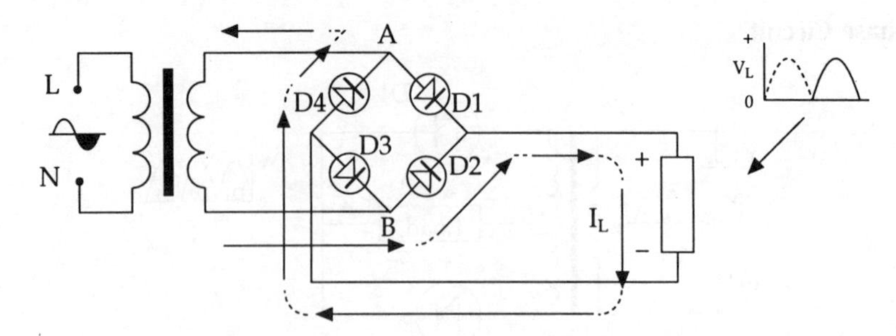

Current flow in a bridge circuit on the other half cycle

On the next half cycle point B will be positive with respect to A. Diodes D2, D4 conduct, while D1, D3 block current. For either half cycle, this causes current to flow in the same direction through the load, giving the load voltage (V_L) waveform shown:

$$V_L = 0.637\ V_{peak}$$

Advantage of the bridge circuit: no transformer with a centre tapped secondary needed.

Disadvantage of the bridge circuit: increased volt drop because current must flow through two diodes in series. The combined volt drop will be about 1.4V and this must be taken into consideration with low voltage d.c. supplies.

Smoothing

The output voltage of a simple rectifier circuit is a direct current only in the sense that it does not reverse its direction. It is not however a smooth, continuous voltage such as that obtained from a battery.

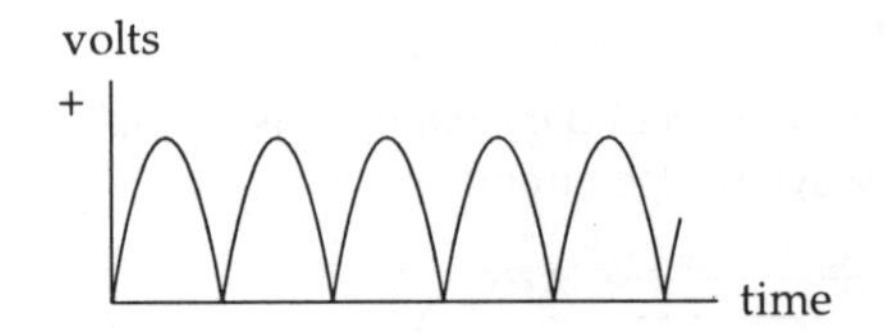

Output waveform of unsmoothed full wave rectifier

This unsmoothed output is acceptable for battery charging or electroplating, but it is not good enough for radio, television or digital electronic circuits.

For these circuits, a smoother output is required, and this is obtained by use of a reservoir capacitor which stores charge and so 'fills in the gaps' of the rectifier's output.

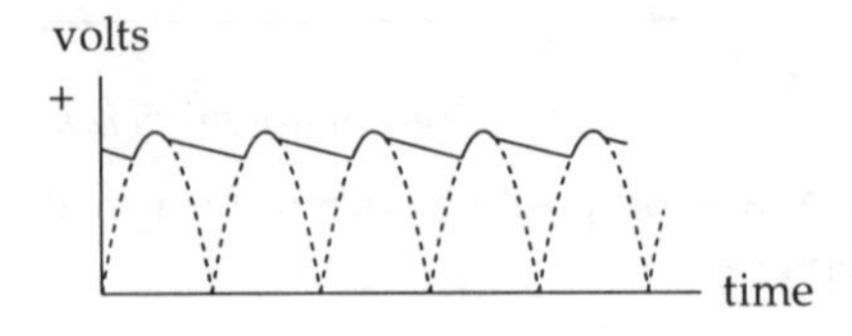

Use of reservoir capacitor to smooth the rectifier output

Action of the Reservoir Capacitor in a Half Wave Circuit

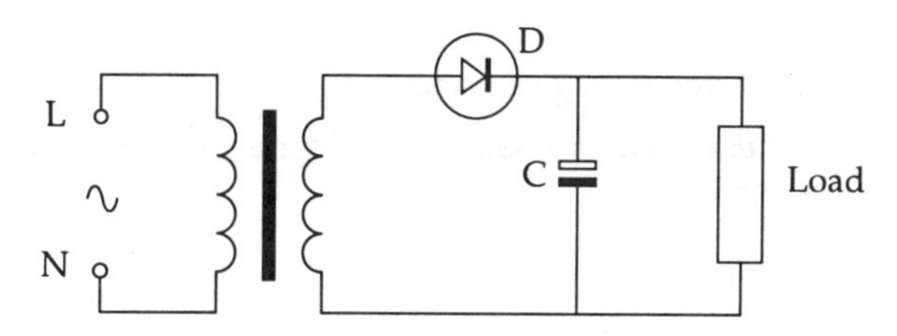

The Half Wave Rectifier Circuit

The reservoir capacitor is connected across the load as shown. Because the load and the reservoir capacitor are connected in parallel it does not matter which comes first in the circuit: the load or the capacitor.

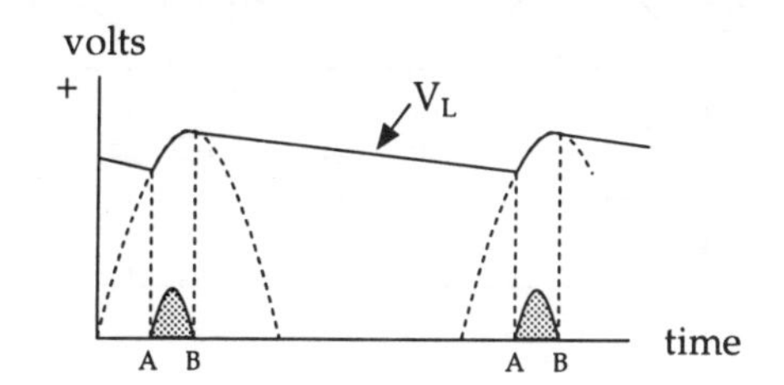

The output waveform of the half wave rectifier circuit smoothed by a reservoir capacitor

The diode conducts from A to B, charging the reservoir capacitor C to the peak of the supply voltage. At B, the **diode stops conducting**.

The capacitor has now been charged to the **peak value** of the secondary voltage. Throughout the remainder of the cycle, the energy stored in the capacitor supplies current to the load.

The capacitor therefore discharges and its voltage falls. The voltage across the load, V_L, will therefore fall until the next positive half cycle charges the capacitor again. This charging and discharging of the capacitor gives rise to a small **ripple** on the output voltage.

The Action of the Reservoir Capacitor on a Full Wave Circuit

When a reservoir capacitor is used with a full wave rectifier, it is charged twice as often as in the half wave circuit.

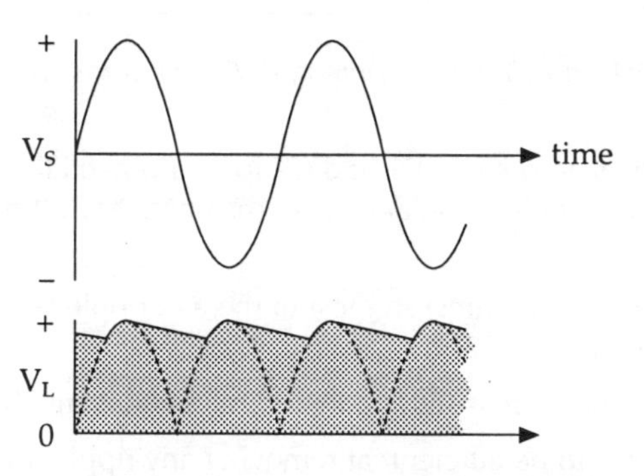

With a full wave circuit the reservoir capacitor is also charged on the negative half cycle of the supply.

This has two effects:

1) The ripple frequency is doubled.
2) The ripple voltage is halved for a given size of capacitor.

   (Alternatively, a smaller capacitor could be used for a given maximum value of ripple voltage that can be tolerated.)

The Ripple Filter

To reduce ripple voltage further, a ripple filter may be used. In the diagram below, C2 and R1 form the capacitance-resistance (CR) filter

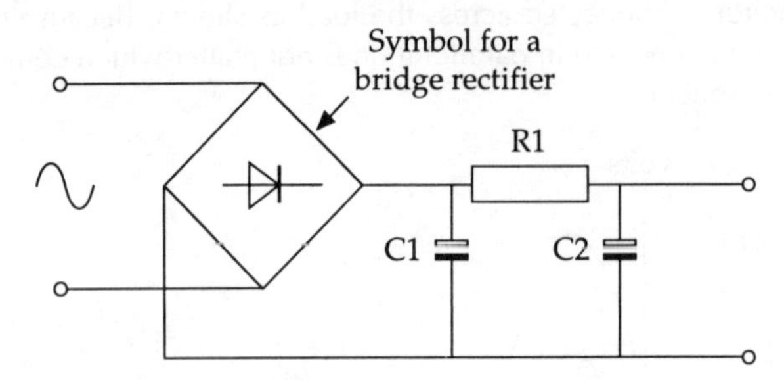

The CR ripple filter

In the diagram of the ripple filter:

C1 is the reservoir capacitor.

R1 is a resistor that attenuates (cuts down) the ripple voltage.

C2 is the smoothing capacitor; this offers a low impedance (low a.c. resistance) to the ripple voltage.

The Ripple Filter Action

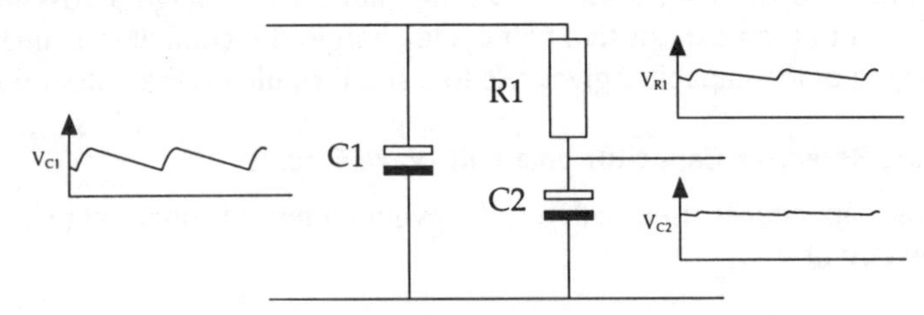

R1 and C2 form a potential divider across the reservoir capacitor C1 to cut down the a.c. ripple.

The ripple voltage appears across C1; R1 and C2 form a potential divider across C1 to the a.c. ripple voltage. The capacitive reactance (a.c. resistance) of C2 to the a.c. ripple is very low (a few ohms).

R1 is much larger (a few tens of ohms) so most of the a.c. ripple will be dropped across R1 and very little will appear across C2.

The d.c. output is taken from across C2 and therefore contains much less ripple.

R1 needs to be large if it is to be efficient at removing any ripple, but it also needs to be as low as possible to prevent a significant d.c. volt drop, which will reduce the d.c. output voltage.

With low voltage, large current power supplies (e.g. 5V, 10A), it is not possible to have any resistance at all!

Even with R1 = 0.5 ohms, then with 10A of load current all the output voltage would be dropped across this filter resistor.

Instead this type of supply uses only a very large capacitor (C1 plus C2 = 5000μF or more) or employs additional circuitry to provide a Regulated Power Supply.

Voltage Regulation

Practical power supplies will generally have a range of specifications, but one of the most important is the Voltage Regulation of the power supply.

Voltage Regulation is a measure of a power supply's ability to maintain a constant output voltage even if the current taken from it varies (i.e. different loads are connected).

Voltage Regulation (V.R.) is normally expressed as a percentage and is given by:

$$\text{V.R.} = \frac{\text{No load voltage} - \text{full load voltage} \times 100\%}{\text{No load voltage}}$$

Where:

No load voltage: is the output voltage with the load open circuit.
Full load voltage: is the output voltage when the rectifier circuit is supplying the maximum load current for which it was designed.

For example, suppose the output voltage changes by 100mV from a no load voltage of 10V, then:

$$\text{V.R.} = \frac{(10 - 9.9) \times 100\%}{10} = 1\%$$

The lower the V.R. figure the better the power supply will be at maintaining a constant output voltage.

Why should the d.c. voltage change as the current supplied to the load changes?

Firstly, the load current must flow through the ripple resistance R if fitted. As the current increases, the voltage drop across R will increase so giving a lower output voltage.

Secondly, since the secondary winding of the transformer has resistance, a varying load current will produce a varying voltage drop.

Thirdly, when the load current increases, the reservoir capacitor voltage will fall further from its peak value. This causes the average value of the output voltage to fall. (the a.c. ripple will also increase).

Finally, the forward volt drop across the diode changes as the current through the diode changes.

Summary of Half-wave and Full-wave Rectifier Circuits

- The half wave circuit utilises only one half cycle of the mains input waveform. The full wave circuit utilises both halves.
- There are two types of full wave circuit: the bi-phase, which uses two diodes and a centre tapped transformer, and the bridge, which uses a conventional transformer but four diodes.

- ❐ A large reservoir capacitor is used to smooth the pulses of uni-directional voltage. For the same load current a larger capacitor is required with the half wave circuit compared to the full wave to reduce the ripple to the same level.
- ❐ A ripple filter may be employed, but the resistance of this filter is usually too large for low voltage (5V), large current (3A) supplies. For these supplies a much larger reservoir capacitor is used (4700μF or more).

Self-assessment questions *(answers page 146)*

1. The function of a rectifier circuit is to:
   a) act as a reservoir of current.
   b) produce direct current from an a.c. supply.
   c) produce a smooth direct current.
   d) charge batteries.

2. The average output voltage obtained from a half wave circuit is:
   a) half of the peak voltage of the a.c. supply.
   b) equal to one peak only of the a.c. supply.
   c) more than the output of a full wave circuit, because there is only one diode in the series with the current.
   d) half that of a full wave circuit, because only half of the wave is rectified.

3. In order to reverse the polarity of the d.c. output voltage from the Biphase rectifier power pack, we must:
   a) reverse the secondary connections on the transformer.
   b) reverse all diodes in the circuit.
   c) reverse the L and N connections.
   d) reverse all the positive and negative cycles of the mains.

4. The advantage of a full wave rectifier circuit over a half wave circuit is that:
   a) it uses more diodes.
   b) a centre tapped transformer can be used.
   c) the average output voltage is greater.
   d) it gives a perfectly smooth output.

5. A full wave rectifier gives an unsmoothed output voltage. If the peak value of the sine wave input is 200V, the average d.c. output voltage would be:
   a) 31.8V b) 50V c) 100V d) 127V

6. An advantage of the bridge circuit over the bi–phase circuit for full wave rectification is that
   a) it gives a larger output.
   b) it gives a smoother output.
   c) no centre tapped transformer is necessary.
   d) it uses more diodes.

7. The function of a reservoir capacitor in a power pack is:
   a) to give more output current.
   b) to give a more input current.
   c) to give a more continuous output voltage.
   d) to improve the ripple content of the output.

Answer the remaining questions as either true or false.

8. If a larger reservoir capacitor is used in a power pack, the amplitude of the ripple voltage is reduced.

   True or false? _______

9. An electrolytic capacitor would be ideal for a reservoir capacitor because it is always polarised, and the alternating current which it passes is small.

   True or false? _______

10. The use of a reservoir capacitor in a power pack means that the diodes have to withstand a greater reverse voltage (called the peak inverse voltage) so their peak inverse voltage rating must be increased.

    True or false? _______

11. If the load current is increased then the voltage across the reservoir capacitor will be lower but the amplitude of the ripple will be unaltered.

    True or false? _______

12. Because the ripple frequency is doubled in a full wave circuit the reservoir capacitor must be increased in value.

    True or false? _______

13. To reduce the ripple voltage still further, a CR ripple filter may be used. The resistor has a higher resistance to the ripple voltage compared to the capacitive reactance (opposition to the a.c. ripple) of the capacitor.

    True or false? _______

14. The smoothing capacitor which is used in the ripple filter has a low impedance to a.c. and a low resistance to the d.c.

    True or false? _______

15. Where a large current low voltage supply is required, even a resistor cannot be used to form a CR ripple filter, because the volt drop across the resistor is too large.

    True or false? _______

16. The voltage at A with respect to point B is 10V positive. Therefore the voltage at C with respect to B is 10V negative.

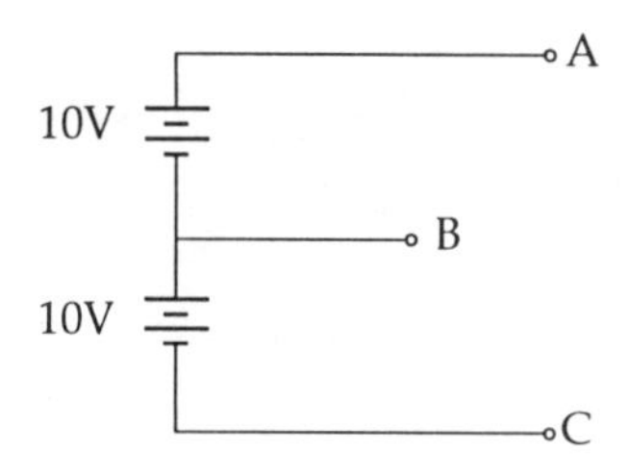

True or false? _______

17. The voltage at A with respect to B at any instant of time is of equal magnitude and opposite polarity to the voltage at C with respect to B.

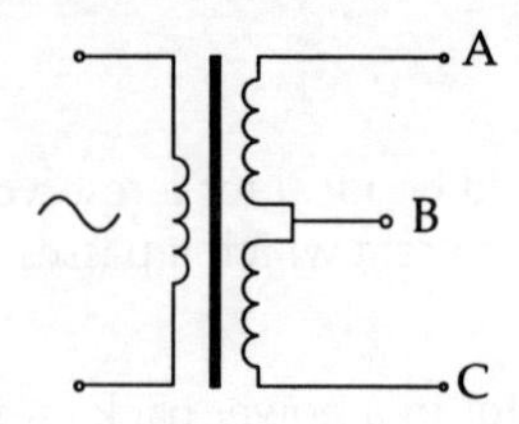

True or false? _______

Basic principles and operation of transformers

Introduction

The mains electricity supply is an a.c. supply because this is an efficient way of transmitting electrical power, and it is easy to change from one voltage to another using a transformer.

Nearly all electronic equipment needs a low voltage d.c. supply to operate the semiconductor devices. A **transformer** is therefore needed to **step down** the voltage to a suitable level and to **isolate** the equipment from the mains supply.

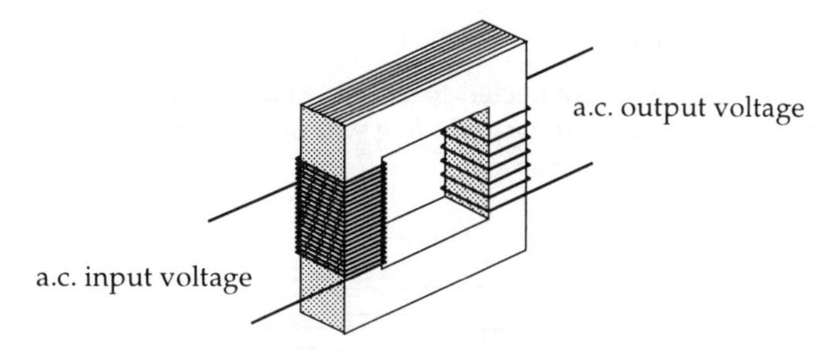

The Transformer

The mains transformer has a primary winding and a secondary winding. The secondary winding has fewer turns of wire than the primary winding and the transformer **steps down** the voltage.

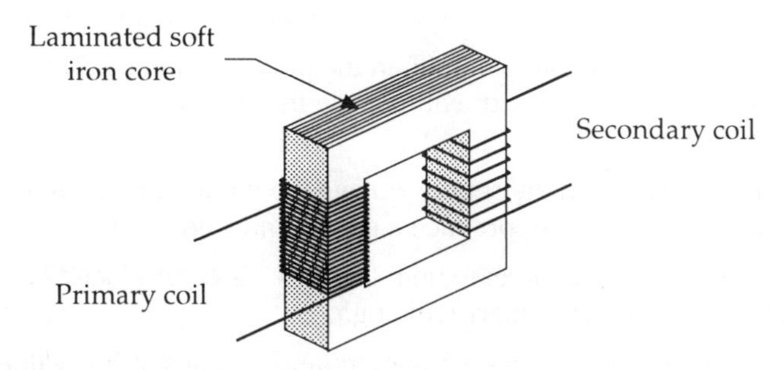

The Step Down transformer

A current flows in the primary winding and sets up a magnetic field. This magnetic field is concentrated in the iron core upon which the windings are wound, and surrounds the secondary winding.

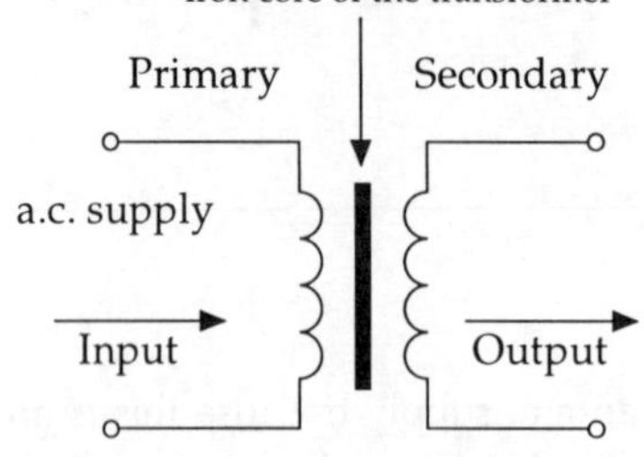

Transformer symbol

Because the input a.c. current constantly changes, the magnetic field also changes and this induces a voltage in the secondary winding. With fewer turns of wire on the secondary than on the primary a lower a.c. voltage is produced at the secondary.

Transformer action NO Load

When the primary winding is connected to an a.c. supply, an alternating current flows. This a.c. current flowing in the primary winding will produce an alternating magnetic flux in the soft iron core of the transformer.

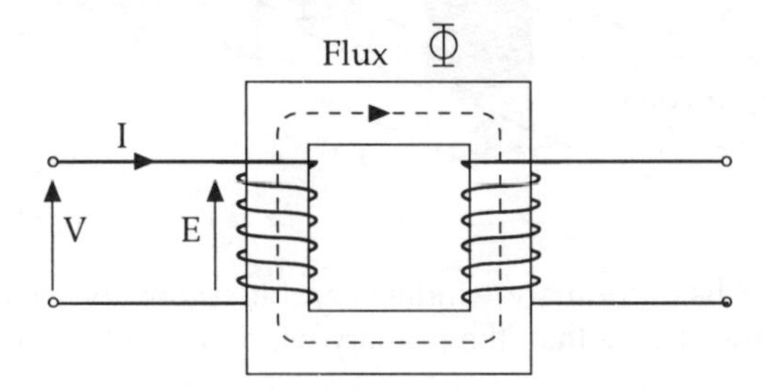

The primary current sets up a magnetic flux in the core

The alternating flux will induce an e.m.f. (E) in the primary winding. This induced e.m.f. is opposite in polarity to the supply and will oppose the supply voltage. For this reason it is called the back e.m.f.

If this back e.m.f. (E) is less than the supply voltage the input current will increase, increasing the flux in the core, and in turn produce a greater induced e.m.f.

The input current will therefore increase until the back e.m.f. produced is equal and opposite to the supply voltage in the primary winding.

The input current cannot then increase any more and remains at this value, called the no load current (I_0). The no load current is relatively low compared to the full rated current of the transformer.

Because the magnetic flux is common to the primary and the secondary windings an e.m.f. is also induced in the secondary winding.

The same e.m.f. will be induced in **each** turn of **both** the primary and secondary windings and so the secondary e.m.f. will depend upon the ratio of the primary turns to the secondary turns.

$$V_P/V_S = N_P/N_S$$

where:

V_P = primary voltage
V_S = secondary voltage
N_P = primary turns
N_S = secondary turns

The ratio N_P/N_S or $N_P : N_S$ is called the turns ratio

By varying the turns on the windings the transformer can be made to convert the a.c. input voltage to a higher (step up) or lower (step down) voltage.

Caution! Never connect the transformer to d.c.

The action of the transformer depends upon an alternating magnetic flux. A transformer therefore cannot be connected directly to a d.c. supply. It can, however, be used on a square wave, i.e. switched d.c. supply.

When connected to a constant d.c. supply only the resistance of the primary winding limits the current. On d.c. there is no **induced back e.m.f.** to limit the flow of primary current. Since the primary resistance is low, simply the resistance of the copper wire, the current that would flow would normally be sufficient to burn out the winding.

Transformer Action On Load

By connecting a load across the secondary winding, current will be drawn from the secondary. This current flowing through the secondary winding tends to reduce the magnetic flux in the core.

This reduction will reduce the e.m.f. induced in the primary winding and allow more primary current to flow from the mains supply.

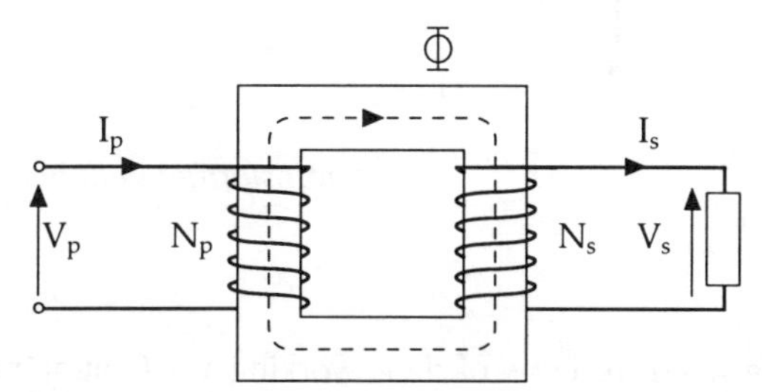

Transformer action 'on load'

The result is that the input primary current is increased to compensate for the current taken from the secondary.

With an increase in primary current the flux level is restored and remains constant. Therefore the secondary voltage remains constant.

In a good transformer losses are small (95–98% efficient). In this case the power in the primary equals the power in the secondary.

The power rating of a transformer is usually quoted in Volt-Amperes (VA).

Therefore for a good transformer:

$$\textbf{Input VA = Output VA}$$

Therefore:

$$V_P \times I_P = V_S \times I_S$$

From which:

$$I_S/I_P = V_P/V_S = N_P/N_S$$

$$\text{The voltage ratio} = \text{the turns ratio} = \frac{1}{\text{Current ratio}}$$

This means that a step down transformer will step **down** the voltage but step **up** the current. Similarly, a step up transformer will step **up** the voltage but step **down** the current.

Transformer Secondaries

A transformer may have a number of secondary windings.

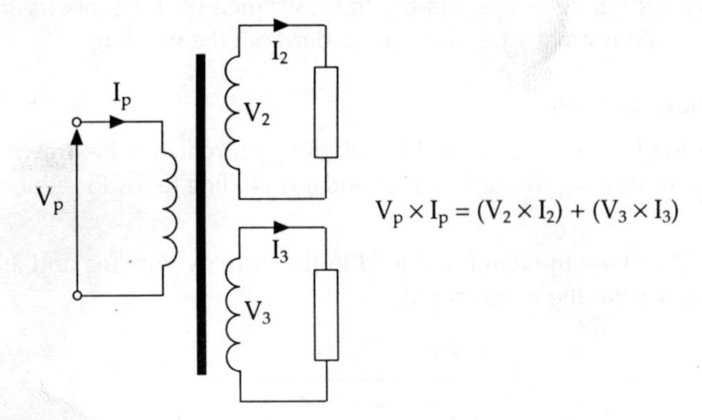

A transformer with two secondary windings

Power Rating

Power transformers are rated in terms of their working VA (volt-amperes). The VA for a transformer can be estimated by calculating the total power consumed by each secondary and multiplying this by 1.1 (this is to allow for losses in the transformer).

Example

If in the above figure the maximum voltage and current rating for the two secondaries are as follows:

Secondary 1: $V_2 = 12V$, $I_2 = 0.5A$

Secondary 2: $V_3 = 6V$, $I_3 = 2A$

then the maximum power that can be provided by the secondaries will be:

$$(12 \times 0.5) + (6 \times 2) = 18W$$

The VA rating of the transformer would therefore be:

$$18 \times 1.1 = 19.8$$

This would be classed as a 20VA transformer.

Note: all the voltages quoted are r.m.s. values.

Regulation

If the output voltage of a secondary winding falls significantly when it is supplying its full rated secondary current, then the regulation of the transformer is poor. (The d.c. resistance of the wire which forms the windings is one of the reasons why the output voltage falls under load.)

The regulation of a transformer is therefore a measure of its ability to maintain its rated output voltage under load.

The regulation is usually quoted as a percentage. Typically 10%.

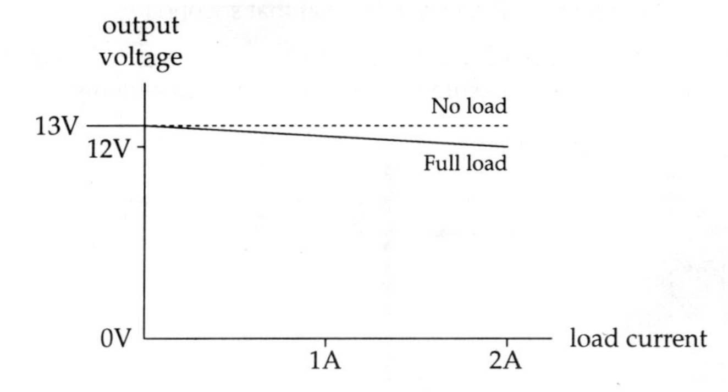

The regulation figure indicates the percentage fall in the output voltage between no-load and the full rated load current being drawn.

The transformer is usually designed to provide the quoted secondary voltage under full load conditions. Therefore off load, the secondary voltage may be 10% higher.

Resistance of Transformer Windings

Primary winding resistance

This will depend upon the VA rating of the transformer. The smaller the power rating then the smaller the diameter of wire that can be used for the primary winding; the smaller the physical size of the transformer. This will give a relatively high resistance for the primary winding.

The primary resistance typically can vary from a few tens of ohms to a few hundred ohms, depending upon its VA rating.

Secondary Winding Resistance

In a step down transformer the secondary winding resistances are considerably lower than the primary winding.

Again the resistance will depend upon the current rating of the secondary under test. The higher the rating the lower the resistance for a given secondary voltage rating. Typically for low voltage high current secondaries the resistance will be less than 1 ohm.

Transformer Faults

The transformer is a very efficient and reliable component and normally gives very little trouble. However if it does fail then it is usually a catastrophic failure.

Caution: Never test a transformer under conditions where the primary winding terminals are exposed and live! This should only be attempted **after** you have had suitable training in working on live mains equipment.

Open Circuit (O/C) Windings

This is a very easy fault to find. If the primary winding is O/C then there will be no secondary voltages.

If a secondary is O/C then there will be no voltage at that secondary.

TO TEST: **disconnect the mains plug** from the 13A socket outlet. Using the resistance range of a multirange test meter, measure the resistance of each winding.

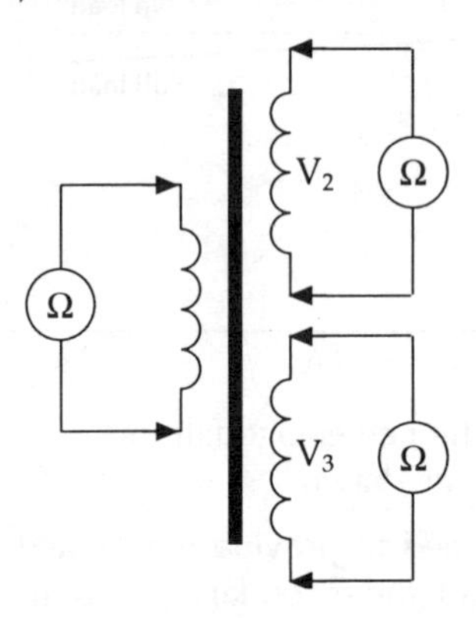

Measuring the primary and secondary winding resistances

Short Circuit (S/C) Windings

If the primary winding is S/C then the fuse will blow in the primary circuit.

Similarly if a secondary winding is S/C then the large current that flows will be reflected back into the primary (by the transformer action) and again cause the fuse in the primary circuit to blow.

TO TEST: **disconnect the mains supply**. The primary winding resistance is easily measured and any S/C readings are obvious.

The secondary windings, however, because they can normally be less than 1 ohm, need to be checked more accurately: i.e. is the resistance 0.5 ohms or is it zero ohms?

Another fault that may blow the fuse would be a breakdown in the insulation resistance between the primary winding and the core of the transformer, which is earthed. All transformers that are to be used in equipment must be tested, between the primary and the core, with a high voltage insulation resistance tester.

Short Circuited Turns

The insulation between adjacent turns of wire may break down. This has the effect of producing a short circuited turn or turns which will cause large circulating currents to flow and draw (by the transformer action) more current from the supply. The possible symptoms of this will be a slight reduction in the output voltages and/or the transformer will be running hotter than normal. (It may actually smell and there may be a wax-like substance dripping from the windings of the transformer.)

There is no remedy for this condition other than to replace the transformer.

Series Connection of the Secondary Windings

The secondary windings may be connected together to produce alternative secondary voltages. The windings may be connected as series-aiding to produce a higher voltage or has series-opposing to lower the secondary voltage.

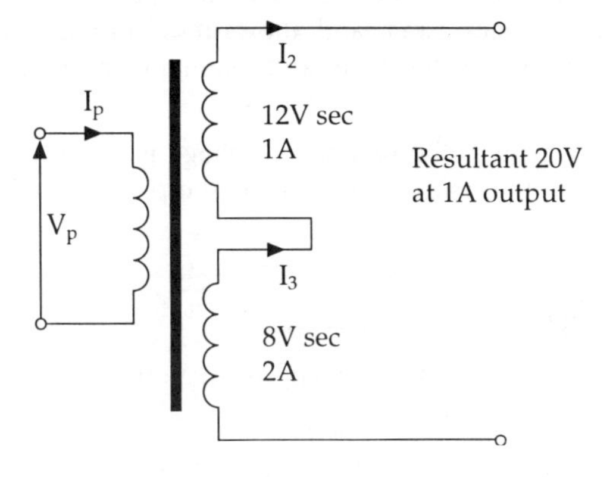

Series-aiding connection

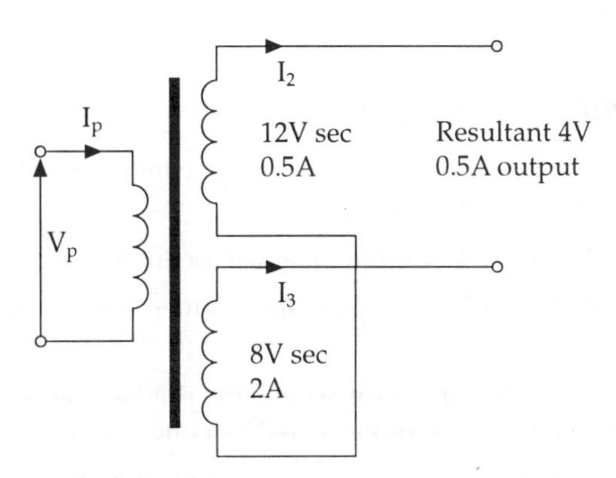

Series-opposing connection

Because the windings are in series, the output current supplied cannot be greater than the smaller winding is capable of providing!

Remember that the VA of the transformer is unchanged!

Summary of Basic Principles and Operation of Transformers

- ❒ A transformer is needed to step down the voltage and to isolate equipment from the mains supply.
- ❒ The transformer consists of primary and secondary windings wound around a laminated iron core. For a step down transformer there are fewer turns on the secondary than on the primary.
- ❒ Current flowing in the primary winding sets up a magnetic field (flux) in the iron core. This flux will change in sympathy with the a.c. input voltage and current.
- ❒ The changing flux produces (induces) a voltage in the primary winding which is opposite and almost equal to the applied mains voltage. This is called a back e.m.f. and limits the off load primary current to a very small value. This same changing flux will also produce (induce) a voltage in the secondary winding.
- ❒ When a load current is drawn off from the secondary, the magnetic flux in the core is weakened. This weakens the back e.m.f. and allows more primary current to flow. This in turn strengthens the flux in the core, and maintains the output voltage at a constant level.
- ❒ The ratio of the primary voltage to the secondary voltage is called the voltage ratio. This voltage ratio is equal to the ratio of the turns on the primary to the turns on the secondary.

$$V_P/V_S = N_P/N_S$$

- ❒ Most transformers are very efficient, and as an approximation:

  Input VA = Output VA.
- ❒ Transformer secondaries may be connected in series-aiding or series-opposing to provide other secondary output voltages.

Self-assessment questions *(answers page 147)*

1) A transformer with a turns ratio of 4:1 has a primary winding of 1000 turns. How many turns are there on the secondary winding?
2) If the primary voltage in Question 1 is 240V, what will be the secondary voltage?
3) Explain how you would determine a step-up transformer from a step-down transformer.
4) A transformer has a 9V secondary and a 6V secondary; draw a diagram to show how you would connect the secondaries to produce a 3V secondary.
5) A transformer has two 9V secondaries: draw a diagram to show how you would connect the secondaries to produce a 18V secondary.
6) A small transformer has a primary resistance of 40 ohms whereas a much larger transformer has a resistance of only 24 ohms. Is the larger transformer faulty or normal?

   If you consider it to be normal give your reasons.
7) A transformer has two secondaries rated as follows:

   Sec A 12V at 0.5A; Sec B 6V at 2A.

What would be the VA rating of the transformer?

8) Explain what would be the effect of connecting the primary winding of a mains transformer to 240V d.c.

Practical Exercise

1) Obtain two or more different mains transformers: i.e. different sizes (differing VA). Use manufacturers' data/catalogues to identify them and determine their voltage, current and VA rating. Use an ohmmeter to measure the resistances of the windings.

2) Obtain a transformer for which you have no data. Use the Ohmmeter to identify the primary and secondary windings. From these results and the information regarding the transformers in (1) above, estimate the VA rating of the transformer and the voltage and current rating of the secondaries.

3) This is to be attempted only if you have access to a completely enclosed (**NO LIVE MAINS TERMINALS**) mains step down transformer with several secondary terminals which are accessible. The transformer must only be connected to the mains supply via a 13A plug top, fitted with a correctly rated fuse.

   i) Apply the mains to the transformer and measure the output voltage at each secondary (meter on a.c. volts range!).

   ii) **Disconnect the mains supply**: connect the secondaries in series-aiding. Re-connect the mains and measure the combined output voltage.

   iii) **Disconnect the mains supply**: connect two secondaries in series-opposing. Re-connect the mains and measure the output voltage. (For this test it would be helpful if the two secondaries had different voltage ratings. If they are both the same then the output voltage theoretically should be zero.)

Capacitance and capacitors

Introduction

Capacitors are used as a store of electric charge and energy (they can store electrical energy in the form of an electric field). There are many types of capacitor all of which have different properties.

A capacitor is created by two parallel conducting plates separated by an insulator, called a *dielectric*.

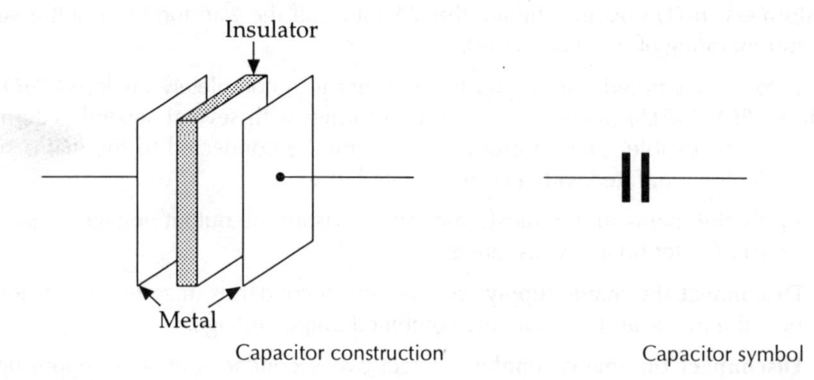

Capacitor construction Capacitor symbol

To construct a reasonable size of capacitor then the area of the plates must be large. The insulator must be a good dielectric and should be as thin as possible for a given working voltage.

To create a large capacitance, two strips of aluminium foil are placed either side of an insulator (dielectric) and then rolled up to form a tube. This will produce a large plate area in a physically small volume.

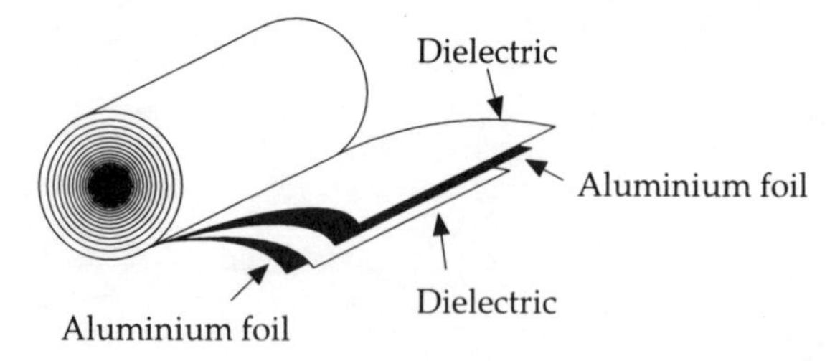

Capacitor construction with a large plate area

Charging a Capacitor

When the plates of a capacitor are connected to a source of e.m.f., electrons flow until the capacitor is charged. The plate connected to the negative pole of the battery receives electrons and becomes negatively charged. The other plate connected to the positive pole of the battery loses an equal number of electrons, making this plate positively charged.

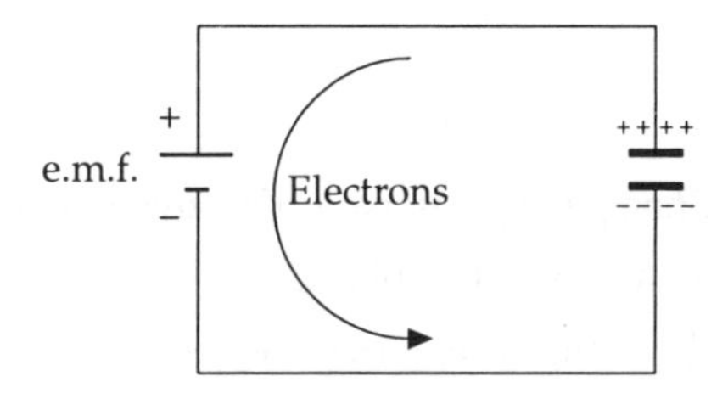

Electron movement to charge a capacitor

The charge on the plates of the capacitor creates a potential difference between the plates. Current continues to flow until the potential difference (p.d.) across the capacitor is equal to the applied e.m.f. This charging process usually takes a very short time, after which there is no more movement of electrons. The charging current has stopped flowing.

The charge (Q) stored by a capacitor is measured in coulombs (abbreviation C). One coulomb of charge is stored when a current of one ampere (6.28×10^{18} electrons) flows for one second.

$$\mathbf{Q = i \times t \text{ coulombs}}$$

The charge storage capacity of a capacitor is called its capacitance. The unit of capacitance is the Farad (abbreviation F).

A capacitor has a capacitance of 1 Farad when a charge of 1 coulomb gives a p.d. between the plates of 1 Volt.

$$\mathbf{Q = C \times V \text{ coulombs}}$$

Where:

Q = charge in coulombs
C = capacitance in farads
V = potential difference in volts

Energy Storage

A charged capacitor stores energy. The amount of energy is small compared with that of a battery, but nevertheless it is capable of doing useful work as in a camera flash, where the energy stored in a charged capacitor is rapidly discharged through a flash tube.

Energy Stored.

The energy stored (W) is given by:

$$\mathbf{W = \tfrac{1}{2} CV^2 \text{ joules}}$$

Example

A 2000μF capacitor charged to a voltage of 50V will store a charge of:

$$\frac{2000}{2} \times 10^{-6} \times 50^2 = 1000 \times 2500 \times 10^{-6} = 2.5 \text{ joules.}$$

Note: a μF is one millionth of a Farad (see Metric prefixes, page 136).

Applications

Capacitors are mainly used:

a) as a *store* of charge or energy: e.g. reservoir of charge in a power pack.

b) to *couple* a.c. signals in and out of amplifiers.

c) to *decouple* unwanted a.c. signals to ground: e.g. mains supply interference filters.

d) in conjunction with resistors to form *timing circuits* or *frequency selective circuits*.

e) in conjunction with coils to form *tuned circuits*.

Types of capacitor

There are many different types of capacitor. When choosing a capacitor some of the important features are:

- *Capacitance*: the farad is too large a unit for most applications and so capacitor values are given in μF, nF of pF.
- *Voltage rating*: the maximum voltage that can be continuously applied to the capacitor is called its working voltage. A higher voltage than this will cause the capacitor dielectric to break down.
- *Tolerance*: the maximum or minimum values allowed from the nominal value: i.e. the spread of values above and below the nominal value. For example, the nominal value of 100μF plus 10% (110μF) and minus 20% (80μF).
- Physical size

The Dielectric

The most important part of a capacitor, and the part which affects all of these properties, is its dielectric. So a capacitor is usually named after the type of dielectric used in its construction. The characteristics of the dielectric will not only determine the capacitance, working voltage, tolerance, size and price but also:

a) stability (the change in capacitance over the life of the capacitor)

b) insulation resistance (the leakage resistance across the dielectric)

c) temperature coefficient (the change in capacitance with a change in temperature).

Types of Capacitor in Common Use

Polyester
Polycarbonate
Polystyrene
Polypropylene
Mixed Dielectric (paper and polyester film)
Ceramic
Electrolytic *
Tantalum *
Silvered Mica
Waxed Paper
Metallised film

* **Note**: electrolytic and tantalum capacitors are polarised, which means that they must always be connected to the correct supply polarity (+ and –).

Capacitor Faults

a) **Short Circuit** – insulation breaks down and the resistance between the plates falls to a low value (a few ohms).

To test: an ohmmeter can be used to test for short circuits.

b) **Open Circuit** – the connection to one of the plates breaks down internally.

To test: many modern digital multirange instruments now have a capacitance range and this can be used to measure the capacitance of the capacitor. Alternatively the suspect capacitor may be bridged with a known good capacitor.

c) **Leaky Capacitor** – the insulation resistance has partially broken down so the capacitor has a resistance of a few thousand ohms. This causes a high leakage current to flow. This fault is common in electrolytic capacitors.

To test: because this fault may only show up at the full rated working voltage of the capacitor, high voltage capacitors need to have their insulation resistance checked using a high voltage insulation tester. Alternatively the suspect capacitor may be temporarily substituted with a known good capacitor.

d) **Out of tolerance** – the capacitance value may change due to age and be outside its tolerance rating. This fault will be noticeable in timing circuits or tuned circuits.

To test: measure its capacitance or substitute with a known good capacitor.

Replacing Capacitors

The different types of capacitors have widely differing characteristics. Therefore when servicing electronic equipment always replace, where possible, like for like: e.g. replace a polyester with another polyester having the same capacitance and working voltage.

Caution!!!

1) Before handling large capacitors (several thousands of microfarads) that have previously been charged, for example in a power supply, the capacitors should first be discharged.

Do not however discharge these capacitors by shorting out the terminals with a screwdriver. There is sufficient energy stored to blow pieces of metal out of the screwdriver and damage your eyes. Also the large current that flows may cause the capacitor to fail.

To discharge the capacitor, carefully clip a resistor (1kΩ) across the terminals and leave it connected for at least a minute. Then, and only then, the capacitor may be **completely discharged** by shorting out the terminals.

2) Polarised capacitors (electrolytic), when connected between two points in a circuit, must be connected with the **terminal marked + to the more positive point, and the terminal marked – to the least positive point**.

Failure to do this will cause the capacitor to pass an excessive current, overheat and explode.

3) When fitting or substituting electrolytic capacitors in a circuit **always** ensure that the **working voltage** of the capacitor is **high enough** for the circuit under test.

Summary of Capacitance and Capacitors

- A capacitor is formed from two conducting plates separated by an insulator called a dielectric.
- When a capacitor is connected across a battery then the positive potential of the battery will attract electrons off one plate which then becomes positive. The negative terminal of the battery repels (pushes) electrons onto the other plate which then becomes negative.
- During this process electrons are moving and constitute an electrical current, called the charging current. The charge (Q) on the capacitor therefore depends upon the current flowing (i) multiplied by the time (t) for which it is flowing.

$$Q = i \times t \text{ coulombs}$$

- The amount of current flowing depends upon the applied voltage and upon the area of the metal plates or capacitance of the capacitor.

  The charge Q is therefore also equal to the voltage applied (V) times the capacitance of the capacitor (C).

$$Q = C \times V \text{ coulombs}$$

- If a capacitor is charged from a supply, which is then disconnected, then there is energy stored in the electric field of the capacitor.

  The amount of energy stored: $W = \frac{1}{2} CV^2$ joules.

  With a good capacitor this energy will be stored in the capacitor for many hours.
- The dielectric of a capacitor determines such things as: capacitance, voltage rating, tolerance, physical size, stability, insulation resistance and temperature coefficient. The capacitor is therefore named after its dielectric: e.g. polyester, ceramic, mica, etc.
- Capacitors should always be replaced with one of the same type, particular care being given to the capacitance value, the working voltage and the tolerance.
- Polarised capacitors must always be connected to the correct polarity (observe the markings on the case). Failure to do this can cause them to explode.
- Care should be taken when handling equipment with large capacitors that are fully charged. Before handling, switch off the equipment and discharge the capacitors slowly via a resistor.

Practical Exercise

1) Obtain at least six different capacitor types. Use manufacturers' data/catalogues to identify each capacitor and its characteristics. Use the data to construct a table for each capacitor as shown below.

Example

Capacitor type	polyester
Range of values	0.001 to 10μF
Working voltage	750V
Tolerance	±20%
Maximum operating temperature	125°C
Polarised	No

2) Use a capacitance meter to measure the value of each of the capacitors. Determine whether or not each capacitor is within its tolerance rating.

3) Connect two capacitors of the same capacitance value in:

a) series with each other

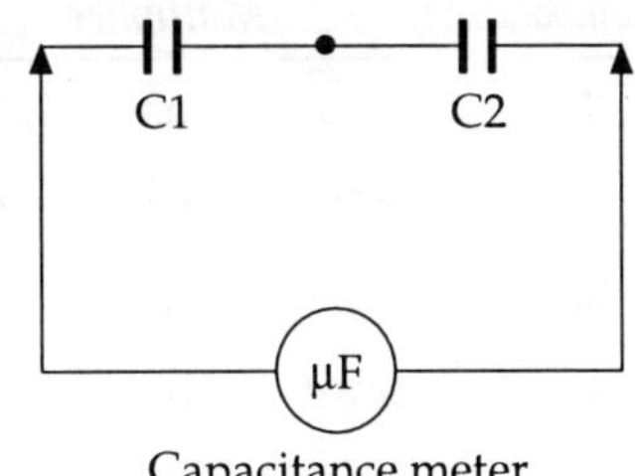

Series connection

b) Parallel with each other

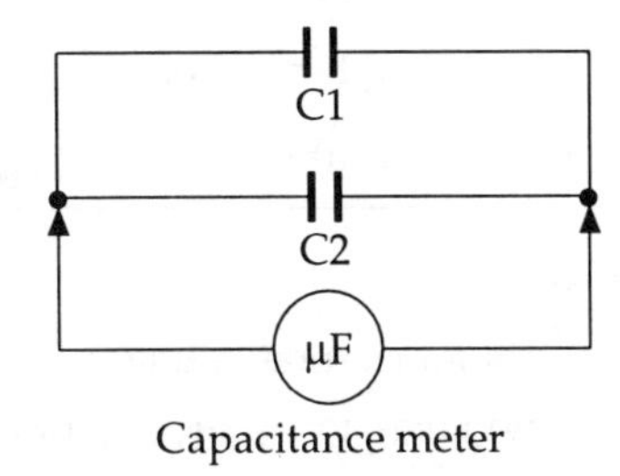

Parallel connection

Use the capacitance meter to determine the resultant capacitance.

Metric prefixes

Introduction

The range of numbers involved in Electrical Engineering is very wide. Capacitors of a few millionths of one farad; resistors of a few million ohms; frequencies of a few million hertz; time intervals of a few micro or nanoseconds.

To make writing and handling these numbers more convenient, we use metric multipliers which increase in steps of one thousand.

Prefix	*Symbol*	*Multiplier*		
Tera	T	$\times 10^{12}$	or	1,000,000,000,000
Giga	G	$\times 10^{9}$	or	1,000,000,000
Mega	M	$\times 10^{6}$	or	1,000,000
Kilo	k	$\times 10^{3}$	or	1,000
Basic unit		$\times 1$		
Milli	m	$\times 10^{-3}$	or	$\frac{1}{1,000}$
Micro	µ	$\times 10^{-6}$	or	$\frac{1}{1,000,000}$
Nano	n	$\times 10^{-9}$	or	$\frac{1}{1,000,000,000}$
Pico	p	$\times 10^{-12}$	or	$\frac{1}{1,000,000,000,000}$

Quantities are normally written as:

Number, prefix, unit

For example: 0.000,000,47 Farads = 470nF or 0.47µF

i.e. 470(number) n(prefix) F(unit)

150,000,000 ohms = 150M ohms

However: on circuit diagrams where the value of a component is written alongside the circuit symbol, the unit is not shown and the prefix is placed where the decimal point would normally be.

For instance resistance:

2.2k ohms would be written 2k2
220 ohms would be written 220R
3.3M ohms would be written 3M3

Similarly capacitance:

4.7 nano farads becomes 4n7
100 pico farads becomes 100p
2.2 microfarads becomes 2µ2

Conversion

It is often necessary to express quantities with the most convenient prefix.

For example which is the more convenient?

a) 0.000 000 47 farads

b) 0.47uF

c) 470nF

Answer: (c) is the more convenient because there are no decimal points involved.

We therefore need to be able to express the same quantity with more than one metric prefix.

0.000, 000, 1 seconds = 0.000, 1ms = 0.1µs = 100ns

Again the most convenient is 100ns but we must know that it is also 0.1µs.

Converting to a Smaller Unit

For example:

seconds to ms	Move decimal point 3 places right
ms to µs	Move decimal point 3 places right
µs to ns	Move decimal point 3 places right

Converting to a Larger Unit

For example;

ns to µs	Move decimal point 3 places left
µs to ms	Move decimal point 3 places left
ms to s	Move decimal point 3 places left

Exercises *(answers page 147)*

Write down the quantities in the units indicated.

1) 2200 ohms in k ohms

2) 470 000 ohms in k ohms

3) 470 000 ohms in M ohms

4) 1500 000 ohms in M ohms

5) 0.01A in mA

6) 15mA in amps

7) 1.8mA in amps

8) 0.000 001 farads in µF

9) 0.000 000 15 farads in µF

10) 0.000 000 15 farads in nF

11) 2200pF in nF

12) 500pF in nF

13) 50mV in volts
14) 900μA in amps
15) 0.068nF in pF

Note: Metric Multipliers and Ohm's Law

V = I x R

1mA x 1k ohm = 1 volt
1μA x 1M ohm = 1 volt
1mA x 1 ohm = 1mV
1μA x 1k ohm = 1mV

Self-assessment questions *(answers page 148)*

1) Convert the following time periods into milliseconds:
   i) 0.001s   ii) 100μs   iii) 10ns
2) Express the following values in more convenient units:
   a) 6.8×10^{3} volts   b) 1.2×10^{-11} farads
   c) 7.6×10^{-4} amps   d) 8.2×10^{-10} amps
3) From 0.256A take 4,300μA and give the answer in mA.
4) Add the following time periods together and express the answer in μs:
   0.000, 000, 1s   0.000, 001s   0.000, 01s.
5) Add the following voltages together and express the answer in volts.
   10,000,000uV   10,000mV   0.24kV.
6) Express the following values in more convenient units:
   a) 0.0053A   b) 18,925W
   c) 19,500,000 ohms   d) 0.000, 006 25 ohms

Voltage regulator employing discrete components

Study Note

Before studying this helpline you will need to study **Part 2 Section 2 Information and Skills Bank** The bi-polar transistor p.209; The common emitter conection p.217: and single stage audio amplifier, p.229.

Introduction

Instead of using an integrated circuit voltage regulator such as a 7805, it is possible to create a regulator circuit using two transistors and a zener diode.

Basic Principle

One of the transistors is connected in series with the supply to the load and forms the series regulator. The other transistor is used as an error amplifier, to provide an output proportional to the difference between the output voltage and a reference voltage provided by the zener diode.

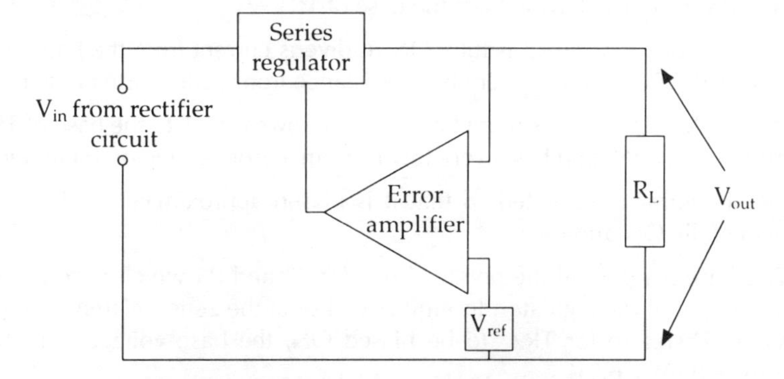

Comparator method of voltage regulation

Principle of Operation

If the load current increases then the output voltage will fall.

The difference between the output voltage Vout and the reference voltage Vref falls.

The output from the error amplifier then has to increase the current through the series regulator and reduce the voltage dropped across it: i.e. the resistance of the series regulator has to fall.

If the load current decreases then the output voltage will tend to rise. The error amplifier must now reduce the current through the series regulator and increase the volt drop across it: i.e. the resistance of the series regulator must rise.

The Circuit

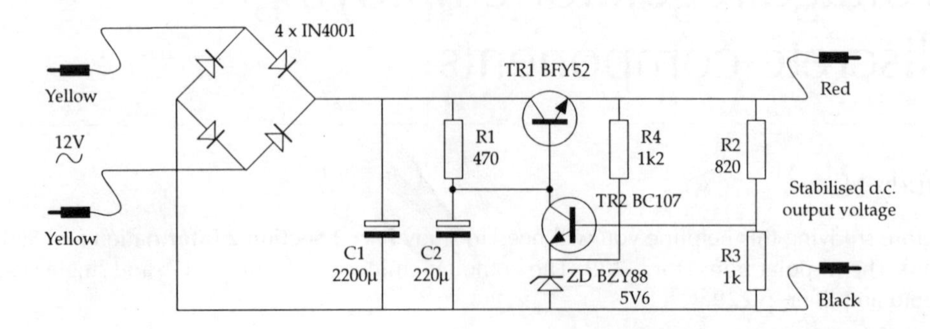

Voltage regulator using discrete components.

Function of Components

TR1 — TR1 is the series regulator and behaves as a variable resistance between the input and output of the circuit.

TR1 is connected as an emitter follower and its resistance between the collector and emitter will vary as its base current varies.

R1 — R1 feeds base current to TR1 to bias the transistor ON. Its value is chosen to bias TR1 fully ON.

TR2 — TR2 forms the error amplifier and controls the current into the base of TR1: that is, it diverts current away from the base of TR1.

For example, when TR2 is fully ON, it diverts current from the base of TR1. TR1 is turned OFF and has a very high resistance from collector to emitter.

When TR2 is OFF, maximum base current flows via R1 to the base of TR1. TR1 is turned fully ON and has a very low resistance from collector to emitter.

TR1 is normally operated so that it is biased approximately half way between being fully ON and OFF.

R2/R3 — TR2 base is fed from the potential divider R2 and R3 which is connected across the output of the regulator. Its emitter will sit at the zener reference voltage of 5.6 Volts. Therefore for TR2 to be biased ON, the base voltage must be at least 5.6V + 0.6V = 6.2V.

R2 and R3 are chosen so that TR2 is biased ON.

R4 — This is used to bias the zener diode into its breakdown region and produce a constant reference voltage on the emitter of TR2.

C2 — This is a decoupling capacitor to ensure that no a.c. ripple appears on the base of TR1 via R1, becomes amplified by TR1, and appears at the output.

Circuit Action

Suppose that the output voltage tends to increase. The proportion of the output voltage developed across R3 will also increase. This will increase the base potential of TR2 which in turn will conduct more, drawing more current through R1.

The base current of TR1 will reduce and its collector-emitter resistance will increase. More voltage will be dropped across TR1 and the output voltage will fall.

Similarly if the output voltage tends to fall, the proportion of the output voltage developed across R3 will also decrease. This will reduce the base potential of TR2 which in turn will conduct less, drawing less current through R1.

The base current of TR1 will increase and its collector-emitter resistance will decrease. Less voltage will be dropped across TR1 and the output voltage will rise.

Summary

- A voltage regulator is employed to ensure that the output of a power supply will remain constant and independent of changes in the load current.
- In the series regulator the control element is a transistor connected as an emitter follower.

  The resistance of the transistor (collector to emitter) is controlled by the base current.
- The output voltage is monitored and compared with a reference voltage. The difference between the two produces an output from an error amplifier which in turn controls the base current of the series regulator transistor.
- If the output voltage increases, the difference between V_{out} and V_{ref} increases. This causes the error amplifier to pass more current which diverts current away from the series regulator and increases its resistance, restoring the output voltage to its original level.
- If the output voltage decreases, the difference between V_{out} and V_{ref} decreases. This causes the error amplifier to pass less current, which allows more current to flow to the series regulator and reduce its resistance, restoring the output voltage to its original level.

Self Assessment Questions *(answers page 148)*

Questions 1 to 12 refer to the circuit diagram of the voltage regulator circuit on p.140.

1) Which transistor carries the main load current of the stabiliser?
   Trace on the circuit diagram the path of the load current.

2) Which component/s provide/s the voltage reference for the regulator?
   a) R1 & R2 b) TR2 c) ZD

3) Which component is used as a store of energy?
   a) C1 b) C2 c) TR1

4) Which component is used as a de-coupling capacitor?
   a) C1 b) C2 c) ZD

5) What is the function of R4?
   a) to provide emitter current for TR2.
   b) it is the regulator in conjunction with ZD.
   c) to bias ZD into reverse breakdown.

6) What fault symptom would you expect if R1 became open circuit?
   a) the output voltage would be a rise.
   b) the output voltage would be 0V.
   c) the output voltage would be half.
7) What fault symptom would you expect if R2 became open circuit?
   a) the output voltage would rise.
   b) the output voltage would be 0V.
   c) the output voltage would be half.
8) What fault symptom would you expect if ZD became open circuit?
   a) the output voltage would rise.
   b) the output voltage would be 0V.
   c) the output voltage would be half.
9) What approximate voltage do you expect to obtain at the output of this regulator circuit?
   a) 11V   b) 15.6V   c) 17V
10) What voltage would you expect to measure across C1 with a digital multirange instrument on a d.c. range?
   a) 12V   b) 15.6V   c) 17V
11) How could the regulator output be increased by a small amount?
   a) by changing the mains transformer.
   b) by changing the series regulator.
   c) by putting a small variable resistor in series with R2.
12) Is the power dissipated in TR1 at maximum:
   a) when the regulator is OFF load?
   b) when it is ON load?
   c) when it is supplying half full load?

Answers to questions in Part 1: Power Supplies

Section 1

Task c: Investigation of rectifier circuits

1) Output negative to produce a negative supply rail
2) 1 pulse in 20ms compared to 2 pulses in tests 4 & 5.
3) 10ms: Half
4) The peak should be 0.7V less in test 5 due to extra diode volt drop.
5) It only uses 2 diodes and gives less volt drop
6) It requires a centre tapped transformer.
7) 50Hz
8) 100Hz
9) Test 6 should produce a higher (0.7V) output voltage.
10) Uses both half cycles of the mains; has a higher ripple frequency which is easier to filter; For a given size of ripple you can use a smaller reservoir capacitor.
11) CRO indicates pk -pk; digital meter indicates r.m.s. (r.m.s. = Peak Voltage × 0.707)
12) Yes

Section 2

a.c. waveforms

1) A) P.T. = 0.025s F = 40Hz
   B) P.T. = 1ms F = 1kHz
   C) P.T = 100ms F = 10Hz
2) P.T. = 10ms; F = 100Hz; pk = 10A; r.m.s. = 7.07A
3) a) 100Hz b) 10MHz c) 50kHz
4) a) 2μs b) 3.3μs c) 50μs
5) a) 14.14V b) 42.42V c) 282.8V
6) a) 70.7mV b) 2.12A c) 35.35V

a.c. waveforms – phase

1) False; I leads V
2) True.
3) True.
4) True.
5) False; anti-clockwise.
6) False; V leads I
7) False; I leads V
8) True. Back e.m.f. opposes initial flow of current.
9) False; current must flow before the voltage can build up on the plates.

10) True.
11) False; they will go in and out of phase.
12) False; they will be out of phase.
13) True.
14) True.
15) True.

Using the cathode ray oscilloscope

Peak and r.m.s. values

	pk-pk	*pk*	*r.m.s.*
a)	17.5V	8.75V	6.186V
b)	0.175V	87.5mV	61.8mV
c)	2.25mV	1.125mV	0.795mV
d)	37.5mV	18.75mV	13.25mV

Periodic Time & Frequency

	P.T	*Frequency*
a)	17.5ms	57Hz
b)	175μs	5.7kHz
c)	22.5ms	44.4Hz
d)	3.75ms	266.6Hz

Function generators

1) c
2) b
3) d
4) a
5) a
6) a
7) b
8) a
9) b
10) d

Simplified semi-conductor theory

1) b) & c)
2) b) & d)
3) c) There are equal numbers of holes and electrons in pure semiconductor material.
4) a) P-type material has positive holes introduced.
5) b) N-type material has free electrons introduced
6) a) Electrons are available due to temperature. (hole-electron pairs)
7) a) Holes are available due to temperature.
8) d) Each bond consists of two electrons, one from each atom.
9) b) A hole is introduced into the lattice for each impurity atom.
10) a) A free electron is introduced into the lattice for each impurity atom.

Semi-conductor diodes

1) a) A PN junction has a high resistance in one direction and a low resistance in the other.
2) a) Forward bias breaks down the potential barrier.
3) d) Minority carriers (holes on the N side and electrons on the P side) are attracted across the junction by the reverse bias. This forms the reverse current.
4) b) Reverse biasing assists the potential barrier
5) a) Leakage current doubles for every few degrees C rise in temperature.
6) a) Lower forward volt drop than silicon.
7) b) Lower reverse current than germanium.
8) a) Current in the 3 ohm resistor = 2A; Current in the 2 ohm resistor = 0A
   b) Current in the 3 ohm resistor = 0A; Current in the 2 ohm resistor = 2.5A
9) 655.36mA
10) A1 = 1.111A, A2 = 0, A3 = 0.8333A

Zener diodes

1) b) The diode is broken down in the reverse direction.
2) d) If the maximum power of the diode is exceeded then the diode is destroyed.
3) b) Temperature will affect the volt drop.
4) b) The 10 ohm series resistor has to drop 15V therefore the current in the zener is 1.5A
5) b) Current = 10V/10ohms = 1A.
6) b) The zener unloaded takes 1.5A therefore if a load takes 0.5A then 1A continues to flow in the zener.
7) b) The load voltage is still 10V.
8) d) Load resistance lowers, load current increases, current in zener reduces.

Multirange test instruments

1) b) Sum of the volt drops around the circuit equals the supply.
2) c)
3) a)
4) c) The resistance (R1 + R2) and supply voltage are known; the series current can be determined and hence the volt drop across each resistor.
5) b) Sum of the branch current equals the supply current.
6) c)
7) a)
8) d)

Resistors and resistor colour coding

1) Top left. 470R 10% Top right. 5R6 5%
Bottom left 3k9 10% Bottom right 1M8 20%

2)

R	*Max*	*Min*
100k	110k	90k
3.3k	3465	3135
150R	165	135
2R7	2.727	2.673

3) a) 0.33 ohm 20% b) 4700 ohms 1%
c) 6,800,000 ohms 20% d) 22000 ohms 10%

4) a) Yellow, Violet, Gold, Silver
b) Brown, Red, Gold, Red
c) Yellow, Violet, Silver, Gold

Power supply specifications

1) c)
2) a)
3) d
4) a)
5) d) If the smoothing is effective the a.c. ripple should be small.
6) d)
7) c)
8) d) Most transistors and integrated circuits require a small d.c. voltage to operate them.

Half-wave and full-wave rectifier circuits

1) b)
2) d)
3) b)
4) c)
5) d) Average = peak x 0.636
6) c)
7) c)
8) True
9) True
10) True
11) False; ripple will also increase.
12) False; The reservoir can be reduced in value
13) True
14) False; capacitor has infinite resistance to d.c.
15) True
16) True
17) True

Basic principles and operation of transformers

1) 1000/4 = 250
2) 240/4 = 60V
3) The secondary resistance will be less than the primary resistance because there are fewer turns on the secondary of a step down transformer, and vice verse for a step up transformer.
4)

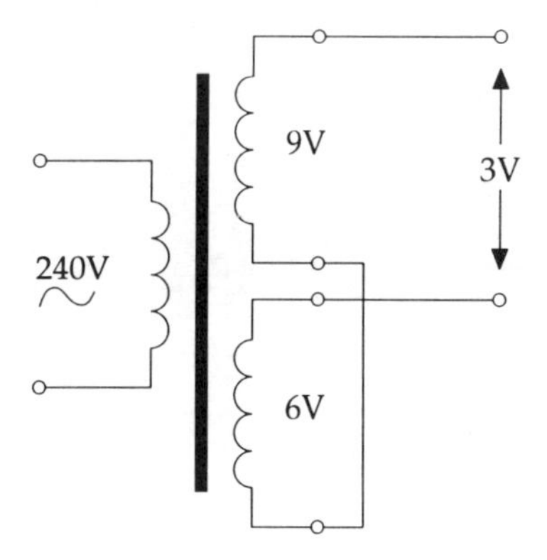

5)

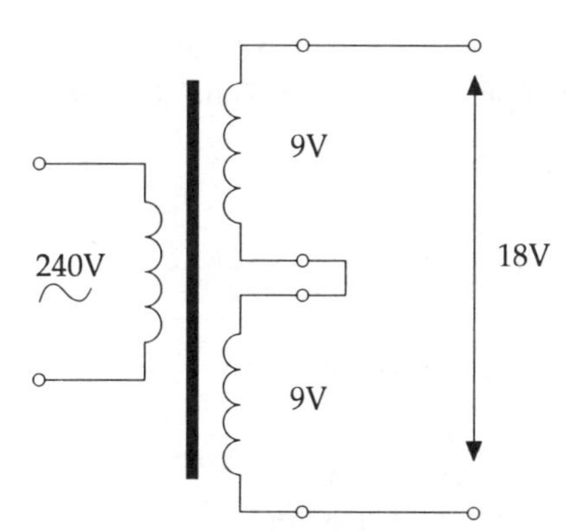

6) Normal: The larger transformer will have a greater VA rating. This means that the primary current will be larger and therefore the diameter of the wire used for the primary will need to be greater. The resistance for the same number of primary turns will therefore be less.
7) Secondary Power = (12 x 0.5) + (6 x 2) = 18W
   Allowing for losses the VA rating will be: 18 x 1.1 = 19.8
   Therefore a suitable transformer would be a 20VA.
8) With d.c. applied to the transformer there will be no changing flux and no back e.m.f. to limit the primary current. Either a fuse will blow or the primary winding will burn out.

Metric Prefixes

Exercises:

1) 2.2k
2) 470k
3) 0.47M
4) 1.5M
5) 10mA
6) 0.015A
7) 0.0018A
8) 1μF
9) 0.15μF
10) 150nF
11) 2.2nF
12) 0.5nF
13) 0.05V
14) 0.0009A
15) 68pF

Self Assessment Questions

1)	i) 1ms	ii) 0.1ms	iii) 0.00001ms	
2)	a) 6.8kV	b) 16pF	c) 0.76mA (760μA)	d) 0.82nA (820pA)
3)	251.7mA			
4)	11.1μs			
5)	260V			
6)	a) 5.3mA	b) 18.925kV	c) 19.5MΩ	d) 6.25μΩ

Voltage Regulator Discrete Components

1)		TR1 is the series control transistor.
2)	c)	the zener diode provides Vref
3)	a)	C1 is the main reservoir capacitor
4)	b)	C2 is used to decouple the base of TR1
5)	c)	R4 biases the zener into reverse breakdown
6)	b)	With R1 open circuit TR1 has no base current and is OFF
7)	a)	If R2 is open circuit TR2 is OFF. TR1 receives max base current, TR1 fully ON and the output voltage will rise.
8)	a)	With ZD open circuit TR2 emitter voltage will rise and turn TR2 OFF. TR1 receives max base current, TR1 fully ON and the output voltage will rise.
9)	a)	The voltage on the base of TR2 will be 0.6V above 5.6V i.e. 6.2V. This voltage is across R3. Using the potential divider rule the voltage across R2 and R3 will be 11.28V approx.*
10	b)	12V x 1.414 minus the volt drop across the diodes (1.4V)
11	c)	Altering the voltage on the base of TR2 will alter the output voltage.
12	b)	Max power under full load.

* The potential divider rule is covered in **Part 2 Section 2 Information and Skills Bank** Single stage audio amplifier, p.229.

Part 2: Amplifiers

Contents

How to use this part

This project assumes that you have studied the previous Power Supplies project and undertaken the practical tasks.

Readers with little prior electrical knowledge

It is suggested that you look at the first practical task (bipolar transistor characteristics) and then study in depth the associated Information and Skills Bank references for that task. When you have completed the self assessment questions then attempt Task a. Continue in this manner until you have completed Tasks a, b, c and d, by which time you will have designed, constructed and tested a single stage bi-polar transistor amplifier!

You may then undertake tasks e, f and g and design, construct and test a FET amplifier.

Readers with considerable previous experience.

These readers may progress straight to the Amplifiers design project (tasks d and g). Completion of the project will demonstrate their knowledge and competence for all the BTEC objectives shown on the syllabus/topic coverage sheet for this project.

Syllabus/Topic Coverage Chart

Electronics NII Syllabus topic	**Tasks Section 1**	**Information and Skills bank Section 2**
B-2-a Describes the d.c. operation of bi-polar transistors B-2-c States base/emitter voltage	*Task a* *Bi-polar Characteristics*	Amplifier specifications Bi-polar transistor Common emitter circuit Common emitter characteristics
C-3-a Describes operation of bi-polar amplifier	*Task b* *Measure gain and bandwidth of a common emitter amplifier*	Bi-polar amplifier Gain and bandwidth of common emitter amplifier
C-3-e Defines bandwidth	*Task c* *Tests on bipolar amplifier*	Fault finding on a bi-polar amplifier
C-3-b Describes g.m. model of transistor C-3-c States g.m. = $I_C/25$ mS C-3-d Calculates gain of bi-polar amplifier C-3-f Uses manufacturer's data C-3-g Constructs and tests bi-polar amplifier B-2-d Describes switching action of transistor	*Task d* *Design and test bipolar amplifier*	Load lines Equivalent circuits Capacitive reactance
B-2-a Describes the d.c. operation of uni-polar transistors	*Task e* *FET*	Amplifier specifications FET transistor
C-3-a Describes operation of uni-polar amplifier	*Task f* *Measure gain and bandwidth of FET amplifier*	FET amplifier Gain and bandwidth of FET amplifier.
C-3-e Defines bandwidth	*Task f* *Tests on FET amplifier*	Fault finding on a FET amplifier
C-3-d Calculates gain of FET amplifier C-3-f Uses manufacturers' data C-3-g Constructs and tests FET amplifier B-2-d Describes switching action of transistor	*Task g* *Design and test FET amplifier*	Load lines Equivalent circuits Capacitive reactance

Study Note: Information and Skills Bank references; Bi-polar transistor (p.209) and Common emitter circuit (p.217), need to be studied to achieve the BTEC syllabus topic D-4-b objectives (regulated power supply using transistors and zener diode).

D-4-b	Series voltage regulator using 2 transistors & zener diode	Transistor/zener stabiliser Bi-polar transistor Common emitter

Outline of Amplifiers project

You have now been with your new small electronics firm (BIT & BYTE Ltd) for 12 weeks. You have designed your power supply and your training supervisor now gives you a another project to develop your knowledge and practical skills with amplifiers.

You are given the following information about the design project together with access to a skills and information bank, electronic components, manufacturer's data, catalogues, and test equipment.

Design project

S.A.S. Alarms Ltd were delighted with the power supplies you designed for their alarm panels. The S.A.S. alarm panel has now has two supplies; the 8.3V supply you previously designed and a 24V supply at 1A.

S.A.S. Alarms are developing a total home security panel which detects intruders, flame and smoke. As an added feature it also will provide sound monitoring of any selected room within the house from any other room. This can to be used to monitor babies, the old, infirm or any sick person as well as any room that may contain valuables, while the occupiers are resident in the house.

The supplies in the control panel are therefore required to power several sound detectors each requiring a small low frequency audio amplifier to drive an alarm sounder.

The alarm sounder is basically a small loud speaker with a built in power amplifier and a volume control.

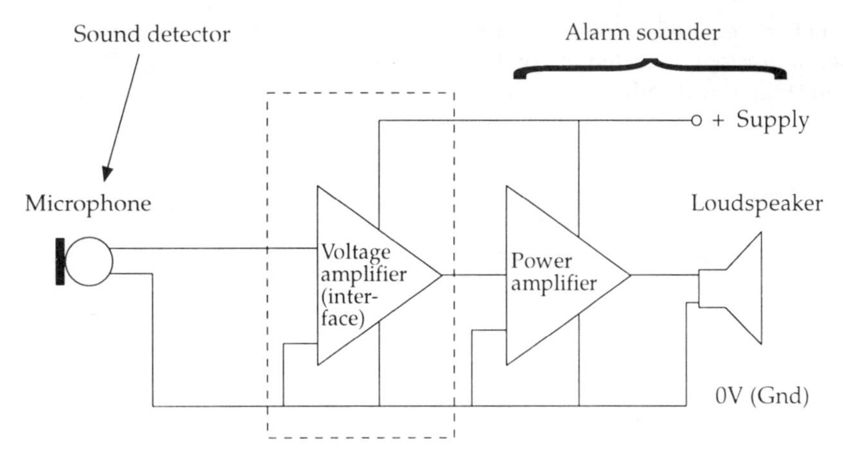

S.A.S. sound detectors and alarm sounders

There are two types of sound detectors available:

- ❒ Type A has a low internal resistance (1k ohms approx.) and provides an output of 10mV r.m.s.
- ❒ Type B a very high internal resistance (1M ohm) and provides an output voltage of 0.35V r.m.s.

There are also two types of alarm sounder:

- ❒ Type X, which has an input resistance of 10k ohms and requires a minimum input of 500mV r.m.s.
- ❒ Type Y, which has an input resistance of 10k ohms and requires a minimum input of 2.8V r.m.s.

S.A.S. alarms would like you to design, build and test suitable amplifiers to interface the sound detectors to the alarm sounders. They also require a full report of your designs in order that they can cost and choose the most suitable sound detector and alarm sounder to fit in their alarm panels.

To interface each type of sound detector to an alarm sounder you will need to design two different amplifiers.

Project 1

Design and test a single stage amplifier to interface **type A** detector to an alarm sounder **type X**.

Amplifier Specification
A single stage resistive loaded amplifier
Supply voltage 8.3V
Stage gain 60 or more
Frequency range 20Hz to 20kHz
Output resistance 10k or more
Input resistance 1k ohms maximum
External Load 10k ohms
Input signal 10mV r.m.s.

The amplifier required to interface type A detector to type X sounder is ideally suited to a bi-polar transistor connected in common emitter configuration. It has a low input impedance and will provide a large gain to a small input signal of a few mV (Task d).

Project 2

Design and test a single stage amplifier to interface a **type B** detector to an alarm sounder **type Y**.

Amplifier Specification
A single stage resistive loaded amplifier
Supply voltage 24V
Stage gain 8 or more
Frequency range 20Hz to 20kHz
Output resistance greater than 40k ohms
Input resistance 1M ohms minimum
External load 10k ohms
Input signal 0.35V r.m.s.

The amplifier required to interface a type B detector to a type Y sounder is ideally suited to a Junction FET transistor connected in a common source configuration. The FET has a large input impedance, provides a voltage gain and can accept the larger input signals (Task g).

Part 2: Amplifiers

Section 1
Design project: tasks and investigations

Below is a summary of the practical tasks for this project, together with the associated cross-references to the **Section 2 Information and Skills Bank**. The practical tasks follow a logical order of progression but, depending on previous knowledge and experience, completing all the tasks may not be necessary (unless your lecturer instructs you to do so).

Summary of practical tasks

Remember!
If you already have some knowledge and experience of transistors and amplifiers then it may only be necessary for you to complete the two main tasks (d) and (g) above. If you have little previous experience, you must study the relevant parts of **Section 2 Information and Skills Bank** before attempting any practical tasks.

Task a

Common emitter static characteristics

Refer to ***Section 2 Information and Skills Bank*** *references: The common emitter connection, p.217; Common emitter static characteristics, p.223.*

Aims

To plot the transfer and output characteristics of transistors connected in common emitter. To determine from the characteristics the current gain h_{fe}, the output resistance and the mutual conductance g.m.

Equipment

A low voltage power unit providing a variable d.c. output voltage 0–30V, up to 500mA.
Three digital multirange meters.
A breadboard and suitable connecting wires.

Components

Resistor: 68k; Transistor; BC108; 1k ohm potentiometer; 9V battery.

Tools

A pair of wire cutters/strippers and snipe nosed pliers.

Circuit Diagram

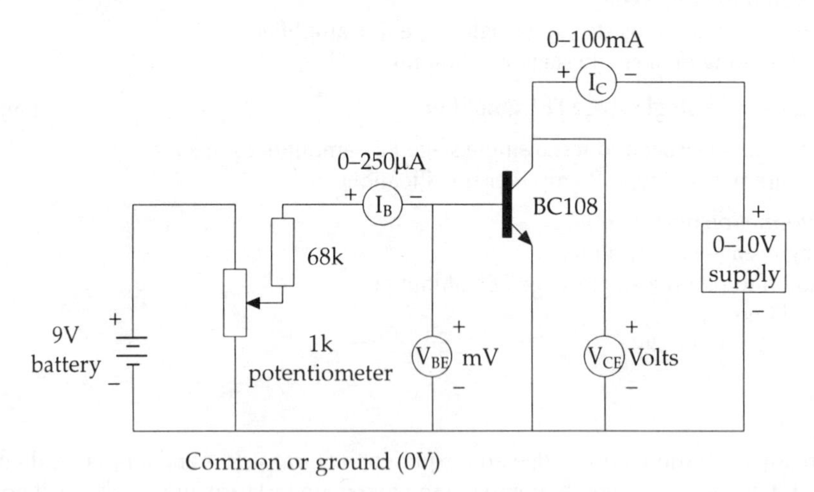

Circuit for obtaining the common emitter characteristics

Note: If you have two variable d.c. supplies then you may use one to replace the 9V battery and the potentiometer. The 68k should be increased to 220k however (see the test circuit in **Section 2 Information and Skills Bank** Common emitter static characteristics, p.223.). Similarly if you have access to only one digital multirange meter then you will have to connect and disconnect it to take all the readings around the circuit.

Approach

Breadboard the circuit: use the breadboard pattern below to construct the circuit. (If you are not sure how to use a breadboard see **Part 1 Section 2 Information and Skills Bank** How to use breadboards, p.96).

Practical Tips

1) Position the equipment and circuit breadboard exactly as it appears on the circuit diagram, **keeping all meter and circuit breadboard leads as short as possible**. (With all transistor amplifiers, long leads may pick up stray signals or cause oscillation, which may affect your readings.)

2) Connect up the potential divider network, including ammeter I_B and voltmeter V_{BE}. Check that it works (provides a variable voltage), reduce V_{BE} to zero, switch off the battery and then connect it to the base of the transistor (**keep all leads as short as possible**).

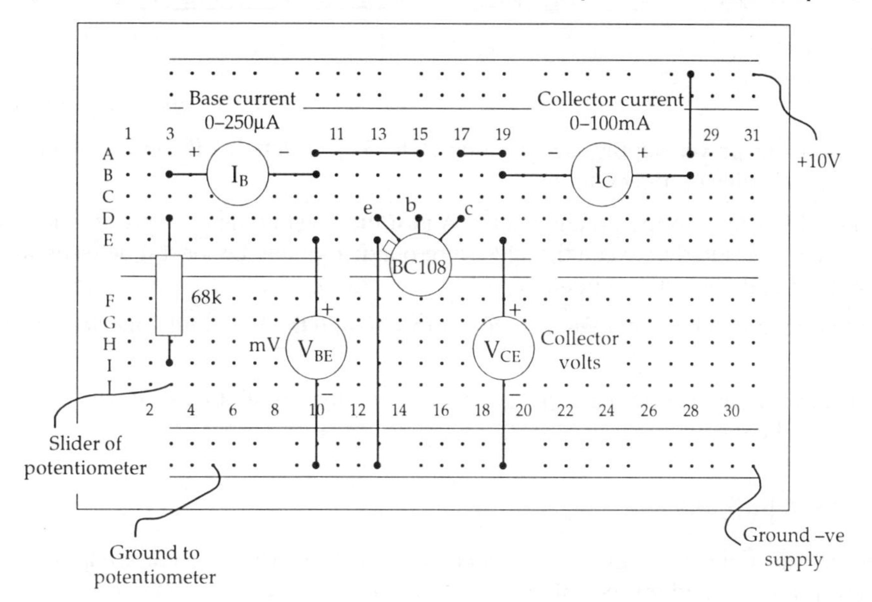

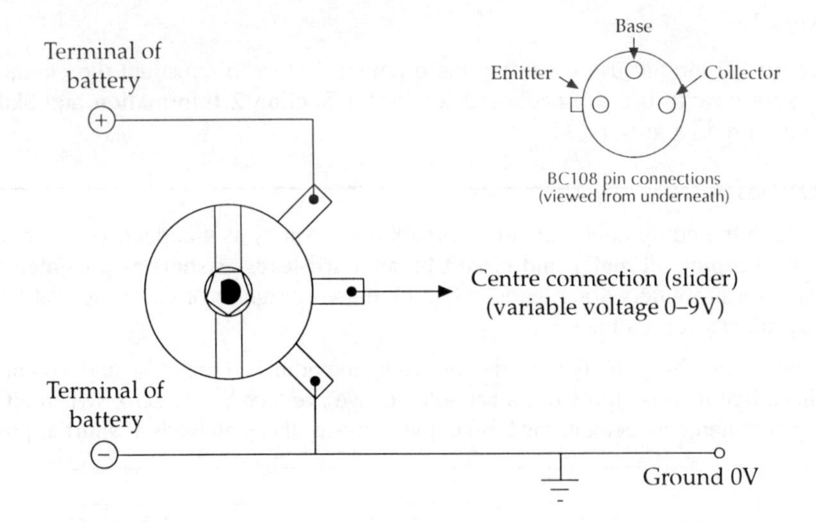

Connections to potentiometer

Switch on the power supply, set the output to 5V, **switch it off**, and then connect it between the collector and common.

Connect the battery to the potentiometer. Adjust the potentiometer to give minimum output voltage (measured between the slider to ground) and then connect it between the base and ground of the transistor circuit.

Wire two of the multirange meters in circuit to measure the base and collector currents as shown.

Check all your connections carefully and when you are satisfied that your wiring is correct, switch on all the equipment.

Transfer Characteristic

In this test the collector voltage V_{CE} is kept constant at 5V. The base current I_B is increased in steps from zero and for the value of each base current given in the table you record the collector current I_C flowing.

Results Table

I_B (μA)	0	5	10	20	30	40	50	60	70	80	90	100
I_C (mA)												

Now plot a graph of I_C (vertical axis) against I_B (horizontal axis). From the graph determine the current gain h_{fe}.

$$h_{fe} = \frac{\text{a change of } I_C}{\text{corresponding change of } I_B}$$

Output Characteristic

To obtain the output characteristic, the base current is fixed at some value, the collector voltage is increased in steps and the collector current recorded at each step. The base current is then increased to a new value and the procedure repeated to obtain a family of curves.

Using the same circuit as before, reduce the collector current to zero. Adjust the base current to 20μA and ensure that the base current remains constant at this value for the first test. Now increase the collector voltage in the steps shown in the table below. At each step measure and record the value of collector current flowing.

Results Tables

	Collector voltage (volts)													
I_B = 20μA	0.1	0.2	0.3	0.4	0.5	0.6	0.7	0.8	0.9	1.0	2.0	3.0	4.0	5.0
I_C (mA)														

Now reduce the collector voltage to zero, increase the base current to 40μA and take another set of readings.

	Collector voltage (volts)													
I_B = 40μA	0.1	0.2	0.3	0.4	0.5	0.6	0.7	0.8	0.9	1.0	2.0	3.0	4.0	5.0
I_C (mA)														

Now carry on the procedure and complete all the tables.

	Collector voltage (volts)													
I_B = 60μA	0.1	0.2	0.3	0.4	0.5	0.6	0.7	0.8	0.9	1.0	2.0	3.0	4.0	5.0
I_C (mA)														

	Collector voltage (volts)													
I_B = 80μA	0.1	0.2	0.3	0.4	0.5	0.6	0.7	0.8	0.9	1.0	2.0	3.0	4.0	5.0
I_C (mA)														

	Collector voltage (volts)													
I_B = 100μA	0.1	0.2	0.3	0.4	0.5	0.6	0.7	0.8	0.9	1.0	2.0	3.0	4.0	5.0
I_C (mA)														

Now plot a graph of collector current I_C (vertical axis) against collector voltage V_{CE} (horizontal axis) for each fixed value of base current I_B.

Determination of the Output Resistance R_{out}

From the output characteristic, determine the output resistance of the transistor. A suitable point on the graph would be where the collector voltage is 3V and the base current is 80μA. In other words, take a change in collector voltage around 3V and using the I_B = 80μA curve estimate the change in collector current (see **Section 2 Information and Skills Bank** Common emitter static characteristics, p.223.)

$$R_{out} = \frac{\text{a change in collector voltage}}{\text{the corresponding change in collector current}}$$

Determination of h_{fe} from the Output Characteristic

$$\text{The current gain } h_{fe} = \frac{\text{a change in collector current}}{\text{the corresponding change in } I_B}$$

for a constant collector voltage V_{CE}.

If we draw a vertical line up the graph from the X axis say at a V_{CE} of 3V then all the way up that line the voltage is **constant** at 3V.

If we now project across to the Y axis where the vertical line intersects the 60μA and 80μA curves, for example, then we can read off the change in collector current for this change (60 to 80μA) in base current (see **Section 2 Information and Skills Bank** Common emitter static characteristics, p.223.)

Use your graph to determine the h_{fe}. Compare this value with the value obtained from the transfer characteristic.

Determination of the Mutual Conductance g.m.

The output characteristic can also be a plot of collector current against collector voltage for a fixed base-emitter voltage V_{BE}, instead of base current I_B. In other words we use the same output characteristic as before, but replace the fixed base currents with the base voltages V_{BE} which created them.

To determine the base voltage for each base current we need to take another set of measurements. Using the same procedure that we used for the output characteristic, set the base current to all the values used in the test and for each value of base current measure and record the base voltage V_{BE}.

I_B (μA)	20	40	60	80	100
I_C (mA)					

When you have completed the table insert the values for V_{BE} on your output characteristic alongside the corresponding value of base current. In other words, if a base voltage of 620mV causes a base current of 60μA, then alongside the 60μA curve write 620mV. (see **Section 2 Information and Skills Bank** Common emitter static characteristics, p.223).

Now:

$$\text{The g.m.} = \frac{\text{change in collector current } I_C}{\text{change in } V_{BE} \text{ causing change in } I_C} \quad \text{(mS)}$$

at a fixed value of collector-emitter voltage.

To determine the change in collector current (I_C), draw a vertical line on your graph from a point around the middle of the voltage axis – say 4V. Mark this line where it intersects the 610mV and the 630mV curves. From these two marks, project across horizontally to the Y-axis and read off the corresponding values of collector current. Now, subtracting the smaller value of collector current from the larger will provide the change in collector current, for a change in base voltage (V_{BE}) of 610mV to 630mV.

Now determine the g.m. for your transistor!

Tip: Keep the output characteristic you have just plotted, it will be useful when you are designing your single stage bi-polar amplifier (Task d).

Task b

Determination of the gain and bandwidth of a common emitter amplifier

Refer to ***Information and Skills Bank*** *The single stage audio amplifier, p.229; Stage gain of a common emitter amplifier, p.241.*

Aims

To measure and calculate the stage gain of a resistive loaded common emitter amplifier and determine the bandwidth.

Equipment

Low voltage power unit providing a variable d.c. output voltage 0-30V, up to 500mA.
A dual trace CRO and a function/signal generator.
A digital voltmeter (capable of measuring a few mV at a frequency of 1kHz).
A breadboard and suitable connecting wires.

Components

Resistors: 82k; 22k; 2 x 3k3; 2 x 1k. Capacitors: 100μ; 2 x 10μ; 2n2; Transistor; BC108.

Tools

A pair of wire cutters/strippers and snipe nosed pliers.

Approach

Breadboard the circuit shown below.

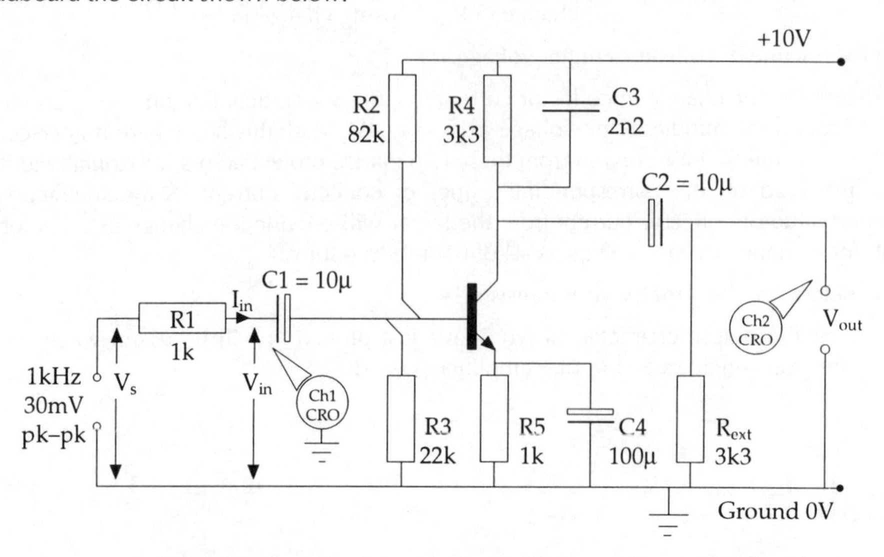

Circuit for measuring gain and bandwidth of a common emitter amplifier

Switch on the power supply, set the output to 10V, **switch off**, and connect the power supply to the circuit.

Connect the CRO and the function generator as indicated on the diagram (take care to connect the correct CRO and generator leads to ground).

Suggested Breadboard Layout

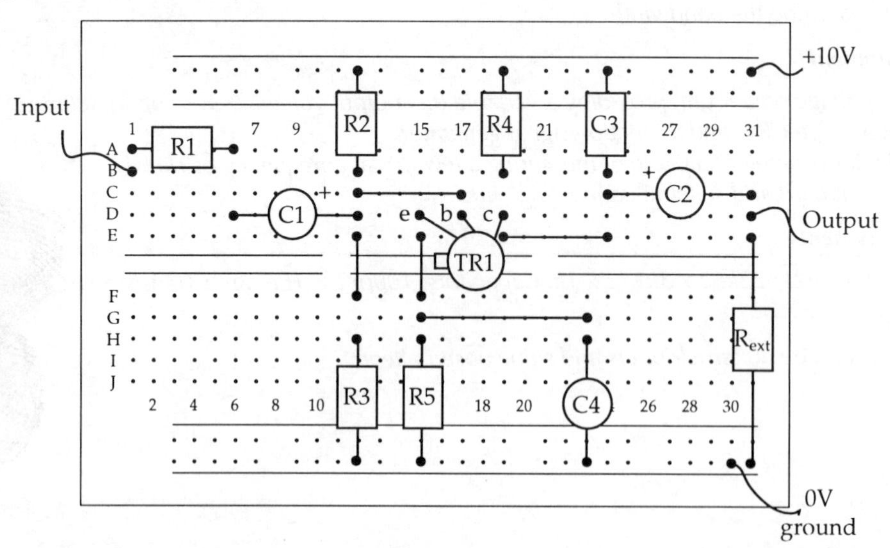

Check all your connections carefully and when you are satisfied that your wiring is correct, switch on all the equipment. Using channel 1 of the CRO, which is monitoring the input, set the function/signal generator to provide an input voltage V_{in} of 30mV peak to peak at 1kHz.

Use channel 2 of the CRO to observe and measure the output waveform. Determine the voltage gain (A_v).

Calculation of Voltage Gain

The common emitter amplifier provides a current and voltage gain. There is phase inversion and the following approximate relationships give the voltage and power gain.

Voltage Gain

$$A_v = g.m. \times R_L = 40 \times I_E \times R_L$$

(since $g.m. = 40 \times I_E$)

or:

$$A_v = \frac{h_{fe} \times R_L}{R_{in}}$$

Power Gain

$$A_p = \frac{h_{fe}^2 \times R_L}{R_{in}}$$

Determination of g.m.

Measure the voltage across the emitter resistor (R5) with the voltmeter and determine the emitter current I_E.

Calculate the g.m. from: **g.m. = 40 I_E**

Determination of the Load R_L

The resistor R_{ext} represents the resistance of an external load. In other words, when the amplifier output is connected to feed another stage then the input resistance of this next stage will load up the collector resistor R4. To a.c. frequencies the coupling capacitor C2 has a very low reactance (low opposition to a.c.). Therefore the resistor R_{ext} is effectively in parallel with R4 in terms of a.c.

Remember the +10V and the 0V rails are at the same potential in terms of a.c. because of the low internal resistance of the power supply to a.c. (the large reservoir capacitor in the power supply has a very low reactance).

The capacitor C3 represents the input capacitance of the circuit the amplifier is driving and the stray and wiring capacitances of the amplifier circuit. The reactance of this capacitor will be large at 1kHz compared to the equivalent value of the load and although it is also in parallel with the load, it can be ignored at mid and low frequencies.

Therefore the effective load of the amplifier R_L is R4 in parallel with R_{ext}

$$R_L = \frac{R4 \times R_{ext}}{R4 + R_{ext}} \qquad \frac{\text{Product}}{\text{Sum}}$$

Calculate the load R_L and the stage gain: (A_v = g.m. × R_L).

Study Note: You will probably find that the measured value of gain and the calculated value of gain may differ. This is because the calculated value using the g.m. relies on the relationship 40 × I_E at a specific temperature, and assumes that components have no losses, and that all resistors have no tolerance. The measured value of the voltage gain will also include any losses in the coupling capacitors, the tolerance of components, and any inaccuracies in the readings obtained due to the instruments and observation error. If you are relying on the CRO to determine the input and output voltages then the measurements must be taken very carefully.

From the relationship:

$$\mathbf{A_v = \frac{h_{fe} \times R_L}{R_{in}}}$$

We can determine the h_{fe} if we know R_{in}.

Determination of R_{in}

In order to determine R_{in} we first need to calculate I_{in}

$$I_{in} = \frac{V_s - V_{in}}{R1} \qquad \text{where R1 = 1k}$$

With V_{in} set to 20mV peak to peak, use the digital voltmeter on the 200mV a.c. range to measure V_s and V_{in}.

Now calculate I_{in}!

$$\text{The input resistance } R_{in} = \frac{V_{in}}{I_{in}}$$

Now calculate R_{in} and then calculate h_{fe}!

$$Av = \frac{h_{fe} \times R_L}{R_{in}}$$

$$\text{Therefore: } h_{fe} = \frac{A_v \times R_{in}}{R_L}$$

Use the measured value of A_v in the formula.

Power Gain A_p

Using the values previously calculated for: h_{fe}; R_{in}; and R_L determine the power gain from the relationship:

Power Gain:

$$A_P = \frac{h_{fe}^2 \times R_L}{R_{in}}$$

Determining the Bandwidth

The gain of an amplifier falls off at high and low frequencies. The fall-off at low frequencies is mainly due to the rising capacitive reactance of the coupling capacitors and the source by-pass capacitor. The coupling capacitors therefore couple less signal in and out of the amplifier and the rising reactance of the by-pass capacitor allows NFB to lower the gain.

The fall-off at high frequencies is largely due to the input capacitance of the circuit the amplifier is driving. Because the amplifier under test is not driving a normal load C3 has been included to simulate this effect.

When the gain falls to half power then this is the limit of the lower and upper frequency that can be handled by the amplifier.

To find the upper and lower half power points the input frequency is altered over the range 20Hz to 100kHz.

It is important that the amplitude of the test signal from the function generator is **kept constant at all frequencies** (i.e. it must be kept at 30mV peak to peak).

Step 1: Use channel 1 on the CRO to monitor the amplitude of a 1kHz input signal and use channel 2 to measure the amplitude of the output at 1kHz. Determine the gain; this is then the mid frequency gain.

Step 2: Now reduce the frequency of the function generator until the output falls to 0.707 of the mid frequency gain. This is the low frequency half power point F_{low}

Step 3: Now increase the frequency from the mid point frequency until the output once again falls to 0.707 of the mid frequency gain. This is the high frequency half power point F_{high}

The bandwidth of the amplifier is the difference between these two frequencies.

$$\text{Bandwidth} = F_{high} - F_{low} \text{ Hz}$$

Determine the bandwidth of your amplifier.

Task c

Measuring, Testing and Fault Diagnosis on a Single Stage Transistor Amplifier

*Refer to **Section 2 Information and Skills Bank**: The common emitter connection, p.217; The single-stage audio amplifier, p.229; Determination of the stage gain of a common emitter amplifier, p.241; Fault-finding on a single-stage bi-polar transistor amplifier, p.246)*

Aims

To observe the effects of fault conditions on a single stage bi-polar transistor amplifier.

Equipment

Low voltage power unit providing a variable d.c. output voltage 0–30V, up to 500mA.
A dual trace CRO and a function/signal generator.
A digital voltmeter.
A breadboard and suitable connecting wires.

Components

Resistors: 82k, 22k, 2 × 3k3, 2 × 1k. Capacitors: 100μ, 2 × 10μ, 2n2; Transistor: BC108.

Tools

A pair of wire cutters/strippers and snipe nosed pliers.

Approach

Breadboard the circuit shown below. Switch on the power supply, set the output to 10V, **switch off**, and connect the power supply to the circuit.

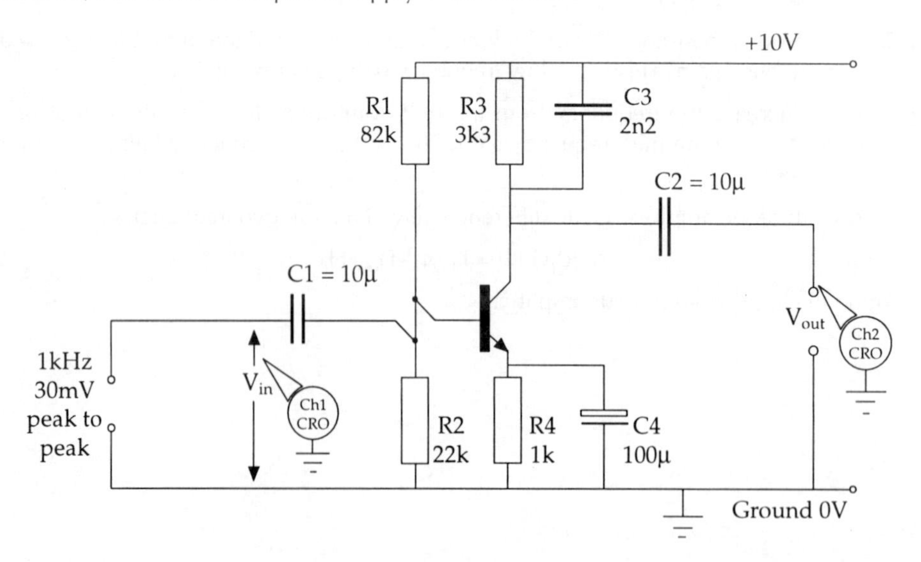

Circuit for Fault Diagnosis in a Common Emitter Amplifier

Suggested Breadboard Layout

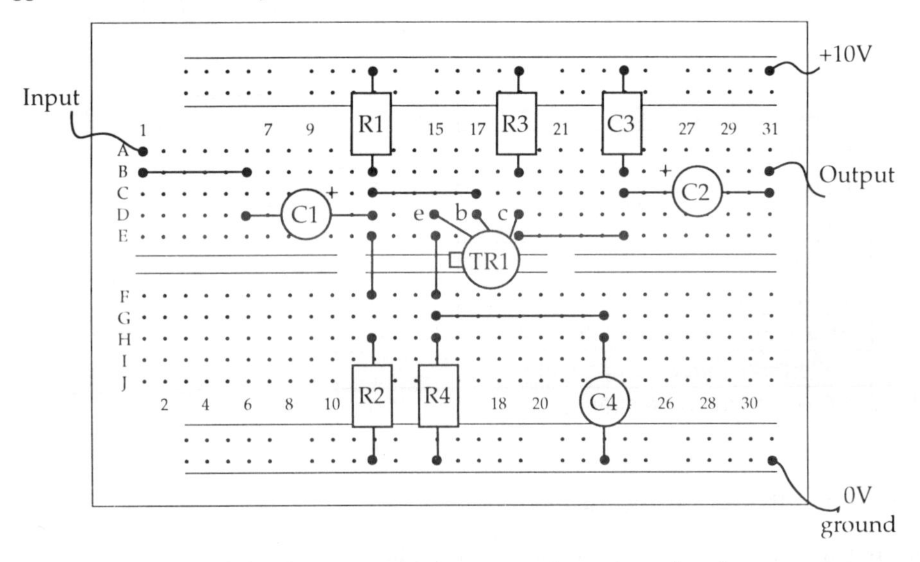

Connect the CRO and the function generator as indicated on the diagram. (Take care to connect the correct CRO and generator leads to ground.) Check all your connections carefully and when you are satisfied that your wiring is correct, switch on all the equipment.

Using channel 1 of the CRO, which is monitoring the input, set the function/signal generator to provide an input voltage V_{in} of 30mV peak to peak at 1kHz. Use channel 2 of the CRO to observe and measure the output waveform. Determine the voltage gain.

Observation of Fault Conditions

Create each of faults given in the table below one at a time. Observe the effect and complete the table, then rectify the fault (remove the fault) before creating the next fault condition.

Simulate the faults by unplugging one end of a resistor, unplugging the appropriate lead on the transistor and placing short circuits across the leads of the transistor.

At all times when creating or rectifying faults switch OFF the power supply.

Faults on the PDER Circuit (d.c. conditions)

For each fault in the table below predict what effect the fault will have. Measure the actual voltages and compare the results with your prediction.

Fault	*Base voltage*	*Collector voltage*	*Emitter voltage*
No fault			
R1 o/c			
R2 o/c			
R3 o/c			
R4 o/c			
Collector o/c internally			
Emitter o/c internally			
Base/emitter s/c internally			
Collector/emitter s/c internally			
Base/collector s/c internally			

Dynamic Faults (a.c. conditions)

These are faults that affect the circuit's ability to amplify a.c. signals.

Although the fault may be caused by incorrect bias its effect may only be noticeable when the amplifier is handling a signal, i.e. dynamic conditions.

Create each fault from the table below, in the circuit, one at a time. Observe the effect on the output waveform and/or gain of the amplifier. Rectify the fault, check that the circuit is working, then create the next fault in the table.

At all times when creating or rectifying faults switch OFF the power supply.

For each fault in the table below predict what effect the fault will have. Observe the actual outputs, sketch all waveforms, record voltages or gains measured, and compare the results with your prediction.

Fault	*Observations (sketch all waveforms)*
No fault	Measure the gain of the stage
Increase output signal to 500mV	Observe output waveform. Restore the input to 30mV
R3 increased to 10k	Measure the peak to peak waveform
R1 increased to 180k	Determine the stage gain
R4 increased 10k	Measure voltages on base, emitter, collector. Determine the stage gain
C1 open circuit	Observe the output
C1 short circuit	Measure voltages on base and collector. Observe output waveform
C2 open circuit	Observe any output waveform

Faultfinding

At this stage you now need access to previously faulted circuit boards. The faults should be inserted on the boards in such a way that they cannot be found simply by visual inspection. Connect all the test equipment, power up the boards and observe the symtoms. Try to estimate in your observations if the fault is dynamic (the circuit is unable to handle an a.c. signal) or static (d.c. conditions on the transistor).

Use the results of all the tests above to determine the fault. Rectify the fault and test the circuit.

Task d

Designing a single stage bi-polar amplifier

Refer to ***Section 2 Information and Skills Bank:*** *Amplifier specifications, p.199; The bi-polar transistor, p.209; The common emitter connection, p.217; Common emitter static connections, p.223; The single-stage audio amplifier, p.229; Fault-finding on a single-stage bi-polar transmitter amplifier, p.246; Load lines, p.273; Equivalent circuits, p.287; Capacitive reactance, p.296).*

Aims

To design, construct and test a resistive loaded single-stage bi-polar amplifier to the following specifications:
Supply voltage 8.3V; Stage gain 80 or more; Frequency range 20Hz to 20kHz; Input resistance 1k ohms minimum; Output load: an alarm sounder of 10k ohms; Input signal 10mV r.m.s.

Equipment

A dual trace oscilloscope.
A multirange digital voltmeter.
A low voltage power supply providing a variable d.c. output voltage range; 0–30V, up to 200mA.
A signal/function generator; range 0 to 100kHz, output; a few mv to 1V.
A breadboard and suitable connecting wires.

Components

A range of resistors: 1k to 100k.. A range of capacitors: 1nF to 500μF; A BC108 transistor or near equivalent.

Tools
A pair of wire cutters/strippers and snipe nosed pliers.

Approach

Study the worked example on how to design a single stage amplifier given below. Design your own circuit to the specification given in the design project. Then build and test your design to see that it meets the specification.

Worked Example

Before we can begin to design the amplifier then we need to study the specification to determine the supply voltage, the input signal, and the gain required. In order to illustrate the design approach consider the following specification.

Specification for the Worked Example

Supply voltage 9V; Stage gain 60 ; Frequency range 100Hz to 10kHz; Input resistance 1k ohms minimum; Output load: an alarm sounder of 2k ohms; Input signal 10mV r.m.s.

Choice of Transistor

From the specification we will need a transistor that will: operate from 9V, have a low input resistance, be driven by a 10mV signal, and provide a good current gain (h_{fe}) over the specified frequency range. For this specification a general purpose, low power bi-polar transistor such as a BC108 or its equivalent, connected in common emitter configuration, would be suitable. From the manufacturer's data, the BC108 typically has an h_{fe} of 180 and an h_{ie} (R_{in}) of 2k ohms.

A suitable common emitter circuit diagram with potential divider bias and emitter stabilisation is shown below.

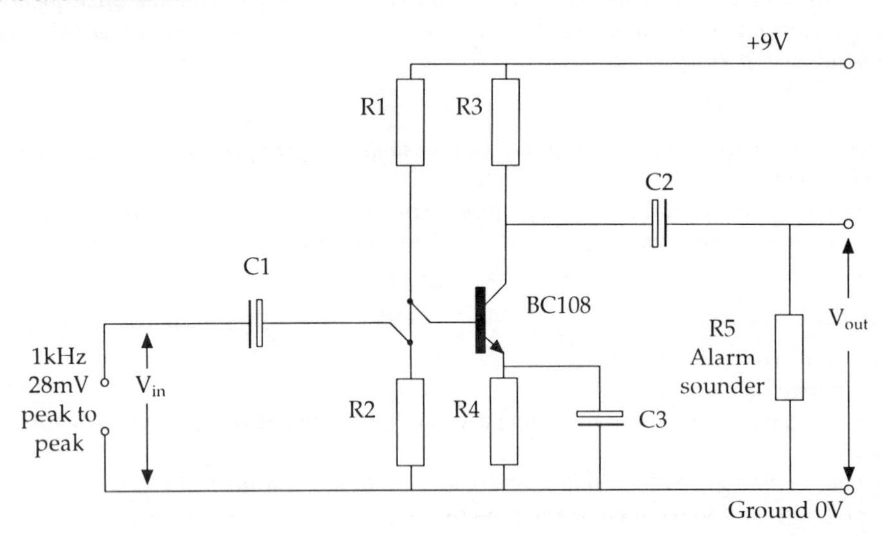

Circuit diagram for the worked example

Next, we need to set the standing collector voltage that will allow the maximum output voltage swing without distortion.

Setting the Standing (Quiescent) Collector Voltage

The supply voltage is 9V and we will need at least 10% of this voltage (0.9V say 1V) to be dropped across the emitter resistor to provide emitter stabilisation. The remaining 8V will be the maximum peak to peak swing of the collector voltage.

Maximum Output Voltage Swing

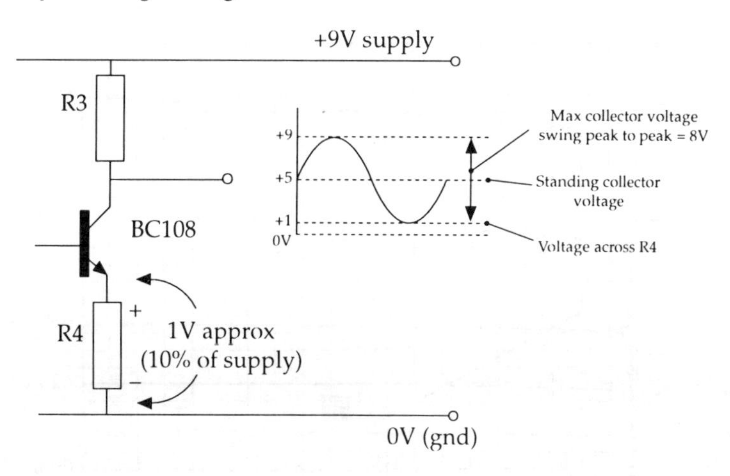

Setting the Standing Collector Voltage

Therefore with 1V across the emitter resistor the collector will sit at 5V with respect to ground. The maximum swing of the output voltage without clipping will be plus or minus 4V.

Choosing the Operating Point

We can now use the output characteristic for the BC108 to determine the operating point and plot the load line.

We have now fixed two points on the X axis of the BC108 output characteristic: the supply voltage is 9V and the standing collector voltage is 5V. The operating point will have to lie on a line projecting upwards from the 5V point on the X axis!

If we now choose a suitable base bias voltage (say V_{BE} = 610mV) that will give a standing collector current between 1 to 2mA, then where this curve cuts the vertical 5V line is the operating point (P).

Remember: if we choose too high a base bias voltage, then the standing collector current and the power dissipated in the transistor will be too high, and the transistor will overheat.

Having determined these two points then the d.c. load line can be plotted on the output characteristic

Plotting the Load Line

To plot the load line we draw a point from the 9V point on the X axis, through the operating point P, to the Y axis. This will be the d.c. load line.

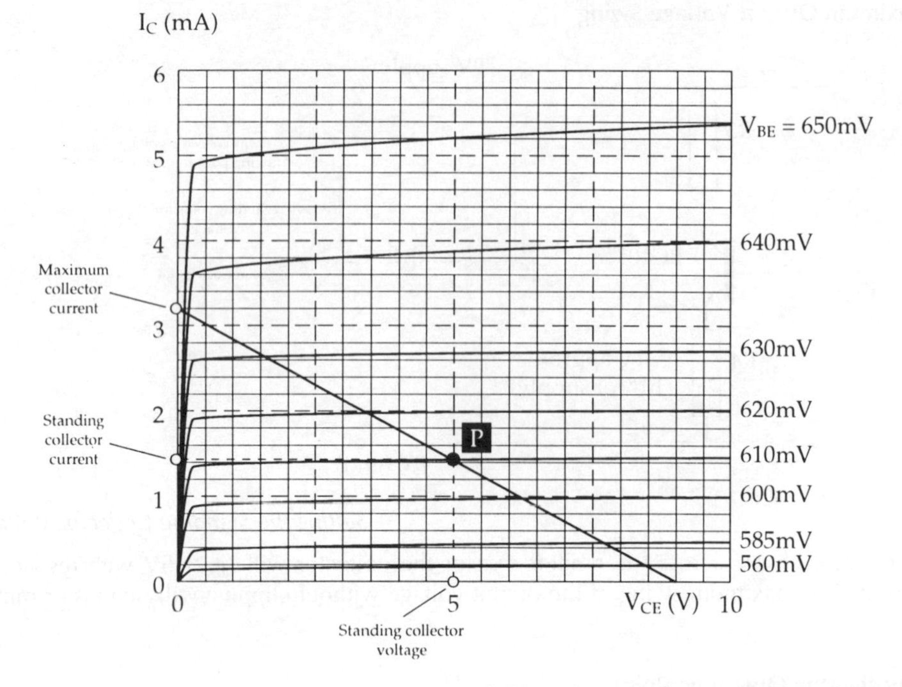

BC108 Output Characteristic and load line

Where this line cuts the Y axis is the maximum collector current that can flow when the transistor is fully turned ON. This current is determined by the supply voltage divided by R3 + R4. From the graph this current is 3.2mA.

$$\text{Now } I_{CMAX} = \frac{V_{supply}}{R3 + R4}. \text{ Therefore } R3 + R4 = \frac{9V}{3.2} = \mathbf{2.8k\ ohms}$$

We now need to calculate the value of R4, and this value subtracted from 2.8k will give the value of the load resistor R3.

Determining the Standing Collector Current

By projecting horizontally from the operating point P across to the Y axis, the value of the standing collector current can be determined. This is the value of current that will flow in the collector before an a.c. signal is applied.

From the graph the standing collector current is 1.45mA.

Determining the value of the emitter resistor (R4)

The voltage drop across the supply is typically 10% of the supply voltage. In this example that will be 0.9V. If we round this up to 1V then the value of the emitter resistance will be 1V divided by the standing (quiescent) emitter current.

The quiescent emitter current is virtually the same as the collector current; therefore if we use the standing collector current the emitter resistor R4 will be:

$$\text{Now } R4 = \frac{\text{Volt drop across R4}}{\text{Standing collector current}} = \frac{1V}{1.45} \text{ k ohms} = 0.689\text{k ohms or } \mathbf{680\ ohms}$$

(nearest preferred value).

Calculating the Load Resistor (R3)

The total resistance obtained from the d.c. load line (2.8k) less the value of R4 will give the load resistance R3.

Therefore R3 = 2,800 – 680 = 2120 ohms or **R3 = 2k2** (nearest preferred value).

Next we have to calculate the value of the bias resistors R1 and R2.

Calculating the value of R1

As a rule of thumb, the current through R1, must be at least 11 times the required base current, so that changes in base current will not affect the voltage dropped across R1 significantly. If the average h_{fe} is 180 for a BC108 then knowing the steady collector current (1.45mA) we can calculate the required base current.

$$h_{fe} = \frac{I_C}{I_B} \quad \text{Therefore } I_B = \frac{I_C}{h_{fe}} = \frac{1{,}450}{180}\,\mu A$$

$$\mathbf{I_B = 8.0\mu A}$$

Therefore the current flowing in R1 = 11 × I_B = **88.6μA** (round this up to 100μA).

The voltage across R1 is the supply voltage less the base bias voltage (610mV) and the emitter voltage V_E.

$$V_{R1} = 9V - 1.61V = 7.39V$$

$$\text{From which R1} = \frac{7.39}{0.1} \text{ k ohms} = 73.9\text{k}$$

Or **R1 = 75k** (nearest preferred value E24 series).

Calculating the Value of R2

The current through R2 will be the current in R1 less the base current since I_B flows into the base of the transistor. i.e. 100μA – I_B = 92μA.

The voltage across R2 will be $V_E + V_{BE}$.

$$\text{Therefore the value of R2} = \frac{1 + 0.61}{92 \times 10^{-6}} = \frac{1.61}{0.092} = 17.5 \text{ k ohms or } \mathbf{18k} \text{ (E12 series)}$$

Calculating the value of the coupling capacitors C1 and C2

As a rule of thumb, the reactance of the coupling capacitor, at the lowest frequency of operation, must not be more than one tenth of the value of the effective input resistance of the amplifier.

The effective a.c. input resistance is R1 and R2 and R*in* all in parallel.

$$\text{R1 and R2 in parallel} = \frac{R1.R2}{R1 + R2} = \frac{75 \times 18}{75 + 18} = 14.5\text{k}$$

R1 and R2 in parallel with R_{in} (h_{ie}) gives:

$$\frac{14.5 \times 2}{14.5 + 2} = 1.76k$$

Therefore the capacitive reactance at the lowest frequency (100Hz in the worked example specification) must not be more than a tenth of this value: i.e. 176 ohms or less.

The capacitive reactance is given by: (see **Section 2 Information and Skills Bank** Capacitive reactance, p.296):

$$X_C = \frac{1}{2\pi \times f \times C} \text{ ohms}$$

$$\text{Therefore, } C = \frac{1}{2\pi \times f \times X_C} \text{ Farads} = \frac{1{,}000{,}000}{2\pi \times 100 \times 176} \mu F$$

$$= \mathbf{9.09\mu F} \text{ (10}\mu\text{F preferred value)}$$

Since the output resistance is very much larger than the input resistance this value will also be suitable for C2.

Remember: these values are the minimum values we can get away with: it is quite in order to fit a larger value if you wish!

Calculation of the Emitter By-pass Capacitor

Using the same rule of thumb, the reactance of the by-pass capacitor (C3) must not be more than one tenth of the emitter resistor R4.

R4 = 680 ohms. Therefore the reactance must be 68 ohms or less at 100Hz.

$$X_C = \frac{1}{2\pi \times f \times C} \text{ ohms}$$

$$\text{Therefore, } C = \frac{1}{2\pi \times f \times X_C} \text{ Farads} = \frac{1{,}000{,}000}{2\pi \times 100 \times 68} \mu F$$

$$= 23.53\mu F \text{ (taking } 2\pi = 6.25\text{)}$$

$$\text{or } \mathbf{C3 = 22\mu F} \text{ (nearest preferred value)}$$

Determination of the Stage Gain from the Output Characteristic

On the BC108 output characteristic on page 172 plot the a.c. load line.

Remember: the effective a.c.load is R3 and R5 in parallel (see **Section 2 Information and Skills Bank** Load Lines, p.273).

Determine the output swing of voltage for an input of 10mV rms. (approximately 28mV peak to peak), and calculate the stage gain.

Now you are ready to attempt one of the design projects, adopt a similar procedure with the new design specification.

Design Project

Now calculate all the component values for the design specification:

Supply voltage 8.3V; Stage gain 80 or more; Frequency range 20Hz to 20kHz; Input resistance 1k ohms minimum; Output load: an alarm sounder of 10k ohms; Input signal 10mV r.m.s.

Use the output characteristic provided at the end of this task to plot your load line.

Construction of Your Design

Insert all your calculated values on the diagram below.

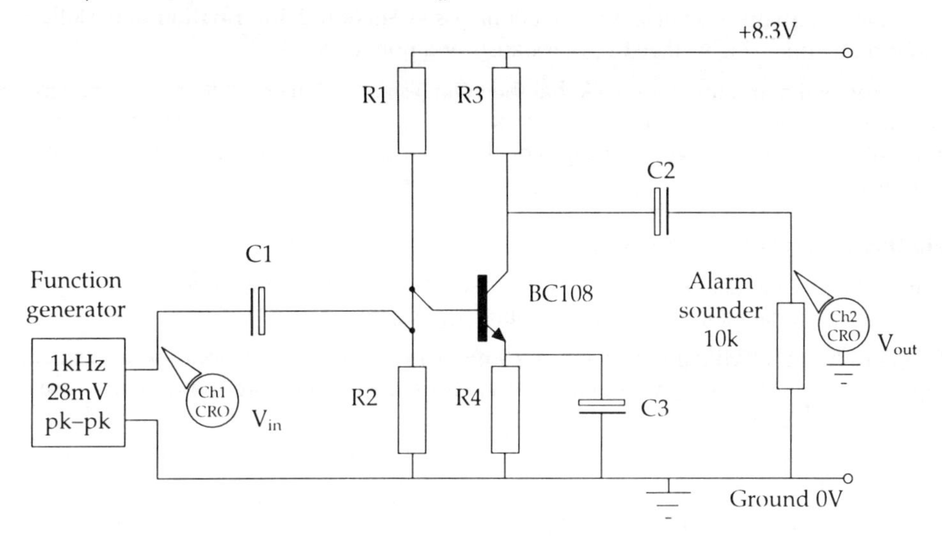

Circuit of a Common Emitter Amplifier

Now breadboard your circuit.

Suggested Breadboard Layout

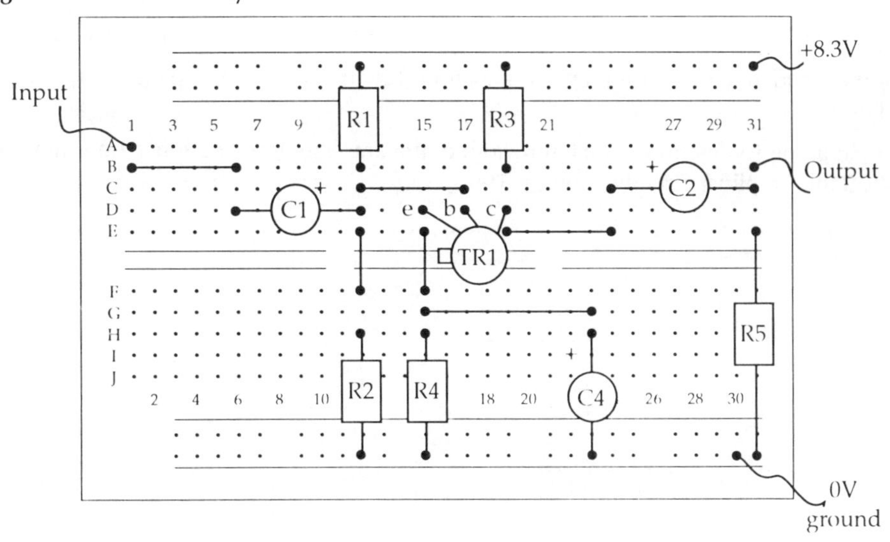

Connect the CRO and the function generator as indicated on the diagram (take care to connect the correct CRO and generator leads to ground).

Testing Your Design

Check all your connections carefully and when you are satisfied that your wiring is correct, switch on all the equipment.

Using channel 1 of the CRO, which is monitoring the input, set the function/signal generator to provide an input voltage V_{in} of approximately 30mV peak to peak at 1kHz. Use channel 2 of the CRO to observe and measure the output waveform. Determine the voltage gain.

If at this stage your circuit is not functioning see **Section 2 Information and Skills Bank** Faultfinding on a single-stage bi-polar transistor amplifier, p.246.

Using the digital multimeter check that the voltages around the circuit are the same as your calculated values. (Remember the components have tolerances and in many cases the nearest preferred value was chosen. For these reasons the measured values may well be different!)

Plotting the Frequency Response

Connect a 2n2 capacitor across the 10k resistor that represents the alarm sounder (this is to simulate the capacitance of the alarm sounder).

Keeping the input voltage constant at 20mV peak to peak throughout the test, inject the frequencies given in the table below. At each frequency measure the output voltage and determine the voltage gain.

	Frequency Hz															
	20	40	50	70	100	200	500	1k	2k	5k	7k	10k	20k	50k	70k	100k
Output																
Gain																

Frequency Gain table

Plot the frequency response graph in pencil on the log-linear graph paper provided at the end of this task.

Use the graph to determine the bandwidth of the amplifier (see **Section 2 Information and Skills Bank** Amplifier Specifications, p.199).

BC108 Output Characteristic

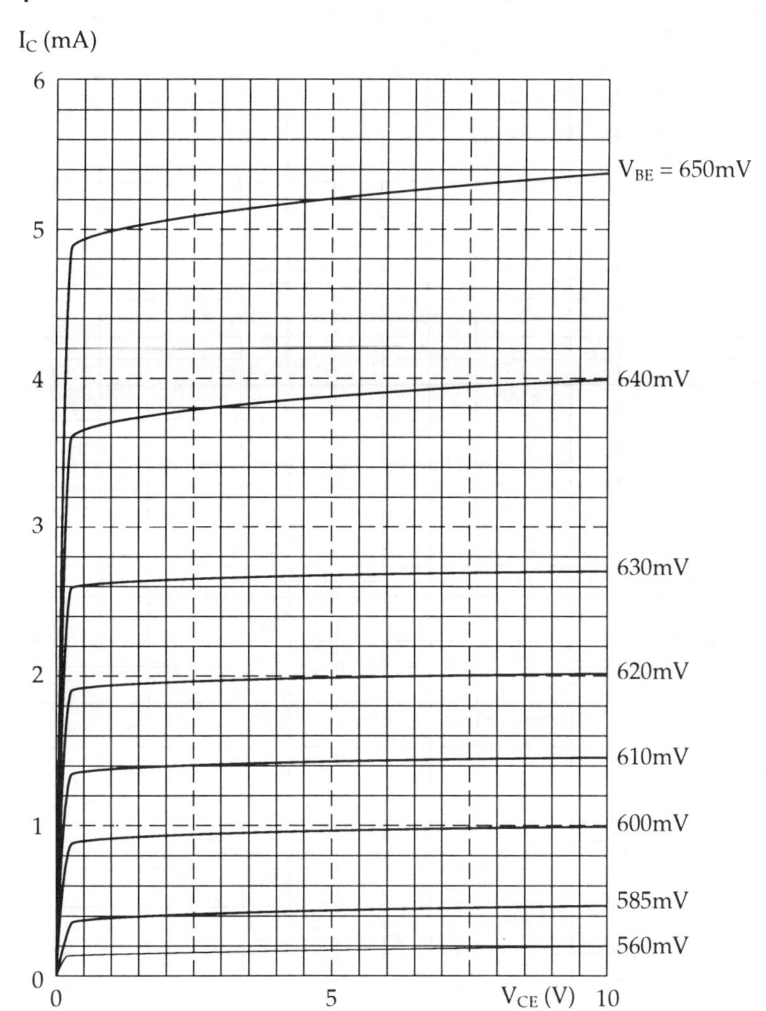

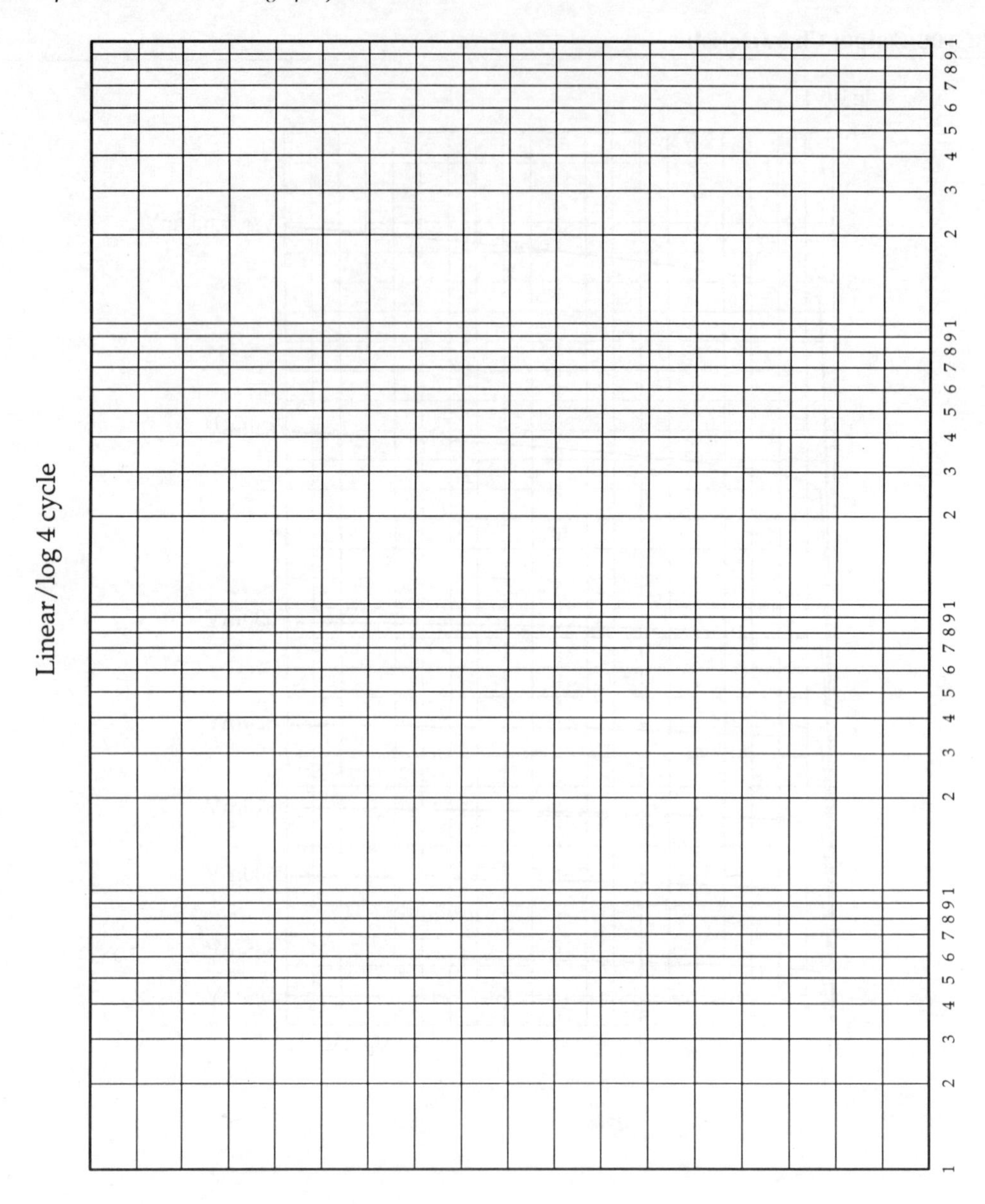

You have now completed design project 1!

Task e

Plotting the output characteristics of a junction FET

Refer to ***Section 2 Information and Skills Bank:*** *The field effect transistor, p.255).*

Aims

To plot the output characteristic of a junction FET transistor connected in common source.
To determine the mutual conductance g.m. from the output characteristics.

Equipment

Two low voltage power unit providing a variable d.c. output voltage 0–30V, up to 500mA
Three digital multirange meters.
A breadboard and suitable connecting wires.

Components

Transistor 2N3819.

Tools

A pair of wire cutters/strippers and snipe nosed pliers.

Circuit Diagram

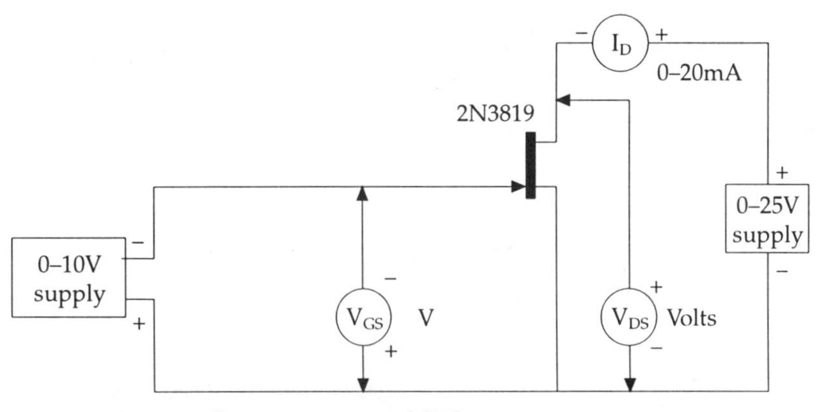

Circuit for obtaining the common source static characteristics

Note: if you only have access to one digital multirange meter then you will have to connect and disconnect it to take all the readings around the circuit. If you have access to two, then connect one in the drain circuit to measure current and use the other switch between measuring the gate and drain voltage.

Approach

Breadboard the circuit (a suggested layout for your breadboard is shown below). Switch on the power supplies, set their outputs to 0V, **switch them off**, and then connect one between the drain and source and the other between the gate and source.

Wire one of the multirange meters in circuit to measure the drain current as shown (0 to 20mA range). Remember to keep all connecting leads as short as possible.

Check all your connections carefully and when you are satisfied that your wiring is correct, switch on all the equipment.

Suggested Breadboard Layout

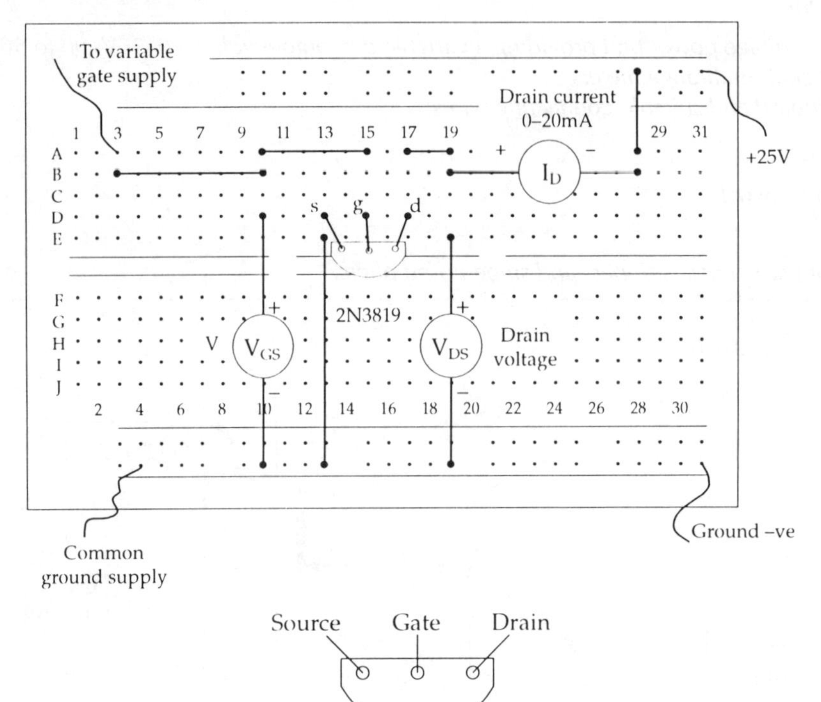

2N3819 pin out (viewed from underneath)

Output Characteristic

To obtain the output characteristic, the gate-source voltage is fixed at some value, the drain voltage is increased in steps and the drain current recorded at each step. The gate-source voltage (V_{GS}) is then increased to a new value and the procedure repeated to obtain a family of curves.

Step 1: Adjust the gate-source voltage to 0V and ensure that the gate voltage remains constant at this value throughout the first test.

Step 2: Now increase the drain voltage in the steps shown in the table below. At each step measure and record the value of drain current flowing.

Step 3: Now reduce the drain voltage to zero, increase the gate voltage to –0.5V and take another set of collector voltage and corresponding drain current readings.

Results Tables

	Drain voltage (volts)											
V_{GS} = 0V	1.0	2.0	3.0	4.0	5.0	6.0	7.0	8.0	10	15	20	25
I_D (mA)												

	Drain voltage (volts)											
V_{GS} = –0.5V	1.0	2.0	3.0	4.0	5.0	6.0	7.0	8.0	10	15	20	25
I_D (mA)												

	Drain voltage (volts)											
V_{GS} = –1.0V	1.0	2.0	3.0	4.0	5.0	6.0	7.0	8.0	10	15	20	25
I_D (mA)												

	Drain voltage (volts)											
V_{GS} = –1.5V	1.0	2.0	3.0	4.0	5.0	6.0	7.0	8.0	10	15	20	25
I_D (mA)												

	Drain voltage (volts)											
V_{GS} = –2.0V	1.0	2.0	3.0	4.0	5.0	6.0	7.0	8.0	10	15	20	25
I_D (mA)												

Now plot a family of curves of drain current I_D (vertical axis) against drain voltage V_{DS} (horizontal axis) for each fixed value of gate voltage V_{GS}.

Temperature Effects

You may have experienced a reduction in drain current as the drain voltage increased. This is due to the effect of temperature. The velocity with which the majority charge carriers travel through the channel depends upon the drain source voltage and the temperature within the channel of the FET.

An increase in temperature reduces the charge carrier velocity and this effectively reduces the drain current. In other words, for a given gate-source voltage as the drain voltage is increased during the test, the power dissipated within the FET is increased, increasing the temperature and reducing the drain current slightly.

When plotting the graph, therefore, ignore any reduction in drain current as the drain voltage is increased and plot a flat graph at the maximum value of drain current for that particular test.

Determination of the Output Resistance R_{out}

Because of temperature effects, it is not easy to obtain the output resistance from the characteristics because they are often very flat (horizontal line).

Unless we can compensate for the temperature effects it is not possible to accurately determine the output resistance from the output characteristic. The output resistance for a junction FET is however several tens of k ohms.

Determination of the Mutual Conductance g.m.

The output characteristic is a plot of drain current against drain voltage for a fixed gate-source voltage V_{GS}.

$$\text{Now: the g.m.} = \frac{\text{change in drain current } I_D}{\text{change in } V_{GS} \text{ causing the change in } I_D} \text{ mS}$$ **at a fixed value of drain-source voltage.**

Therefore, if we draw a vertical line up the graph, at say 8V, then all the way up this line will represent a fixed drain-source voltage of 8V. Where this line intersects the –1.0V and–2.0V graphs project across to the Y axis. This will give the change in drain current for a change in gate-source voltage of 1V.

Now read off the corresponding values of collector current from the Y axis.

Now determine the g.m. for your transistor!

Plotting the Mutual Characteristic

The mutual characteristic is a plot of drain current against gate-source voltage.

Using the same circuit as before, we record the drain current flowing for various values of gate-source voltage, at a fixed value of drain-source voltage.

Complete the table below.

	Drain voltage = 10V			
Gate source voltage V_{GS} = 0V	–0.5V	–1.0V	–1.5V	–2.0V
Drain current I_D (mA)				

Now plot a graph of drain current (vertically) against gate-source voltage (horizontally) for the readings obtained.

Use the same method outlined in **Section 2 Information and Skills Bank** Common emitter static characteristics (p.223) to determine the g.m.

Designing a FET Amplifier

Keep the graph you have just plotted as it will be useful when you undertake Task g (Designing a single stage FET amplifier).

Task f

1) Determination of the gain and bandwidth of a common source amplifier

Refer to ***Section 2 Information and Skills Bank:*** *The field effect transistor, p.255; Fault-finding on a single-stage FET amplifier, p.268.*

2) Measuring, testing and fault diagnosis on a single-stage junction FET amplifier.

Aims

1) To measure and calculate the stage gain of a resistive loaded common source amplifier and determine the bandwidth.
2) To observe the effects of fault conditions on a single-stage junction FET amplifier.

Equipment

Low voltage power unit providing a variable d.c. output voltage 0 - 30V, up to 500mA
A dual trace CRO
A function/signal generator.
A digital voltmeter (capable of measuring a few mV at a frequency of 1kHz)
A breadboard and suitable connecting wires.

Components

Resistors: 3k3, 390R, 1M, 33k, 2 x 10k. Capacitors: 25μ: 2 x 100n, 2n2; Transistor: 2N3819.

Tools

A pair of wire cutters/strippers and snipe nosed pliers.

Approach

Breadboard the circuit shown below. Switch on the power supply, set the output to 24V, **switch off**, and connect the power supply to the circuit.

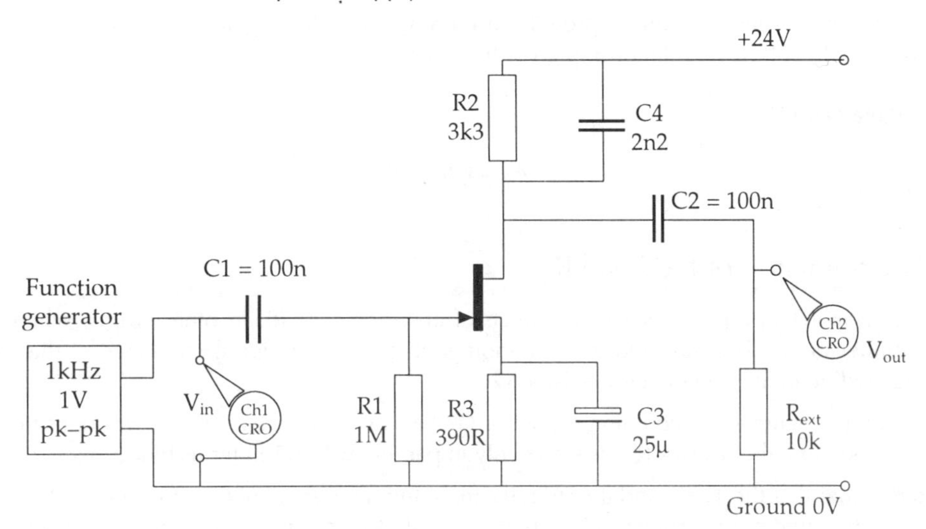

Circuit for measuring gain and bandwidth of a common source amplifier

Connect the CRO and the function generator as indicated on the diagram (take care to connect the correct CRO and generator leads to ground).

Suggested Breadboard Layout

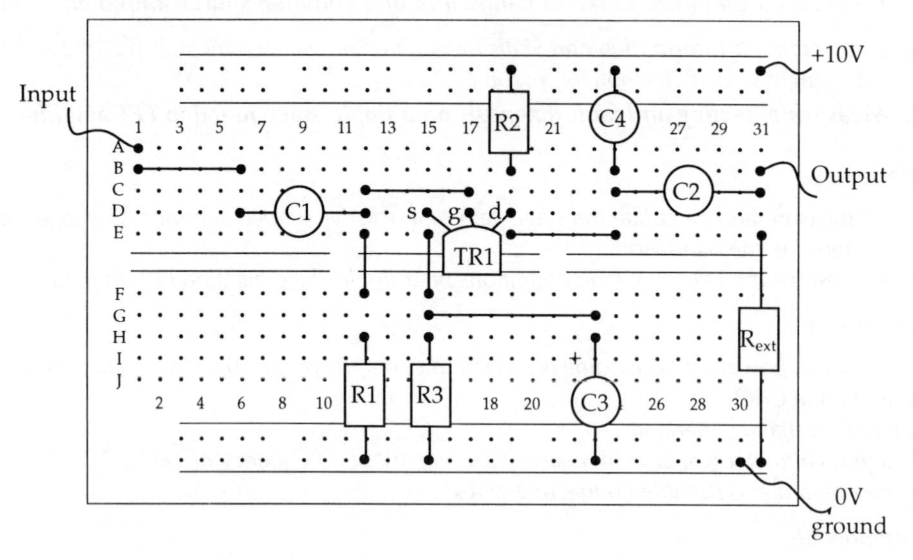

Check all your connections carefully and when you are satisfied that your wiring is correct, switch on all the equipment.

Using channel 1 of the CRO, which is monitoring the input, set the function/signal generator to provide an input voltage V_{in} of 1V peak to peak at 1kHz. Use channel 2 of the CRO to observe and measure the output waveform. Determine the voltage gain (A_v).

Calculation of Voltage Gain

The common source amplifier provides a current and voltage gain. There is phase inversion and the following relationship gives the voltage gain.

Voltage Gain

$$A_v = g.m. \times R_L$$

Determination of the Load R_L

The resistor R_{ext} represents the resistance of an external load. In other words, when the amplifier output is connected to feed another stage then the input resistance of this next stage will load up the collector resistor R2.

To a.c. frequencies the coupling capacitor C2 has a very low reactance (low opposition to a.c.), therefore the resistor R_{ext} is effectively in parallel with R2 in terms of a.c.

Remember that the +24V and the 0V rails are at the same potential in terms of a.c. because of the low internal resistance of the power supply to a.c. (the large reservoir capacitor in the power supply, and power rail decoupling capacitors, have a very low reactance to a.c.). The capacitor C4 represents the input capacitance of the circuit the amplifier is driving and the stray and wiring capacitances of the amplifier circuit.

The reactance of this capacitor will be large at 1kHz compared to the equivalent value of the load and although it is also in parallel with the load, it can be ignored at mid and low frequencies.

Therefore the effective load of the amplifier R_L, is R2 in parallel with R_{ext}:

$$R_L = \frac{R2 \times R_{ext}}{R2 + R_{ext}} \quad \text{i.e.} \quad \frac{\text{(product)}}{\text{(sum)}}$$

Calculate the load R_L and the stage gain: (A_v = g.m. × R_L)

Study Note. Use the value for g.m. that you found in Task f.

Determining the Bandwidth

The gain of an amplifier falls off at high and low frequencies.

The fall off at low frequencies is mainly due to the rising capacitive reactance of the coupling capacitors and the source by-pass capacitor. The coupling capacitors couple less signal in and out of the amplifier and the rising reactance of the by-pass capacitor allows NFB to lower the gain.

The fall off at high frequencies is largely due to the input capacitance of the circuit the amplifier is driving. Because the amplifier under test is not driving a normal load C4 has been included to simulate this effect.

When the output falls to half power then this is the limit of the lower and upper frequency that can be handled by the amplifier.

To find the upper and lower half power points the input frequency is altered over the range 20Hz to 100kHz.

It is important that the amplitude of the test signal from the function generator is kept **constant at all frequencies** (i.e. it must be kept at 1V peak to peak).

Step 1: Use channel 1 on the CRO to monitor the amplitude of a 1kHz input signal and use channel 2 to measure the amplitude of the output at 1kHz. Determine the gain; this is then the mid frequency gain.

Step 2: Now reduce the frequency of the function generator until the output falls to 0.707 of the mid frequency gain. This is the low frequency half power point F_{low}

Step 3: Now increase the frequency from the mid point frequency until the output once again falls to 0.707 of the mid frequency gain. This is the high frequency half power point F_{high}

The bandwidth of the amplifier is the difference between these two frequencies.

$$\text{Bandwidth} = F_{high} - F_{low} \text{ Hz}$$

Determine the bandwidth of your amplifier.

Observation of Fault Conditions

Create each of faults given in the table below one at a time. Observe the effect and complete the table, then rectify the fault (remove the fault) before creating the next fault condition. Simulate the faults by unplugging one end of a resistor, unplugging the appropriate lead on the transistor and placing short circuits across the leads of the transistor.

At all times when creating or rectifying faults switch OFF the power supply.

d.c. Faults (Static Tests)

For each fault in the table below predict what effect the fault will have. Measure the actual voltages and compare the results with your prediction.

Fault	*Gate voltage*	*Drain voltage*	*Source voltage*
No fault			
R1 o/c			
R2 o/c			
R3 o/c			
R3 = 10k			
R2 = 33k			
Source o/c internally			
Drain/source s/c internally			
Drain o/c internally			
Gate/source s/c internally			

a.c. Conditions (Dynamic Faults)

These are faults that affect the circuit's ability to amplify a.c. signals.

Although the fault may be caused, for example, by incorrect bias (a d.c. condition) its effect may only be noticeable when the amplifier is handling a signal, i.e. dynamic conditions.

Create each fault from the table below, in the circuit, one at a time. Observe the effect on the output waveform and/or gain of the amplifier. Rectify the fault, check that the circuit is working, then create the next fault in the table.

At all times when creating or rectifying faults switch OFF the power supply.

For each fault in the table below predict what effect the fault will have. Observe the actual outputs, sketch all waveforms, record voltages or gains measured, and compare the results with your prediction.

Fault	*Observations (sketch all waveforms)*
No fault	Measure the gain of the stage
Increase input signal to 3V peak to peak	Observe output waveform. Restore input to 1V
R2 increased to 33k	Observe output waveforms
R1 reduced to 10k	Determine the stage gain
R3 increased 10k	Observe output waveform
C1 open circuit	Observe output
C3 open circuit	Determine the stage gain
C2 open circuit	Observe any output waveform

Faultfinding

At this stage you now need access to previously faulted circuit boards. The faults should be inserted on the boards in such a way that they cannot be found simply by visual inspection.

Connect all the test equipment, power up the boards and observe the symtoms. Try to estimate in your observations if the fault is dynamic, (the circuit is unable to handle an a.c. signal) or static (d.c. conditions on the transistor).

Use the results of all the tests above to determine the fault. Rectify the fault and test the circuit.

Task g

Designing a single stage bi-polar amplifier

Refer to ***Section 2 Information and Skills Bank:*** *Amplifier specifications, p.199; The field effect transistor, p.255; Fault-finding on a single-stage FET amplifier, p.268; Load Lines, p.273; Capacitive reactances, p.296.*

Aims

To design, construct and test a resistive loaded single stage junction FET amplifier to the following specifications:
Supply voltage 24V
Stage gain 8 or more
Frequency range 20Hz to 20kHz
Input resistance 1M ohms minimum
Output load: an alarm sounder of 10k ohms
Input signal 0.35V r.m.s.

Equipment

A dual trace oscilloscope.
A multirange digital voltmeter.

A low voltage power supply providing a variable d.c. output voltage range; 0-30V, up to 200mA.
A signal/function generator: range 0 to 100kHz, output: a few mv to 1V.
A breadboard and suitable connecting wires.

Components

A range of resistors: 1k to 100k. A range of capacitors: 1nF to 50μF. A 2N3819 transistor or near equivalent.

Tools

A pair of wire cutters/strippers and snipe nosed pliers.

Approach

Study the worked example on how to design a single stage amplifier given below. Design your own circuit to the specification given in the design project. Then build and test your design to see that it meets the specification.

Worked Example

Before we can begin to design the amplifier, we need to study the specification to determine the supply voltage, the input signal, and the gain required. In order to illustrate the design approach consider the following specification.

Specification for the Worked Example

Supply voltage 25V; Stage gain 5; Frequency range 200Hz to 10kHz; Input resistance 1M ohms minimum; Output load: an alarm sounder of 3k ohms; Input signal 1V peak to peak.

Choice of Transistor

From the specification we will need a transistor that will: operate from 25V, have a high input resistance, be driven by a 1V signal, and provide a voltage gain of 5 over the specified frequency range.

For this specification a general purpose, low power junction FET transistor such as a 2N3819 or its equivalent, connected in common source configuration, would be suitable.

The next step is to to set the standing drain voltage that will allow a reasonable output voltage swing without distortion.

Setting the Standing (Quiescent) Drain Voltage.

From the specification: the input signal is 1V peak to peak and the gain required is 5, so the output swing needs to be 5V peak to peak.

The supply voltage is 25V so if we set the drain voltage at 12.5V this will allow a peak swing at the drain of at least 10V. It will also allow for the bias voltage dropped across the source resistor, which is the next thing we have to decide upon.

Choosing a Suitable Bias Voltage

From the specification we need a stage gain of 5 and since the stage gain is determined by g.m. × R_L, the effective a.c. load must be reasonably high. This means that we need to choose a gate bias voltage that will give a low standing drain current.

Remember: with a smaller gate voltage the standing drain current will be higher. This will mean that the drain load resistor must be smaller and so the stage gain will also be lower.

If we therefore choose a gate bias voltage of say V_{GS} = –1.5V, this will give a low standing drain current of 3 to 4mA.

We can now use the output characteristic below for the 2N3819 to determine the operating point and plot the d.c. load line.

Locating the Operating Point

We have now fixed two points on the X-axis of the 2N3819 output characteristic: the supply voltage is 25V and the standing collector voltage is 12.5V.

The operating point will have to lie on a line projecting upwards from the 12.5V point on the X axis!

Where the –1.5V V_{GS} curve cuts this 12.5V vertical line is the operating point P!

Having determined these two points, the d.c. load line can be plotted on the output characteristic as shown below.

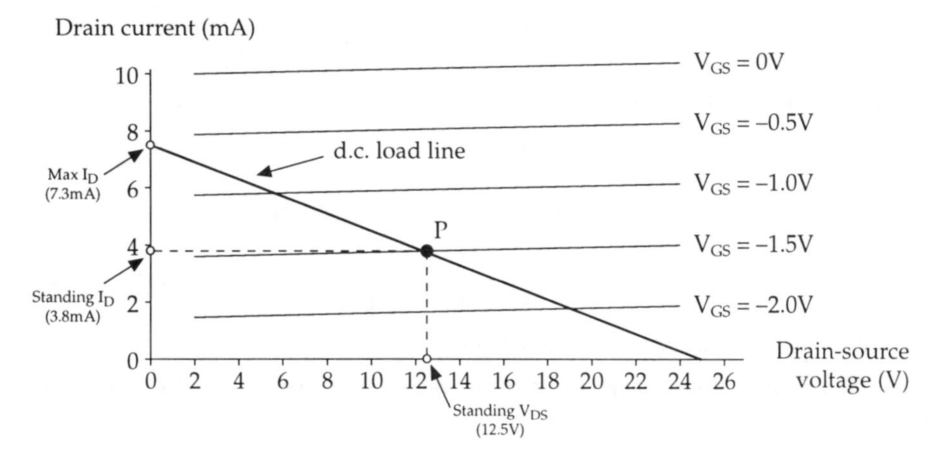

Output characteristic for 2N3819

Plotting the Load Line

To plot the load line draw a point from the 25V point on the X axis, through the operating point P, to the Y axis. This will be the d.c. load line. Where this line cuts the Y axis is the maximum drain current that can flow when the transistor is fully turned ON.

This current is determined by the supply voltage divided by R2 + R3. From the graph this current is 7.3mA.

$$\text{Now: } I_{Cmax} = \frac{\text{Vsupply}}{\text{R2 + R3}} \text{. Therefore R2 + R3} = \frac{25\text{V}}{7.3} \text{ k ohms}$$

R2 + R3 = 3.43k ohms

We now need to calculate the value of R3, and this value subtracted from 3.43k will give the value of the load resistor R2.

Determining the Standing Drain Current

By projecting horizontally from the operating point P across to the Y axis the value of the standing drain current can be determined. This is the value of current that will flow in the drain before an a.c. signal is applied.

From the graph the standing collector current is 3.8mA.

Determining the Value of the Source Resistor (R3)

The voltage drop across R3 must be 1.5V to provide the bias required by our chosen bias point. The value of the source resistor will be 1.5V divided by the standing (quiescent) drain current (3.8mA).

$$\text{R3} = \frac{\text{volt drop across R3}}{\text{standing collector current}} = \frac{1.5\text{V}}{3.8} \text{ k ohms}$$

From which R3 = 0.395k ohms or **390 ohms** (nearest preferred value).

Calculating the Load Resistor (R3)

The total resistance obtained from the d.c. load line (3.43k) less the value of R3 will give the load resistance R2.

Therefore R2 = 3430 – 390 = 3040 ohms

Or nearest preferred values: (2k7; 3k3)

R2 = **3k3** in the E12 series.

(The higher value is chosen to give more gain.)

A common source circuit diagram with the values we have so far calculated is shown below.

Circuit Diagram

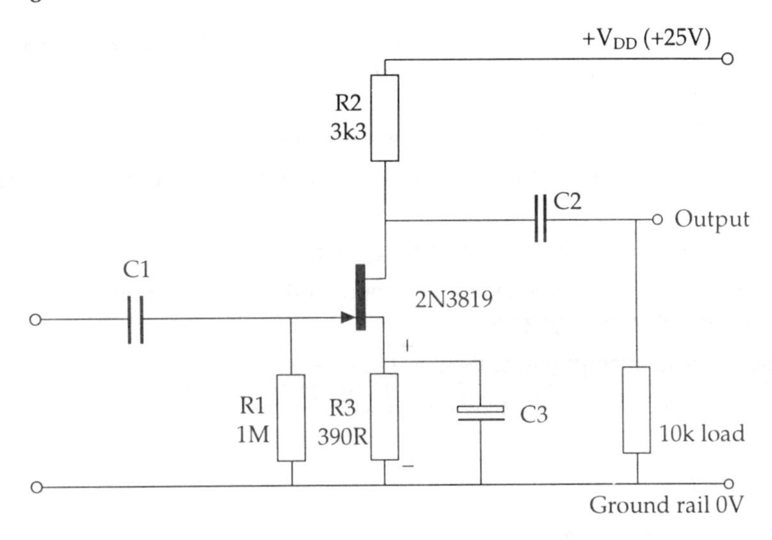

Circuit diagram for design of a FET amplifier

We now need to calculate the values of the capacitors C1, C2 and C3.

Calculating the Value of the Coupling Capacitors C1 and C2

As a rule of thumb, the reactance of the coupling capacitor C2, at the lowest frequency of operation, must not be more than one tenth of the value of the load resistance on the amplifier. The load resistance on the amplifier is 10k ohms.

The capacitive reactance of C2 at the lowest frequency (200Hz in the worked example specification) must not be more than a tenth of this value, i.e. 1k ohms or less.

The capacitive reactance is given by:

$$X_C = \frac{1}{2\pi \times f \times C} \text{ ohms}$$

$$\text{Therefore, } C = \frac{1 \times 10^9}{2\pi \times f \times X_C} \text{ nFarads} = \frac{10^4}{2\pi} \text{ nF} = 800\text{nF (1.0}\mu\text{F preferred value)}$$

(See **Section 2 Information and Skills Bank** Capacitive reactance, p.296.)

A similar calculation for an input resistance of 1M gives a value for C1 of 8nF (10nF preferred value).

Calculation of the Source By-pass Capacitor C3

Using the same rule of thumb, the reactance of the by-pass capacitor (C3) must not be more than one tenth of the source resistor R3.

R3 = 390 ohms, therefore the reactance must not be more than 39 ohms at 200Hz.

$$X_C = \frac{1}{2\pi \times f \times C} \text{ ohms}$$

$$\text{Thus, } C = \frac{1 \times 10^6}{2\pi \times f \times X_C} \mu\text{Farads} = \frac{1{,}000{,}000}{2\pi \times 100 \times 39} \mu\text{F} = 20.5\mu\text{F (taking } 2\pi = 6.25)$$

Or C3 = **22μF** (nearest preferred value)

Remember: these values for the capacitors are the minimum values we can get away with. It is quite in order to fit a larger value if you wish!

Determination of the Stage Gain From the Output Characteristic

On the 2N3819 output characteristic plot the a.c. load line (see **Section 2 Information and Skills Bank** Load Lines, p.273).

Remember: The effective a.c. load is R2 and R_{load} in parallel.

Determine the output swing of voltage for an input of 0.35V rms. (approximately 1V peak to peak), and calculate the stage gain.

Now you are ready to attempt one of the design projects.

Adopt a similar procedure with the new design specification.

Design Project

Now calculate all the component values for the design specification:

Supply voltage 24V; Stage gain 8 or more; Frequency range 20Hz to 20kHz; Input resistance 1M ohms minimum; Output load: an alarm sounder of 10k ohms; Input signal 0.35V r.m.s.

2N3819 Output Characteristics

In order to design the amplifier we will need a set of output characteristics for the 2N3819.

You may either use the output characteristic you plotted for Task e or use the set of test results below to plot another output characteristic.

Drain-source voltage V_{DS}	*Drain current I_D (mA)*				
(volts)	V_{GS} = 0V	V_{GS} = −0.5V	V_{GS} = −1.0V	V_{GS} = −1.5V	V_{GS} = −2.0V
1	5.61	4.55	3.56	2.57	1.55
2	8.69	6.78	5.01	3.34	1.83
3	9.89	7.54	5.39	3.61	1.92
4	10.33	7.78	5.57	3.66	1.96
5	10.51	8.0	5.70	3.73	2.01
6	10.52	8.0	5.71	3.74	2.05
7	10.52	8.0	5.72	3.75	2.06
8	10.52	8.0	5.73	3.76	2.07
10	10.52	8.0	5.73	3.76	2.08
15	10.52	8.0	5.73	3.76	2.10
20	10.52	8.0	5.73	3.76	2.13
25	10.52	8.0	5.73	3.76	2.18

Results table for a 2N3819 FET

Construction of Your Design

Insert all your calculated values on the diagram below.

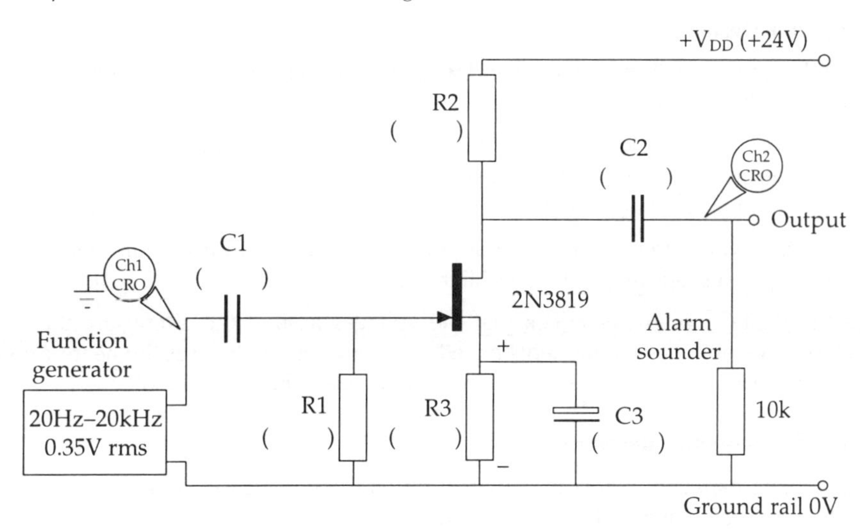

Circuit diagram for the design of the FET amplifier

Now design your own breadboard layout and then breadboard your circuit.

Breadboard Layout

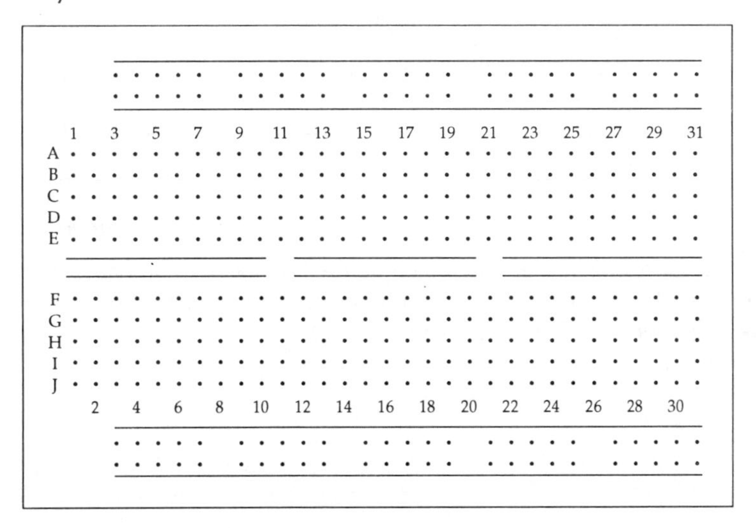

2N3819 connections

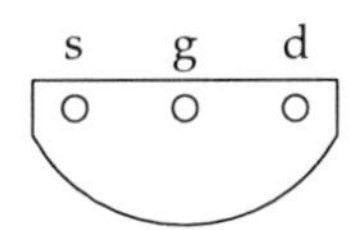

Connect the CRO and the function generator as indicated on the diagram (take care to connect the correct CRO and generator leads to ground).

Testing Your Design

Check all your connections carefully and when you are satisfied that your wiring is correct, switch on all the equipment.

Using channel 1 of the CRO, which is monitoring the input, set the function/signal generator to provide an input voltage V_{in} of approximately 1V peak to peak at 1kHz. Use channel 2 of the CRO to observe and measure the output waveform. Determine the voltage gain.

If at this stage your circuit is not functioning see **Section 2 Information and Skills Bank** Fault-finding on a single-stage FET amplifier, p.268.

Using the digital multimeter check that the voltages around the circuit are the same as expected (remember the components have tolerances and in all cases the nearest preferred value was chosen. For these reasons the measured values may well be different!)

Plotting the Frequency Response

Connect a 2n2 capacitor across the 10k resistor that represents the alarm sounder (this is to simulate the capacitance of the alarm sounder).

Keeping the input voltage constant at 1V peak to peak throughout the test, inject the frequencies given in the table below.

At each frequency measure the output voltage and determine the voltage gain.

	Frequency Hz															
	20	40	50	70	100	200	500	1k	2k	5k	7k	10k	20k	50k	70k	100k
Output																
Gain																

Frequency Gain table

Plot the frequency response graph in pencil on the log-linear graph paper provided below. Use the graph to determine the bandwidth of the amplifier (see **Section 2 Information and Skills Bank** Amplifier specifications, p.199).

Linear/log 4 cycle

You have now completed design project 2!

Part 2: Amplifiers

Section 2
Information and Skills Bank

Contents

Amplifier specifications

Introduction

When we amplify a signal, we make it larger, and an **amplifier** is an electronic circuit that does this job.

If the output from the amplifier is larger than the input signal supplied to the amplifier, we then say that the amplifier has so much **gain**.

Voltage Amplifier

If the output voltage is larger than the input voltage then we have a voltage amplifier.

The gain of the voltage amplifier is determined by:

$$\text{Voltage gain} = \frac{\text{output voltage}}{\text{input voltage}} \quad \text{(no units)}$$

There are no units of gain since it is a voltage ratio.

Example

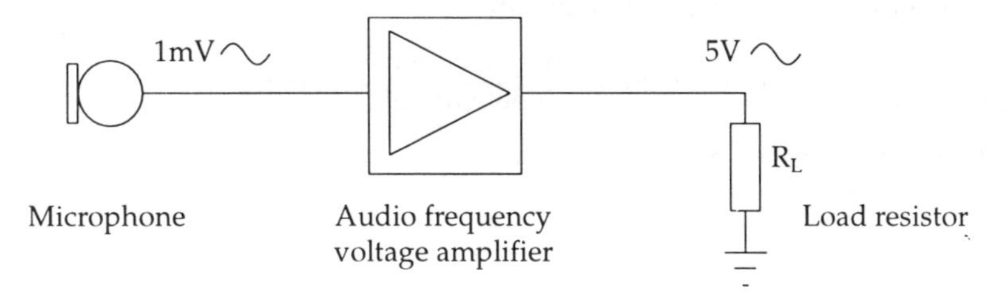

An Audio Frequency Voltage Amplifier

The microphone output is 1mV peak. The voltage amplifier amplifies this signal until it has an amplitude of 5V peak. The output voltage of the amplifier is 5000 times larger than the input voltage; the voltage gain is 5000.

Power Amplifier

The output from the voltage amplifier is often required to drive a loudspeaker. To drive a loudspeaker cone in and out takes power (voltage and current). The voltage amplifier can provide sufficient voltage (5V) but there is very little current available. The power amplifier does not provide very much voltage gain, but does provide a output voltage together with a larger output current.

Example

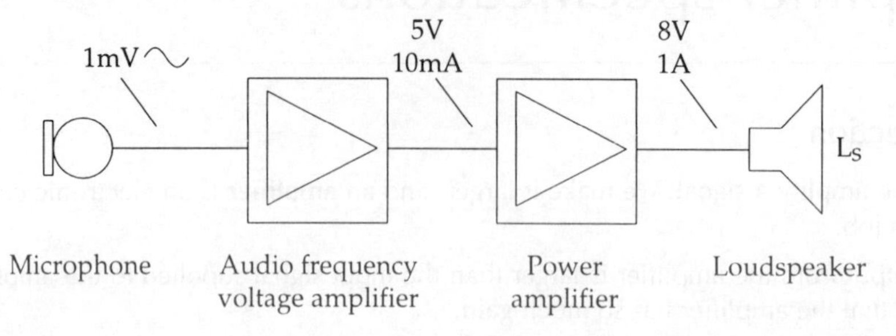

An audio frequency voltage amplifier + power amplifier

The peak inputs to the power amplifier are 5V at 10mA. The peak outputs are 8V at 1A, therefore the power gain:

$$\text{Power gain} = \frac{\text{output power}}{\text{input power}} = \frac{8\text{V} \times 1{,}000\text{mA}}{5\text{V} \times 10\text{mA}} = \frac{8{,}000}{50} = 160 \text{ (no units)}$$

Current Amplifier

If the output current is larger than the input current then we have a current amplifier.

The gain of a current amplifier is determined by:

$$\text{Current gain} = \frac{\text{output current}}{\text{input current}} \text{ (no units)}$$

Again there are no units as it is a current ratio.

Example

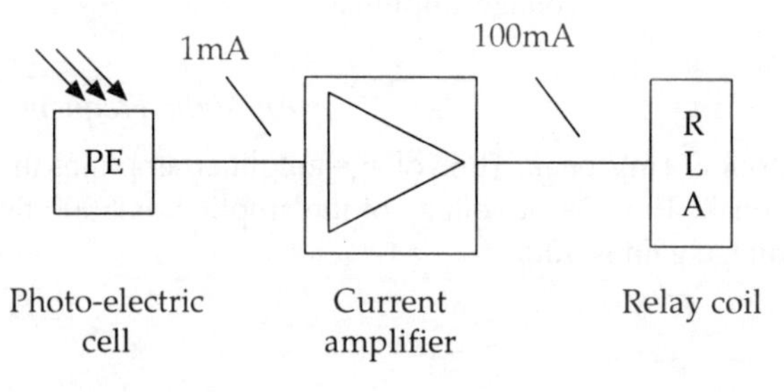

A Current Amplifier

The photoelectric cell gives an input current of 1mA. The relay coil requires a current of 100mA to 'pull-in' the relay contacts. Therefore the current gain of the amplifier must be 100.

Distortion

An amplifier is required to increase the size of a signal without changing its shape. If it does change the shape then distortion has been introduced.

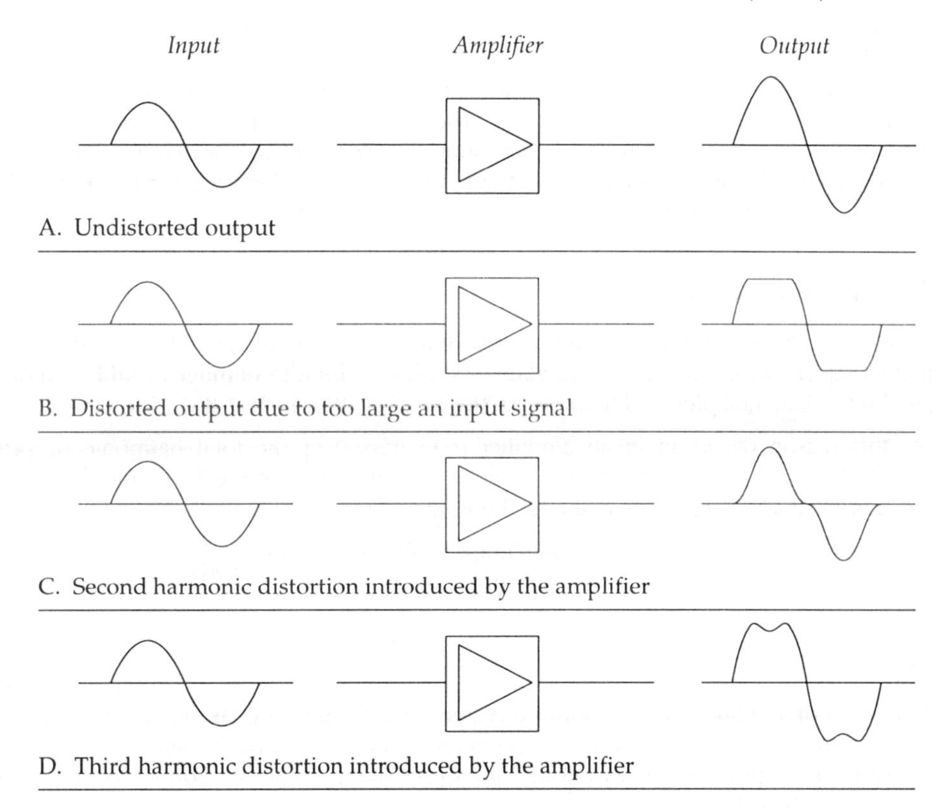

Distortion of a sine wave signal in an audio amplifier

Harmonics

The effect of distorting a sine wave signal is to introduce **harmonics**. These are whole number multiples of the fundamental input frequency.

For example, if the input signal is a 100Hz sine wave then:

2nd harmonic is 200Hz
3rd harmonic is 300Hz
4th harmonic is 400Hz etc.

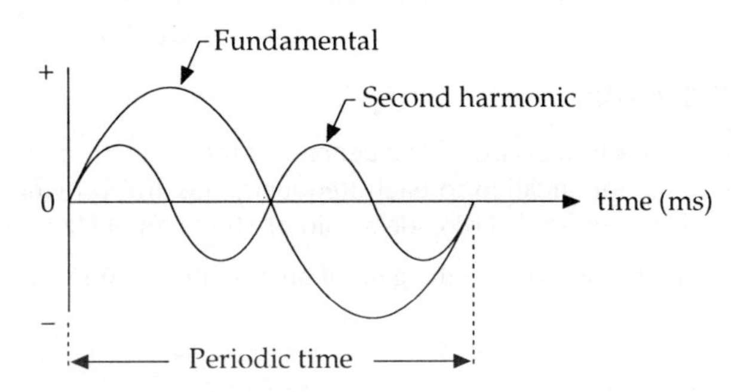

A fundamental and its second harmonic (two cycles in the same periodic time).

Complex Waveforms

Sound waves are examples of complex waveforms. These waveforms consist of a sinewave fundamental plus a number of harmonics. The sound of a complex wave will depend on: the number of harmonics, the type of harmonic, (odd or even) and the amplitude of the harmonics in relation to the fundamental and to each other. Generally the higher the order of the harmonic the smaller is its amplitude.

Total Harmonic Distortion

All electrically complex waves also consist of a sinewave fundamental and harmonics. A perfect squarewave consists of a fundamental plus an **infinite number of odd** harmonics; i.e. the fundamental plus odd harmonics: 3rd, 5th, 7th, 9th, 11th, 13th, 15th, etc.

For this reason distortion in an amplifier is expressed as the total harmonic distortion (THD). That is, from a pure sine wave input, the amplitude of the harmonics introduced are expressed as a percentage of the total output voltage.

$$\text{THD} = \frac{\text{harmonic output voltage (r.m.s.)}}{\text{total output voltage (r.m.s.)}} \times 100\%$$

Phase

As a result of amplification the output waveform is often **out of phase** with the input waveform. With single stage (one transistor) amplifiers then the output waveform will be 180° out of phase with the input waveform. Strictly speaking it is more correct to say that the waveform is **inverted**.

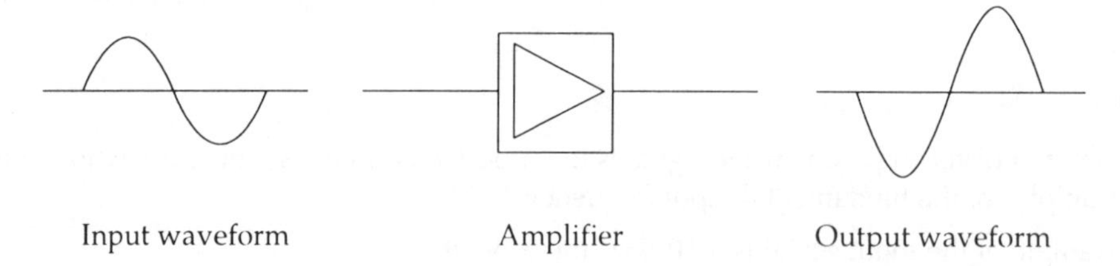

Phase inversion in an audio voltage amplifier

If there are two stages (two transistors) then the difference in phase will be 360°: i.e. the output waveform will be back in phase with the input waveform.

Frequency Response

An audio amplifier has to amplify all frequencies in the range from 20Hz to 20kHz. It has to provide the same amplification to each frequency: i.e. if it provides a gain of 100 to a 1kHz frequency then it must also provide a gain of 100 to an 8kHz frequency.

The **frequency response** is a plot of the gain of an amplifier against frequency.

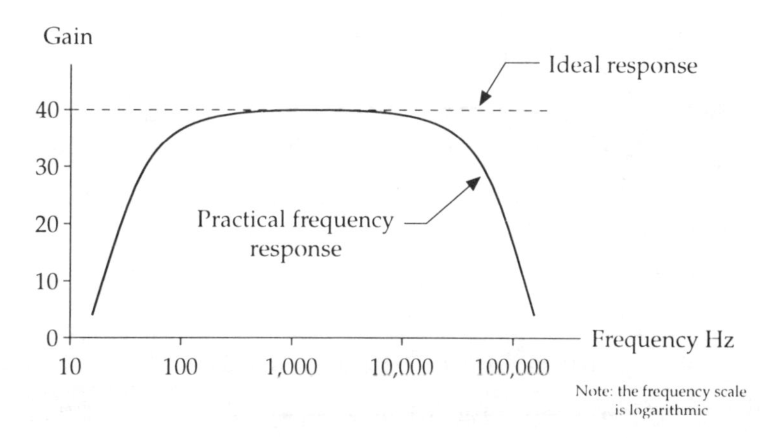

Frequency response of an audio amplifier

Bandwidth

In practice the gain of the amplifier falls off at high and low frequencies. Therefore a practical amplifier will only amplify a **certain band of frequencies**. The **bandwidth** of an amplifier is the band of frequencies it can amplify without the output falling below half power.

Determination of Bandwidth

The bandwidth of an amplifier can be determined from the frequency response curve.

1) The maximum gain of the amplifier is determined from the graph (usually around the mid frequency point 1 kHz).
2) This value is then multiplied by 0.707 and a line drawn across the response curve at this value.
3) Where this line cuts the response curve, lines are projected down to the frequency axis and the frequencies read off from the frequency scale.
4) The lower frequency is then subtracted from the higher frequency and this gives the bandwidth of the amplifier.

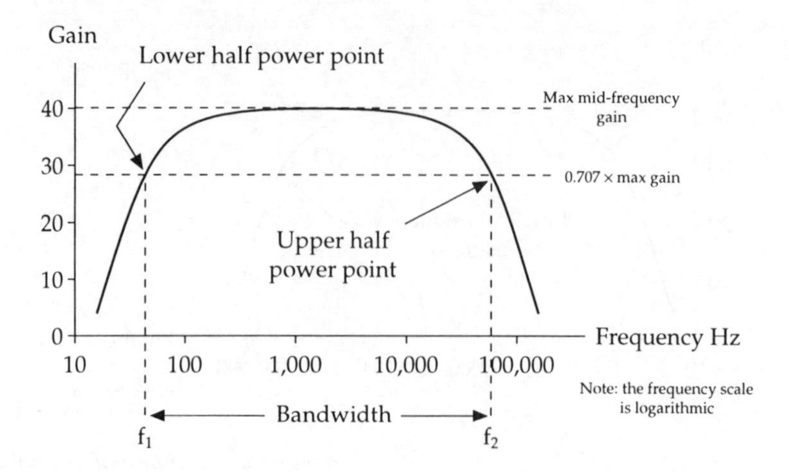

Determination of the bandwith of an audio amplifier

Half Power Points

Power may be obtained from the relationship:

$$\text{Power } P = V^2/R$$

Therefore power is proportional to V^2, and $0.707^2 = 0.5$

Therefore when the output voltage falls to:

$$0.707 \times \text{mid frequency gain},$$

then this corresponds to half power.

Even though the plot above is gain against frequency, the relationship still applies since the gain is the output voltage divided by the input voltage. Because the input voltage must be kept constant at each frequency, when plotting the response curve, the value for the gain is directly proportional to the output voltage.

These points are also called the -3dB points. This is another way of expressing half power.

Matching

When two single stage amplifiers are connected together in tandem, to obtain maximum gain then the output resistance of the first amplifier must be matched to the input resistance of the second amplifier. If they are matched then the total gain is the product of the two individual gains.

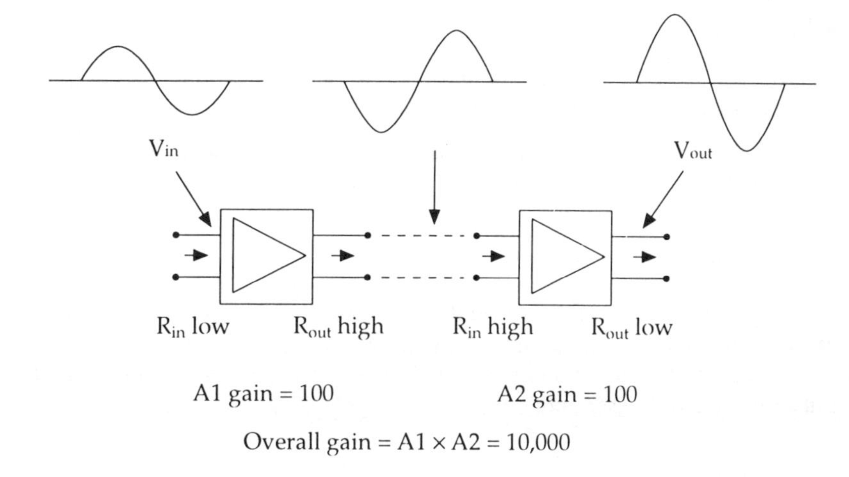

Gain of two matched voltage amplifiers

If the two stages are not matched then the second amplifier can 'load up' the first amplifier and lower its gain.

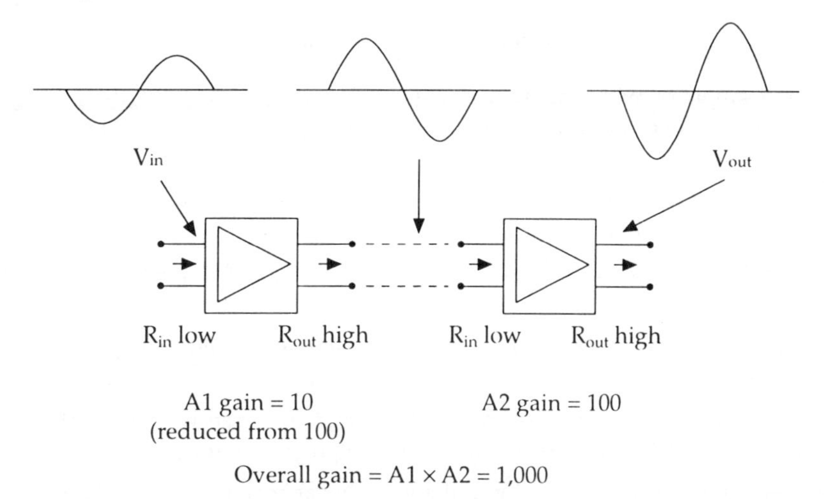

Gain of two unmatched voltage amplifiers

As well as matching between stages, both the input and output to the complete amplifier must also be matched to obtain maximum power at the output. Often however the lower overall gain due to mismatch is overcome by having an additional stage of amplification.

Input Resistance

This is the resistance offered to a signal at the input of the amplifier. Because the input circuit also has a small amount of capacitance the input resistance is really an impedance. (Impedance is the opposition offered to a.c. by the combination of resistive and reactive components.) Since the input capacitance is very small the effect of its capacitive reactance is often ignored (see **Information and Skills Bank** Capacitive reactance, p.296)

Output Resistance

This is the output resistance of the amplifier, and to obtain maximum power transfer to the load, the load resistance should be the same value.

Again the output also has capacitance (and may have inductance) and we should strictly speak of the output impedance (the units of reactance and impedance are ohms).

Buffer Amplifier

The buffer amplifier (also called an emitter follower) is a stage that can be interposed between two unmatched voltage amplifiers above to improve the matching. It has unity voltage gain (a gain of 1) and matches the high output resistance of the first amplifier to the low input resistance of the second.

Summary of Amplifier Specifications

- An amplifier may be used to amplify voltage, current or power. The gain of an amplifier is given by:

$$\text{Gain} = \frac{\text{output}}{\text{input}} \text{ (no units)}$$

- A good amplifier will amplify a signal without introducing distortion. Distortion is where the shape of the waveform (other than being larger) is altered by the amplification process.
- Distortion of a sinewave introduces harmonics. A harmonic is a whole number multiple of the fundamental frequency of the sinewave.
- Complex waves consist of sinewaves; a fundamental and a number of odd and even harmonics.
- A good squarewave consists of a fundamental frequency and an large number of odd harmonics.
- The process of amplification often introduces a phase inversion; the effect is similar to a phase shift of 180° between the input and output waveforms.
- The frequency response is a plot of the gain (linear scale) of the amplifier against frequency (logarithmic scale).
- The bandwidth of an amplifier is the range of frequency that it can amplify without its output falling below half power.
- In order to obtain maximum gain from two or more stages of amplification, the output resistance of the first stage should be equal or matched to the input resistance of the second stage.
- The matching may be obtained from a matching transformer or from a special amplifier called a buffer amplifier.

Self-assessment questions *(answers page 300)*

1) An audio frequency power amplifier may be used to amplify:
   a) d.c. voltage
   b) a.c. voltage
   c) a.c. voltage and current
   d) a.c voltage or a.c. current, but not voltage and current.

2) The power gain of an amplifier may be determined by:

   a) $\text{Gain} = \dfrac{\text{output voltage}}{\text{input voltage}}$

   b) $\text{Gain} = \dfrac{\text{output current}}{\text{input current}}$

   c) $\text{Gain} = \dfrac{\text{output voltage} \times \text{output current}}{\text{input voltage} \times \text{input current}}$

   d) $\text{Gain} = \dfrac{\text{input voltage} \times \text{input current}}{\text{output voltage} \times \text{output current}}$

3) Distortion in an audio amplifier occurs if the output signal is:
   a) the same size as the input signal
   b) smaller than the input signal
   c) bigger than the input signal
   d) a different shape to the input signal

4) The third harmonic of the mains frequency is:
   a) 50Hz b) 100Hz c) 150Hz d) 300Hz

5) If a complex wave consists of a fundamental and its fifth harmonic, then the fifth harmonic compared to the periodic time of the fundamental will have:
   a) 0.5 cycles in one fifth of the periodic time
   b) 1 cycle in the periodic time
   c) 5 cycles in half the periodic time
   d) 2.5 cycles in half the periodic time

6) A good square wave contains a:
   a) fundamental frequency only
   b) fundamental and few odd and even harmonics
   c) fundamental and a large number of even harmonics
   d) fundamental and a large number of odd harmonics

7) A single stage audio amplifier inverts the input signal. This is the same as a phase shift of:
   a) 0° b) 90° c) 180° d) 270°

8) Three stages of amplification will provide an effective phase shift of:
   a) 90° b) 180° c) 270° d) 360°

9) The bandwidth of an amplifier is the range of frequencies that the amplifier can handle. The bandwidth is defined as all the frequencies:
   a) amplified by the amplifier
   b) in the middle frequency range
   c) that are given the same maximum gain
   d) amplified with at least half the power of the middle frequencies

10) The bandwidth of an amplifier can be determined by doing the following on the frequency response curve:
    1) subtracting the two frequencies corresponding to the half power points
    2) multiplying the mid frequency gain by 0.707
    3) reading off the two frequencies corresponding to the half power points
    4) finding the half power points

    The correct order to carry out these steps is:
    a) 2,4,3,1
    b) 4,2,3,1
    c) 1,3,4,2
    d) 2,3,4,1

The bi-polar transistor

Study Note

Before studying this helpline you will need to be familiar with the **Part 1 (Power Supplies) Section 2 Information and Skills Bank** Simplified semiconductor theory, p.65 and Semiconductor diodes, p.75.

Introduction

The bi-polar junction transistor is a three layer, two junction device which is capable of controlling current and can be used to amplify.

Transistors can be PNP or NPN devices. The three layers which have wire connections to them are called Emitter, Base and Collector.

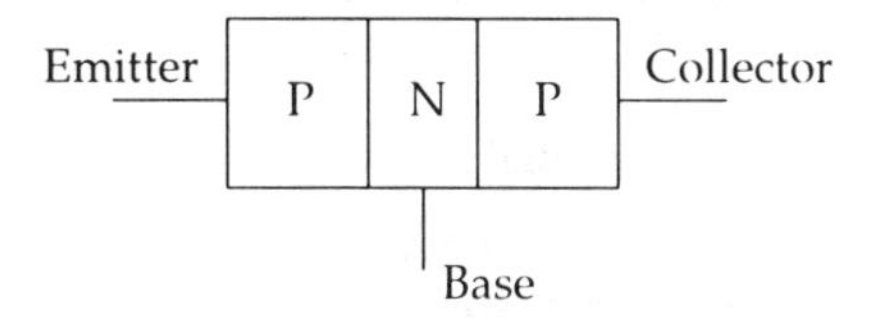

An PNP transistor

In the PNP transistor holes cross the base from the emitter to the collector.

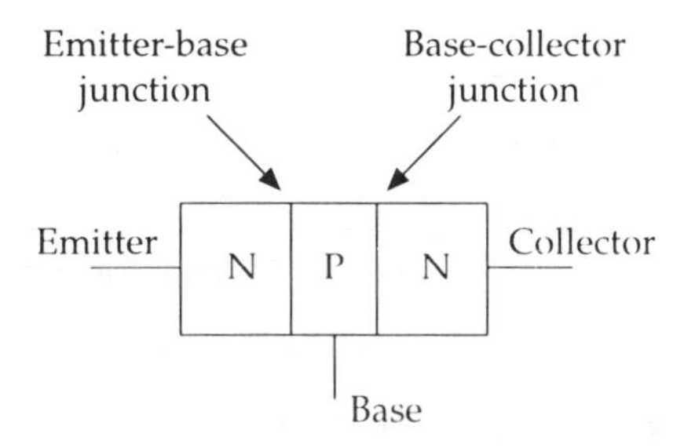

A NPN transistor

In the NPN transistor electrons cross the base from the emitter to the collector. Therefore current directions in the PNP and NPN transistors are opposite to each other.

The more common of the two types is the NPN transistor.

The Common Base Connection

The transistor is connected so that the base is **common** to the **input** and **output** circuits. The input is between emitter and base. The output is between collector and base.

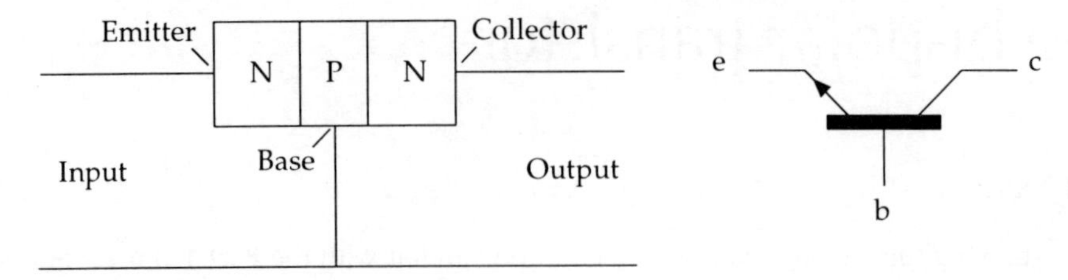

Input is between emitter and base. The output is between collector and base. On the right is the symbol for an NPN transistor

Transistor Action

In use, the collector-base junction is reverse biased; (the applied voltage assists the potential barrier) and the emitter-base junction is forward biased (the applied voltage breaks down the potential barrier).

In an NPN transistor this means that the emitter is made negative and the collector positive in relation to the base.

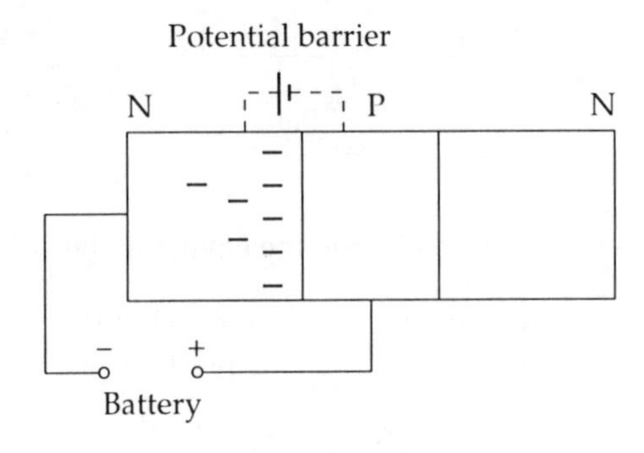

Forward bias *of the emitter-base junction. The negative potential of the battery drives electrons towards the emitter-base junction.*

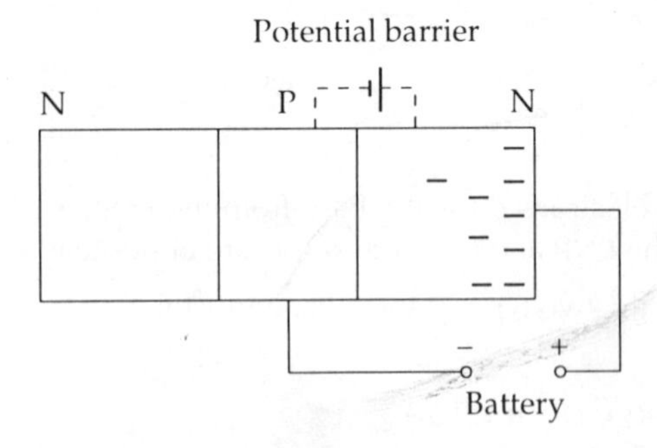

Reverse bias *of the base-collector junction. The positive potential of the battery attracts electrons towards the right-hand side of the collector region*

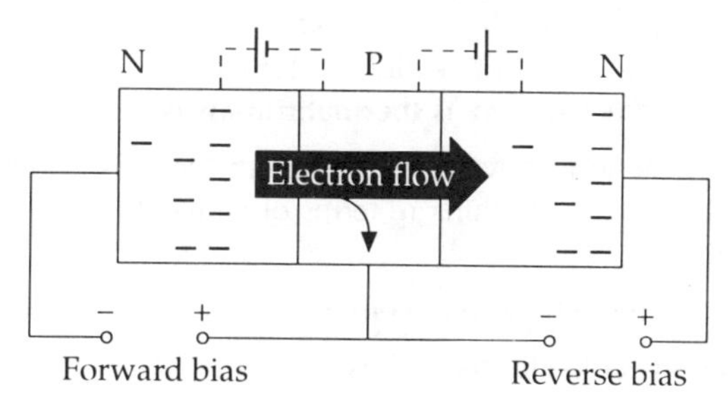

The forward bias gives a negative ***push*** *and the reverse bias a positive* ***pull*** *to the electrons*

When the emitter base region is forward biased, electrons flow from the emitter into the base.The base is a very thin region and these electrons rapidly spread through it by diffusion. The electrons in the base region are attracted by the collector's positive potential; they cross over into the collector region and form the collector current.

Nearly all the electrons injected into the base by the emitter are collected by the collector and only a few electrons, 1% to 2%, flow out of the base and form the base current.

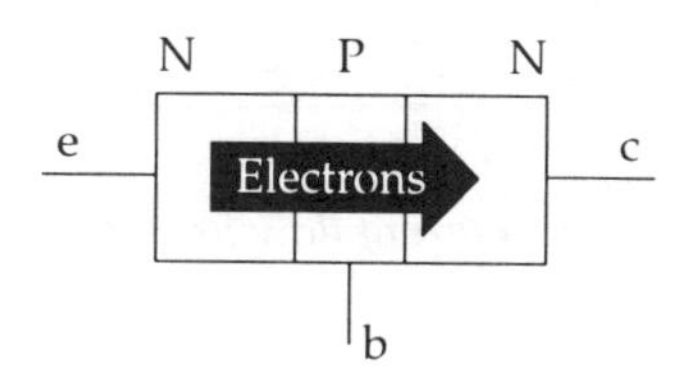

Virtually all the electrons cross the base from the emitter into the collector. The collector current is almost equal to the emitter current.

Collector Leakage Current

When the collector base junction is reverse biased and the emitter is left open circuit, a very small, almost constant, reverse leakage current flows (I_{CBO}). This is due to the presence of minority carriers (electrons) in the P-type base. Any electron which is near the base-collector junction is attracted across the junction by the positive collector potential and forms this very small leakage current.

This leakage current depends upon temperature. The leakage current increases as the temperature increases.

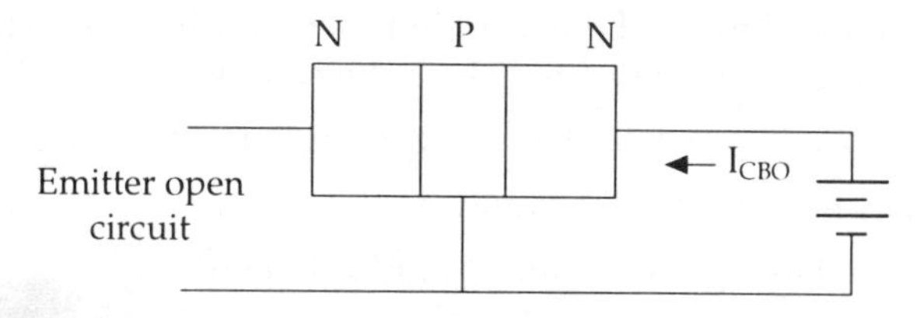

Collector leakage current increases with temperature

The collector leakage is small and does not increase the overall collector current. However if the transistor becomes overheated then this leakage current will increase considerably, further increasing the collector current, which in turn increases the temperature of the device. This can lead to an effect known as **thermal run-away**.

Remember! The transistor action above is described in terms of movement of electrons (electron flow). In practice we tend to think in terms of conventional current flow, which is in the opposite direction.

i.e. Forward bias current from the base to the emitter

Reverse bias current from the collector to the base.

The arrow on the transistor symbol always shows the direction of conventional current flow!

Example

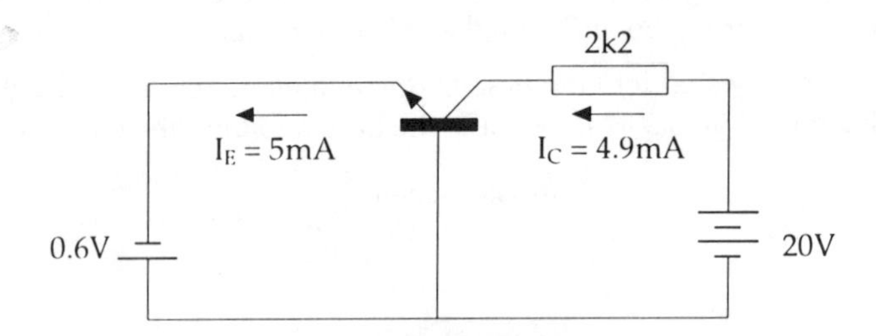

Conventional current flow in a common base transistor amplifier

Current Gain

In the circuit shown above the forward bias voltage of 0.6V produces an emitter current of 5mA; the resulting collector current is 4.9mA and flows through the 2k2 ohm resistor in the output circuit.

$$\text{This gives a current gain of } = \frac{I_C}{I_E} = \frac{4.9}{5} = \mathbf{0.98} \text{ (no units)}$$

Notice that the current gain is always less than 1. This is because the output collector current is always less than the emitter input current in the common base circuit.

The circuit can however provide a power gain.

Power Gain

The collector current flows through the 2k2 resistor and will have a voltage developed across it. The output voltage developed across the 2k2 resistor is:

$$I_C \times R_{2k2} = 4.9 \times 10^{-3} \times 2.2 \times 103 = \mathbf{10.8V}$$

Therefore the output power in the 2k2 load resistor:

$$(I_C \times V_{2k2}) = 4.9\text{mA} \times 10.8\text{V} = \mathbf{52.8mW}$$

The **input** power to the transistor is the forward bias voltage times the forward current:

$$0.6\text{V} \times 5\text{mA} = 3\text{mW}$$

Therefore the power gain of the circuit $= \frac{\text{output power}}{\text{input power}} = \frac{52.8}{3} = $ **17.6 times**

The Transfer Characteristic

A graph of the collector current against the emitter current is called the current transfer characteristic. It shows how I_E controls I_C.

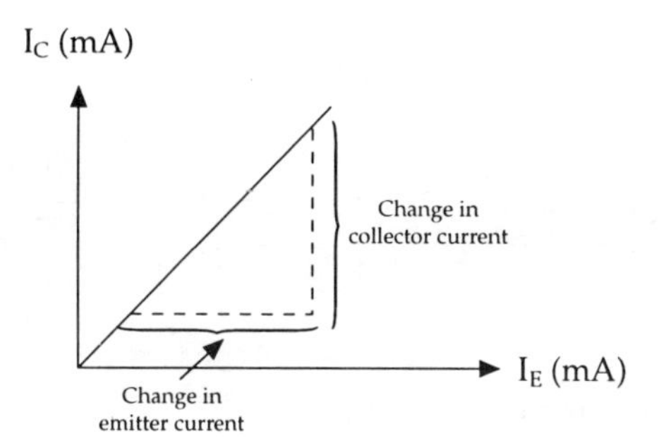

A graph showing how I_E controls I_C

The common base current gain can be determined from the graph. The common base current gain is given the symbol h_{fb} (or the Greek letter α).

Where: h is used to represent the word hybrid (parameter)
f is used to represent forward gain
b is used to represent common base circuit.

$$h_{fb} = \frac{\text{a small change in collector current}}{\text{corresponding change in emitter current}}$$

This ratio is always less than 1!

Summary of the Bi-polar Transistor

- The bi-polar transistor is a sandwich of PNP or NPN semi-conductor material. This produces two diode junctions.
- The base-emitter junction is forward biased and the base-collector junction reverse biased.
- Basic NPN Transistor Action: a small forward bias starts electrons moving into the emitter. Virtually all these electrons cross the base from the emitter to the collector.
- The collector current depends upon, and is controlled by, the emitter current.
- The collector current is almost equal to the emitter current. The base current is very small (μA).
- The collector current plus the very small base current equals the emitter current.

$$\mathbf{I_E = I_C + I_B}$$

- The conventional flow of emitter current flows into a forward biassed NP junction which offers a low resistance. Therefore only a small forward bias voltage is needed to cause it to flow.
- Collector current emerges from a reverse biased PN junction which has a very high resistance.

Self-assessment questions *(answers page 300)*

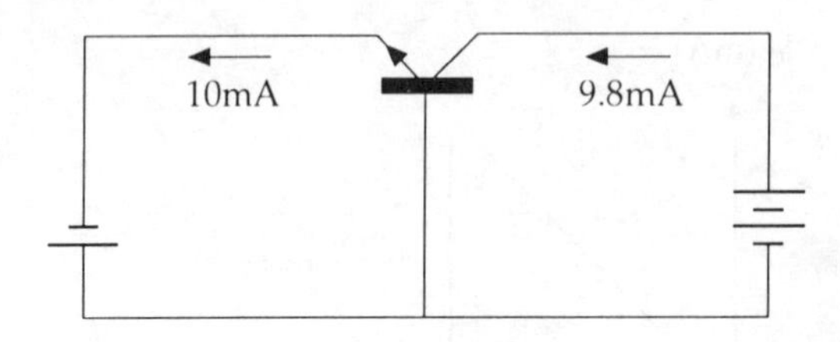

Figure 1

1) The base current in the circuit of Fig. 1 is:
   a) 0.2mA flowing into the base
   b) 0.2mA flowing out of the base
   c) 19.8mA flowing into the base
   d) 29.8ma flowing out of the base

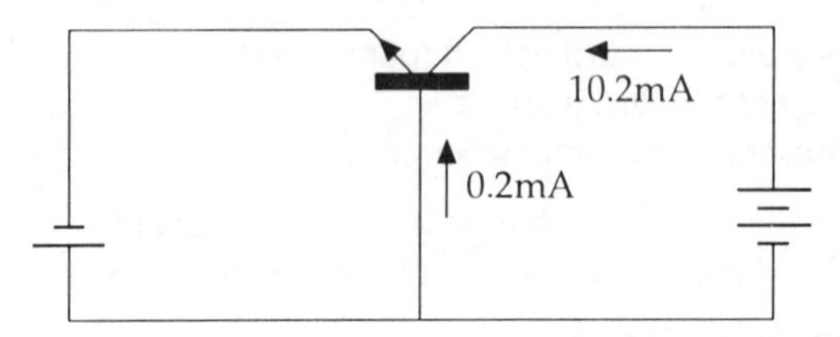

Figure 2

2) The emitter current in the circuit of Fig. 2 is:
   a) 10mA flowing into the transistor
   b) 10mA flowing out of the transistor
   c) 10.4mA flowing into the transistor
   d) 10.4mA flowing out of the transistor

3) When the emitter current of a transistor changes by 6mA its collector current changes by 5.88mA. What is its current gain?
   a) 0.98 b) 5.88 c) 9.8 d) 0.12

4) The current gain of a transistor in common base is:
   a) any value greater than 1 b) equal to 1
   c) equal to or less than 1 d) always less than 1

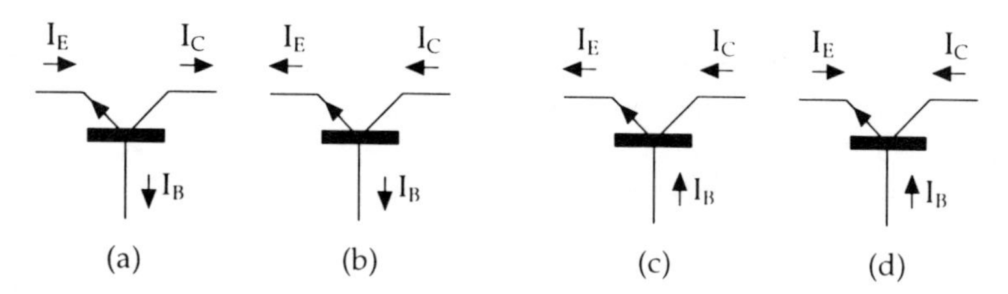

Figure 3

5) Which diagram in Fig. 3 correctly shows the direction of conventional d.c. currents in a NPN transistor?

6) The relationship between the currents in a bi-polar junction transistor is:
   a) $I_C = I_E + I_B$ b) $I_B = I_E + I_C$
   c) $I_E = I_C - I_B$ d) $I_C = I_E - I_B$

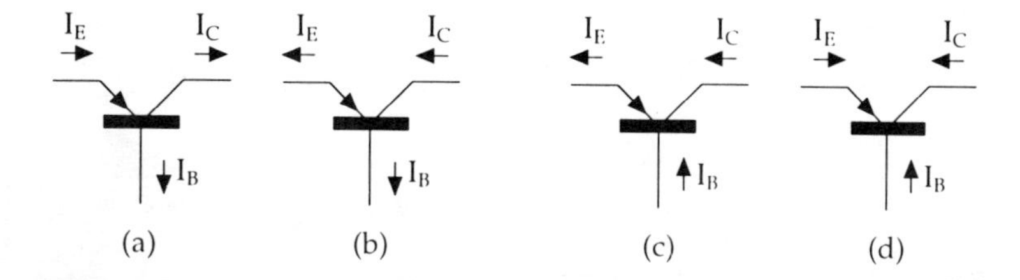

Figure 4

7) Which diagram in Fig. 4 correctly shows the direction of conventional d.c. currents in a PNP transistor?

8) In a common base circuit, when the emitter current is zero then:
   a) the collector current is zero.
   b) the base current and the collector currents are zero.
   c) a small leakage current flows in the base circuit.
   d) a small leakage current flows in the base and collector circuits.

9) Since power is equal to voltage times current, then the common base circuit will provide:
   a) only a power gain
   b) a current and power gain
   c) a voltage and power gain
   d) a voltage, current and power gain.

10) The reverse leakage current will:
   a) increase with an increase in temperature
   b) reduce with an increase in temperature
   c) increase with an reduction in temperature
   d) will not alter as the temperature changes.

11) The voltage that is applied between the emitter and base:
   a) assists the emitter base potential barrier
   b) reduces the emitter base potential barrier
   c) has no effect upon the potential barrier
   d) creates a potential barrier.

12) The voltage that is applied between the collector and base:
    a) assists the collector-base potential barrier
    b) reduces the collector base potential barrier
    c) has no effect upon the potential barrier
    d) creates a potential barrier.

The common emitter connection

Study Note

Before studying this section you will need to be familiar with The bi-polar transistor (p.209).

Introduction

In the common emitter connection the emitter is **common** to the 'input' and 'output' terminals.

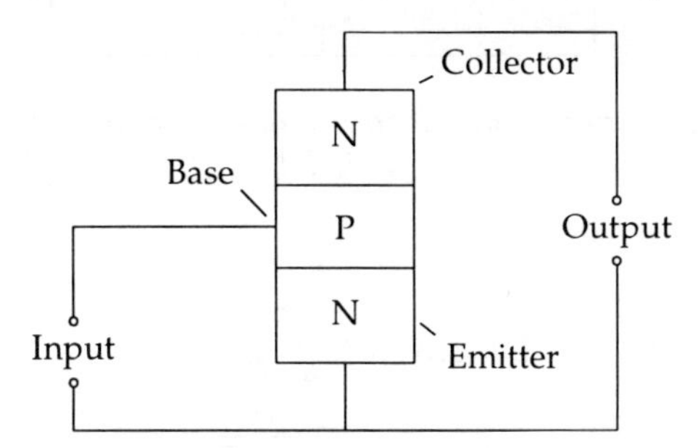

The input is between base and emitter. The output is between base and collector.

The base-emitter junction is forward biased and the base-collector junction is reverse biased.

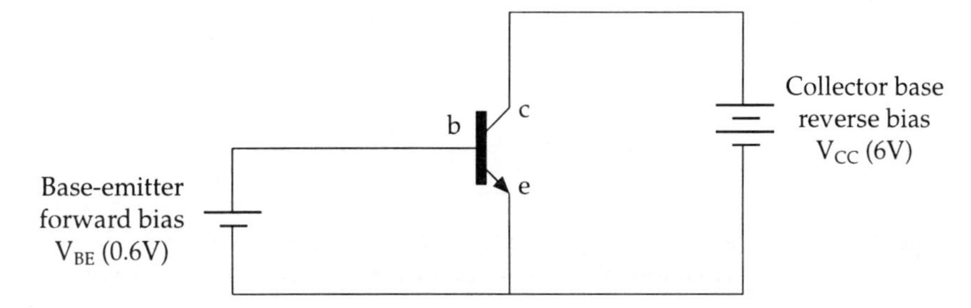

Common Emitter Connection

Note: Because the emitter is now common to both the collector and the base, the reverse bias battery is now connected between collector and emitter and not collector and base as it was in the common base circuit.

However, since with respect to the emitter the collector voltage is +6V and the base voltage is +0.6V, the collector-base potential is still (6 – 0.6) = 5.4V reversed biased: i.e. the collector is 5.4V more positive than the base.

Transistor Action

A common base circuit showing typical voltages and currents is shown below.

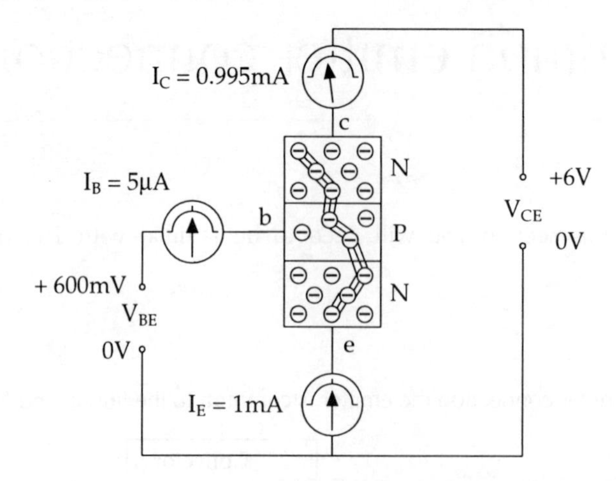

***Note**: A small base current starts electrons moving in the emitter. These electrons are pulled through the base by the positive collector.*

If the base current is now doubled:

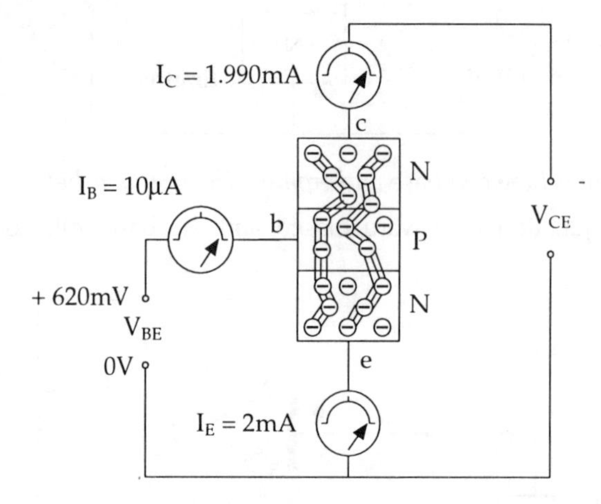

***Note**: A 20mV increase in V_{BE} doubles the base current I_B, which doubles I_C and I_E. Therefore an increase of 5μA in I_B causes an increase of 1mA in I_C.*

Current Gain

Because a change in a very small base current produces a much larger change in the collector current there is now quite a large current gain.

Transfer Characteristic

If we plot collector current against base current, we produce the current transfer characteristic.

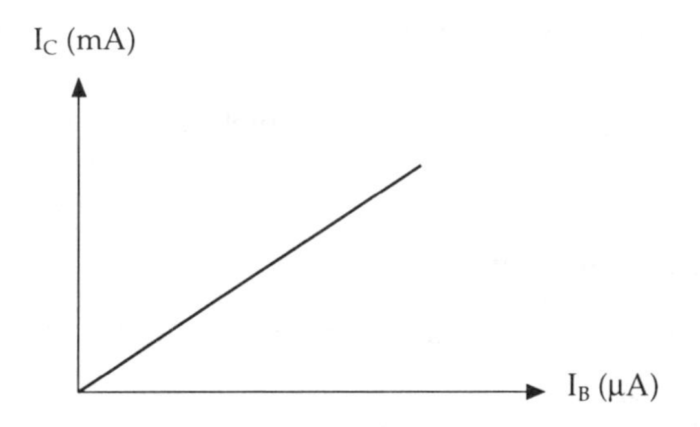

Common emitter current transfer characteristic

This is a straight line graph, and shows that the increase in collector current I_C is directly proportional to the increase in base current I_B.

The current gain in common emitter has the symbol h_{fe} (or β).

Where: h = hybrid parameter.
f = forward gain
e = common emitter

$$h_{fe} = \frac{\text{change in collector current}}{\text{change in base current}} \text{ (no units)}$$

The current gain is greatly dependent on how many electrons re-combine with holes in the base of the transistor. This in turn depends on the width of the base and because the base width cannot be controlled with very precise accuracy, the gains of transistors may vary considerably from one specimen to another, even though they all carry the same type number. For example, the spread of current gain can be from 50 to 250 with a mean of 100 for a given type of transistor.

Leakage Current

If the base of the transistor is left open circuit then a reverse leakage current will flow from the collector, through the base and into the emitter.

I_{CEO} means collector leakage current (I_C) in common emitter (E) with the base open circuit (O).

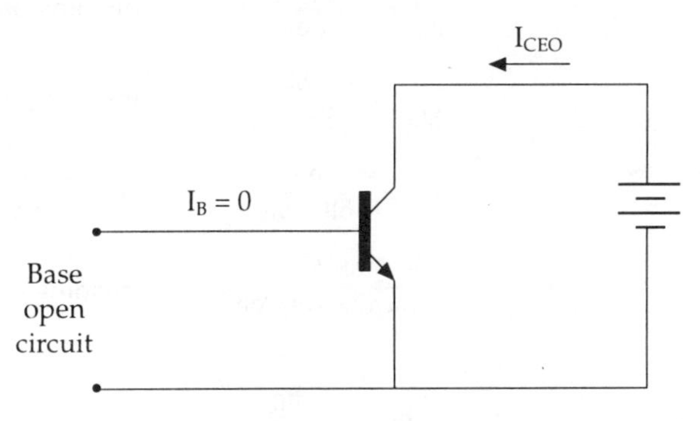

Collector-emitter leakage current I_{CEO}

I_{CEO} is always present and flows in the same direction as any collector current produced by a base current I_B.

I_{CEO} increases with temperature and can cause problems but for most cases it can be ignored.

Summary (Common Emitter Circuit)

- A small base current controls a much larger collector-emitter flow of current.
- As before: $I_E = I_C + I_B$

  But since I_B is always much smaller than I_C and I_E we normally make the approximation that:

  $$I_C = I_E$$
- The forward current gain h_{fe} is large.
- The input current is now base current and the collector current is the output current.
- The collector leakage current is always present and contributes to the total collector current. It increases with temperature.

The Relationship Between the Current Gain in Common Base and the Current Gain in Common Emitter

In common base the forward current gain:

$$h_{fb} = \frac{\text{change in collector current}}{\text{change in emitter current}} = \frac{\delta I_C}{\delta I_E}$$

In common emitter the current gain:

$$h_{fe} = \frac{\text{change in collector current}}{\text{change in base current}} = \frac{\delta I_C}{\delta I_B}$$

Now:

$$\boxed{I_E = I_C + I_B}$$

Therefore: $\delta I_B = \delta I_E - \delta I_C$

and: $\dfrac{\delta I_B}{\delta I_C} = \dfrac{\delta I_E - \delta I_C}{\delta I_C}$ dividing both sides by δI_C

and: $\dfrac{\delta I_C}{\delta I_B} = \dfrac{\delta I_C}{\delta I_E - \delta I_C}$ inverting both sides

and: $h_{fe} = \dfrac{\delta I_C}{\delta I_E - \delta I_C}$ since $\dfrac{\delta I_C}{\delta I_B} = h_{fe}$

and: $= \dfrac{\delta I_C/\delta I_E}{\delta I_E/\delta I_E - \delta I_C/\delta I_E}$ dividing through by δI_E

Which gives:

$$\boxed{\mathbf{h_{fe} = \frac{h_{fb}}{1 - h_{fb}}}}$$

If the current gain is quoted for common base, then the common emitter current gain may easily be determined.

$$\text{For example, if } h_{fb} = 0.99, \text{ then } h_{fe} = \frac{0.99}{1 - 0.99} = 99$$

Self-assessment questions *(answers page 217)*

1) In a common emitter circuit:
    a) The emitter is the input and the collector the output.
    b) The base is the input and the collector is the output.
    c) The base is common to both input and output
    d) The collector is common to both input and output

2) To bias an NPN transistor in the common emitter circuit configuration then, with respect to the emitter:
    a) The base and collector are positive
    b) The base and collector are negative
    c) The base is positive and the collector negative
    d) The base is negative and the collector positive.

3) To bias an PNP transistor in the common emitter circuit configuration then with respect to the emitter:
    a) The base and collector are positive
    b) The base and collector are negative
    c) The base is positive and the collector negative
    d) The base is negative and the collector positive.

4) Which two of the statements i) to iv) below best describes the transistor action in a common emitter NPN transistor?
    i) The base-emitter potential pushes electrons into the emitter.
    ii) The base-emitter potential pulls electrons into the emitter.
    iii) The collector-emitter potential pulls electrons from the base.
    iv) The collector-emitter potential pushes electrons from the base.

    a) i) and iii)
    b) i) and iv)
    c) ii) and iii)
    d) ii) and iv)

5) If the base current in a common emitter circuit is doubled then:
    a) the collector current is doubled
    b) the emitter current is nearly doubled
    c) the collector current is nearly doubled
    d) the collector and emitter currents are doubled.

6) The collector leakage current Iceo:
    a) is independent of temperature
    b) increases with an increase of temperature
    c) decreases with a increase in temperature
    d) increases with a decrease of temperature.

7) The collector leakage current I_{CEO}:
   a) increases the normal collector current
   b) reduces the normal collector current
   c) has no effect on the normal collector current
   d) only exists when the base is open circuit.

8) In a common emitter transistor the current gain h_{fe} equals

   a} $\frac{\text{change in base current}}{\text{change in collector current}}$

   b} $\frac{\text{change in collector current}}{\text{change in base current}}$

   c} $\frac{\text{change in collector current}}{\text{change in emitter current}}$

   d} $\frac{\text{change in emitter current}}{\text{change in collector current}}$.

9) The common emitter transfer characteristic indicates that:
   a) collector current is independent of base current
   b) collector current is inversely proportional to base current
   c) collector current is directly proportional to base current
   d) collector current depends upon temperature.

10) In a common emitter circuit then:
   a) the input and output resistance is low (few tens of ohms)
   b) the input and output resistance is high (few tens of k ohms)
   c) the input resistance is low and the output resistance is high
   d) the input resistance is high and the output resistance is low.

Common emitter static characteristics

Introduction

If we vary the base voltage of a transistor connected in common emitter, then we alter the base current which in turn alters the collector current and collector voltage.

In order to see the way the transistor behaves we can plot graphs to determine how all these quantities vary in relation to each other These characteristics are obtained when the transistor is not handling a signal, and for this reason are called the **static characteristics**.

A typical setup for obtaining these characteristics is shown below.

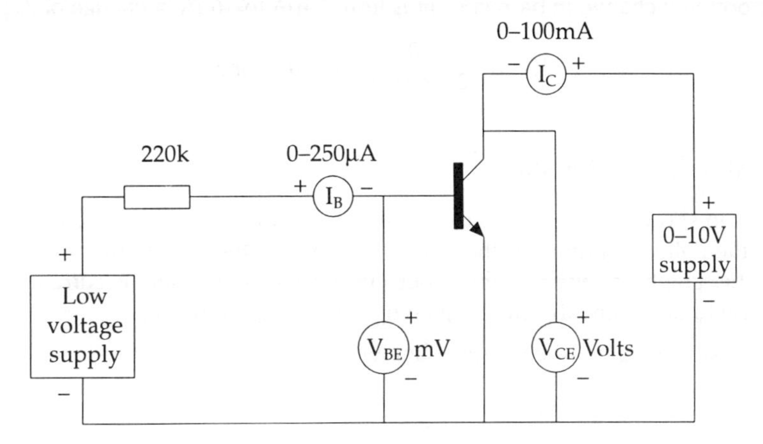

Circuit for obtaining the common emitter static characteristics

The Input Characteristic

This graph is obtained by plotting the variation in base current (I_B) for various values of base emitter voltage (V_{BE}), with the collector voltage (V_{CE}) held at a constant value. From this graph we can obtain the input resistance R_{in} (also known as the hybrid parameter h_{ie}).

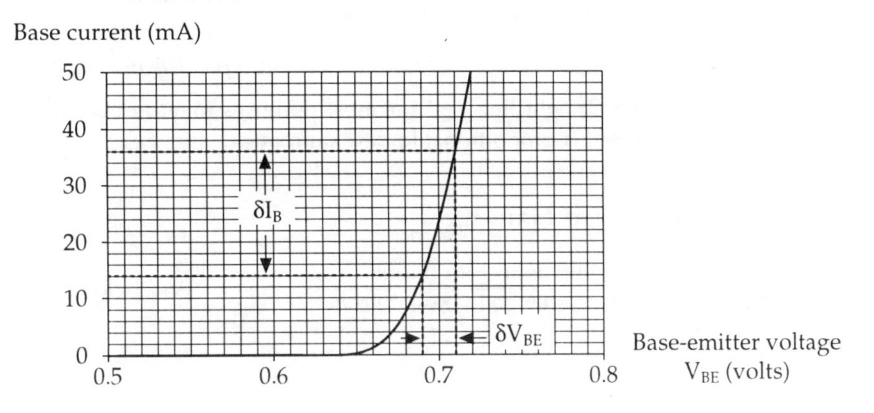

The Input Characteristic

The graph shows that no base current flows until the base voltage is in excess of 0.65 volts. At this point the forward bias is breaking down the potential barrier and the transistor action is starting.

Determination of R_{in} (h_{ie})

$$R_{in} = \frac{\text{a change in the base-emitter voltage}}{\text{the corresponding change in base current}} = \frac{\delta V_{BE}}{\delta I_B} \quad (V_{CE} \text{ held constant})$$

From the graph above:

The base-emitter voltage varies from 690mV to 710mV: a change of 20mV.

The corresponding change in base current is from 14μA to 36μA: a change of 22μA.

$$R_{in} = \frac{0.02}{22 \times 10^{-6}} = \mathbf{909\ ohms}$$

The Transfer Characteristic

This is a graph showing how the collector current (I_C) varies as the base current (I_B) is varied. Since the collector current is the output current and the base current is the input current then this graph shows how the input current varies the output current. This is the graph we use to determine the current gain h_{fe} (or β) of the transistor.

A typical transfer characteristic is shown below.

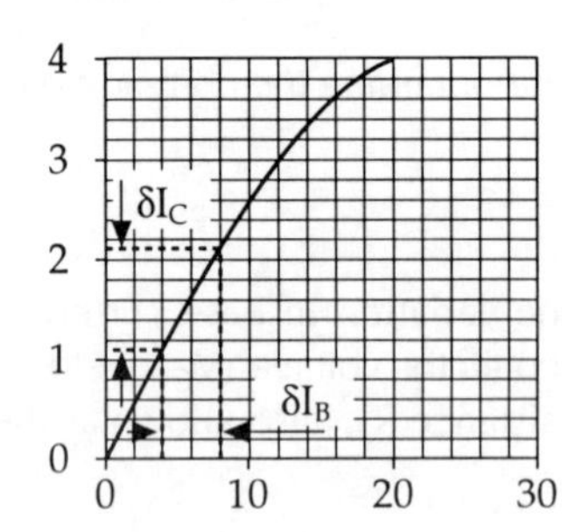

A typical Transfer Characteristic

The graph is very nearly a straight line and for most purposes may be considered so. In other words, collector current is proportional to base current.

Determination of Current Gain h_{fe}

$$h_{fe} = \frac{\text{a change in the collector current}}{\text{the corresponding change in base current}} = \frac{\delta I_C}{\delta I_B} \quad (V_{CE} \text{ held constant})$$

From the graph above:

The collector current varies from 1.1mA to 2.1mA: a change of 1mA.

The corresponding change in base current is from 4μA to 8μA: a change of 4μA.

$$\text{Therefore, } h_{fe} = \frac{1 \times 10^{-3}}{4 \times 10^{-6}} = \mathbf{250}$$

The Output Characteristic

The output characteristic is a graph showing the variation of the collector current (I_C) against the variation of collector voltage (V_{CE}) for fixed values of base voltage (V_{BE}).

The collector voltage is the output voltage and the collector current is the output current; for this reason it is called the output characteristic.

The output characteristic is probably the most useful of all. It can be used to determine the output resistance, the current gain of the transistor; and later on we shall use it to plot a.c. and d.c. load lines.

A typical output characteristic is shown below.

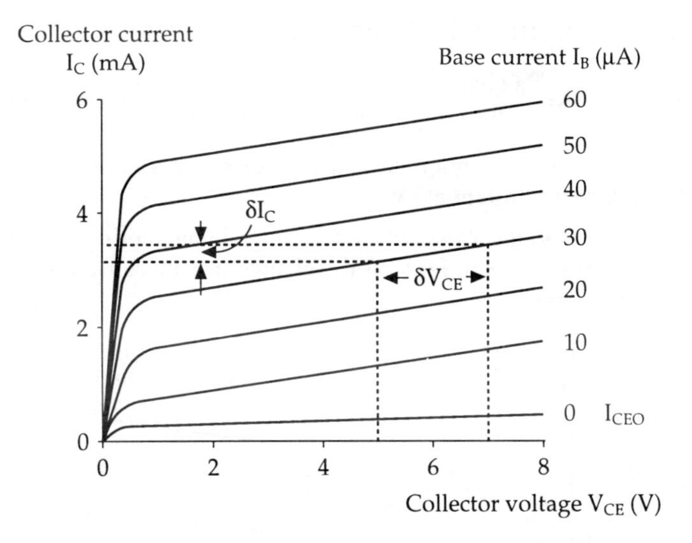

A Typical Output Characteristic

Notice that for each fixed value of base current I_B the curve rises sharply, bends at what is called the **knee** of the curve and then flattens out.

We normally operate the transistor with collector voltages beyond the knee of the characteristic, i.e. on the flat part. Over this part of the graph the collector current does not alter very much even if there is a large change in V_{CE}.

Therefore, for a fixed value of collector voltage the only way the collector current I_C can be changed or controlled is by changing the base current.

The bi-polar transistor is a current operated device.

Finally notice that when $I_B = 0$, the base is open circuit. Only the small leakage current flows I_{CEO}.

Determination of the Output Resistance R_{out}

We use the part of the graph after the knee of the characteristic to calculate the output resistance.

$$R_{out} = \frac{\text{a change in the collector voltage}}{\text{the corresponding change in collector current}} \quad \text{for a constant base current } I_B$$

We determine the changes on one curve only; this ensures that the base current is constant and that the collector current only changes as a result of the change in collector voltage.

Therefore using the graph of $I_B = 30\mu A$:

The change in collector voltage V_{CE} is: 7V – 5V = 2V; the corresponding change in I_C is: 3.3mA – 3.1mA = 0.2mA

$$\text{From which, } R_{out} = \frac{2}{0.2} \text{ k ohms} = \mathbf{10k}$$

Determination of h_{fe} from the Output Characteristic

$$\text{The current gain, } h_{fe} = \frac{\text{a change in the collector current } I_C}{\text{the corresponding change in } I_B} \quad \text{for a constant collector voltage } V_{CE}$$

If we draw a vertical line up the graph from the X axis say at a V_{CE} of 7V, then all the way up that line the voltage is **constant** at 7V.

If we now project across to the Y axis where the vertical line intersects the 30μA and 50μA curves, for example, then we can read off the change in collector current for this change (30 to 50μA) in base current.

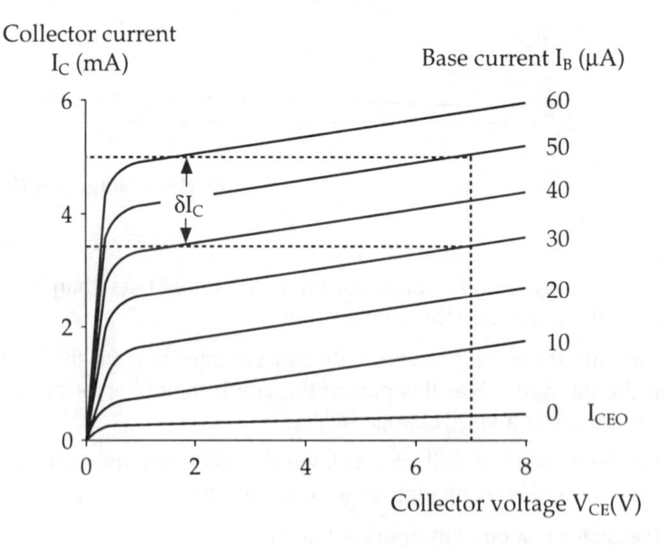

Obtaining h_{fe} from the output characteristic

From the graph: at a fixed collector voltage of 7V:

The change in collector current is: 5.1 – 3.3 = 1.8mA. The corresponding change in I_b is: 50 – 30 = 20μA

$$\text{Therefore, the current gain } h_{fe} = \frac{1.8 \times 10^{-3}}{20 \times 10^{-6}} = \mathbf{90}$$

Note:

1) This is a different value to the value we previously obtained for h_{fe} from the transfer characteristic. This is because the transfer graph and the output graph are typical graphs and are not related to any particular transistor or to each other.
2) Also: the value of h_{fe} depends upon the value of the collector current and can vary by as much as 60% or more. The h_{fe} will be a maximum for a particular value of collector current, but for any further increase the h_{fe} falls off. For this reason the manufacturers will quote the h_{fe} for a range of collector currents.

Mutual Conductance (g.m.)

The mutual conductance of a transistor is a measure of how much the collector-emitter flow of current is affected by the base voltage.

In other words, if the base voltage is changed by a few mV how much does the collector current change?

The mutual conductance is given the symbol g.m.

$$\text{g.m.} = \frac{\text{a change in the collector current } I_C}{\text{change in } V_{BE} \text{ causing change in } I_C} \text{ (S) } V_{CE} \text{ held constant}$$

The units are A/V or Siemens. Since the quantities above are measured in mA and mV the units are Siemens.

Determination of the Mutual Conductance, g.m.

The output characteristic can also be a plot of collector current against collector voltage for a fixed base-emitter voltage V_{BE}, instead of base current I_B.

In other words we use the same output characteristic as before, but replace the fixed base currents with the base voltages V_{BE} which created them.

$$\text{g.m.} = \frac{\text{change in collector current } I_C}{\text{change in } V_{BE} \text{ causing change in } I_C} \text{ (S)}$$

at a **fixed value of collector-emitter voltage**.

Therefore, if we draw a vertical line up the graph, at say 8V, and where this line intersects the 630mV and 610mV graphs, project across to the Y axis, then read off the corresponding values of collector current, we can determine the g.m.

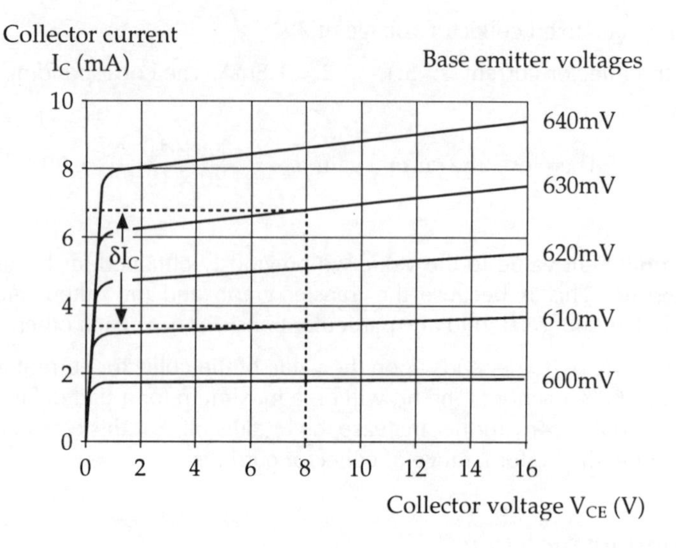

A graph to determine the mutual conductance g.m.

From the graph: the change in base-emitter voltage will be: 630 – 610mV = 20mV and the corresponding change in the collector current I_C is: 6.8mA – 3.4mA = 3.4mA

$$\text{Therefore, g.m.} = \frac{3.4 \times 10^{-3}}{20 \times 10^{-3}} = 0.17\text{S or } 170\text{mS}$$

Summary of Common Emitter Characteristics

- The input characteristic is a plot of base current against base-emitter voltage The input resistance (R_{in}) can be determined from this graph.

$$R_{in} = \frac{\text{a change in the base-emitter voltage}}{\text{the corresponding change in base current}} \quad (V_{CE} \text{ held constant})$$

- The transfer characteristic is a plot collector current against base current. The current gain h_{fe} can be obtained from this graph.

$$h_{fe} = \frac{\text{a change in the collector current}}{\text{the corresponding change in base current}} \quad (V_{CE} \text{ held constant})$$

- The output characteristic is a plot of collector current against collector voltage. The output resistance R_{out} can be obtained from this graph.

$$R_{out} = \frac{\text{a change in the collector voltage}}{\text{the corresponding change in collector current}} \quad \text{with base current constant}$$

  The h_{fe} and the mutual conductance g.m. may also be obtained from the output characteristic.

- The output characteristic is also used to plot d.c. and a.c. load lines (see **Section 2 Information and Skills Bank:** Load lines, p.273)

Now attempt Task a Plotting the common emitter static characteristics.

Single-stage audio amplifier

Introduction

With the common emitter circuit a few mV change in the base voltage produces a small change in the base current. This produces a very much larger change in the collector-emitter flow of current because of the current gain of the transistor.

If the collector current is made to flow through a load resistor (R_L) in the collector circuit, then as the base current changes, this will cause a change in the collector current flowing in this resistor. The changing collector current through this resistor will produce a changing collector voltage across the resistor.

This changing voltage is the useful output voltage of the amplifier. It is an **amplified version** of the change in base voltage.

The circuit has provided **voltage amplification**.

Typical Voltage and Currents in a Resistive Loaded Amplifier

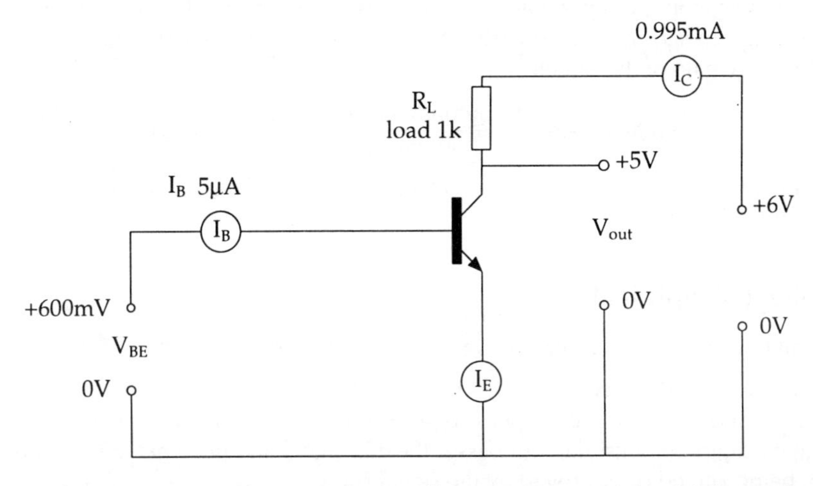

A 5μA base current causes a 0.995mA collector current and the 1V dropped across R_L leaves the collector voltage at approximately 5V

If now the base current is doubled, the collector current doubles and the volt drop across the load resistor doubles.

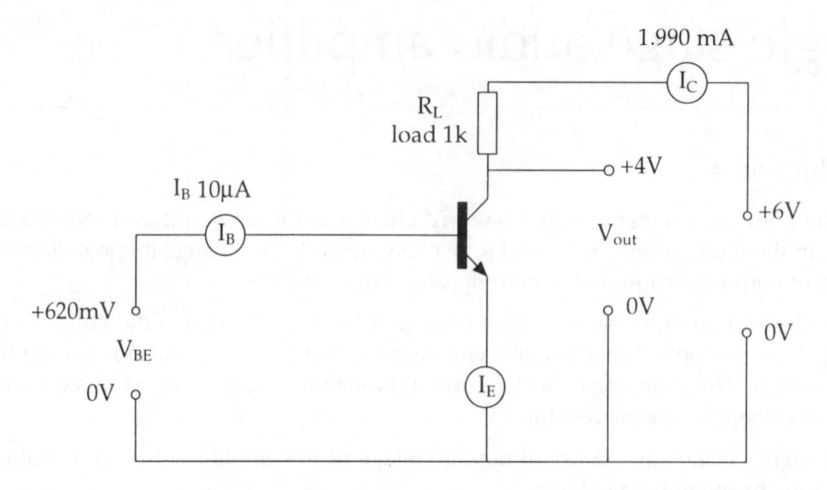

Base voltage increases by 20mV, the base current doubles, the collector current and volt drop across R_L double, leaving the collector voltage at 4V.

The small 20mV change in base voltage caused a change in the volt drop across the load resistor of approximately 1000mV (1.0V).

If the small change in the input voltage, represents the input signal to the amplifier and the corresponding change in voltage across the load resistor, represents the output voltage, then the voltage gain of the circuit is:

$$\text{Voltage gain} = \frac{\text{output voltage}}{\text{input voltage}} = \frac{1{,}000}{20} = \mathbf{50}$$

This means that the change in output voltage is 50 times larger than the change in the input voltage.

a.c. Voltage Amplification

A common emitter configuration may be used as an a.c. voltage amplifier.

A small bias voltage (610mV) is applied between the base and the emitter to start the transistor action. An a.c. signal (20mV peak to peak) is then fed through a capacitor (C1) to the base. The capacitor 'blocks' d.c. but passes the a.c. signal and prevents the small bias voltage from being altered or destroyed by the signal source.

The small a.c. signal, applied between the base and emitter, varies the bias voltage, which varies the base current, which in turn varies the collector current and the volt drop across the load.

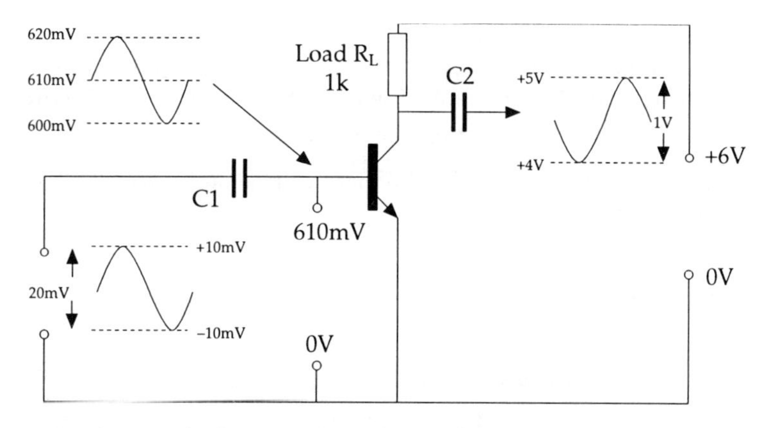

The a.c. signal varies the base voltage, which varies the collector current. This in turn causes a changing volt drop across R_L, making the collector voltage vary in sympathy with the input signal

The output voltage is taken via another capacitor (C2) which blocks the d.c. potential present on the collector and passes only the variation in collector voltage to the output. In other words, it **couples** the a.c. signal out from the amplifier.

Notice that as the base voltage becomes more positive with respect to the emitter, the base and collector currents increase, but the volt drop across the collector resistor also increases, which makes the collector voltage less positive with respect to the emitter. In other words, there is a phase inversion between the input and output signals.

Remember that in a common emitter circuit the output is taken from between the collector and the emitter.

Simple Biasing

It is not very practical to have a small battery of 610mV to provide the base-emitter bias. This small voltage can be obtained from the battery used to supply the collector voltage.

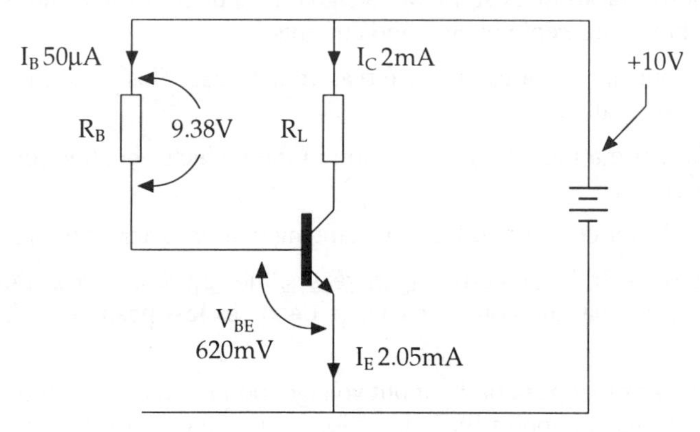

Base resistor providing simple bias

The base is forward biased by the resistor R_B and allows base current to flow, which in turn causes the larger collector current to flow in the load resistor R_L.

When the base emitter is forward biased, then the base emitter voltage V_{BE} will be between 0.6V and 0.7V depending upon how much base current is flowing. In the simple bias circuit shown above V_{BE} is 0.62mV. Therefore the voltage dropped across R_B will be:

$$10V - 0.62V = 9.38V$$

R_B therefore provides the necessary level of base voltage and current to bias the transistor and start the transistor action.

Practical amplifier

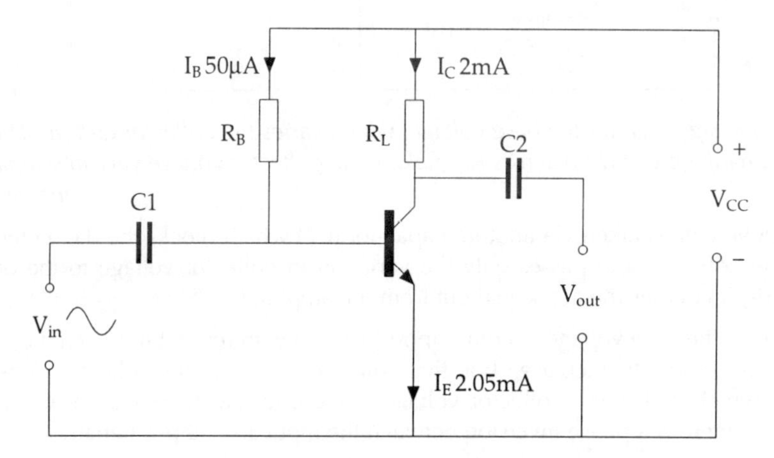

Practical common emitter amplifier

Waveforms in a Practical Common Emitter Amplifier

Look carefully at the waveforms shown on the next page.

The steady base bias voltage produces the steady base and collector currents. These are the d.c. conditions that must exist, before a signal is applied, to start the transistor action. These are called the **quiescent** voltages and currents.

The small a.c. input signal varies the base bias voltage V_{BE}. This causes the base bias current I_B to vary in sympathy.

This in turn will cause a much larger variation in the collector-emitter flow of current due to the transistor action.

The changing collector current produces a changing volt drop across the load resistor R_L.

Remember that as the collector current I_C increases, the volt drop across the collector resistor also increases, so that the collector voltage becomes **less positive with respect to the emitter**.

This means that a positive peak of the input voltage (point a) has produced a negative peak in the output voltage V_{CE} (point b). This action causes the phase inversion between the input and output signals.

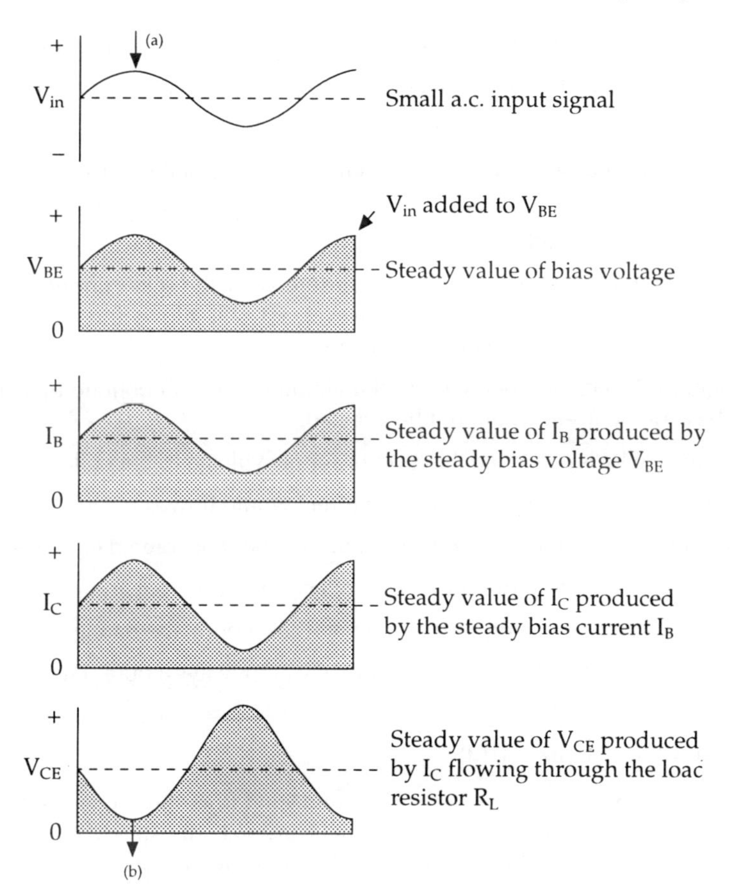

Waveforms in a practical common emitter amplifier

Self-assessment questions (1) (answers page 301)

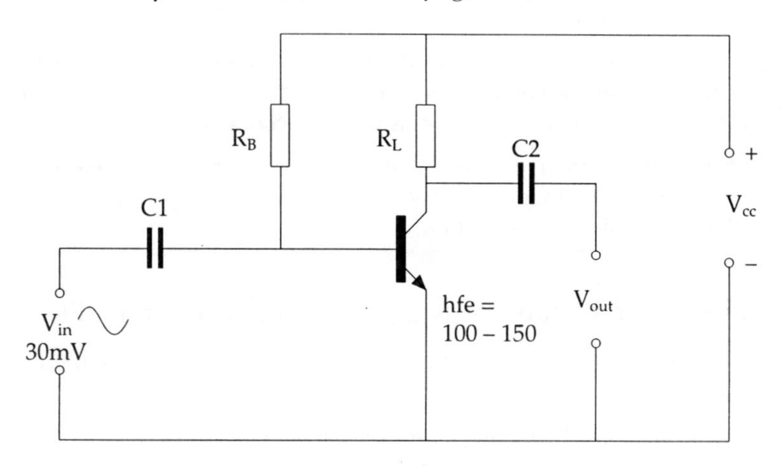

Circuit diagram for questions 1-12

Study the diagram above and decide whether the statements below are **True** or **False**.

1) The input voltage will be amplified and inverted at the output by this circuit.
2) The current gain of the transistor is 100.
3) The resistor R_B feeds a steady direct current to the base of the transistor.
4) If the resistor R_B became open circuit, only the positive half cycle of the input waveform would be amplified.
5) Capacitor C1 blocks direct current and ensures that the d.c. current in R_B can flow only into the base of the transistor. However, C1 still allows the a.c. input signal to vary the base-emitter voltage of the transistor.
6) If capacitor C1 became open circuit, the amplifier would continue to work normally as the base bias current would not be affected.
7) The purpose of R_L is to reduce the level of supply voltage to the collector.
8) If RL became open circuit, no collector current would flow.
9) If R_L became short circuit, the collector current would become dangerously large.
10) Capacitor C2 is used to prevent the d.c. collector voltage appearing at the output.
11) The output voltage on the right-hand side of C2 is a pure alternating voltage.
12) The collector voltage on the left-hand side of C2 is always a pure direct voltage.

Disadvantages of Simple Bias

The simple bias arrangement, using a resistor connected from the supply to the base of the transistor, provides a steady fixed base current. However, the steady collector current it produces due to transistor action, may vary for the following reasons:

1) The leakage current I_{CEO} increases with temperature.

   The leakage current can double for every few degrees rise in temperature of the transistor. This can lead to thermal runaway. In other words, the temperature rises – the leakage current increases – the collector current increases – which further increases the temperature and so on.

2) To have a fixed steady value of collector current, we need a fixed value of base bias current. To fix the base current then we need to fix the base bias voltage. The simple bias arrangement does not fix the base bias voltage accurately enough.

   Also remember: the current gain h_{fe} of the transistor varies from one transistor to another. If R_B is calculated for an average value of h_{fe}, it will be the wrong value for transistors having a different value of h_{fe}.

The Potential Divider, Emitter Resistor, Stabilising Circuit (PDER)

The base-emitter voltage V_{BE} is obtained from a potential divider and this holds the base voltage more constant.

Thermal runaway is prevented by having a resistor in the emitter circuit.

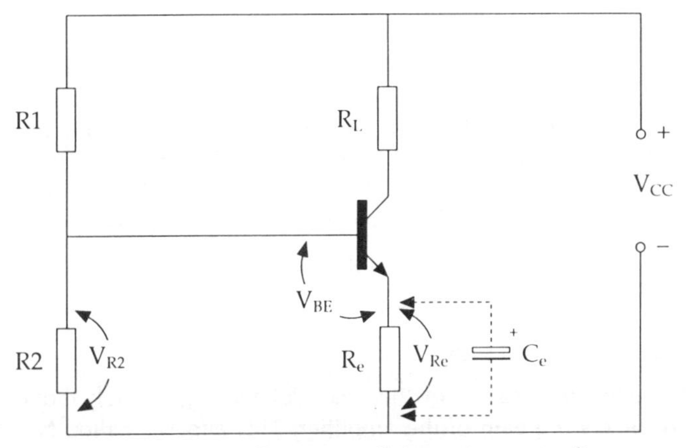

A potential divider, emitter resistor, stabilising circuit

Resistors R1 and R2 form a potential divider chain which holds the base potential constant. The base voltage relative to the common rail is the voltage dropped across R2 (V_{R2}).

The transistor action sets up the collector-emitter flow of current. (Remember I_C is almost equal to I_E.) The emitter current flows through the emitter resistor R_e and produces a volt drop V_{Re}. The emitter voltage relative to the common rail is therefore V_{Re}.

The base-emitter voltage (V_{BE}) is the difference between V_{R2} and V_{Re}.

$$V_{BE} = V_{R2} - V_{Re} \text{ volts}$$

Stabilisation

This arrangement can stabilise against variations in the temperature of the transistor.

If the temperature increases, then both I_C and I_E increase. This will increase the volt drop across R_e.

Since the base voltage V_{BE} is determined by:

$$V_{BE} = V_{R2} - V_{Re} \text{ volts}$$

then as V_{R2} is fixed by the potential divider, V_{BE} must become smaller as V_{Re} increases.

If the base bias voltage is reduced then the base current and the consequent collector current must reduce. This compensates for the initial increase in temperature and holds I_C constant.

a.c. and d.c. Conditions

The transistor amplifier has two sets of conditions and each must be considered separately.

The d.c. conditions are all the voltages and currents necessary to make the transistor work.

The a.c. condition is when the transistor is handling (amplifying) an a.c. signal.

The PDER circuit has now satisfied all the d.c. conditions for the transistor but has upset the a.c. operating condition.

Feedback

Let us now consider the a.c. conditions.

A small change in base voltage due to an a.c. signal produces a change in base current, which in turn produces a much larger change in the collector-emitter flow of current. This will produce a change in the emitter voltage V_{Re}. The amount of change in the collector current depends upon the amount of change in the base emitter voltage V_{BE}.

The base voltage even to an a.c. signal is still determined by:

$$V_{BE} = V_{R2} - V_{Re} \text{ volts}$$

So as V_{R2} is varied by the input signal so also is V_{Re}.

This reduces the effective change in the base voltage V_{BE}, which reduces the effective input signal and lowers the gain of the amplifier. This effect is called **Negative Feedback** (NFB).

Emitter Bypass Capacitor

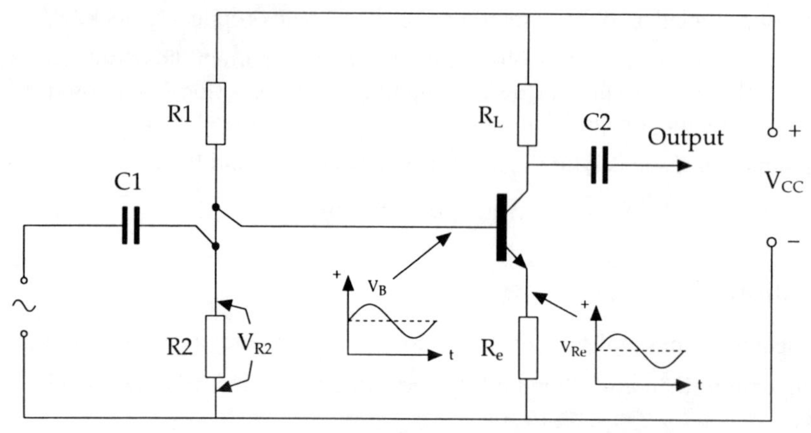

As the signal drives the base voltage positive then the emitter voltage follows the base voltage and also becomes more positive

To overcome this effect the emitter voltage must be held constant even though the emitter current is changing due to the input signal.

A large capacitor if therefore placed across the emitter resistor and holds the emitter voltage constant by its stored charge.

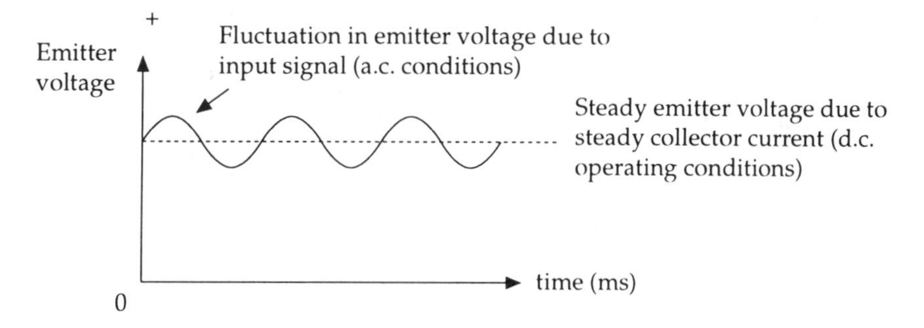

The extra charge given to the capacitor on the positive half cycle is discharged on the negative half cycle. The average charge over one cycle is constant. Capacitor voltage remains constant

Determination of Voltages and Currents in the PDER circuit

Often we need to know the approximate voltages and currents in the circuit.

Consider the circuit below.

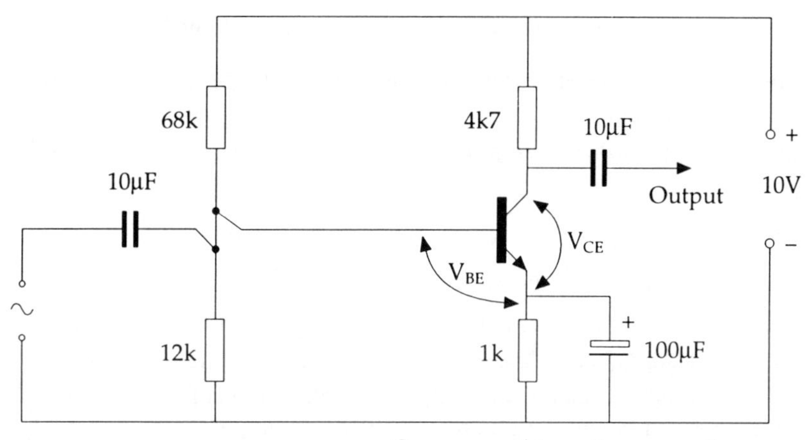

Circuit for worked example

1) **Determine the approximate collector current assuming that V_{BE} = 0.7V.**

   The base voltage will be the voltage across the 12k resistor.

   To find the voltage across this resistor we can use the voltage divider rule.

Voltage divider rule

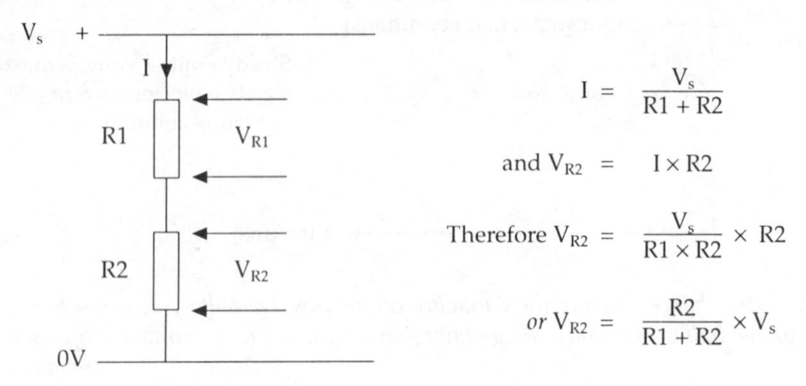

$$\text{Therefore, from the voltage divider rule, } V_{R2} = \frac{12k}{12k + 68k} \times 10V = 1.5V$$

The emitter voltage will be the base voltage less V_{BE}

$$V_E = 1.5 - 0.7 = 0.8V$$

$$\text{The emitter current } I_E = \frac{V_E}{R_e} = \frac{0.8}{1k} = \mathbf{0.8mA}$$

Therefore the collector current = 0.8 mA (since the collector current is approximately equal to the emitter current.)

2) Calculate the collector-emitter voltage V_{CE}

The voltage dropped across the collector load resistor:

$$V_{RL} = I_C \times R_L = 0.8mA \times 5.6k = 4.48V$$

therefore the voltage across the transistor V_{CE}:

$$V_{CE} = V_{supply} - V_{RL} - V_E \text{ (emitter voltage)} = 10 - 4.48 - 0.8 = \mathbf{4.72V}$$

Notice that the supply voltage is shared nearly equally between the transistor and its load resistor. This is so that when a signal is applied, the collector voltage can swing equally above and below the quiescent value, allowing the stage to amplify without distortion.

Summary of Single-stage Audio Amplifier

- The simple bias circuit allows collector current to vary, particularly if the temperature changes.
- Bias stabilisation is the process of holding the quiescent (d.c.) collector current steady.
- The PDER circuit holds the collector current constant and compensates against changes in temperature.
- The capacitor C_E holds the emitter voltage constant when an a.c. signal is applied and prevents NFB lowering the gain of the amplifier.
- The coupling capacitors C1 and C2 allow the a.c. signal to pass but block d.c. and prevent the d.c. conditions on the transistor from being destroyed.

Self assessment questions (2) (answers page 301)

All the questions refer to the circuit below.

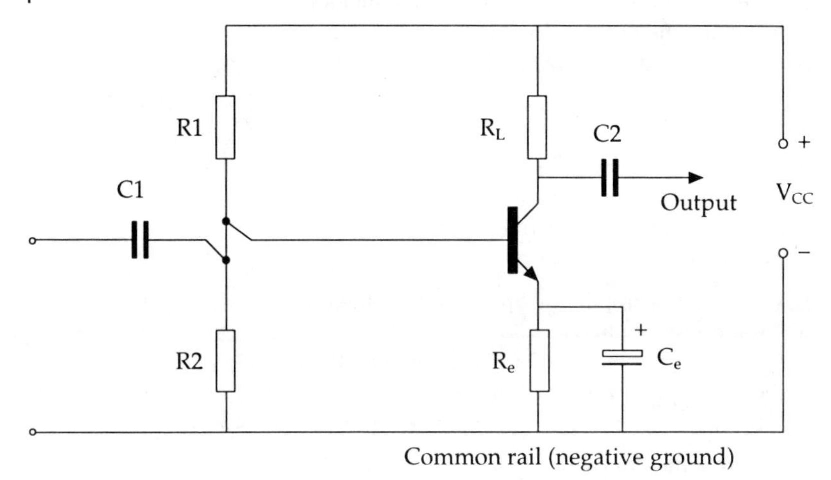

Circuit for self assessment questions

1) The function of the resistors R1 and R2 is:
   a) To hold the base current constant.
   b) to hold the base voltage constant.
   c) to couple in the a.c. signal to be amplified.
   d) to provide a load for the input signal.

2) The function of the emitter resistor R_e is:
   a) To produce a volt drop proportional to the collector current.
   b) To make the emitter positive with respect to the base.
   c) To introduce negative feedback.
   d) To allow current to flow out of the emitter.

3) If R1 became open circuit the effect would be to:
   a) increase the collector current slightly.
   b) reduce the collector current slightly.
   c) cut off the collector current.
   d) greatly increase the collector current.

4) If R2 became open circuit the effect would be to:
   a) increase the collector current.
   b) reduce the collector current.
   c) cut off the collector current.
   d) introduce negative feedback.

5) If C_e became open circuit the effect would be to:
   a) increase the collector current slightly.
   b) reduce the collector current slightly.
   c) cut off the collector current.
   d) introduce negative feedback.

6) The purpose of the resistor R_L is:
   a) to limit the collector current to a safe value.
   b) to reduce the voltage applied to the collector.
   c) to reduce the current flowing through R_e.
   d) to act as a load resistor, to develop the output voltage.

7) If a PNP transistor were employed in the same circuit, which of the following changes would be necessary?
   a) reverse the battery polarity and C_e.
   b) exchange the positions of R1 and R2.
   c) exchange the positions of R_c and R_e.
   d) exchange the connections of collector and emitter.

8) When the circuit is amplifying an a.c. signal, the emitter by-pass capacitor C_e:
   a) has a constant charge.
   b) has no constant charge but is charged by the a.c. signal.
   c) has a varying charge, but over one cycle the average charge is constant.
   d) has no charge, the average charge over one cycle is zero.

9) The purpose of the coupling capacitors C1 and C2 is:
   a) to couple the a.c. signal in and out of the amplifier.
   b) to block the d.c. voltages present in any signal.
   c) to pass the d.c. but block any a.c. component of the signal.
   d) to pass the a.c. but block any d.c. component of the signal.

10) The function of the emitter resistor is:
   a) to reduce the voltage across the transistor.
   b) to limit the emitter current.
   c to produce a compensating voltage that can alter the effective base voltage.
   d) to produce a compensating current that can alter the effective base current.

Stage gain of common emitter amplifier

Introduction

The voltage and power gains of the stage may be calculated using the basic relationships developed below.

Voltage Gain

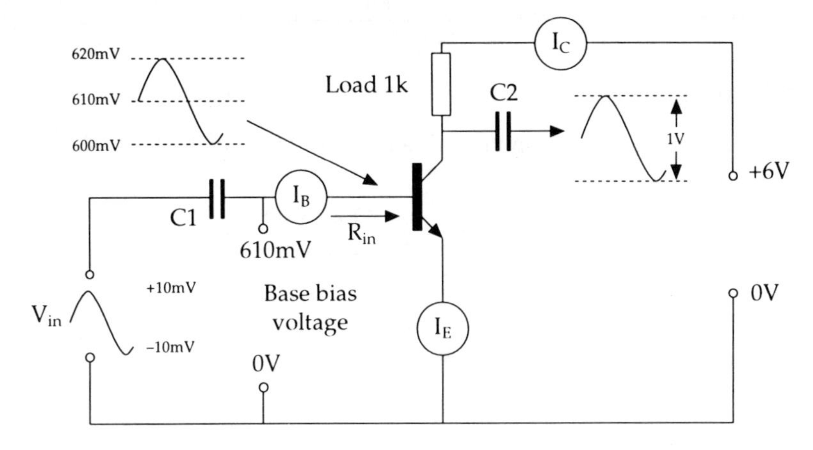

A typical common emitter amplifier

In the circuit above:

$$\text{the collector current } I_C = h_{fe} \times I_B$$

$$\text{the base current } I_B = V_{in}/R_{in}$$

from which:

$$I_C = h_{fe} \times V_{in}/R_{in}$$

This value of current flows through the collector load resistor and develops the output voltage V_{out}.

$$\text{The output voltage } V_{out} = I_C \times R_L$$

$$\text{Therefore } V_{out} = h_{fe} \times V_{in}/R_{in} \times R_L$$

Now V_{out}/V_{in} is the voltage gain, A_v: therefore:

$$\mathbf{A_v = \frac{h_{fe} \times R_L}{R_{in}}}$$

Power Gain

$$\text{Power gain, } A_p = \frac{\text{output power}}{\text{input power}} = \frac{P_{out}}{P_{in}}$$

The input power is given by $I_B^2 \times R_{in}$

The output power is given by $I_C^2 \times R_L$

Therefore:

$$P_{out} = (h_{fe} \times I_B)^2 \times R_L \quad \text{since } I_C = h_{fe} \times I_B$$

from which the power gain:

$$\text{Power gain} = P_{out} = \frac{h_{fe}^2 \times I_B^2 \times R_L}{I_B^2 \times R_{in}}$$

Therefore Power Gain:

$$\boxed{A_p = \frac{h_{fe}^2 \times R_L}{Rin}}$$

Worked Example

If a common emitter transistor amplifier has a load resistance of 2k, R_{in} = 1k, and h_{fe}= 100. Determine the voltage and power gains.

$$A_v = \frac{h_{fe} \times R_L}{R_{in}} = \frac{100 \times 1{,}000}{2{,}000} = \mathbf{50}$$

$$A_p = \frac{h_{fe}^2 \times R_L}{R_{in}} = \frac{100^2 \times 1{,}000}{2{,}000} = \mathbf{5{,}000}$$

Input Resistance R_{in}

The input resistance R_{in} is the input resistance of the transistor. It depends upon the current gain and the **internal emitter resistance of the transistor (r_e)**.

$$\boxed{R_{in} = h_{fe} \times r_e}$$

The value of r_e can be determined from the diode current equation.

$$r_e = \frac{k \times T}{q \times I_e}$$

Where: k = Boltzman's constant (1.374×10^{-23} joules per degree C)
T = Absolute temperature (degrees Kelvin)
q = Charge on an electron (1.59×10^{-19} coulombs)
I_E = emitter current (mA)

The coefficient kT/q has units of voltage and at room temperature (20°C) the internal emitter resistance (r_e) is given by:

$$\boxed{r_e = \frac{25\text{mV}}{I_e\ \text{mA}}}$$

Note:

1) This value of r_e is valid for all transistors.
2) The internal emitter resistance depends upon emitter current.
3) It means that if the emitter current is 1mA then the emitter resistance r_e is 25 ohms.
4) We can use this value of r_e in the voltage gain and power formula (since $R_{in} = h_{fe} \times r_e$).

Mutual Conductance (g.m.)

The mutual conductance of a transistor is a measure of how much the collector-emitter flow of current is affected by the base voltage.

In other words, if the base voltage is changed by a few mV how much does the collector current change?

The mutual conductance is given the symbol g.m.

$$\text{g.m.} = \frac{\text{change in collector current } I_C}{\text{change in } V_{BE} \text{ causing change in } I_C} \quad \text{(Siemens)}$$

The units are A/V or mA/mV or Siemens. If the quantities above are measured in mA/V the units are milli-Siemens.

Determination of the Stage Gain Using g.m.

The voltage gain is given by:

$$\mathbf{A_v = \frac{h_{fe} \times R_L}{R_{in}}}$$

And the input resistance of the transistor is given by:

$$\mathbf{R_{in} = h_{fe} \times r_e}$$

$$\text{Therefore, } A_v = \frac{h_{fe} \times R_L}{h_{fe} \times r_e} = \frac{R_L}{r_e}$$

The emitter resistance r_e is given by:

$$\mathbf{r_e = \frac{25mV}{I_E\, mA}}$$

$$\text{The units for } r_e \text{ are } \frac{mV}{mA}, \text{ therefore } \frac{1}{r_e} = \frac{1}{mV/mA} = \frac{mA}{mV} \text{ or Siemens}$$

$$\text{The g.m. is therefore equal to } \frac{1}{r_e}, \text{ and at room temperature:}$$

$$\text{g.m.} = \frac{1}{r_e} = \frac{1}{25mV / I_E} = \frac{I_E}{25mV}$$

$$\therefore \text{ g.m.} = 0.04 \times I_E \text{ Siemens (since } I_E \text{ is in mA)}$$

As the g.m. is usually given in milli-Siemens, then the g.m. = $40 \times I_E$ mS

The stage gain A_v = g.m. $\times R_L$

Hence:

$$\mathbf{Av = 40 \times I_E \times R_L}$$

or:

$$\mathbf{Av = g.m. \times R_L}$$

This means that if:

I_E = 1mA; then the g.m. = 40 mS (or mA/V)

I_E = 2mA; then the g.m. = 80 mS (or mA/V)

Worked Example

If a transistor with an h_{fe} of 100 is used with a 3k9 load resistor to form a resistive loaded amplifier, then if the emitter current is 1mA, determine the voltage gain:

$$A_v = 3{,}900 \times 40 \times 1 \times 10^{-3} = \mathbf{156}$$

Notice that this method of calculation does not involve the use of h_{fe}, but instead relies upon the actual emitter current flowing. Because of the spread of h_{fe} during manufacture, this approach gives a more accurate result.

Also:

The power gain	=	$h_{fe}^2 \times R_L/R_{in}$
	=	$h_{fe} \times h_{fe} \times R_L/R_{in}$
	=	$h_{fe} \times A_v$ (since $h_{fe} \times R_L/R_{in} = A_v$)

Or Power Gain:

$$\mathbf{A_p = g.m. \times R_L \times h_{fe}}$$

In this example the power gain is:

$$A_p = 100 \times A_v = 15{,}600$$

Summary of Stage Gain of Common Emitter Amplifier

The emitter resistance r_e is:

$$\mathbf{r_e = \frac{25mV}{I_E\ mA}}$$

The mutual conductance g.m. of a bi-polar transistor is:

$$\mathbf{g.m. = 1/r_e = \frac{I_E}{0.25V}\ mS(mA/V)}$$

or:

$$\mathbf{g.m. = 40\ I_E\ (mS)}$$

The voltage gain:

$$A_V = g.m. \times R_L$$

The power gain:

$$A_P = g.m. \times R_L \times h_{fe}$$

Self-assessment questions *(answers page 301)*

1) The mutual conductance of a transistor is the measure of how:
   a) the base current affects the collector current
   b) the base current affects the emitter current
   c) the base voltage affects the emitter current
   d) the base voltage affects the collector current.

2) If the quiescent collector current in a common emitter is set at 2mA the the emitter resistance R_e will be:
   a) 12.5 ohms   b) 25 ohms   c) 12.5k ohms   d) 25k ohms

3) If the current gain h_{fe} for the transistor in question 1 is 100, then the input resistance R_{in} (h_{ie}) will be:
   a) 125 ohms   b) 250 ohms   c) 1250 ohms   d) 12500 ohms

4) The stage gain of a common enitter transistor amplifier with a load resistance of 2k2 ohms is 100. If the h_{fe} is 120, then the input resistance R_{in} (h_{ie}) will be:
   a) 2200 ohms   b) 2400 ohms   c) 2640 ohms   d) 3300 ohms

5) A common emitter amplifier has a standing collector current of 1.5mA. If the collector load resistor is 1k8 ohms then the stage gain A_v will be:
   a) 47.9   b) 108   c) 10800   d) 108000

6) A transistor with an h_{fe} of 180 and an emitter resistance re of 25 ohms is connected as a common emitter amplifier with a collector load resistance of 2k2 ohms. The stage gain A_v will be:
   a) 88   b) 158.4   c) 15840   d) 88000

7) The power gain for the common emitter amplifier in question 5, will be:
   a) 1584   b) 15840   c) 28512   d) 88000

8) A common emitter transistor amplifier has a stage gain A_v of 40 when the standing (quiescent) collector current is 1mA. If the bias is altered to produce a quiescent collector current of 2mA then the stage gain will be:
   a) 40   b) 60   c) 80   d) 120

Fault-finding on a single-stage bi-polar transmitter amplifier

Study Note

Before studying this section you will need to be familiar with The common emitter connection, p.217, and Single-stage audio amplifier, p.229.

Introduction

Before we can begin to find faults in a bi-polar amplifier we will probably need to refresh our memory as to the purpose and function of all the components in the circuit. A typical single stage bi-polar transistor amplifier and a brief description of the circuit is given below.

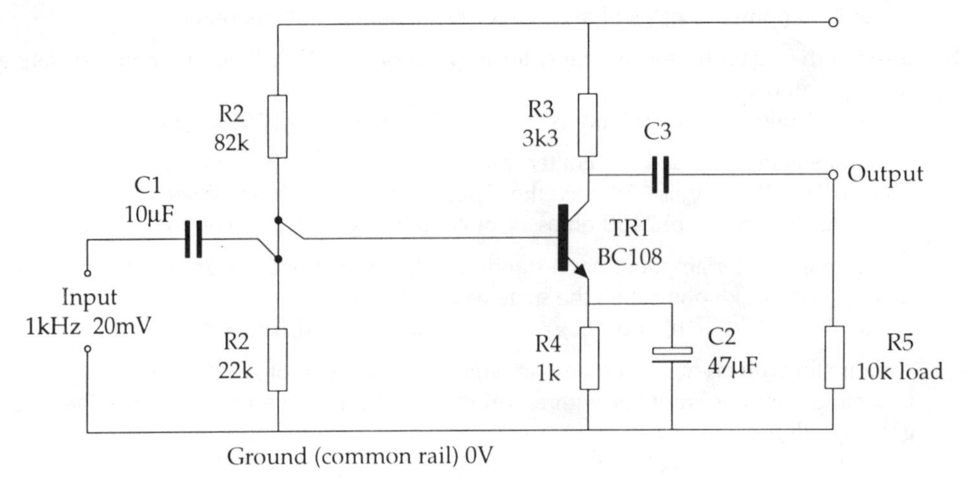

A Bi-polar Single-stage Transistor Amplifier

In the amplifier above:

R1 and R2 are the potential divider bias resistors, providing base bias to start the transistor action. R3 is the collector load resistor, across which the changing collector signal current produces a changing output voltage. R4 is the emitter stabilisation resistor; it works in conjunction with R1 and R2 to stabilise the base bias current against changes in the transistor's temperature.

C2 is the emitter by-pass capacitor; the changing collector-emitter current causes an a.c. variation in the emitter voltage and produces negative feedback, lowering the gain of the stage. C2 is chosen so that it has a low opposition to a.c. (low reactance) and it by-passes any a.c. signal on the emitter to ground, preventing the NFB and increasing the stage gain. C1 and C3 are the input and output coupling capacitors respectively. These couple the a.c. signal in and out of the amplifier, but block d.c. and preserve the d.c. voltages and currents that make the transistor work.

R5 is the load resistor for the amplifier. The a.c. output signal cannot be coupled out of the amplifier by C3 into an open circuit (there must be a d.c. path to each side of a coupling capacitor for it to work).

a.c. and d.c. Conditions

There are two sets of conditions required before any transistor amplifier will amplify:

1) d.c. voltages on the base, emitter and collector to make the transistor work. These are called the **static** conditions.

2) a suitable a.c. signal coupled into and out of the amplifier. These are called the **dynamic** conditions.

Testing a Single Stage Audio Amplifier

When fault-finding on the amplifier these conditions must be investigated separately. That is, the transistor will have to be tested to see if it has all the required d.c. operating conditions and then tested to see if it can handle an a.c. signal.

The first step is to power up the amplifier, apply a suitable input signal and look to see if there is an output signal. A suitable test circuit is shown below.

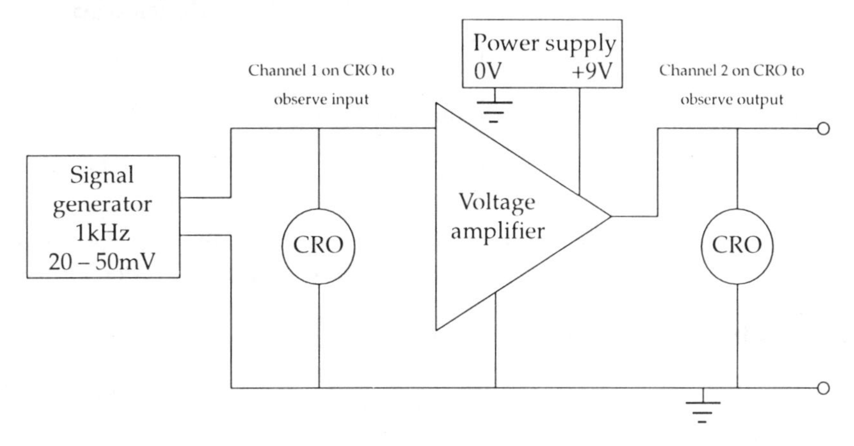

Test Circuit for Single-Stage Audio Amplifier

Symptoms

The first step is to observe the symptoms. For the amplifier above the symptoms and possible causes could be:

1) *No output*: this could be a problem with either the a.c. or d.c. operating conditions.

2) *Low Gain*: use the oscilloscope to measure the voltage gain of the amplifier. If the actual gain of the amplifier is 10 and it should be 100 or more then this fault is easily spotted; if however the actual gain is 80 then it may be overlooked (**observe the symptoms carefully**). This could also be a problem with either the a.c. or d.c. operating conditions.

3) *Distortion*: if the output waveform is not the same shape (remember it should be larger and be 180 degrees out of phase), then distortion has been introduced.

The actual distortion must be carefully observed in order that the cause may be determined. The usual cause of distortion is that the base bias current is incorrect. Typical distorted outputs are shown below.

a) Normal Operation

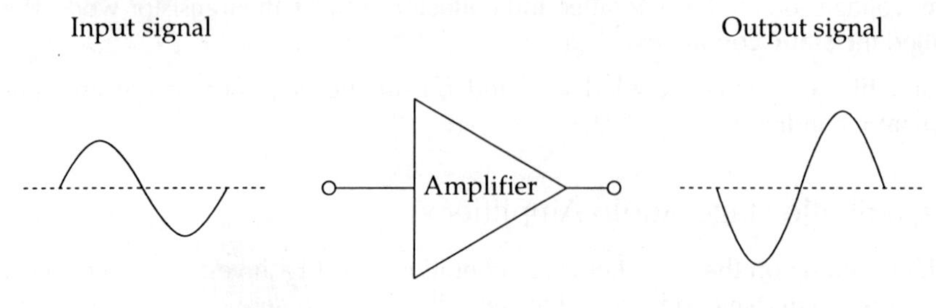

Linear Operation - no distortion

b) Bias too large

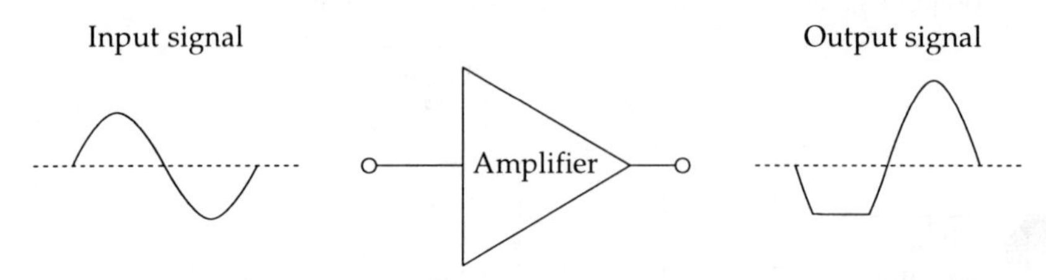

Bias too large. The output waveform is clipped on the negative half cycle. Positive swing of the input signal is driving the transistor into saturation

c) Bias too small

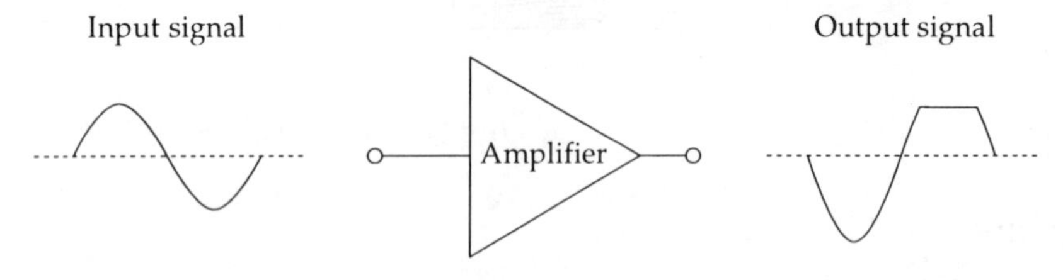

Bias too small. Negative swing of the input signal drives the transistor into cut-off i.e. base-emitter junction is no longer forward biased and the collector current is cut off

d) Too large an input signal.

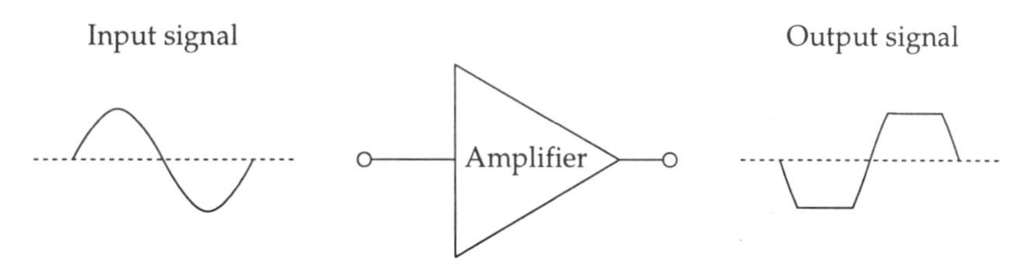

Input signal too large. The transistor is driven into saturation and into cut-off. If the bias is correct then this will result in equal clipping of the positive and negative peaks

Having observed the symptoms we now need to decide if it is the static or dynamic conditions that are at fault.

The d.c. conditions are set up by the PDER circuit (s**ee Section 2 Information and Skills Bank** Single-stage audio amplifier, p.229).

Faultfinding on the PDER circuit

In this section we are only concerned with the d.c. operating (static) conditions for the transistor. Firstly we need to know what the normal voltages should be in a PDER circuit.

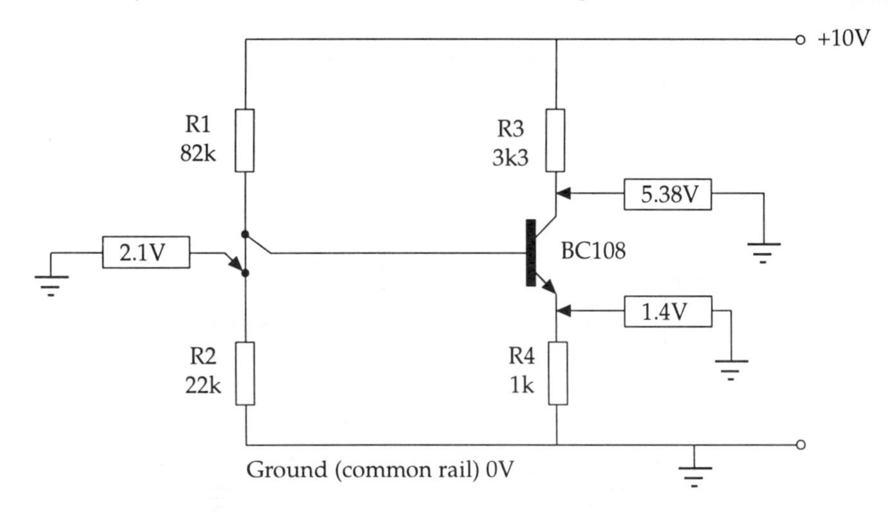

Normal voltages in a PDER circuit

Most of the failures in this circuit will be the result of open circuit resistors or open circuit transistor leads (the most common cause of this is soldered dry joints), or open circuits on the internal transistor leads.

Generally short circuits (s/c) are internal short circuits within the transistor: e.g. base-emitter s/c or base-collector s/c or collector-emittor s/c.

R1 Open Circuit

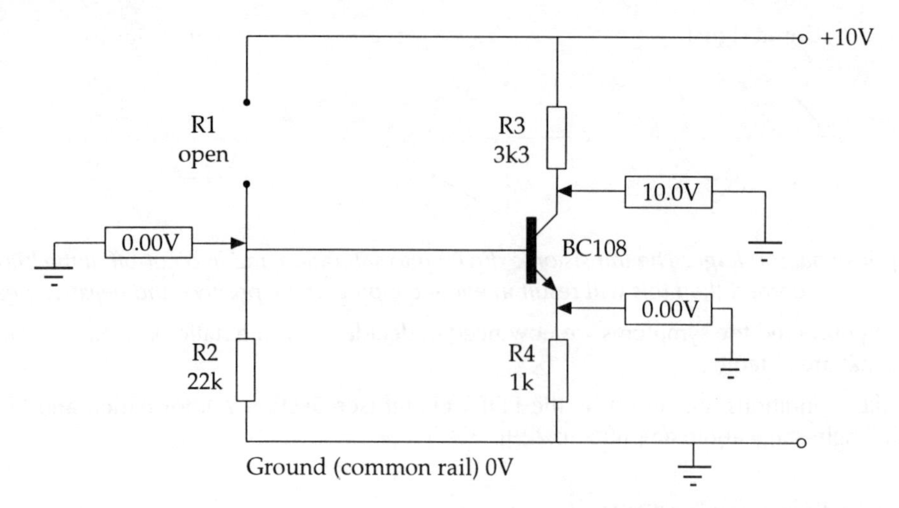

The base is at 0V because R2 is putting the base at ground potential. The transistor is cut off – no base current – no transistor action

R2 Open Circuit

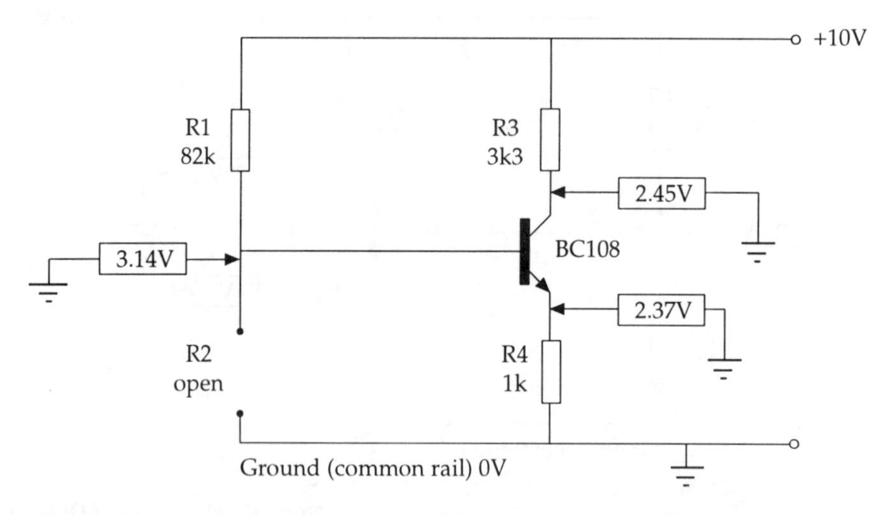

The base voltage can rise and put too much bias on the base. The transistor saturates and R3 and R4 act as a potential divider across the supply. Therefore for the values given the collector voltage is approximately 1/4 of the supply

R3 Open Circuit

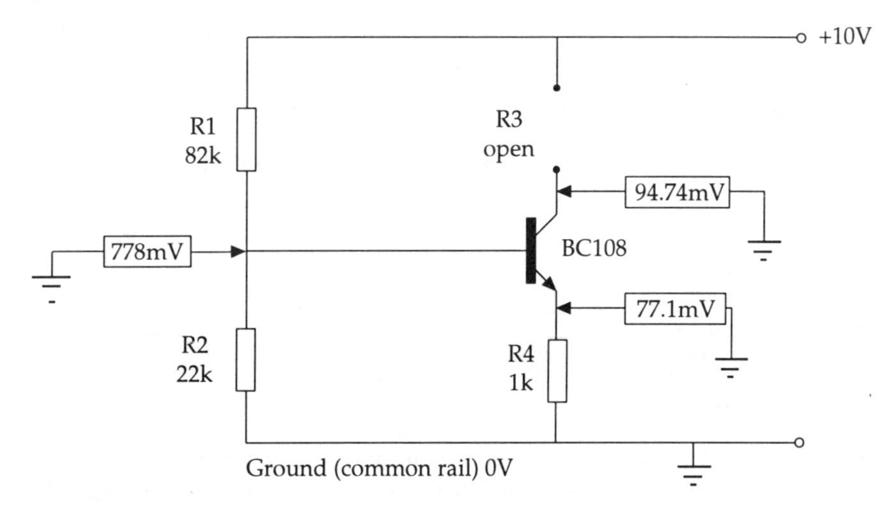

With R3 open circuit then V_C will be approximately 0.7V below V_B. Because there is no transistor action the base voltage falls. The transistor input resistance falls which lowers the effective value of R2. The base is forward biased and when a meter is connected, it completes the circuit from the collector to the ground, and a small voltage is recorded

Collector Open Circuit Internally

This will create the same effect as R3 being open circuit except that the collector voltage will read the full 10V.

R4 Open Circuit.

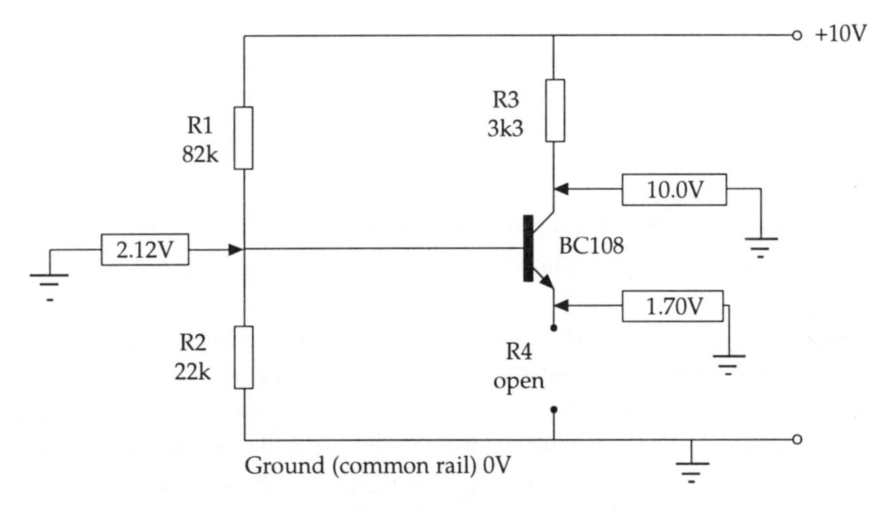

With R4 open current is prevented from flowing in the transistor. The meter completes the circuit to ground and the emitter voltage will be 0.42V below the base with the modern high resistance digital meters. Older instruments will create a volt drop of 0.7V across the base-emitter junction.

Base or Emitter Open Circuit Internally

This will create the same effect as R4 being open circuit except the emitter voltage now reads zero. The meter cannot now complete the circuit.

Internal Short Circuits

With these faults two of the transistor connections will have exactly the same potential. For the circuit given the readings will be:

	V_B	V_E	V_C
Base/collector s/c	2.88V	2.11V	2.88V
Base/emitter s/c	115mV	115mV	10.0V
Emitter/collector s/c	2.12V	2.33V	2.33V

Dynamic Testing

When we are satisfied that the transistor is working and is biased correctly, then the symptoms of no output or low gain must be a dynamic or a.c. fault. To test the circuit dynamically, we need to inject test signals from a signal or function generator.

Coupling the Function Generator

If we inject test signals directly onto the base or collector of the transistor, then, to prevent the d.c. conditions on the transistor from being upset, we should use a coupling capacitor in series with the signal lead of the generator.

The value of this capacitor is important. The reactance (opposition to a.c.) will be greatest at low frequencies. Therefore a large capacitor of at least 80μF will be required to inject a frequency of 20Hz into a typical bi-polar amplifier. For frequencies of 100Hz upwards a 16μF coupling capacitor would be adequate.

Remember however that if the generator coupling capacitor is in series with the circuit's coupling capacitor then the effective capacitance may be half or less, (i.e. an 8μF in series with an 8μF has an overall capacitance of 4μF). This may prevent you coupling in sufficient signal at low frequencies!

Dynamic Faults

On the single-stage amplifier the possible faults may be:

C1 or C2 open circuit (o/c). This will prevent the signal being coupled in or out of the circuit.

To test: either bridge the component (temporarily connect another capacitor in parallel) with a known good component;

OR: Inject a 1kHz signal and use the CRO on either side of C1 to see that the signal is getting through. If C1 is OK, then connect the CRO on either side of C2 to test C2.

If C1 is short circuit (s/c) then this will affect the bias on the transistor and will cause varying degrees of distortion depending upon the signal source.

C2 s/c: depending upon the amplifier load this will lower the gain slightly and will alter the collector potential but the output may only distort when amplifying large input signals.

Emitter by-pass capacitor C3 o/c: this will give rise to a lot of NFB and the gain of the stage will fall to about 3.0.

Emitter by-pass capacitor C3 s/c: if this goes completely s/c then the bias current becomes too large and the transistor saturates.

	V_B	V_C	V_E
For the circuit under test typical voltage would be:	776mV	130mV	000mV

The collector voltage has fallen to Vcesat = 0.13V and for the normal range of input signal the transistor will remain saturated.

Summary of Fault-finding on a Single-stage Bi-polar Amplifier

- The bi-polar amplifier has two sets of conditions: the d.c. conditions required to make the transistor work (static) and the a.c. conditions required to enable it to amplify an a.c. signal (dynamic).
- When testing an amplifier these conditions are investigated separately: a dynamic test will determine if the amplifier can amplify an a.c. signal; the static test will determine whether the d.c. voltages and currents are correct.
- The first most important step when fault-finding is to observe any symptoms (have a good visual inspection of the circuit board first): i.e. low gain, distortion, no output, etc.
- When dynamic testing using a signal or function generator. always use a suitable coupling capacitor (16μF will be suitable for frequencies of 100Hz upwards) in series with the signal lead. The generator may output a small d.c. voltage which can upset the bias conditions.
- With static resistance tests, switch off the power supply and observe the following precautions:
  - i) check that the resistor to be measured is not shunted by (in parallel with) another low resistor or the low resistance of a forward biased diode: e.g. the base-emitter junction of a transistor.
  - ii) when checking diodes or transistors the ohmmeter should provide sufficient voltage to forward bias the diode junction.

Self-assessment questions *(answers page 301)*

A single stage resistive loaded transistor amplifier is constructed using a bi-polar transistor with an h_{fe} of 100 and fed from a 12V supply. The transistor is biased using a PDER circuit. The emitter resistor is by-passed with a 100μF capacitor and the input and output capacitors are 10μF.

The normal voltages on the pins of the transistor are as follows:

$V_B = 2.16V$ $V_C = 7.5V$ $V_E = 1.4V$

In each of the questions below a fault condition exists. State in each case which are the faulty component/s and what the fault is.

The voltages measured on the pins of the transistor are:

	V_B	V_C	V_E
Fault 1	324μV	12V	00.0V
Fault 2	2.97	12V	2.55V
Fault 3	139mV	12V	139mV
fault 4	2.97V	12V	0.00V
Fault 5	3.63V	2.92V	2.85V
Fault 6	3.37V	3.37V	2.6V

Fault 7: All the d.c. voltages are normal but the amplifier has very low gain.

Fault 8: All the d.c. voltages are normal, the input coupling capacitor is OK, but there is no output.

The field effect transistor

Study Note

Before studying this section you will need to have read Part 1 Section 2 Simplified semiconductor theory, p.65.

Introduction

The uni-polar transistor is an alternative to the bi-polar. Its principle of operation is different to the bi-polar and this gives it different properties.

The Junction Gate Field Effect Transistor (JUGFET)

An N–channel JUGFET is made from a bar of N-type semiconductor with heavily doped P regions on two opposite sides. An electrical connection is made between the two P-type regions internally but not through the N-type channel material.

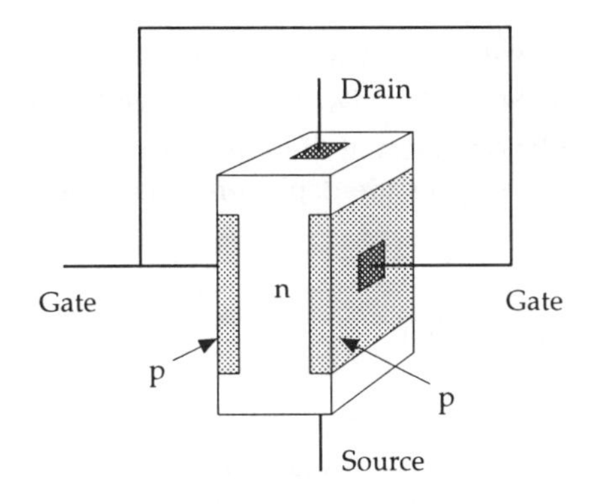

The N-channel FET

Electrical connections are made in three places to create the terminals for the source, drain, and gate.

The N-channel JUGFET

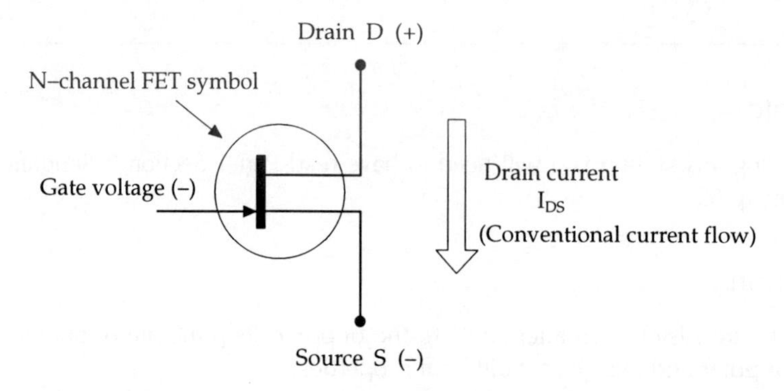

The gate controls the drain-source flow of current

The source (S) is the end of the channel and where majority carriers (electrons in N-type material) enter the channel. The drain (D) is the other end of the channel and where majority carriers leave the channel. The gate (G) consists of two P-type regions on either side of the channel connected together internally. The gate controls the flow of current in the channel.

The drain is made positive with respect to the source and the gate source junctions are always reverse biased in a JUGFET: i.e. negative on the gate relative to source.

The reverse bias forms depletion layers at the P-N junctions. Because of the difference in doping densities between the regions the depletion layer extends deeper into the channel than into the gate regions.

Basic Action of N-channel FET

Gate-Source Voltage Zero

With the drain a few volts positive with respect to the source and the gate voltage zero, a current flows in the N channel. This current is limited by the resistance of the channel and is further controlled by the drain resistor R_D. Typically this current will be a few mA.

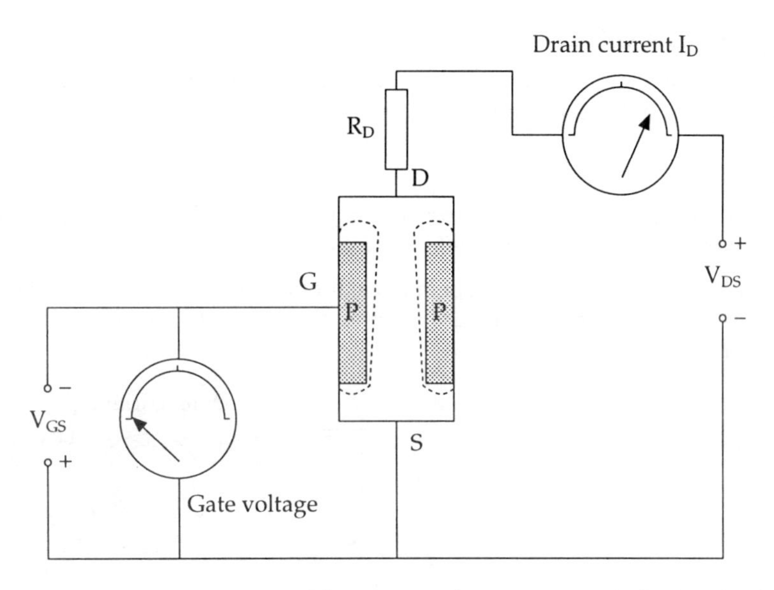

Without gate voltage a current is flowing in the channel

With the gate at the same potential as the source then there is a reverse bias between the gate and the channel. This reverse bias increases as you go from the source towards the drain, widening the depletion area near to the drain.

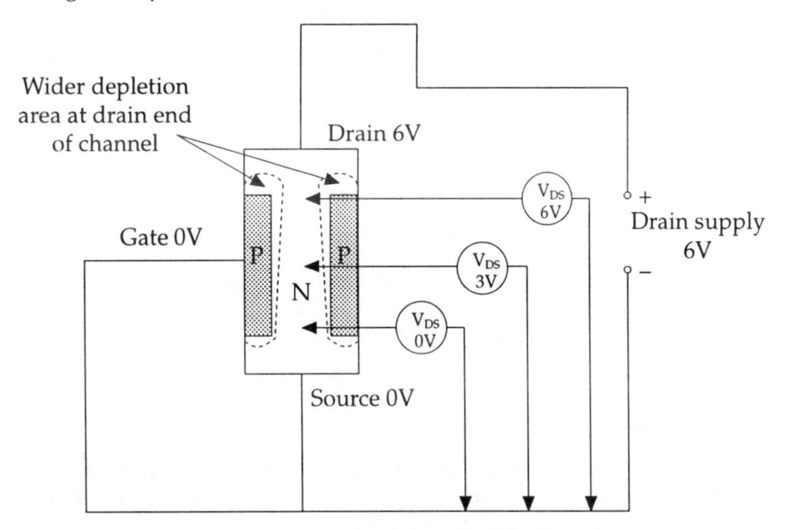

As you go up the channel from source to drain, the N channel becomes more positive with respect to the source, but the whole of the gate area remains at 0V.

This creates a reverse bias between the gate and the channel. (The N-type channel is positive compared to the P-type gate.)

Because in relation to the source the gate is at 0V for its whole length and the channel potential increases going towards the drain, the reverse bias increases as you go towards the drain. This action increases the depletion area at the drain end and restricts the width of the channel.

Increasing the Drain Voltage

If now the drain voltage is increased, the drain current will increase. But the reverse bias and depletion area at the drain end is also increased.

Pinch Off

Eventually the depletion area will extend across the width of the channel and prevent any further increase in drain current. This effect is known as pinch off.

At pinch off the negative electrons repelling each other prevent the channel being closed completely. The drain current however becomes almost constant and independent of any further increase in drain voltage.

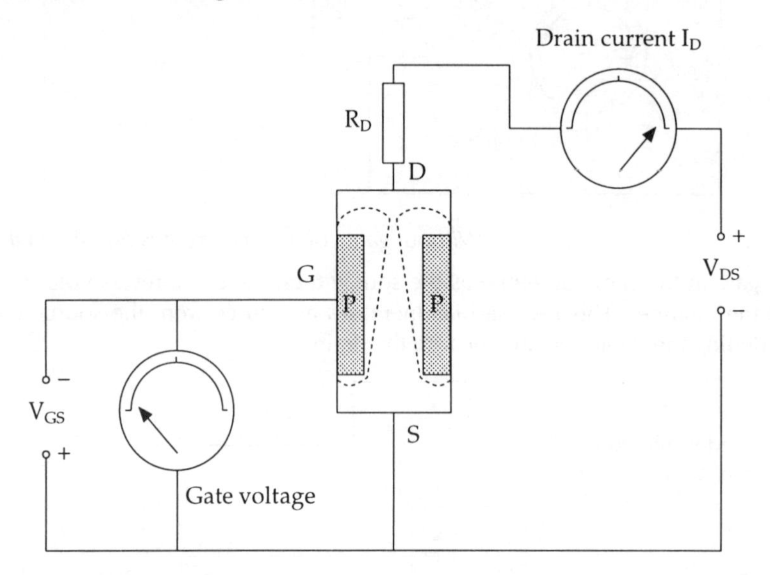

Increasing the drain voltage causes the channel to pinch off. Gate voltage zero

Therefore at pinch off the current through the channel I_D becomes constant and does not depend upon the drain voltage. That is, increasing the drain voltage, after pinch off, will not increase the drain current. If however the drain voltage in increased too much, then the device will break down.

Effect of Gate Voltage

If the drain voltage is kept constant and a negative voltage is applied to the gate, the reverse bias will be increased and the depletion layer extended further into the channel. This reduces the amount of drain current flowing.

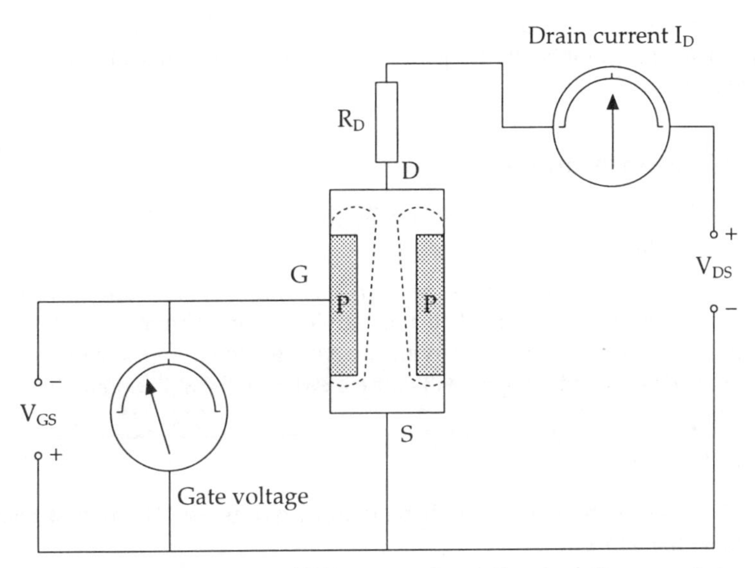

With gate voltage, the depletion area is increased and the channel current is reduced. V_{DS} held constant.

As the gate source voltage is made more negative then the depletion layers extend deeper into the channel. This means that there are fewer majority current carriers (electrons) in the channel and the channel current decreases.

The effect of increasing the negative gate voltage is to cause pinch off to occur at lower values of drain voltage. It also restricts the width of the channel, reducing the maximum drain current that can flow, for a particular value of drain voltage.

A Large Gate-Source Voltage Causes Cut-Off

If the gate voltage is sufficiently negative then the depletion layers meet across the channel and cut the channel off completely!

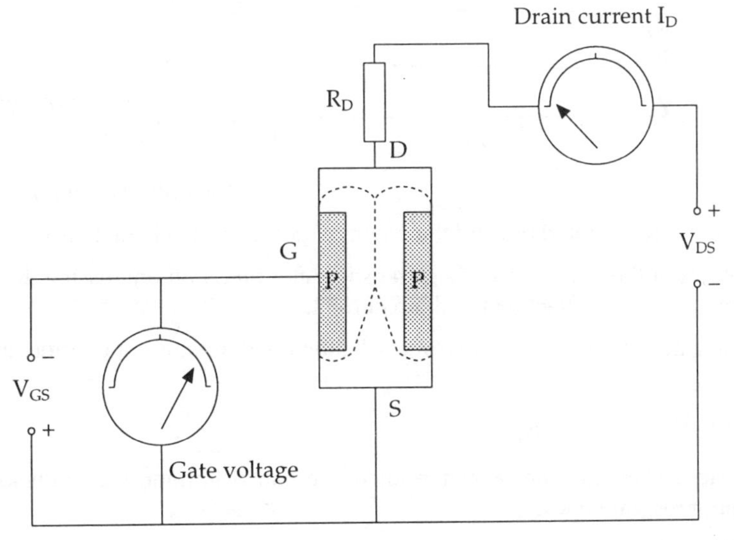

With a large gate voltage the depletion area increases and ***cuts the channel current off***

Because the gate to channel connection is a reversed biased junction the input resistance is very high. Therefore practically no gate current flows into a JUGFET and the device is **voltage operated**.

The JUGFET as an Amplifier

The JUGFET requires a small gate bias voltage to reduce the drain current from a maximum to a steady value.

A small a.c. signal applied to the gate can then vary this bias and cause a variation in the drain current flowing in the channel. If this varying drain current flows through a load resistor R_D in the drain circuit then a varying a.c. voltage will be present on the drain. Typically a swing of 1V or so on the gate will cause a swing of 5V at the drain.

FET Characteristics

With the FET transistor there are two characteristics of interest. The output characteristic and the mutual characteristic.

The Output Characteristic (Drain Characteristic)

This shows the relationship between the drain current and the drain-source voltage for different values of gate-source voltage.

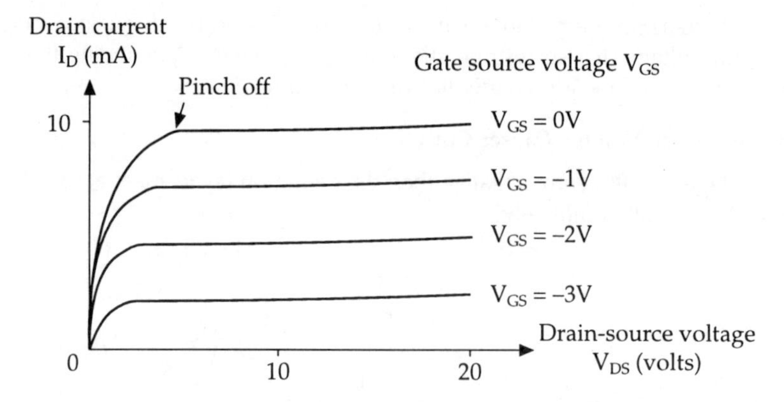

The output or drain characteristic

The FET is normally operated beyond the pinch off point on the characteristic.

Pinch off occurs at the knee of the characteristic. After pinch off, increasing the drain voltage does not have much effect on the drain current.

Changing the gate-source voltage however will alter the drain current considerably.

The Output Resistance R_{DS}

From the output characteristic, after the knee, we can determine the drain-source resistance or output resistance R_{DS}.

$$R_{DS} = \frac{\text{change in drain-source voltage } V_{DS}}{\text{corresponding change in drain current } I_D} \quad \text{(ohms)}$$

$$R_{DS} = \frac{\delta V_{DS}}{\delta I_D} \quad \text{(for a constant value of } V_{GS}\text{)}$$

Typically R_{DS} is between 40k and 1M ohm.

The Mutual Characteristic

This characteristic shows how the drain current varies with gate voltage. The gate voltage is the input voltage and the drain current the output current. This characteristic therefore shows the mutual effect between the input and output of the device.

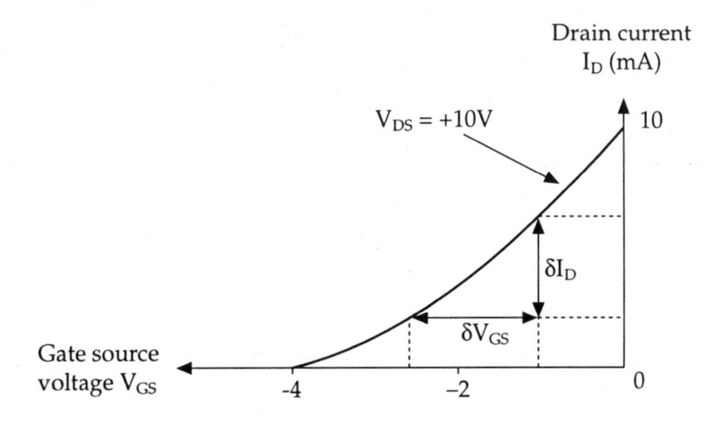

The mutual characteristic

Mutual Conductance g.m.

From the mutual characteristic we can determine the mutual conductance g.m. for the device.

$$\text{g.m.} = \frac{\text{change in drain current } I_D}{\text{corresponding change in } V_{GS}} = \frac{\delta I_D}{\delta V_{GS}} \quad \text{(mA/V or mSiemens)} \quad (V_{DS} \text{ held constant})$$

Typically the g.m. is between 1 to 7 mA/V or mS.

From these two parameters we can calculate a third.

Amplification Factor μ

This parameter is the relationship between the input (V_{GS}) and the output (V_{DS}) voltages.

If we multiply g.m. x R_{DS}:

$$\frac{\delta I_D}{\delta V_{GS}} \times \frac{\delta V_{DS}}{\delta I_D} = \frac{\delta V_{DS}}{\delta V_{GS}} = \mu \text{ (no units)}$$

This is the amplification factor!

For example, if the g.m. is 3mS and R_{DS} is 40k then the amplification factor μ:

$$\mu = 3 \times 10^{-3} \times 40 \times 10^{+3} = 120$$

Input Resistance

The input resistance of a JUGFET is very high due to the reverse biased gate to channel P-N junction. Typically, it is greater than 100M ohms. For this reason practically no input current flows into the FET and the device is purely voltage operated.

In the JUGFET maximum current flows with no voltage on the gate. To use the JUGFET as an amplifier we must reduce the drain current to a steady value before we can apply the a.c. signal. The output characteristic is used to plot a load line and to determine the operating point.

The FET Single-stage Audio Amplifier

To operate the junction FET as an amplifier a small steady bias voltage must be applied to the gate. If there is no bias on the gate then maximum drain current flows. Therefore the N-channel FET is normally operated with its gate negative with respect to the source, to limit the flow of drain current.

Biasing the FET

The junction FET is biased by including a resistor between the source and ground as shown in the diagram.

The source current flowing in resistor R3 produces a volt drop across R3. This makes the source more positive than the ground or common rail.

If the gate is now connected to ground then the gate must be less positive than the source, or in other words the gate must be negative with respect to the source. This small difference in potential between the gate and source provides the necessary bias and reduces the flow of current in the channel.

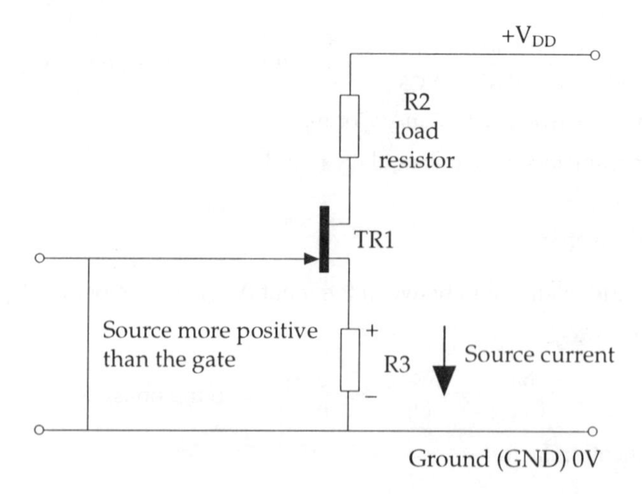

Source resistor R3 provides bias

The input signal is applied between the gate and the ground. Therefore if the gate is simply shorted to ground then this will short out the input signal!

If we connect the gate to the ground via a resistor then the gate will have the same potential as the ground in terms of d.c.

Remember that no current flows into the gate of a FET: it is a voltage operated device, so that although the source and gate are at different potentials, because of the extremely high input resistance, **no current flows**.

If **no current** is flowing in this gate resistor then there can be **no voltage dropped across it**.

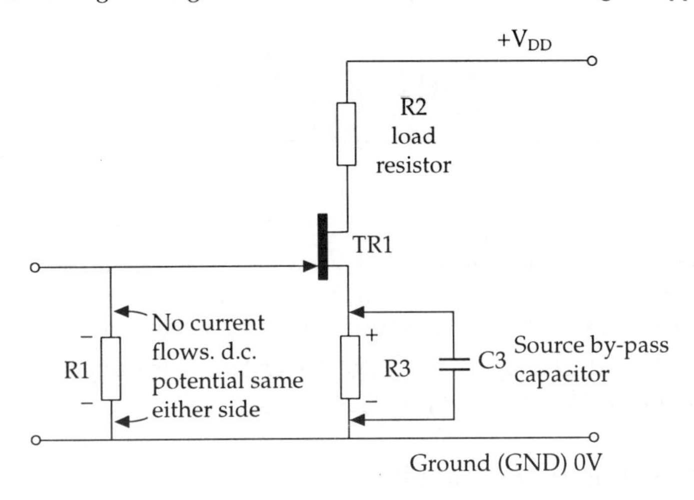

Resistor R1 connects the gate to ground in terms of d.c.

To prevent attenuation (reduction) of the input signal then R1 needs to be a large value and is typically 1M ohm.

Remember, if no current flows, it does not matter how large R1 is, both ends of R1 will still be at the same potential in terms of d.c.

Negative Feedback (NFB)

When the input signal varies the gate voltage, the current in the drain and the source will vary in sympathy. The varying source current will produce a varying source voltage and this has the effect of reducing the effective input signal.

This is exactly the same mechanism that produced negative feedback and lowered the gain of the stage in the common emitter amplifier (see **Section 2 Information and Skills Bank** The Single-stage Audio Amplifier, p.229). Just as in the common emitter amplifier, to overcome the NFB a large capacitor (C3) is connected across the source resistor.

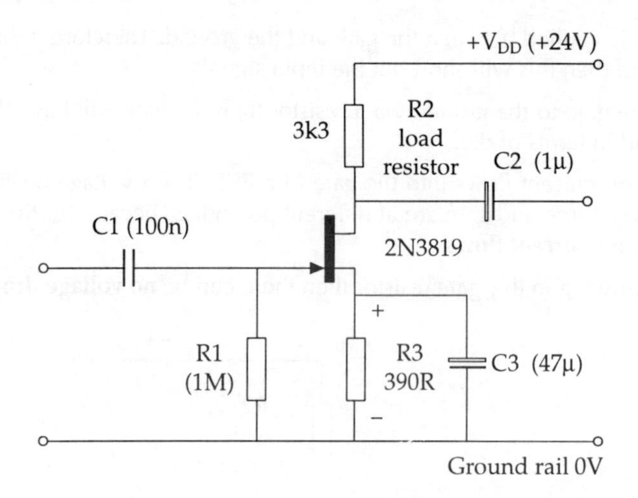

A Typical N-channel junction FET amplifier

A typical junction FET amplifier is shown above with source resistor bias. If improved stabilisation of the operating point is required then a potential divider bias network can be employed instead. This is described in Single stage audio amplifier (p.229).

The capacitors C1 and C2 couple the a.c. signal in and out of the amplifier and prevent the d.c. conditions on the FET from being upset.

The a.c. Operation of the Circuit

R2 is the load resistor. The a.c. input voltage varies the drain current. The varying drain current causes a large variation in the drain voltage. The large voltage swing at the drain is opposite in phase, but varies in sympathy with the input signal.

In other words, a small positive swing at the input produces a larger negative swing in voltage at the output.

Calculation of the Stage Gain of an FET Amplifier

In the circuit below the junction FET has a g.m. of 2.4mS

If an input signal of 1.0V peak to peak is applied to the input, determine:

a) the peak to peak output voltage

b) the stage gain of the amplifier.

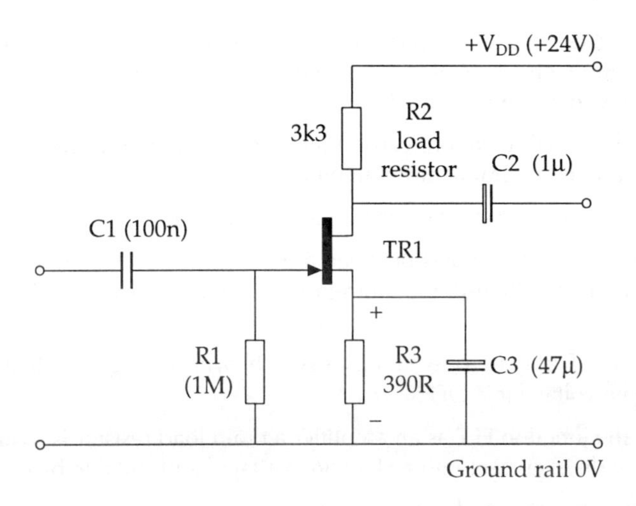

A Typical N-channel junction FET amplifier

Drain Current: the input voltage V_{GS} will cause a peak to peak drain current of:

$$\text{g.m.} \times V_{GS} = 2.4\text{mA/V} \times 1.0\text{V}$$

Therefore the peak to peak swing of drain current will be:

$$\text{peak to peak swing of } I_D = 2.4\text{mA}$$

Drain Voltage: this current flows through the load resistor Rd and therefore the peak to peak voltage produced across R_D will be:

$$2.4 \times 10^{-3} \times 3.3 \times 10^{+3} = 7.92\text{V peak to peak}$$

The stage gain will therefore be:

$$\text{Stage gain} = \frac{\text{the peak-to-peak swing in output voltage (V)}}{\text{the peak-to-peak swing in input voltage (V)}} = \frac{7.92}{1} = \mathbf{7.92}$$

Note: the stage gain can also be found from g.m. × R_L:

$$\text{Therefore: stage gain} = 2.4 \times 10^{-3} \times 3.3 \times 10^{+3} = \mathbf{7.92}$$

Summary of the Field Effect Transistor

- The junction FET is a uni-polar device and conventional current flows between the drain and source in one type of semiconductor material only.
- The junction FET is normally ON and a gate voltage is required to reduce or cut off the drain current.
- With zero gate-source voltage the effect of the drain voltage is to create a channel that narrows at the drain end due to the increasing width of the depletion layer.
- At a particular value of drain voltage the channel is sufficiently narrow to cause pinch off.
- At pinch off the drain current becomes independent of the drain voltage. In other words, increasing the drain voltage will not increase the drain current.

- If a negative gate voltage is applied to an N-channel FET then it will have the effect of increasing the reverse bias along the gate, further increasing the depletion area and reducing the width of the channel.
- The gate voltage will therefore cause pinch off to occur at a lower drain voltage and reduce the amount of drain current flowing.
- With sufficient gate voltage the channel and drain current are cut off completely.
- Because there is a reverse bias between the gate and the channel the input resistance is very high; practically no input current flows into the device and it is voltage operated.
- The junction FET may be operated as a switch; with zero gate voltage it is ON, with sufficient gate voltage it is OFF.
- To operate the junction FET as an amplifier a drain load resistor is required (to convert the changing drain current into a changing voltage) and suitable bias.
- The bias may be obtained using a self or automatic bias circuit. A resistor in the source develops a voltage which makes the source positive with respect to the gate, or conversely the gate negative with respect to the source.
- Self bias is only effective so long as there is a d.c. path between the gate and ground. This may be provided by a 1M ohm resistor. Since no current flows in the gate circuit then there can be no voltage drop across this large resistor, leaving the gate at the same potential as ground.
- The stage gain of a junction FET amplifier is determined by:

$$\text{Stage Gain } A_v = g.m. \times R_L$$

Self-assessment questions (answers page 302)

1) In the P-channel junction FET the main current flow is carried by:
   a) electrons and holes.
   b) electrons only.
   c) holes only.
   d) minority current carriers.

2) The junction FET is a:
   a) voltage operated device.
   b) a current operated device.
   c) a voltage and current operated device.
   d) a voltage or current operated device.

3) The typical input resistance of a junction FET is:
   a) 10k ohms b) 100k ohms c) 500K ohms d) 1M ohms

4) To operate an N-channel JUGFET, with respect to the source:
   a) the drain is positive and the gate negative.
   b) the drain is negative and the gate positive.
   c) both the drain and the gate are negative.
   d) both the drain and the gate are positive.

5) With the correct polarity of drain-source voltage a N-channel JUGFET will:
   a) be conducting and a positive gate voltage turns it off.
   b) be non-conducting and a positive gate voltage turns it on.
   c) be conducting and a negative gate voltage turns it off.
   d) be non-conducting and a negative gate voltage turns it on.

6) If the JUGFET is to be operated as a linear amplifier then:
   a) it is operated before the point of pinch off.
   b) it is operated at the point of pinch off.
   c) it is operated beyond the point of pinch off.
   d) pinch off must not be allowed to occur.

7) The mutual characteristic of a JUGFET is the relationship between:
   a) the gate voltage and the drain current.
   b) the gate voltage and the drain voltage.
   c) the drain voltage and the drain current.
   d) the gate current and the drain current.

8) If the output resistance of a JUGFET transistor is 50k ohms and the g.m. is 2 mS, then the amplification factor μ will be:
   a) 50 b) 100 c) 200 d) 1000

Fault-finding on a single-stage FET amplifier

Study Note

Before studying this section you will need to be familiar with The field effect transistor, p.255. Fault-finding on a bi-polar transistor amplifier, p.246 may also be useful.

Introduction

Before we can begin to find faults in a FET amplifier we will probably need to refresh our memory as to the purpose and function of all the components in the circuit.

A typical single stage FET transistor amplificr and a brief description of the circuit is given below.

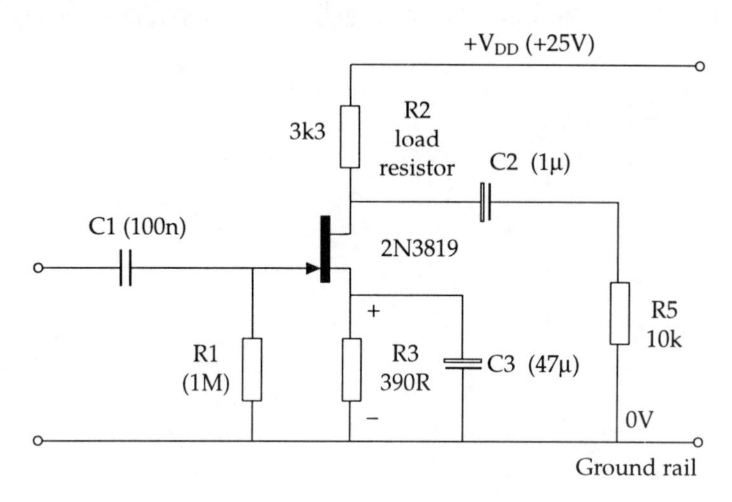

A Typical N-channel junction FET amplifier

In the amplifier above:

R3 is the source bias resistor, providing gate bias to reduce the drain current to a steady (quiscent) value.

R2 is the drain load resistor, across which the changing drain signal current produces a changing output voltage.

C3 is the source by-pass capacitor; the changing drain-source current causes an a.c. variation in the source voltage and produces negative feedback, lowering the gain of the stage. C3 is chosen so that it has a low opposition to a.c. (low reactance) and it by-passes any a.c. signal on the source to ground, preventing the NFB and increasing the stage gain.

C1 and C2 are the input and output coupling capacitors respectively. These couple the a.c. signal in and out of the amplifier, but block d.c. and preserve the d.c. voltages and currents that make the transistor work.

R5 is the load resistor for the amplifier. The a.c. output signal cannot be coupled out of the amplifier by C2 into an open circuit (there must be a d.c. path to each side of a coupling capacitor for it to work).

a.c. and d.c. Conditions

There are two sets of conditions required before any transistor amplifier will amplify:

1) d.c. voltages on the drain, source and gate to make the transistor work. These are called the static conditions.
2) a suitable a.c. signal coupled into and out of the amplifier. These are called the dynamic conditions.

Testing a Single Stage Audio Amplifier

When fault-finding on the amplifier these conditions must be investigated separately. That is, the transistor will have to be tested to see if it has all the required d.c. operating conditions and then tested to see if it can handle an a.c. signal.

The first step is to power up the amplifier, apply a suitable input signal and look to see if there is an output signal. A suitable test circuit is shown below.

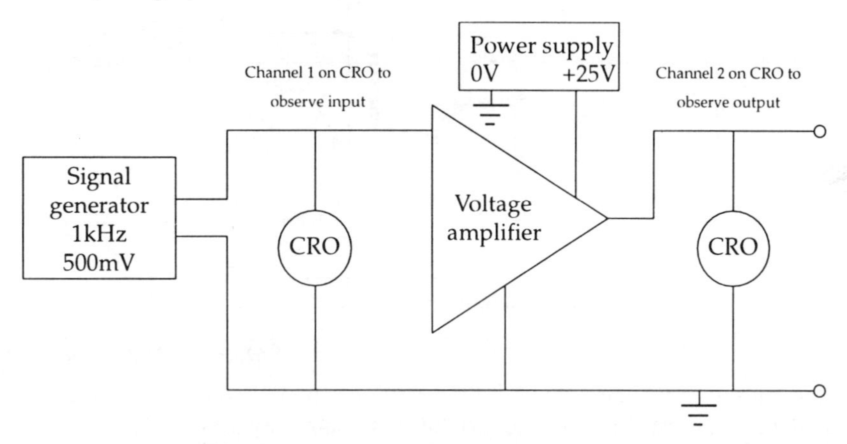

Test Circuit for Single-stage Audio Amplifier

Symptoms

The first step is to observe the symptoms and for the FET amplifier above, the symptoms and possible causes could be:

1) *No output*: this could be a problem with either the a.c. or d.c. operating conditions.
2) *Low Gain*: use the oscilloscope to measure the voltage gain of the amplifier. If the actual gain of the FET amplifier is 2 and it should be 10 or more, then this fault is easily spotted; if however the actual gain is 8 then it may be overlooked (**observe the symptoms carefully**). This could also be a problem with either the a.c. or d.c. operating conditions.

3) *Distortion*: if the output waveform is not the same shape (remember it should be larger and be 180° out of phase), then distortion has been introduced.

   The actual distortion must be carefully observed in order that the cause may be determined. The usual cause of distortion is that the base bias voltage is incorrect. Typical distorted outputs are shown in Faultfinding on a single-stage bi-polar transistor amplifier, p.246.

Having observed the symptoms we now need to decide if it is the static or dynamic conditions that are at fault.

The d.c. conditions are set up by the source resistor (see The field effect transistor, p.255).

Faults Involving the d.c. Conditions.

In this section we are only concerned with the d.c. operating (static) conditions for the FET transistor. Firstly we need to know what the normal voltages should be.

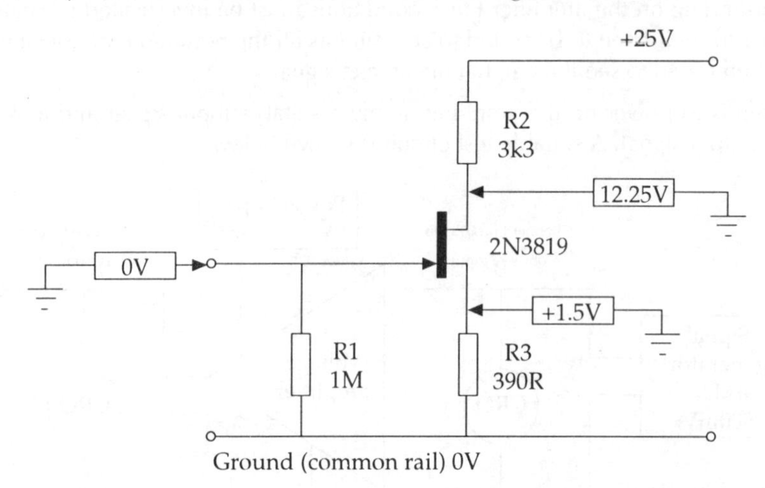

Typical Voltages in a FET circuit

Most of the failures in this circuit will be the result of open circuit resistors or open circuit transistor leads (the most common cause of this is soldered dry joints).

If the source resistor is open circuit then there will be no drain current; the drain voltage will be equal to the supply. When measuring the source voltage the measuring instrument will complete the source to ground circuit and a higher than normal reading can be obtained.

If the drain resistor is open circuit then again there will be no drain-source current. The drain, source and gate voltages will be zero.

If the source by-pass capacitor is open circuit the gain of the stage will be lowered due to NFB.

If the source by-pass capacitor is short circuit then the bias is destroyed. The drain current will be a maximum and the negative half cycles of the output signal will be clipped and severely distorted.

If R1 is open circuit then the output may be normal for very small input signals. For larger input signals however distortion occurs.

Dynamic Testing

When we are satisfied that the FET transistor is biased correctly then the symptoms of no output or low gain must be a dynamic or a.c. fault. On the single stage amplifier then the possible faults will be:

- *C1 or C2 open circuit (o/c).* This will prevent the signal being coupled in or out of the circuit.

  To test: either bridge the component (temporarily connect another capacitor in parallel) with a known good component, OR inject a 1kHz signal and use the CRO on either side of C1 to see that the signal is getting through. If C1 is OK then connect the CRO on either side of C2 to test C2.

- *If C1 is short circuit (s/c)* then this will affect the bias voltage on the gate of the transistor and will cause varying degrees of distortion depending upon the signal source.

- *C2 s/c*: depending upon the amplifier load this will lower the gain slightly and will alter the drain potential but the output may only distort when amplifying large input signals.

Summary of Fault-finding

- The FET amplifier has two sets of conditions: the d.c. conditions required to make the transistor work (static) and the a.c. conditions required to enable it to amplify an a.c. signal (dynamic).
- When testing a FET amplifier these conditions are investigated separately; a dynamic test will determine if the amplifier can amplify an a.c. signal; the static test will determine whether the d.c. voltages and currents are correct.
- The first most important step when fault-finding is to observe any symptoms (have a good visual inspection of the circuit board first): i.e. low gain, distortion, no output etc.
- When dynamic testing using a signal or function generator, always use a suitable coupling capacitor (16nF will be suitable for frequencies of 100Hz upwards) in series with the signal lead. The generator may output a small d.c. voltage which can upset the bias conditions.
- With static resistance tests, switch off the power supply and observe the following precautions:
  i) check that the resistor to be measured is not shunted by (in parallel with) another low resistor;
  ii) when checking diodes the ohmmeter should provide sufficient voltage to forward bias the diode junction.

Self-assessment questions *(answers page 302)*

A single stage resistive loaded transistor amplifier is constructed using a junction FET transistor and fed from a 25V supply. The transistor is biased using a self bias circuit. The source resistor is by-passed with a 25μF capacitor and the input and output capacitors are 100nF.

The normal voltages on the pins of the transistor with respect to ground are as follows.

$V_G = 0V$ $V_D = 12.37V$ $V_S = 1.5V$

In each of the questions below a fault condition exists. State in each case which are the faulty component/s and what the fault is.

The voltages measured on the pins of the transistor are:

	V_D	V_S	V_G
Fault 1	25V	0V	0V
Fault 2	25V	3.6V	0V
Fault 3	12.11V	1.53V	0V
Fault 4	3.69V	2.53V	2.53V
Fault 5	2.65V	2.65V	0V
Fault 6	25V	0V	0V

Fault 7 All the d.c. voltages are normal but the amplifier has low gain.

Fault 8 All the d.c. voltages are normal, the output coupling capacitor is OK, but there is no output.

Load lines

Introduction

The voltage gain of a FET or the current gain of a bi-polar transistor amplifier can be determined by using a load line drawn on the output characteristic of the devices.

All that is required to plot the load line is two points on the output characteristic. If we consider the transistor as a switch then these two points may easily be found.

The Transistor as a Switch

The transistor is also used as a switch. It has the advantage of having no moving parts and is therefore called a solid state switch. Because there are no moving parts, no bouncing contacts, the average transistor can be switched at very high rates of several MHz.

To operate the transistor as a switch then we either bias the device to be fully ON or OFF and ensure that it is not biased anywhere in between these two conditions.

The bi-polar transistor is OFF without any base bias current to start the transistor action. It is fully ON (saturated) with sufficient bias current flowing into the base.

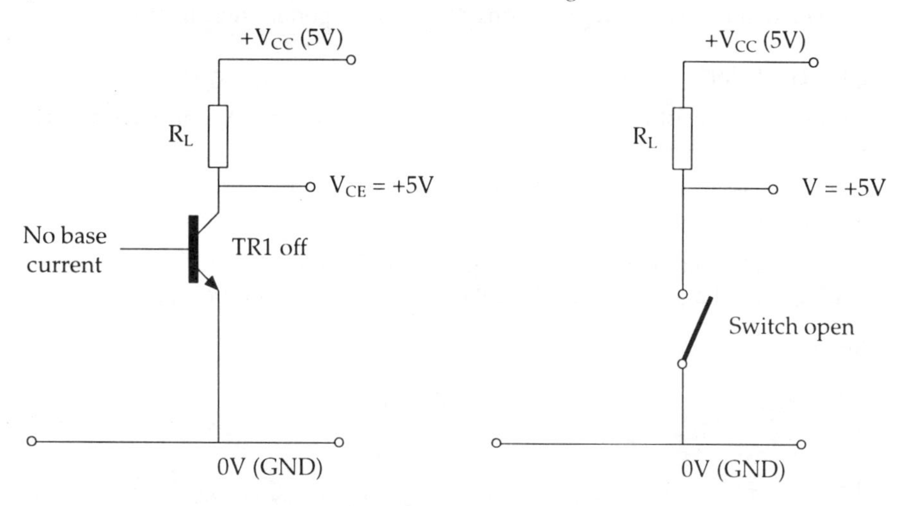

No Bias, TR1 is OFF *The equivalent switch circuit*

The bipolar transistor is fully ON (saturated) with sufficient bias current flowing into the base.

The base is connected to +5V via the base resistor R_b. The value of R_b must be chosen to fully saturate TR1.

With a large base current flowing the transistor is fully saturated and the large collector current produces a large voltage drop across the load resistor. The collector voltage therefore falls to a very low saturation value $V_{CE_{SAT}}$ (approximately 0.2V). For practical purposes this can be considered as 0V.

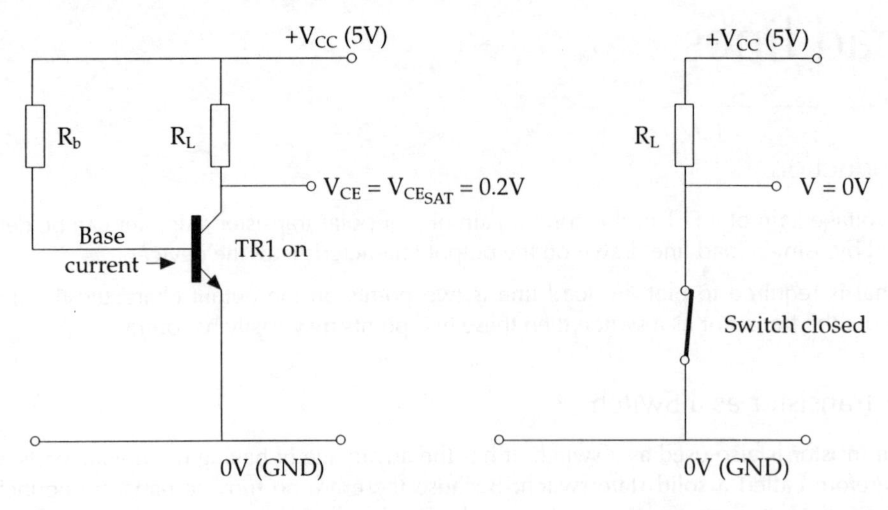

With Bias, TR1 is ON *The equivalent switch circuit*

Similarly the FET may also be used as a switch. The N-channel FET will be fully ON with no gate bias voltage (switch closed) and can be turned fully OFF with sufficient negative gate voltage. The transistor does not behave as a perfect switch because the ON resistance is not zero ohms and there is a leakage current in the OFF condition. For the purpose of plotting the load line however we will consider it to be a perfect switch.

Plotting the Load Line

If we consider the transistor as a perfect switch, supplied from 5V and having a 1k ohm load then, *when the device is OFF*:

a) the collector voltage is equal to the supply voltage (+5V).
b) the collector current is zero.

This now enables the plotting of one of the points on the X axis. That is, under these conditions if the supply voltage is 5V then the collector voltage will also be 5V.

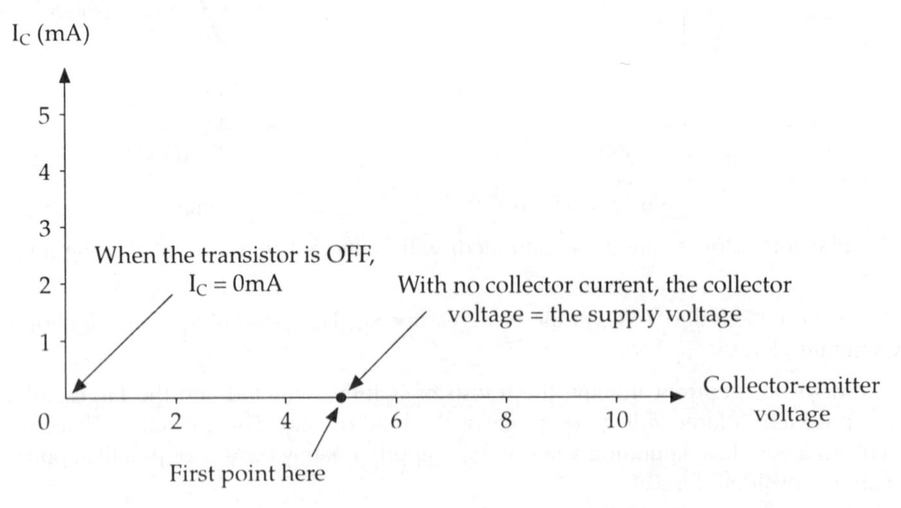

With transistor off, V_{CE} = supply

When the device is ON then:

a) the collector voltage is 0V (assume perfect switch).
b) the collector current will be a maximum. (Again if we assume a perfect switch then the current flowing will only be limited by the load resistor.)

Under these conditions the value of the collector current will be:

$$I_{Cmax} = \frac{\text{supply voltage}}{\text{load resistance}} = \frac{V_{CC}}{R_L} \text{ mA} = \frac{5V}{1k} = 5mA$$

This value of current can now be plotted as the other point on the characteristic.

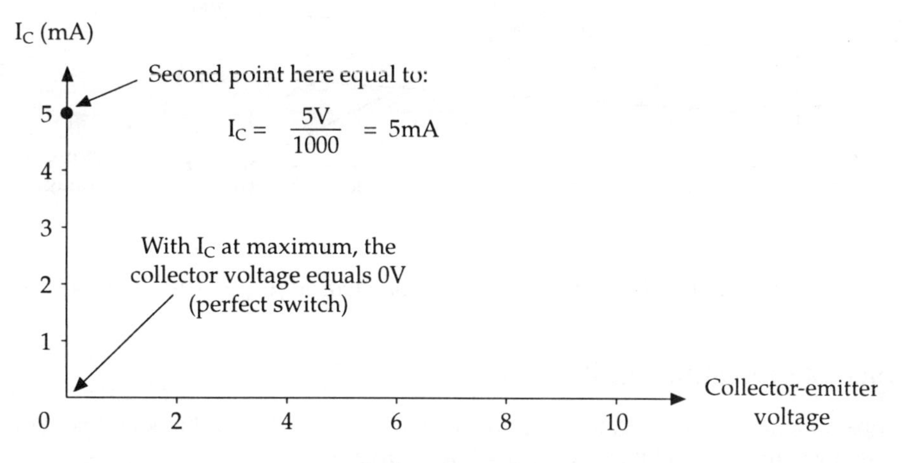

With the transmitter on, I_C will be a maximum; V_{CE} = 0V

We now simply join up these two points to construct the load line for a transistor having a load resistance of 1000 ohms.

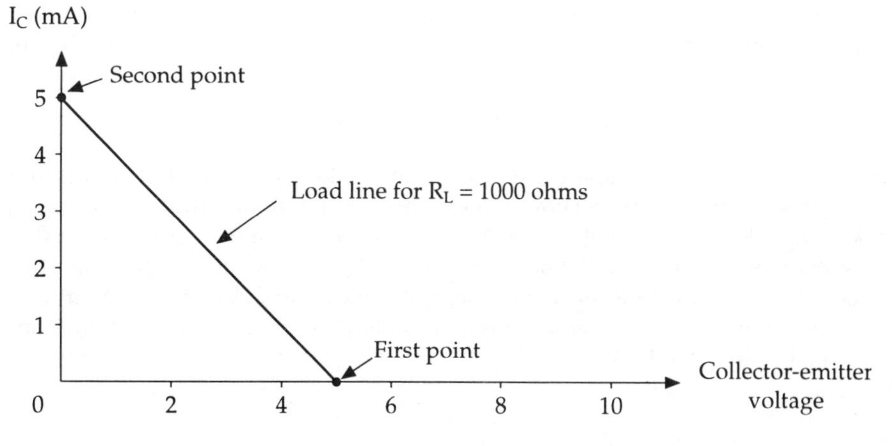

The two points are joined to plot the load line

Plotting the Load Line on the Output Characteristic

By plotting the load line on the output characteristic an operating point can be chosen to give the maximum output voltage swing without distortion occurring.

Once this point has been chosen the maximum current, voltage swings and the stage gain of the circuit can be determined.

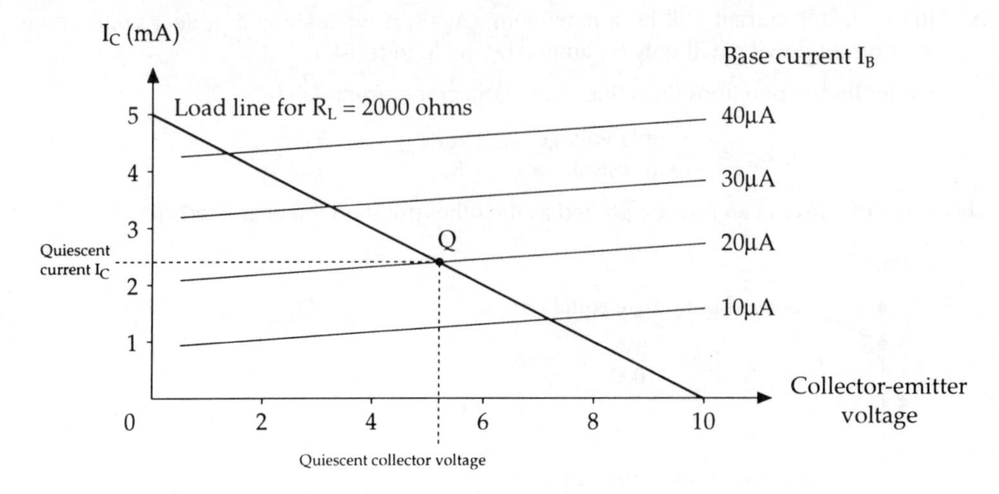

A load line plotted on the output characteristic (V_{CC} = 10V; R_L = 2,000 ohms)

Choice of Operating Point

The operating point is chosen to lie approximately in the middle of the load line. Point Q is chosen where the load line cuts the I_B = 20µA curve.

We can now determine from the graph the quiescent collector current and the quiescent collector voltage, with the operating point Q. We simply project down and across to the X and Y axis, to read off the standing collector voltage and current respectively.

Remember: the quiescent condition is the standing collector voltage and current due to the d.c. conditions existing on the transistor, i.e. before an a.c. input signal has been applied.

Determining the Current Gain of the Stage (A_i)

We can now show on the graph the effect of varying the base current. If we assume that the input signal causes the base current to vary by plus and minus 10µA, i.e. a 20µA peak to peak swing. This variation will cause the operating point to move up and down the load line as shown (by the thick black line) in the next diagram. The input signal will cause the base current to increase to a maximum (peak) of 30µA and fall to a minimum of 10µA. This in turn will cause the collector current to rise to a maximum value of 3.4mA and fall to a minimum value of 1.4mA.

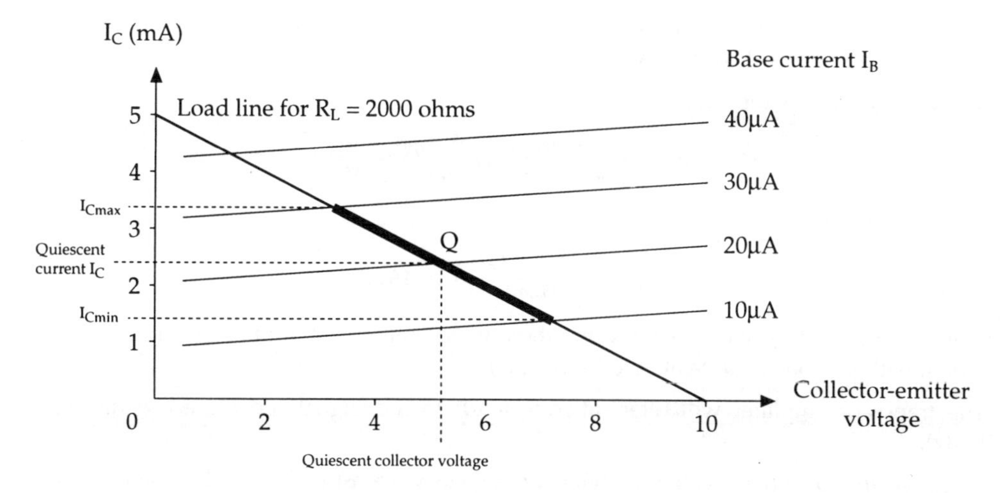

This change in collector current can be determined by projecting to the Y axis of the graph from the load line as shown above.

The current gain of the stage:

$$A_i = \frac{\text{peak to peak change in } I_C}{\text{peak to peak change in } I_B}$$

From the graph:

$$A_i = \frac{(3.4 - 1.4) \times 10^{-3}}{20 \times 10^{-6}} = \frac{2{,}000}{20} = \mathbf{100}$$

Determining the Voltage Gain of the Stage (A_v)

If we use the output characteristic that has the base voltage rather than the base current, then the voltage gain of the stage may be obtained in a similar way.

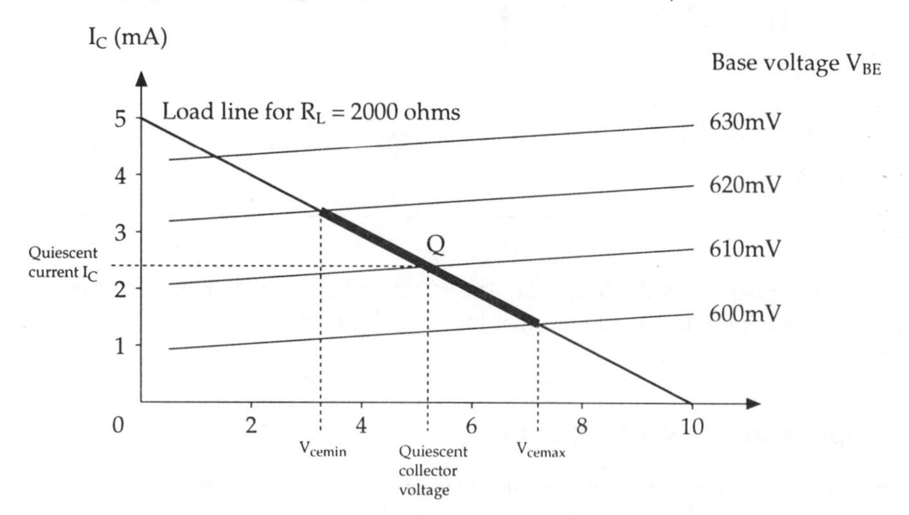

Finding the voltage stage gain: V_{CC} = 10V, R_L = 2,000 ohms

The change in collector voltage can be determined by projecting to the X axis of the graph from the load line as shown above.

The voltage Gain of the Stage:

$$A_v = \frac{\text{peak to peak change in } V_{ce}}{\text{peak to peak change in } V_{be}}$$

From the graph:

$$A_v = \frac{(7.2 - 3.25)}{20 \times 10^{-3}} = \mathbf{197.5}$$

This value of voltage gain assumes that there is no emitter stabilisation resistor, and even then it will only be accurate at zero hertz (d.c.).

The transistor amplifier will offer differing loads to a.c. signals compared to d.c. (zero hertz).

Since we are concerned with amplifying a.c. signals we must plot a d.c. load line to set up the operating point, and another a.c. load line to determine the stage gain.

The effect of the a.c. Signal on the Common Emitter Amplifier

The emitter by-pass capacitor C3, in the circuit below, offers very little opposition to the a.c. signals but a very large opposition to d.c. In other words the emitter by-pass capacitor will shunt R4 at a.c. frequencies but not at d.c.

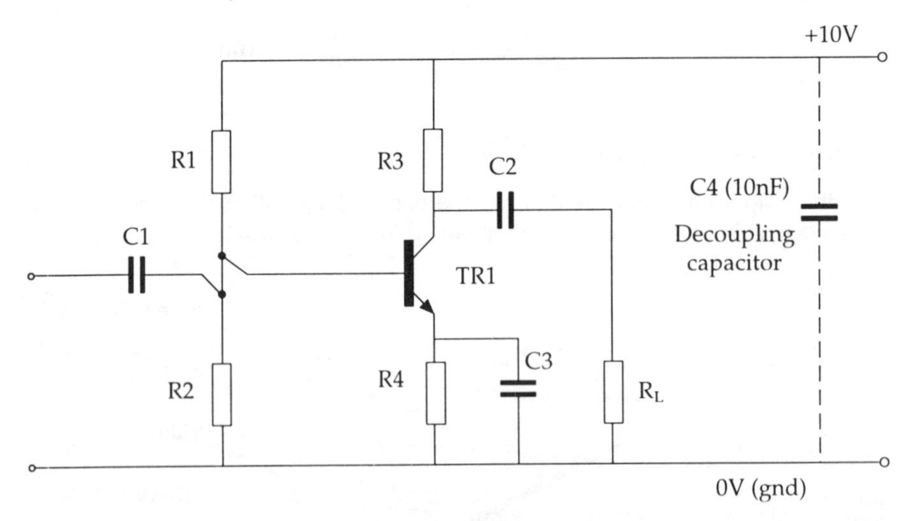

This alters the loading on the transistor. The effective load at zero hertz (d.c.) is R3 + R4 and a d.c. load line must be plotted that includes R4, to determine the correct operating point Q.

Decoupling the Power Supply Rails

It is important to prevent a.c. signals from being coupled into the amplifier via the supply rail and both supply rails should be at 0V in terms of a.c.

The large capacitor in the power supply, because of its construction (rolled up into a tube rather like a coil), has inductance and prevents the two ground rails from being at the same potential to an a.c. signal.

Remember: inductive reactance increases with frequency and this will cause the two power rails to be at differing potentials to a.c. signals.

A decoupling capacitor is often fitted across the power rails (e.g. a 10nF low inductance type) to provide a.c. grounding.

Therefore to a.c. signals:

R3 is connected between the collector and ground and R_L is also connected between the collector and ground. To a.c. therefore, the effective load is R3 in parallel with R_L!

In other words if we assume that C2 and the decoupling capacitor C4 have a low reactance at the mid frequencies then the circuit may be redrawn as shown below.

Remember: this is how the circuit responds to a.c. signals. The d.c. conditions must still be preserved to make the transistor work!

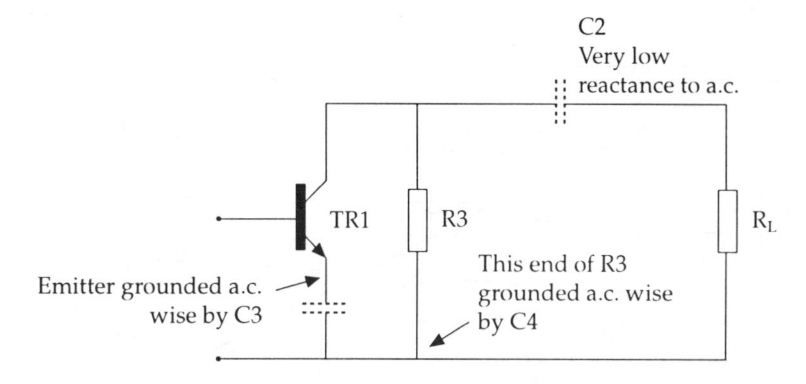

The a.c. load on the transistor TR1

Because of the low reactance of C4 the two power supply rails are at the same potential (0V) to a.c. signals. From a d.c. point of view they are still 10V apart!

The actual loading on the amplifier however will depend upon the frequency.

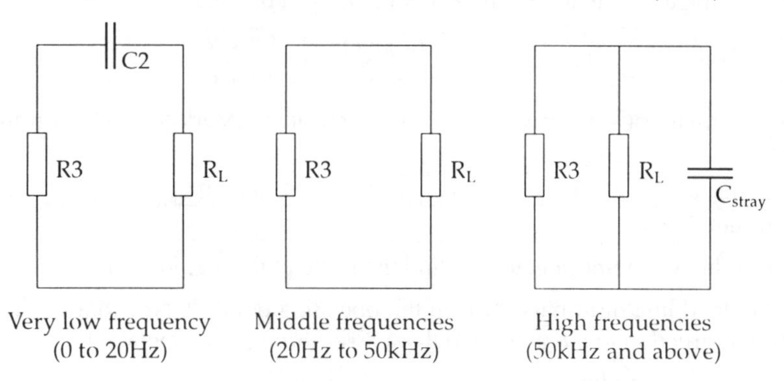

How frequency affects the loading on an audio amplifier

At low frequencies the reactance of C2 is large and the effective load on the amplifier will be: the reactance of C2 in series with R_L, with both C2 and R_L in parallel with R3.

At mid frequencies the reactance of C2 is small and can be ignored; the effective load is R3 in parallel with R_L.

At high frequencies stray capacitance (wiring capacitance and the capacitance of the load circuit) effectively shunt the load (R3 in parallel with R_L) and lower the gain of the stage.

The a.c. Load Line

A d.c. load line must first be plotted on the output characteristic in order to set the operating point Q.

Then an a.c. load line can be plotted on the same output characteristic equivalent to R3 in parallel with R_L.

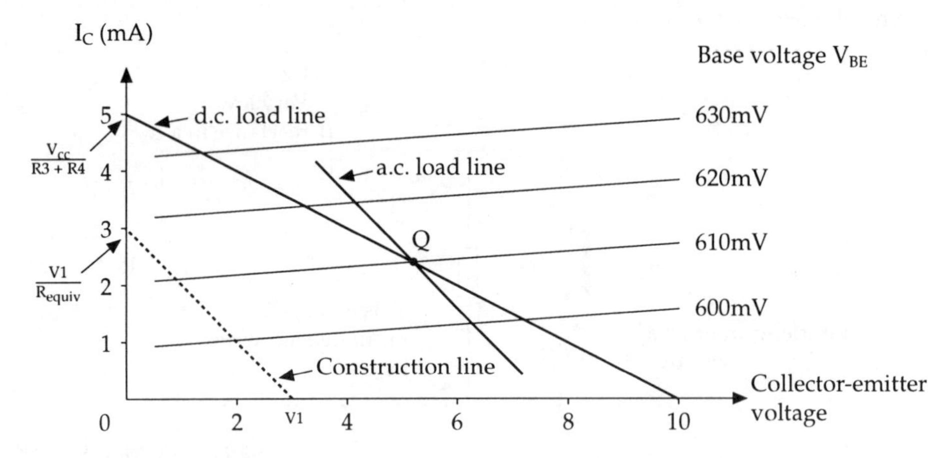

Plotting the a.c. and d.c. load lines

How To Plot the a.c. Load Line

1) Calculate the equivalent resistance of R3 and R_L in parallel.

$$\text{i.e.: } R_{equiv} = \frac{\text{product}}{\text{sum}} = \frac{R3 \times R_L}{R3 + R_L}$$

2) Chose a small convenient value of collector voltage. Mark this value on the voltage (X) axis.

3) Divide this value of voltage by the equivalent resistance R_{equiv} and mark this point on the current (Y) axis.

4) Connect the two points together to find the slope of the a.c. load line.

5) The a.c. load line must pass through the operating point Q. Now draw a line parallel to the construction line drawn in 4) that passes through the operating point Q.

This is now the a.c. load line!

We can now use this a.c. load line to determine the stage gain: i.e. if the input base voltage changes, determine from the a.c. load line how much the collector voltage Vce will change.

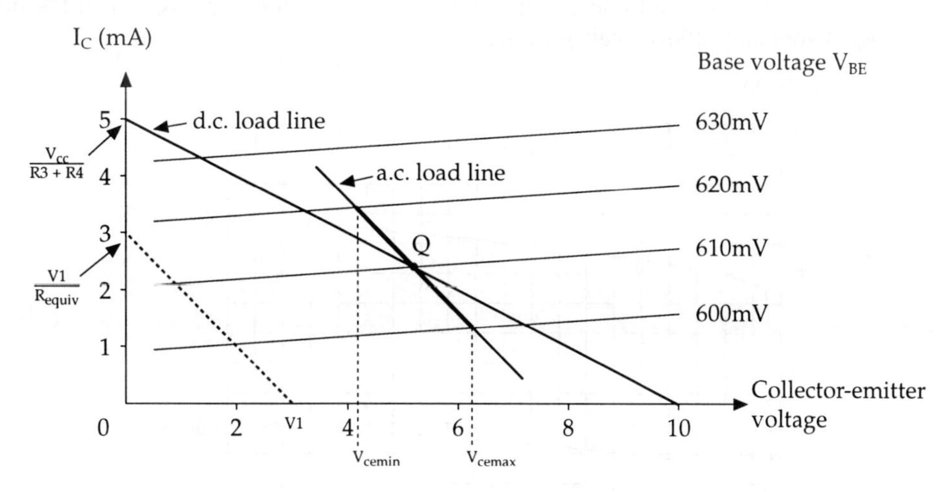

Determining the stage gain (A_v) from the a.c. load line

From the graph:

The change in collector voltage for a 20mV change in base voltage is 6.25 – 4.25 = 2V

$$\text{Therefore, the stage gain } A_v = \frac{2000\text{mV}}{20\text{mV}} = \mathbf{100}$$

Worked Example

Plotting a.c. and d.c. load lines on a bi-polar transistor output characteristic

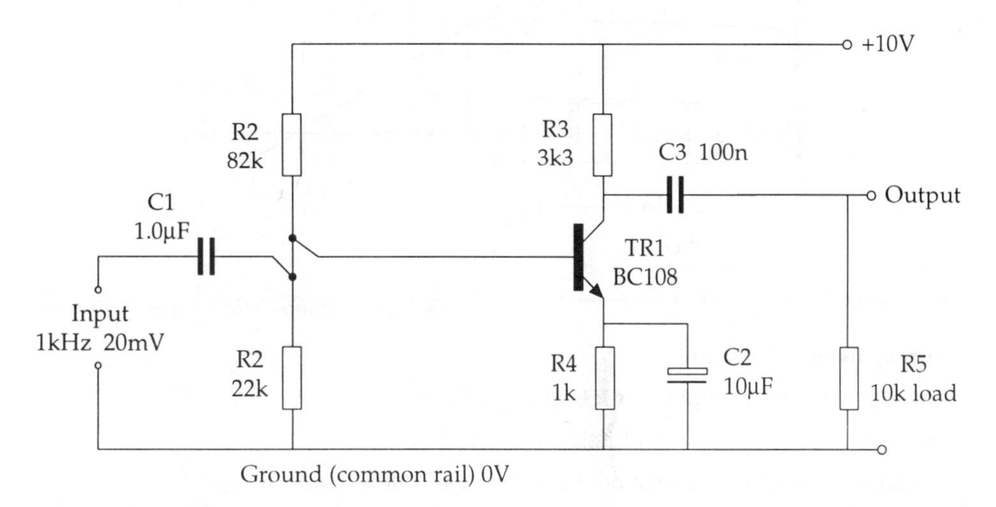

Circuit for worked example

The transistor used in the circuit above has the output characteristic shown below.

1) Plot the d.c. load line on the characteristic and chose a suitable operating point.
2) Next plot the a.c. load line and determine the gain of the stage (A_v) when the input signal applied is 20mV peak to peak at 1kHz.

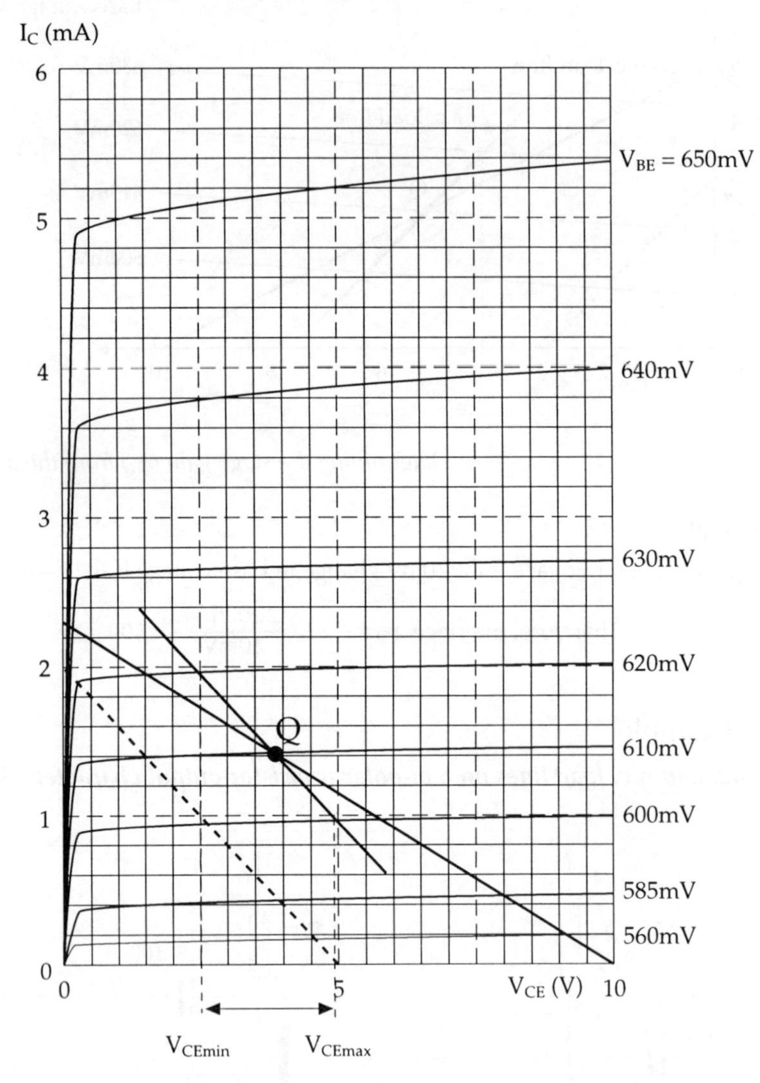

Output characteristic for a BC108 transistor

Plotting the d.c. Load Line

The d.c. load on the amplifier is R3 + R4 = 4.3k ohms.

The supply voltage is 10V so plot this point on the X axis.

The other point can be plotted on the Y axis at a value equal to:

$$\frac{\text{supply voltage}}{\text{d.c. load}} = \frac{10\text{V}}{4.3} \text{ mA} = \mathbf{2.33mA}$$

Now we join up these two points to plot the d.c. load line.

Selecting the Operating Point

From the graph a suitable operating point would be where the d.c. load line cuts the 610mV curve. This will allow a swing of plus or minus 10mV to be easily plotted. Mark this point as point Q.

Plotting the a.c. Load Line

The a.c. load on the amplifier is R3 in parallel with the load RL.

$$\text{The equivalent resistance is: } \frac{R3 \times R_L}{R3 + R_L} = \frac{3.3 \times 10}{3.3 + 10} = \mathbf{2.48k\ ohms}$$

If we chose 5V as the one point for our construction line then we mark this on the X axis.

For the second point of our construction line we divide 5V by 2.48k to find the current.

$$I_c = \frac{5V}{2.48} = 2.016\text{mA, say } \mathbf{2.0mA}$$

We plot this point on the Y axis and draw the construction line for the a.c. load line.

Now we draw a line parallel to the a.c. construction line that passes through the operating point Q. This is the a.c. load line.

Determination of the Stage Gain A_v

To determine the stage gain we use the a.c. load line.

Mark on the output characteristic where the a.c. load line cuts the 600mV curve and the 620mV curve. (This represents the peak to peak swing of 20mV.)

Project down from these two points to the X axis and read off the value of V_{CE} at the two points.

From the graph: $V_{CEmax} = 4.8V$ and $V_{CEmin} = 2.5V$

$$\text{Thererfore the stage gain } A_v = \frac{4.8 - 2.5}{0.02} = \frac{2.3}{0.02} = \mathbf{115}$$

Summary of Load Lines

- The bi-polar and uni-polar transistors may be operated as a switch.
- When the transistor is OFF the collector (or drain) voltage will be equal to the supply voltage. The collector (or drain) current will be zero (apart from leakage).
- When the transistor is ON the collector (or drain) current will be a maximum and the collector (or drain) voltage a minimum. This current will be equal to the supply voltage divided by the load resistor.
- These two conditions may be used to plot a d.c. load line on the output characteristic.
- The load line may then be used to choose a suitable operating (bias) point and from that determine the standing (quiescent) collector (or drain) current and the standing collector (or drain) voltage.

- The transistor circuits offer different properties to a.c. signals and therefore an a.c. load line can be plotted.
- The current and voltage gain of the stage can be determined from the load line.
- The d.c. power lines must both be at ground in terms of a.c. to prevent signals being fed back via these power lines and causing oscillation. This process is called de-coupling.
- Because the d.c. power lines are both at ground to a.c. signals it means that the load on the amplifier is in parallel with the collector (or drain) load resistor. This reduces the effective load of the amplifier and the gain is also lowered.

Self-assessment questions *(answers page 303)*

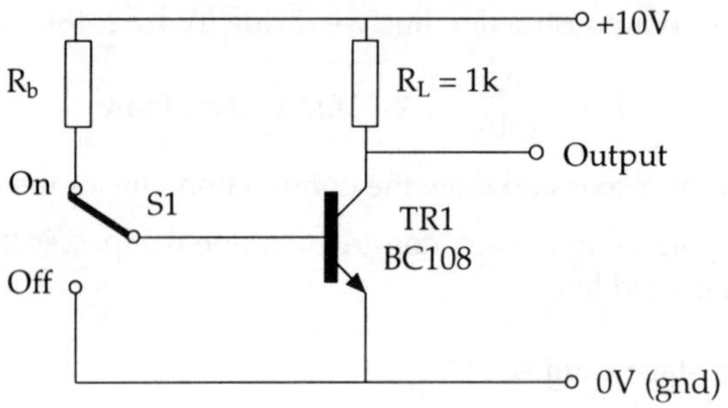

Diagram for questions 1 to 9

1) The bi-polar transistor in the circuit shown is used as a switch. When the transistor is ON then the maximum collector current will be:
   a) 10mA b) 9.8mA c) less than 9.8mA

2) When the transistor is turned off then the collector current will be;
   a) 100μA b) 50μA c) less than 10μA

3) The transistor has a current gain (h_{fe}) of 100. When the transistor is ON with maximum current flowing then the base current will be:
   a) 100μA b)98μA c) less than 98μA

4) With the transistor fully ON then the voltage on the base of the transistor will be:
   a) 10V b)1V c) 0.7V

5) If the base current from question 3 is divided into the supply voltage – base voltage M from question 4, then this will give the value of the base resistor R_b. The nearest preferred value of this resistor will be:
   a) 91k b) 95k c) 100k

6) When the transistor is fully ON then the output voltage will be:
   a) 10V b) 1V c) 0.2V

7) When the transistor is OFF then the output voltage will be:
   a) 10V b) 9.8V c) 0.2V

8) The bi-polar transistor used in the circuit above could be replaced with an N-channel FET if:
   a) the polarity of the supply is reversed
   b) R_b is removed and R_L re-calculated
   c) the supply is reversed and the R_L re-calculated
   d) R_b is removed and a negative 10V is available

9) The difference between the bi-polar and the FET when used as a switch is the:
   a) FET is voltage operated and consumes more power
   b) FET is current operated but consumes less power
   c) bipolar is voltage operated and consumes less power
   d) bipolar is current operated and consumes more power.

10) The circuit of a FET amplifier is shown below together with the results of a test on a 2N3819 FET.
From these results plot the output characteristic.
From the information provided on the circuit diagram, plot the d.c. and a.c. load lines for this amplifier on the output characteristic and then determine the gain of the stage (A_v) from the a.c. load line.

Note: The d.c. and a.c. load lines for a FET amplifier may be drawn in exactly the same way as for the bipolar transistor.

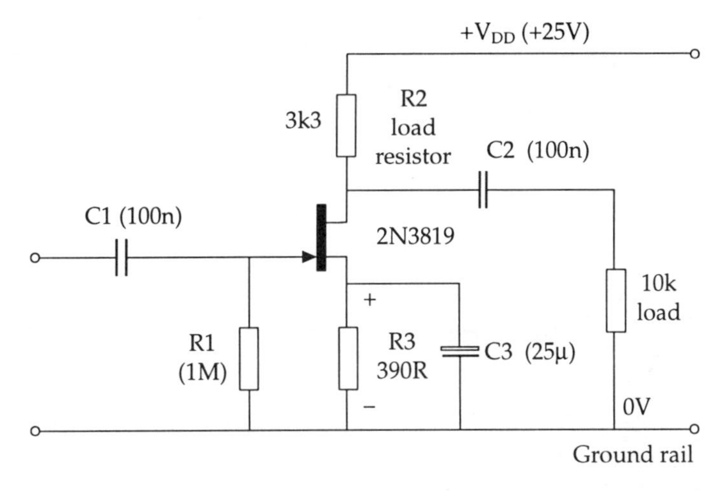

Circuit diagram for question 10

Drain-source voltage V_{DS} (volts)	*Drain current I_D (mA)*				
	$V_{GS} = 0V$	$V_{GS} = -0.5V$	$V_{GS} = -1.0V$	$V_{GS} = -1.5V$	$V_{GS} = -2.0V$
1	5.61	4.55	3.56	2.57	1.55
2	8.69	6.78	5.01	3.34	1.83
3	9.89	7.54	5.39	3.61	1.92
4	10.33	7.78	5.57	3.66	1.96
5	10.51	8.0	5.70	3.73	2.01
6	10.52	8.0	5.71	3.74	2.05
7	10.52	8.0	5.72	3.75	2.06
8	10.52	8.0	5.73	3.76	2.07
10	10.52	8.0	5.73	3.76	2.08
15	10.52	8.0	5.73	3.76	2.10
20	10.52	8.0	5.73	3.76	2.13
25	10.52	8.0	5.73	3.76	2.18

Results table for a 2N3819 FET

Equivalent circuits

Introduction

The voltage gain of a transistor amplifier can be calculated using the a.c. equivalent circuit of the transistor amplifier.

a.c. Equivalent Circuit

The equivalent circuit shows all the physical components of the circuit (resistors, transistors, diodes, etc.) but it also includes the working parameters of these devices.

Remember: a diode will only have a low forward resistance when it is forward biassed! If we wish to measure this low resistance then the ohmmeter must apply sufficient voltage to forward bias the diode. This is the working or dynamic resistance!

Typical working parameters of a transistor are: the input and output resistances and the current gain or g.m. With all the important parameters included in the diagram it makes it easier to calculate the overall gain and to determine the effects of the signal or source resistance on the circuit.

An equivalent circuit behaves in exactly the same way to a.c. signals as the circuit it represents. It shows all the a.c. conditions but does not include any of the d.c. conditions that are required to make the transistor work.

Four terminal Black Box

In the early days when the model of the transistor was none too exact, the transistor was treated as a four-terminal black box.

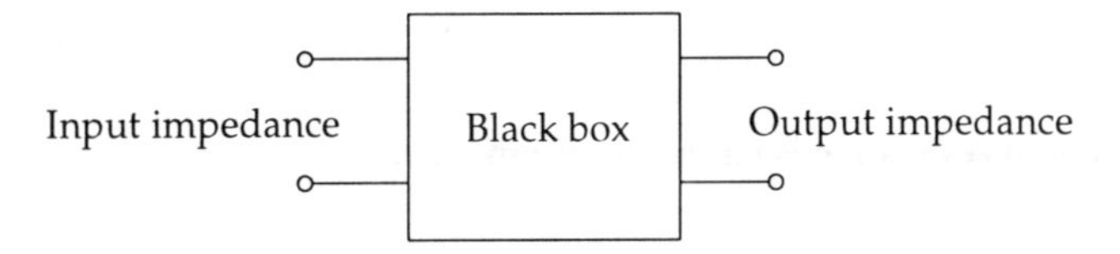

Four-terminal black box model of a transistor

The characteristics of the box were described using hybrid or h parameters.

h-Parameters

These parameters are the input, output and transfer characteristics of the black box. The parameters will of course depend upon whether the transistor is connected in common base, common collector or common emitter.

Common Emitter h-Parameters

h_{ie}; this is the h parameter (h) input impedance (i) in common emitter (e) measured when the collector voltage (V_{CE}) is held constant. (**Note**: In Stage gain of common emitter amplifier, p.241, this was called R_{in}.)

h_{fe}; this is the h parameter (h) for the small signal forward (f) current gain, in common emitter (e).

h_{oe}; this is the h parameter (h) output admittance (o) in common emitter (e) measured when the base current I_B is held constant.

Input Impedance (Resistance)

The input to the bi-polar transistor is a forward biased P-N junction which is mainly resistive. The base-emitter junction does however have capacitance!

The diode junction has two areas with current charges (similar to metal plates) separated by an area depleted of charge carriers (insulator). The junction therefore behaves as a charged capacitor!

To a.c. signals the base emitter junction will offer a combination of resistance and capacitive reactance. **A combination of resistance and reactance is called impedance (units ohms).**

For low and middle frequencies the reactive component is very small and therefore it is quite acceptable to talk of h_{ie} being the input resistance and not impedance.

Output Admittance

The output resistance of the transistor can be determined from the **slope** of the output characteristic (see Common emitter static characteristics, p.223).

$$R_{out} = \frac{\text{change in collector current}}{\text{corresponding change in collector voltage}}$$

This is current divided by voltage or the reciprocal of resistance!

$$\text{i.e.: } R = \frac{V}{I} \text{ therefore } \frac{I}{V} = \frac{1}{R} = \text{admittance (Siemens)}$$

The reciprocal of resistance is called admittance!

$$h_{oe} = \text{output admittance} = \frac{1}{\text{output resistance}} = \frac{1}{R_{out}}$$

$$\boxed{h_{oe} = \frac{1}{R_{out}} \text{ S}}$$

or

$$\boxed{R_{out} = \frac{1}{h_{oe}} \text{ ohms}}$$

The Bi-polar Transistor h-parameter Equivalent Circuit.

The input to the transistor is a voltage driving a current I_b into an input resistance of h_{ie} (or R_{in}).

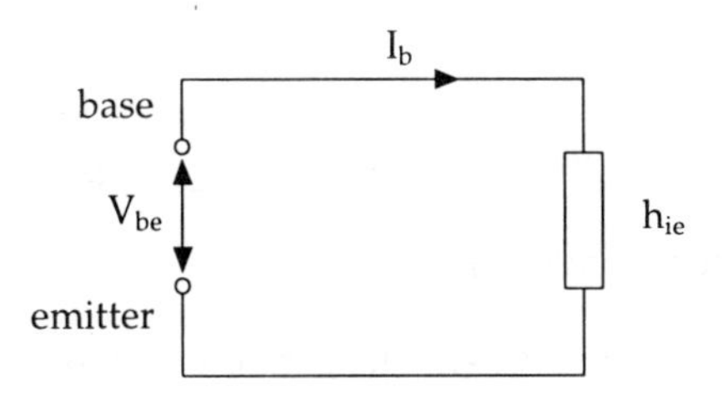

The input a.c. equivalent circuit

The transistor is represented as a constant current generator which generates a current of $I_b \times h_{fe}$: i.e. the collector current is the base current multiplied by the current gain.

The current generator has an output resistance of $1/h_{oe}$ in parallel.

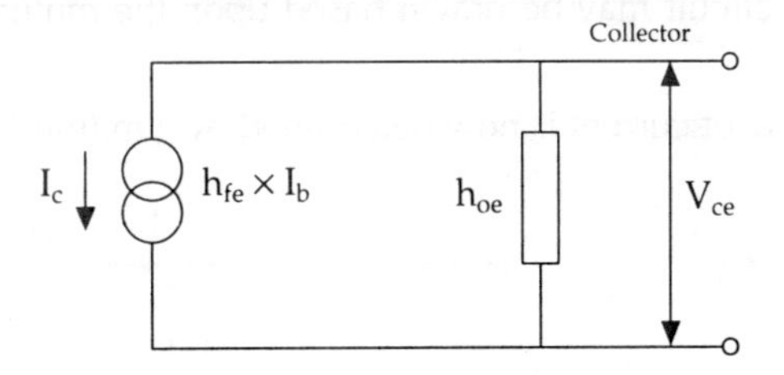

The output a.c. equivalent circuit

The complete model will therefore be:

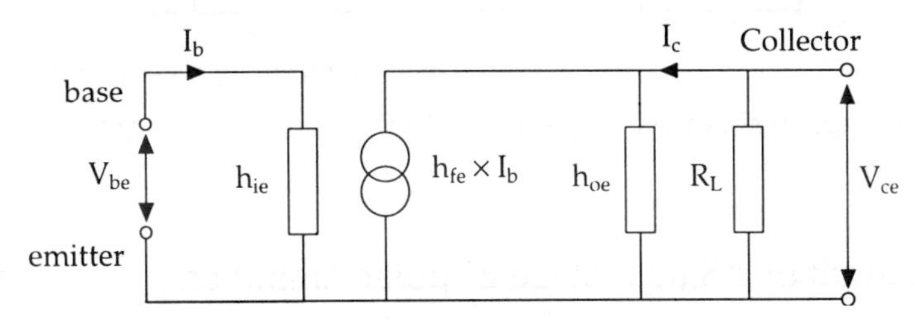

The a.c. equivalent circuit of the transistor and load resistor

The load resistor is in parallel with h_{oe} as far as a.c. signals are concerned and in practice the output resistance ($1/h_{oe}$) is very much larger than R_L, which means the effective resistance is that of R_L. Therefore h_{oe} can be ignored.

Remember: if a 100k ohm resistor is in parallel with a 1k ohm resistor then the effective resistance is approximately 1k ohms.

Output Voltage

The output voltage Vout is the collector current × the load resistor.

$$V_{out} = I_b \times h_{fe} \times R_L$$

Voltage (stage) Gain

The input voltage $V_{in} = V_{be} = I_b \times h_{ie}$

The voltage gain is therefore:

$$A_v = \frac{V_{out}}{V_{in}} = \frac{I_b.h_{fe}.R_L}{I_b.h_{ie}} = \frac{h_{fe}.R_L}{h_{ie}}$$

The ratio $\frac{h_{fe}}{h_{ie}}$ is the mutual conductance g.m.

Therefore the stage gain A_v = g.m. × R_L, and the g.m. is equal to 40 × I_e (mS).

(See Stage gain of common emitter amplifier, p.240.)

The g.m. Equivalent Circuit

The value of g.m. is mainly determined by the d.c. emitter current I_E and is more constant than h_{fe}. An equivalent circuit may be drawn based upon the mutual conductance rather than h-parameters.

The collector-emitter flow of current is now determined by g.m.(mA/V) × V_{be} (volts).

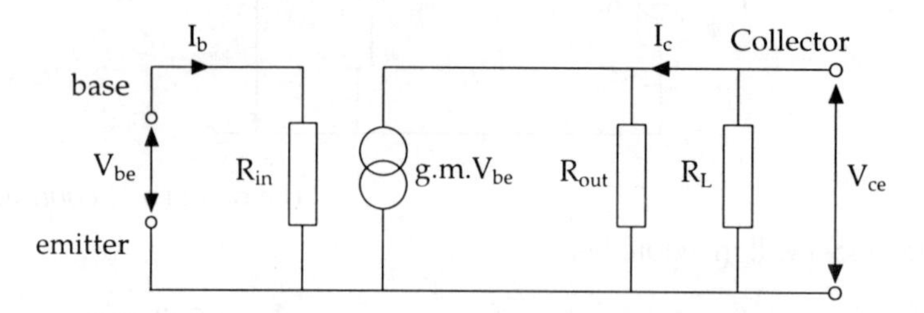

The g.m. model of the a.c. equivalent circuit

Once again the output resistance is in parallel with R_L and if R_{out} is very large compared to R_L it may be ignored.

Equivalent circuit of a Single-Stage Bi-polar Amplifier.

A typical single-stage bi-polar transistor amplifier is shown below. The transistor has an h_{ie} (R_{in}) of 2000 ohms and an h_{fe} of 120.

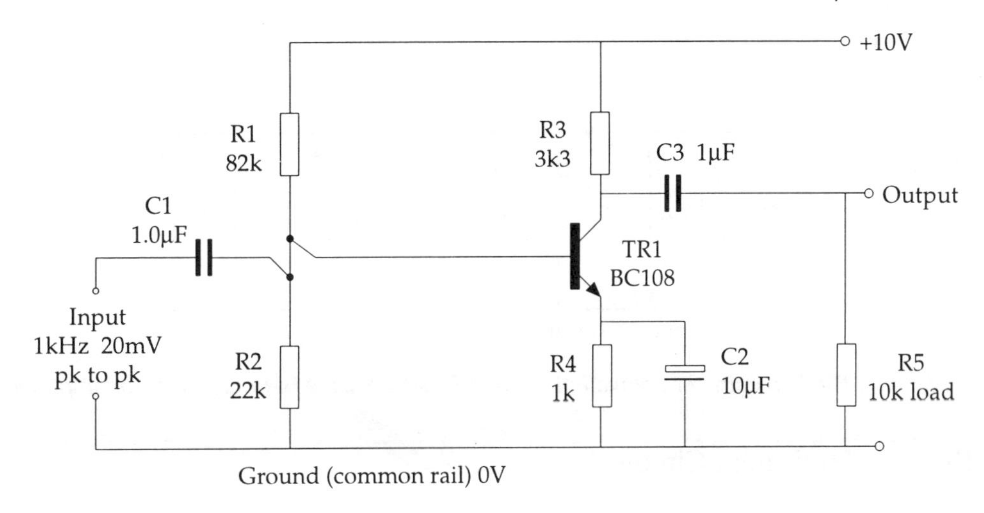

A Bi-polar Single-stage Transistor Amplifier

Remember: in the amplifier above:

R1 and R2 are the potential divider bias resistors, providing base bias to start the transistor action.

R3 is the collector load resistor, across which the changing collector signal current produces a changing output voltage.

R4 is the emitter stabilisation resistor; it works in conjunction with R1 and R2 to stabilise the base bias current against changes in the transistor's temperature.

C2 is the emitter **by-pass** capacitor; because of its low opposition to a.c. signals (low reactance) it simply by-passes any a.c. signal on the emitter to ground. Hence its name, by-pass capacitor.

C1 and C3 are the input and output coupling capacitors respectively. These couple the a.c. signal in and out of the amplifier, but block d.c. and preserve the d.c. voltages and currents that make the transistor work.

R5 is the load resistor for the amplifier. The a.c. output signal cannot be coupled out of the amplifier by C3 into an open circuit (there must be a d.c. path to each side of a coupling capacitor for it to work).

The Equivalent Circuit

To a.c. frequencies that are within the bandwidth of the amplifier, the three capacitors have very small reactances and can be considered virtually short circuit to a.c.

The supply rail is also at ground potential in terms of a.c. (see decoupling capacitors in Load lines, p.273). Therefore the bias resistors R1 and R2 are in parallel with each other and the base emitter resistance (R_{in} or h_{ie}) of the transistor.

These bias resistors therefore shunt the input resistance of the transistor and reduce the input signal. They should have a high value to reduce this effect.

The emitter resistor R4 is shorted out in terms of a.c. by C2.

Equivalent h–Parameter Circuit

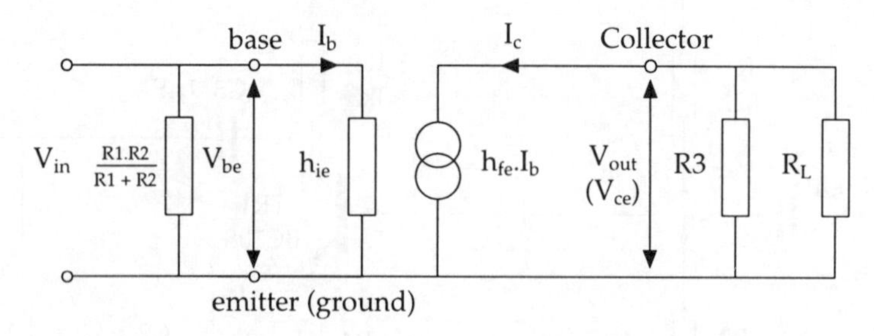

The h–parameter equivalent circuit of a single-stage bi-polar transistor amplifier

The g.m. Equivalent Circuit

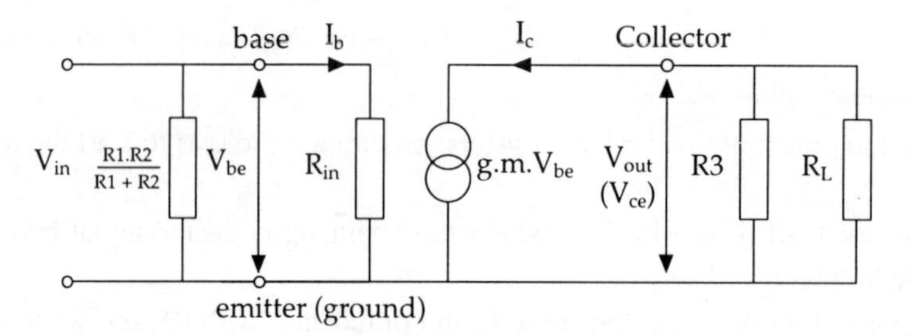

The g.m. equivalent circuit of a single-stage bi-polar transistor amplifier

The Signal Source Resistance

The input signal is obtained from a signal source which will have an internal resistance. This will have the effect of lowering the effective signal entering the amplifier. If we assume that the signal source has a typical internal resistance of 1k ohms then the input circuit will be as follows:

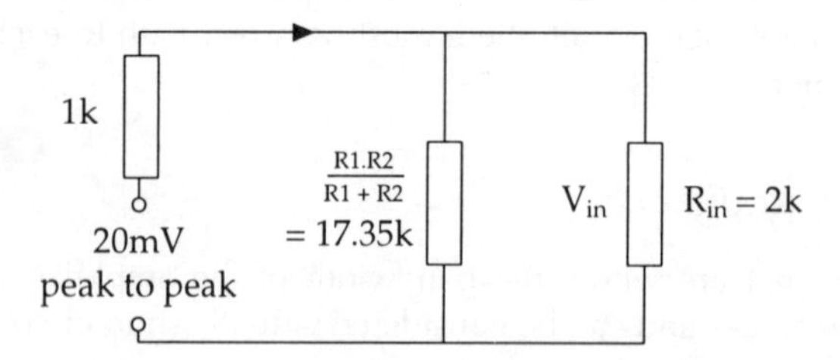

The effective resistance of R1 and R2 in parallel

The effect of R1 and R2 is to lower the input resistance R_{in} to 1.793k ohms.

The effective input signal V_{in} (assuming there is no loss in the coupling capacitor C1) therefore becomes:

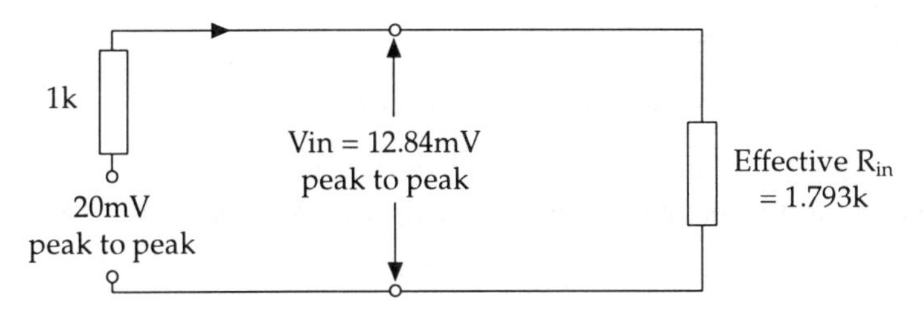

The source resistance and the bias resistors lower the effective input signal by 7.16mV

Equivalent Circuit of a FET Amplifier

The input resistance of a FET amplifier is extremely large and unless potential divider bias is employed the input resistance will be at least 1M ohm.

Remember: there is no input current, the device is voltage operated, so the signal source resistance can be high without loss of signal.

The drain current can be represented by a current generator of g.m.× the input signal voltage on the gate (V_{gs}).

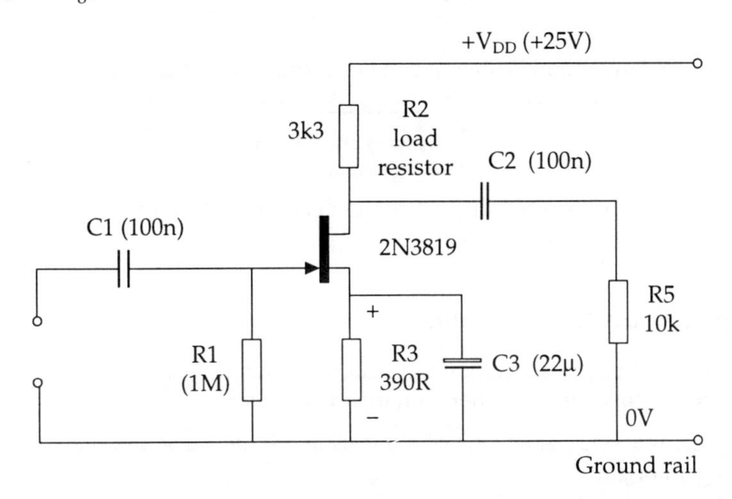

A Typical N-channel junction FET amplifier

The equivalent circuit uses the g.m. model.

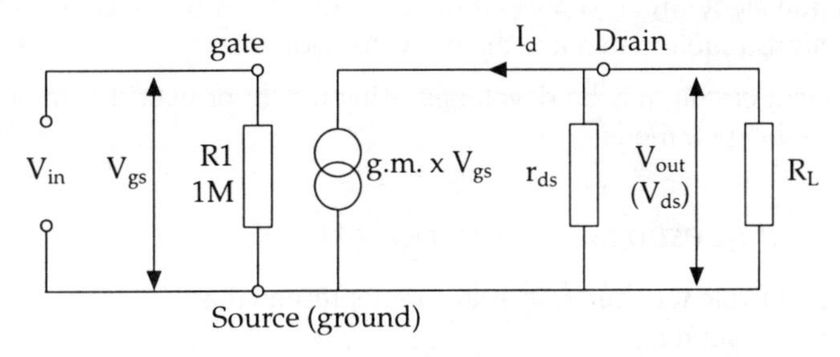

The g.m. equivalent circuit of a single stage FET amplifier

FET Amplifier Voltage Gain (Av)

The voltage gain is determined by the a.c. values of I_d and V_{gs}.

The drain current, $I_d = g.m. \times V_{gs}$

The resistance r_{ds} is the output resistance of the device and is equal to:

$$r_{ds} = \frac{V_{ds}}{I_d}$$

This output resistance and the load resistance are in parallel with each other; the effective load is therefore:

$$\frac{R_L \times r_{ds}}{R_L + r_{ds}}$$

Now the output voltage V_{ds} is given by Id x effective load:

$$V_{ds} = \frac{g.m. \times V_{gs} \times R_L \times r_{ds}}{R_L + r_{ds}}$$

Therefore the voltage gain:

$$A_v = \frac{V_{ds}}{V_{gs}} = \frac{g.m. \times R_L \times r_{ds}}{R_L + r_{ds}}$$

If r_{ds} is very much larger than R_L then it can be ignored, which gives:

$$\mathbf{A_V = g.m. \times R_L}$$

Note: the g.m. of a FET is not constant but depends upon the drain current. The chosen operating point will determine the value of the g.m.

Summary of Equivalent Circuits

- ❒ The equivalent circuit shows the physical components of the circuit such as resistors, diodes, transistors, etc., together with their parameters.
- ❒ The typical working parameters of a transistor are the input and output resistance, the current gain or g.m.
- ❒ With all the parameters included on the circuit it makes it easier to calculate the stage gain and to determine the effects of source and load resistances.
- ❒ The a.c equivalent circuit shows all the a.c. conditions but does not show any of the d.c. conditions required to make the transistor work.
- ❒ An equivalent circuit may be developed using the 'h' or hybrid parameters or alternatively using the g.m. model.

Self-assessment questions *(answers page 303)*

1) In the a.c. equivalent circuit transistors are represented as:
   a) a voltage source
   b) a current source
   c) a constant current generator.

2) In the a.c. equivalent circuit capacitors:
   a) need not be included.
   b) must be included.
   c) are included if they have significant reactance.

3) In the a.c. equivalent circuit resistors:
   a) need not always be included.
   b) must always be included.
   c) are not included if they have a large resistance.

4) In the a.c. equivalent circuit the output resistance of the transistor:
   a) need not be included.
   b) must always be included.
   c) need not be included if it is a very large resistance.

5) In the a.c. equivalent circuit the signal source resistance:
   a) is in series with the input resistance of the transistor.
   b) is in parallel with the input resistance of the transistor.
   c) has no effect on the input resistance and need not be considered.

6) In the a.c. equivalent circuit the d.c. power rails are not shown because:
   a) they do not affect the a.c. circuit.
   b) the a.c. and d.c conditions must be treated separately.
   c) both rails are at ground to a.c. signals and do not exist.

7) In the a.c. equivalent circuit the emitter (or source) resistor is not usually shown because:
   a) It is usually a low resistance compared to the load resistor.
   b) the emitter (source) by-pass capacitor has a very low reactance.
   c) the emitter (source) by-pass capacitor has a very low reactance at the lowest frequency of operation.

8) The a.c. equivalent circuit is used because;
   a) It shows all the physical components.
   b) It shows all the physical components and the parameters of the transistors.
   c) It simplifies the circuit by not including the d.c. power rails.

Capacitive reactance

Introduction

Capacitive reactance is the opposition to the a.c. flow of current offered by the capacitance of a circuit. Capacitive reactance is measured in ohms and is given the symbol X_C.

When a d.c. supply is applied to a capacitor, current only flows to charge or discharge the capacitor. Once the capacitor is fully charged then no more current flows. The capacitive reactance to d.c. is then very large (several M ohms).

a.c. continuously varies in value and polarity; therefore the capacitor is continually charging, discharging, and re-charging, resulting in a continuous flow of current.

The opposition to the flow of a.c. current (capacitive reactance X_C) is therefore very much less.

The Effect of the a.c. Frequency

Consider a circuit where a capacitor is alternately being charged in one direction and then the other.

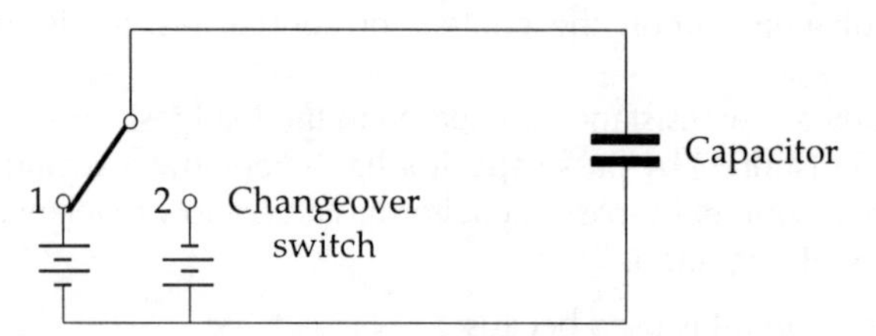

When the switch is moved from 1 to 2, the capacitor discharges and then recharges to the opposite polarity. The same happens when the switch is moved from 2 to 1

While the capacitor is charging or discharging then a current must be flowing.

For a given value of capacitance then the amount of current flowing will depend upon the rate of switching or frequency of switching.

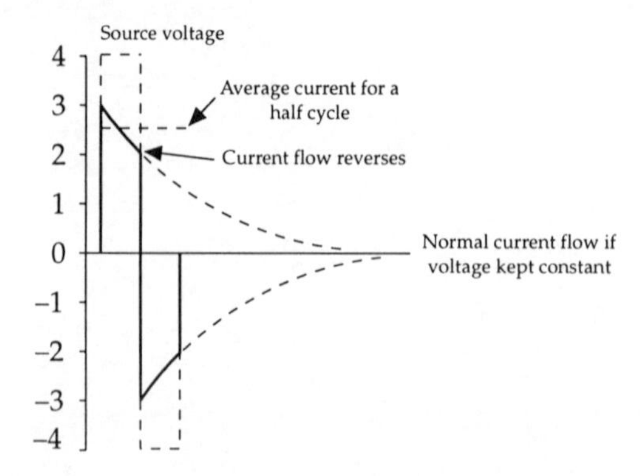

High switch rate = high average current

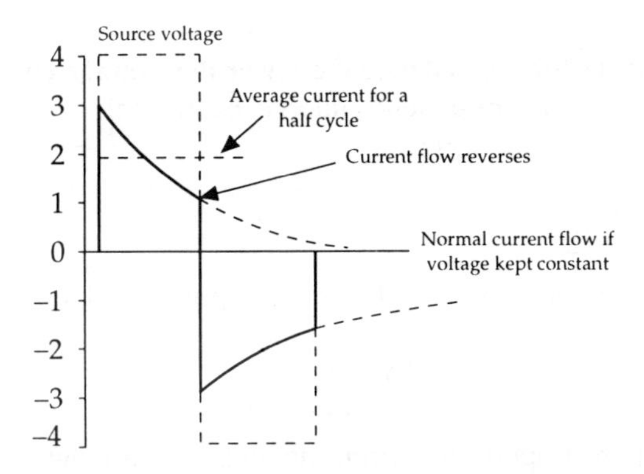

Lower switch rate = lower average current

The lower the frequency the lower the average current.
Current depends upon frequency!

The Effect of the Size of Capacitor

Consider that a voltage is applied to a capacitor for a short period of time and then removed.

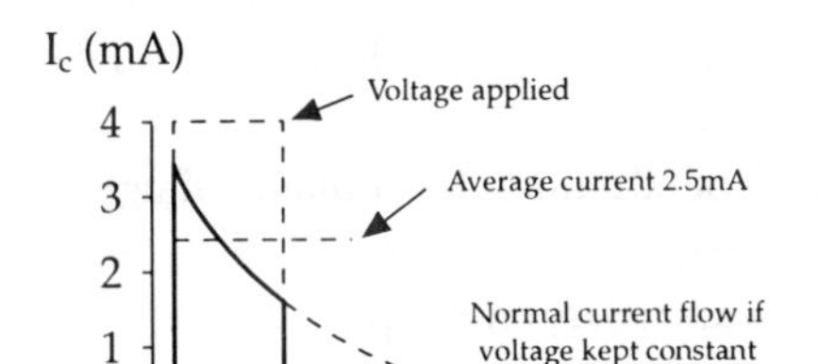

A large initial current flows and decreases rapidly as the small capacitor charges

If the same voltage is applied for the same time to a much larger capacitor:

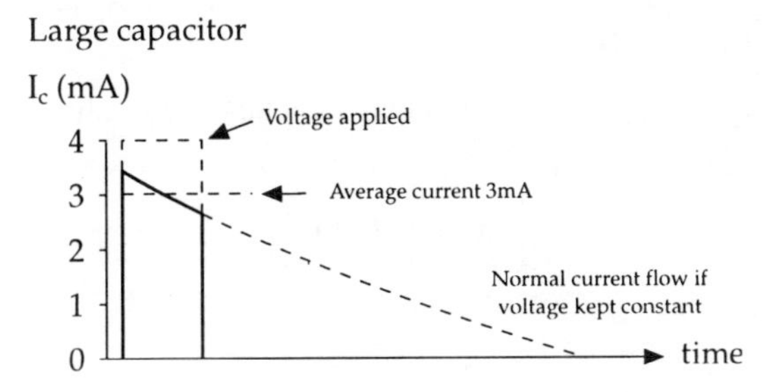

A large initial current flows and decreases less rapidly as the larger capacitor charges

The smaller the capacitance the lower the average current.
Current depends upon capacitance!

The effect of the voltage

The charge on a capacitor depends upon the voltage applied and the capacitance.

$$Q = C \times V$$

The charge Q also depends upon the current flowing and amount of time the current is flowing.

$$Q = i \times T$$

Therefore:

$$i \times T = C \times V$$

And:

$$i = (C \times V)/T$$

For a given time and capacitance the current flowing depends upon the voltage applied!

The current depends upon the applied voltage!

Therefore the Current Flow in a Capacitor Depends Upon:

a) frequency, b) capacitance and c) voltage applied

The a.c. current, I is proportional to $f \times C \times V$

Therefore $I = 2\pi.f.C.V$

Now V/I = opposition to current flow (ohms)

Therefore re-arranging:

$$\frac{V}{I} = \frac{1}{2\pi.f.C} = \text{Capacitive Reactance, } X_c \text{ (ohms)}$$

$$X_c = 1/2\pi.f.C$$

Self-assessment questions *(answers page 303)*

Calculate the capacitive reactance of the following capacitors at the frequency given.

(Take 2π as being 6.25.)

1) 20nF at a frequency of 1.5 kHz
   a) 530 ohms   b) 5.3k ohms   c) 53k ohms

2) 300pF at a frequency of 2MHz
   a) 26 ohms   b) 266 ohms   c) 2,666 ohms

3) 100nF at a frequency of 50 Hz
   a) 3k2 ohms   b) 32k ohms   c) 320k ohms

4) 100nF at a frequency of 800 kHz
   a) 2 ohms   b) 20 ohms   c) 200 ohms

5) 16μF at a frequency of 50Hz
   a) 2 ohms   b) 20 ohms   c) 200 ohms

6) 5nF at a frequency of 80 kHz
   a) 40 ohms   b) 400 ohms   c) 4k ohms

7) 50pF at a frequency of 1MHz
   a) 32 ohms   b) 320 ohms   c) 3200 ohms

8) 250pF at a frequency of 160kHz
   a) 40 ohms   b) 400ohms   c) 4k ohms

Answers to questions in Part 2: Section 2

Amplifier specifications

Q1 (c) A power amplifier must amplify voltage and current.
Q2 (c) Power gain is output power over input power.
Q3 (d) If the shape is altered then distortion has been introduced.
Q4 (c) 3 x 50 = 150
Q5 (d) There will be 5 cycles in the periodic time of the fundamental.
Q6 (d) A perfect squarewave contains an infinite number of odd harmonics.
Q7 (c) Phase inversion is similar to a phase shift of 180°.
Q8 (b) Two stages gives 360° (same as 0°), so a further stage gives another 180°.
Q9 (d) the bandwidth is measured between the half power points.
Q10 (a)

The bi-polar transistor

Q1 (a) The emitter current is equal to the collector and base current.
Q2 (d) $I_E = I_C + I_B$
Q3 (a) current gain = collector current ÷ emitter current
Q4 (d) The emitter current is always larger than the collector current.
Q5 (c) The arrow on the symbol shows the direction of conventional current flow.
Q6 (d) $I_E = I_C + I_B$.
Q7 (a) If I_E flows in I_C and I_B flow out.
Q8 (d) Leakage current flows across the collector base junction.
Q9 (c) The common base does not provide a current gain.
Q10 (a) Leakage current increases with temperature.
Q11 (b) The base/emitter junction is forward biased.
Q12 (a) The base/collector junction is reverse biased.

The common emitter connection

Q1 (b) The emitter is common to input and output.
Q2 (a) The emitter is connected to the negative rail which is grounded.
Q3 (b) The emitter is connected to the positive rail which is grounded.
Q4 (a) The emitter being negative pushes electrons towards the base and the collector potential pulls electrons from the base.
Q5 (d) The collector and emitter currents are doubled but the collector current is slightly smaller than the emitter current.
Q6 (b) Leakage current increases with temperature.
Q7 (a) The leakage current increases the collector current.
Q8 (b) The collector current is the output current, the base the input current.
Q9 (c) The graph is almost a straight line.
Q10 (c) The input is a forward biassed diode junction and the output is a reversed biassed diode.

The single-stage audio amplifier (1)

Q1 true.
Q2 true. The hfe is 100–150.
Q3 true. It provides base bias.
Q4 false. The input signal is not large enough to bias the transistor and start the transistor action.
Q5 true. It preserves the d.c. conditions on the transistor.
Q6 false. The a.c. signal would not be coupled into the amplifier.
Q7 false. R_L has the output signal developed across it.
Q8 true. There would be no d.c. path to the collector.
Q9 false. The collector current depends upon the base voltage and is largely independent of collector voltage.
Q10 true. It blocks the d.c. collector voltage.
Q11 true. Only a.c. will appear on the right hand plate of C2.
Q12 false. The collector voltage is a d.c. onto which the a.c. signal is superimposed.

The single-stage audio amplifier (2)

Q1 (b) To hold the base bias voltage constant.
Q2 (a) It forms part of the d.c. stabilisation network together with the potential divider R1 and R2.
Q3 (c) With no base voltage there will be no transistor action.
Q4 (a) The base voltage would rise increasing the collector current.
Q5 (d) The emitter voltage would follow the base voltage reducing the effective input signal and lowering the gain of the stage.
Q6 (d) The changing collector current is converted into a changing voltage.
Q7 (a) Reverse the battery and C_e
Q8 (c) The charge it receives on one half cycle is given up on the other half.
Q9 (d) They have to block d.c. and pass the a.c. signals.
Q10 (c) Any long term increase in the voltage across Re reduces the effective base bias voltage and consequently the collector current.

Stage gain of a common emitter amplifier

Q1 (d) Q2 (a) Q3 (c) Q4 (c)
Q5 (b) Q6 (a) Q7(b) Q8 (c)

Fault-finding on a bi-polar transistor amplifier

Fault 1 R1 open circuit:-
There is a small base voltage due to leakage from the collector.

Fault 2 R4 open circuit:
The base voltage is higher. There is an emitter voltage due to the measuring instrument completing the circuit. The value of this voltage depends upon the instrument.

Fault 3 Base/emitter short circuit:-
The base and emitter voltages are the same.

Fault 4 Emitter open circuit internally:-
No emitter voltage.

Fault 5 — R2 open circuit:-
Base voltage rises, collector current rises giving a low collector voltage.

Fault 6 — Base/collector short circuit internally:-
The base and the collector voltages are the same.

Fault 7 — Emitter by-pass capacitor open circuit.
N.F.B. lowers the stage gain to about 3.

Fault 8 — The output coupling capacitor is open circuit.

The field effect transistor

Q1 (c) Holes are the majority current carriers in P-type material.

Q2 (a) The JUGFET has a very high input resistance and practically no input current flows.

Q3 (d) The gate to channel is a reversed biased PN junction giving a very large input resistance.

Q4 (a) The gate is made negative to increase the width of the depletion area in the channel.

Q5 (c) The negative gate reduces the channel width and the drain current.

Q6 (c) The normal operation is beyond pinch off; the output resistance is large and the drain current becomes almost independent of the drain voltage.

Q7 (a) This is the relationship between the input (gate voltage) and output (drain current).

Q8 (b) μ = g.m. x rds.

Fault-finding on a single-stage FET amplifier

Fault 1 — R2 open circuit:-
There is no drain current and therefore no source voltage.

Fault 2 — R3 open circuit:-
With no drain current the drain voltage equals the supply voltage. The voltage measured at the source is due to the measuring instrument completing the circuit. The value of voltage measured depends upon the resistance of the instrument.

Fault 3 — R1 open circuit:-
The bias is removed and the drain current increases producing a slightly greater volt drop across the drain load resistor. The voltage across the source resistor has also increased slightly.

Fault 4 — Gate /source short circuit:-
The source and gate voltages are identical.

Fault 5 — Drain/source short circuit:-
The drain and source voltages are indentical.

Fault 6 — Source open circuit internally:-
The measuring instrument cannot now complete the circuit and so no voltage is measured at the source.

Fault 7 — source by-pass capacitor open circuit.
N.F.B. lowers the stage gain to about half.

Fault 8 — The input coupling capacitor is open circuit.

Load lines

Q1 (b) $V_{RL} = 10V - V_{ce_{sat}} = 10 - 0.2V = 9.8V$

Q2 (c) for a low to medium power transistor at room temperature.

Q3 (b) Ic_{max} / h_{fe}.

Q4 (c) forward volt drop of the base/emitter diode.

Q5 (c) the nearest preferred value is 100K

Q6 (c) $V_{ce_{sat}} = 0.2V$

Q7 (a) equal to the supply

Q8 (d) the device is normally ON and a negative supply is needed to turn it off.

Q9 (d) the bi-polar requires base current to start the transistor action

Q10. The solution to this exercise is given in the worked example within Task 2-g (Designing a single-stage FET amplifier).

The d.c. load line is plotted and because the source resistor is small (390R) compared to the drain load resistor (3300R) there is not very much difference between the a.c. and d.c. load lines. You should expect to get a stage gain of between 13 and 14.

Equivalent circuits

Q1 (c) constant current source

Q2 (c) At low frequencies the capacitor reactance may be significant. At high frequencies stray or self capacitance of the circuit may have a significantly low reactance.

Q3 (a) if the resistor is bypassed by a capacitor, e.g. emitter bypass capacitor.

Q4 (c) not included if the collector or drain load resistor is much smaller.

Q5 (a) is in series.

Q6 (c) Both rails should be at ground to a.c. signals.

Q7 (c) The by-pass capacitor shunts the resistor.

Q8 (b) The physical components and their parameters.

Capacitive reactance

Q1 (b) Q2 (b) Q3 (b) Q4 (a)
Q5 (c) Q6 (b) Q7 (c) Q8 (c)

Part 3: Digital Electronics

Contents

How to use this part

Part 3 may be studied at any time, the only previous knowledge required will be found in **Section 2 Information and Skills Bank** Bipolar transistor (p.209) and Field effect transistor (p.255).

Students with little prior electrical knowledge

For these readers a study route is shown overleaf. It is suggested that you look at the first practical task (build and test simple logic circuits) and then study in depth the associated Skills and Information Bank sections for that task. When you have completed the self assessment questions then attempt the first practical task. Continue in this manner until you have completed all the tasks.

Students with considerable previous experience

These readers may progress straight to the design and investigation projects for Part 3. Completion of the project will demonstrate their knowledge and competence for all the BTEC objectives shown on the syllabus/topic coverage sheet for Part 3.

Syllabus/Topic Coverage Chart

Electronics NII Syllabus topic		Tasks Section 1	Information and Skills Bank Section 2
E-5-a/b a & b	Logic symbols, truth tables & Boolean expressions	*Task a* Investigation of basic logic gates	Introduction to digital control Number systems Logic gates
E-5-c/d/e	Logic I.C's; logic families.	*Task b* Investigation of logic gates and truth tables	Truth tables Logic families & logic levels
E-5-f/g	Using logic gates	*Task c* Investigation of de-Morgan's rules; Using NAND/NOR gates	Simplification of Boolean equations de-Morgan's rules
F-6-a/b/c	Designs constructs & tests simple logic circuit	*Task d* Design of a combinational logic circuit	Karnaugh mapping Simplification of Boolean equations de-Morgan's rules
G-7-a/b	Constructs & tests behaviour of flip-flops	*Task e* Tests on a discrete R.S. bi-stable	The bi-stable
G-7-c	Investigates I.C, J.K & D type bi-stables Investigates D & J.K. bi-stables.	*Task f* Tests on R.S. and J.K. bi-stables.	R.S. and J.K. bi-stables
G-7-d	Constructs and tests the operation of counters	*Task g* Test a three-stage binary counter	Binary counters
G-7-d	Constructs and tests the operation of a shift register	*Task h* Test a four-stage shift register	Shift registers
		Task i The design project	All Section 2 **Information and Skills Bank** sections

Outline of Digital Electronics project

You have now been with your new small electronics firm (BIT & BYTE Ltd) for six months. Your training supervisor now gives you a another project to develop your knowledge and practical skills on digital electronics.

You are given the following information about the design project together with access to a skills and information bank, electronic components, manufacturers' data, catalogues, and test equipment.

Design project

S.A.S. Alarms have been asked to provide a security system for a firm of solicitors in the centre of a city. The building has a small underground car park, accessed via a tunnel, large enough for seven cars. Unfortunately the access is very restricted, with poor visibility and there is great difficulty should two cars meet, head on, in the tunnel.

S.A.S. Alarms client requires access control to the car park. This is to take the form of entrance barrier, operated by a magnetic key switch at the entrance to the tunnel and an exit barrier operated by a load cell buried in the road surface of the tunnel.

Access to the tunnel is to be controlled by traffic lights operated by the security personnel at the front of the building. The security personnel are to have a small display on their office counter to indicate either that the car park is empty, has spaces, or is full.Your task is to construct a suitable logic system to process (de-bounce) the input signals to an Up-Down binary counter and construct the counter which has to have a maximum count of seven. The count is to be incremented when 'IN' barrier is raised, and decremented when the 'OUT' barrier is raised.

The count is to be displayed on a seven segment LED display and it is to, display an E for empty, a S for space and an F for full. Your task is to design the logic gate system for the decoder and the drive for the LED display.

The basic system is shown overleaf.

The project therefore involves combinational and sequential logic devices. In order to help you gain knowledge and practical experience with digital electronics, the project once again has been broken down into a number of smaller practical tasks.

Associated with each task is a number of helplines to help you gain the background or under pinning knowledge.

Remember: If you have little previous experience with digital electronics then it will be necessary for you to do all the practical tasks and study all the Information and Skills Bank sections.

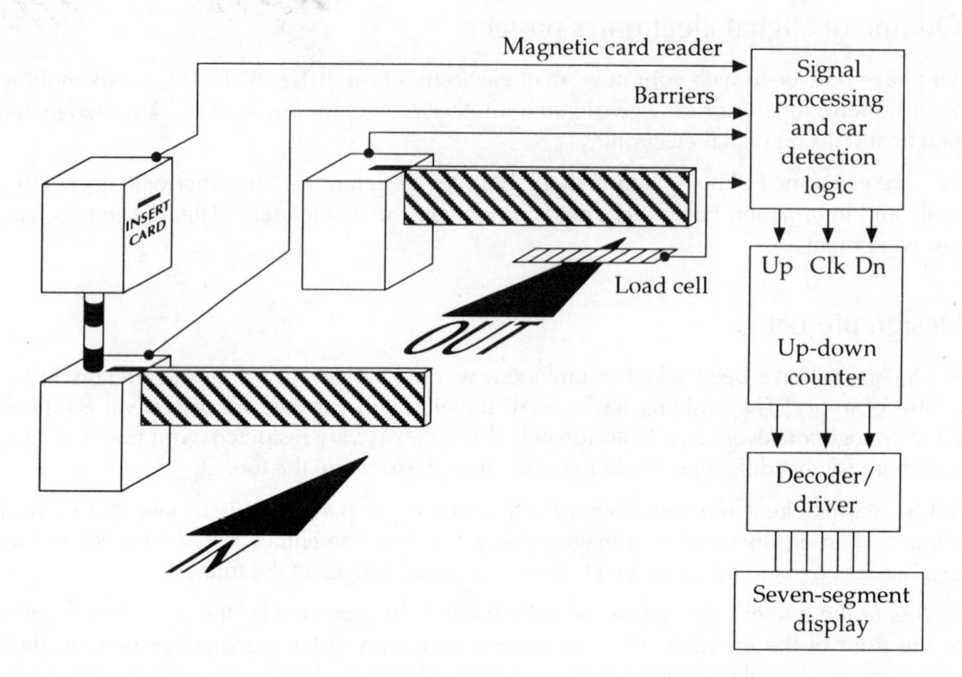

The car park system

Part 3: Digital Electronics

Section 1
Design project: tasks and investigations

Below is a list of practical tasks for part 3 together with the associated cross references to the **Section 2 Information and Skills Bank**. The practical tasks follow a logical order of progression, but depending upon your previous knowledge and experience, completing all the tasks may not be necessary (unless your teacher instructs you to do so).

Summary of practical tasks

Task a

Investigation of diode resistor AND gates, OR gates and a bipolar transistor NOT gate

Refer to ***Section 2 Information and Skills Bank*** *Introduction to digital control, p.357; Number systems, p.363; Logic gates, p.373.*

Study note*: If you have not previously studied Part 1 (Power Supplies) then you may need to refer to the following* ***Information and Skills Bank*** *references before attempting this task: Semi-conductor diodes, p.75; Multirange test instruments, p.89; How to use breadboards, p.96; Resistors and resistor colour coding, p.99; and from Part 2 (Amplifiers) The bi-polar transistor, p.209.*

Aims

To build and test AND, OR and NOT gates.

Equipment

A 5V stabilised power supply. A small breadboard and suitable wires. A digital multimeter.

Components

Two silicon diodes, IN4001 (or similar). Resistors: 10k, 4k7, 1k, 330. Two ultra miniature single pole change-over switches (suitable for plugging in to the breadboard). BC108 transistor, LED.

Approach

Construct the circuits below on a breadboard one at a time and test their operation.

AND Gate

Connect up the diode resistor AND gate as shown in the diagrams below.

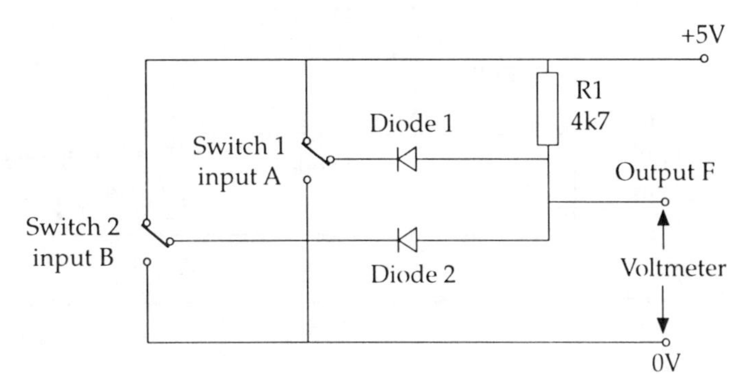

Diode resistor AND gate

Suggested Breadboard Layout

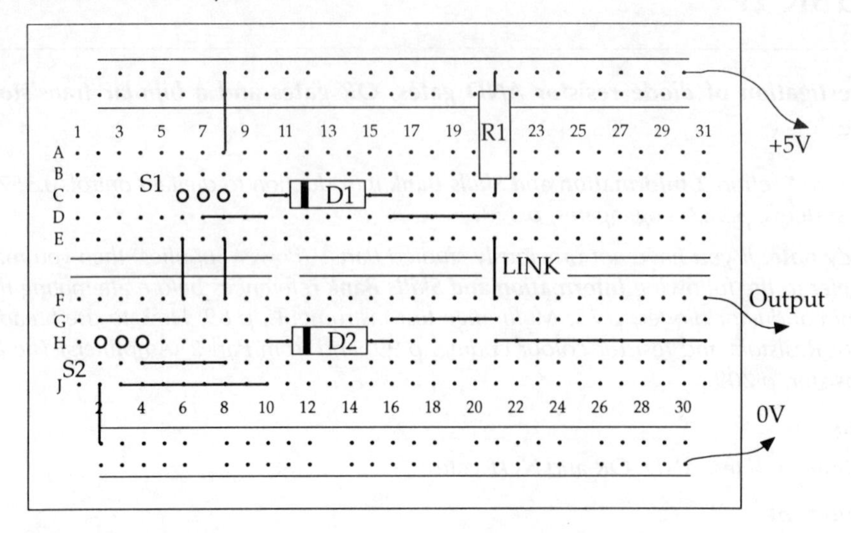

Measure the voltage at the output for the following conditions:

a) S1 and S2 both switched LOW

b) S1 LOW; S2 HIGH

c) S1 HIGH; S2 LOW

d) S1 and S2 both switched HIGH.

Expected Result: You should find that the output will be HIGH only when both inputs are switched HIGH (+5V).

OR Gate

Connect up the diode resistor OR gate as shown in the diagrams below.

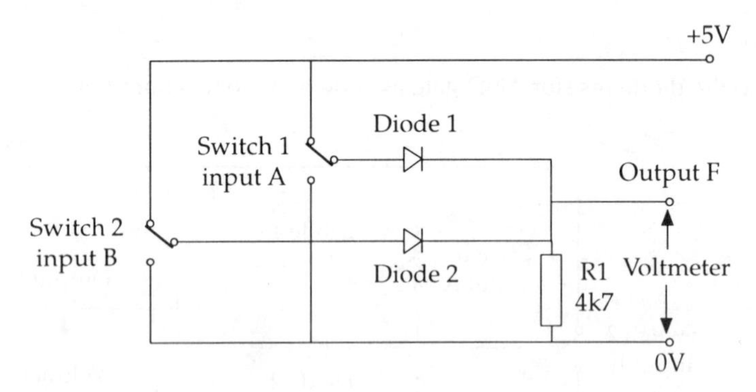

Diode resistor OR gate

Suggested Breadboard Layout

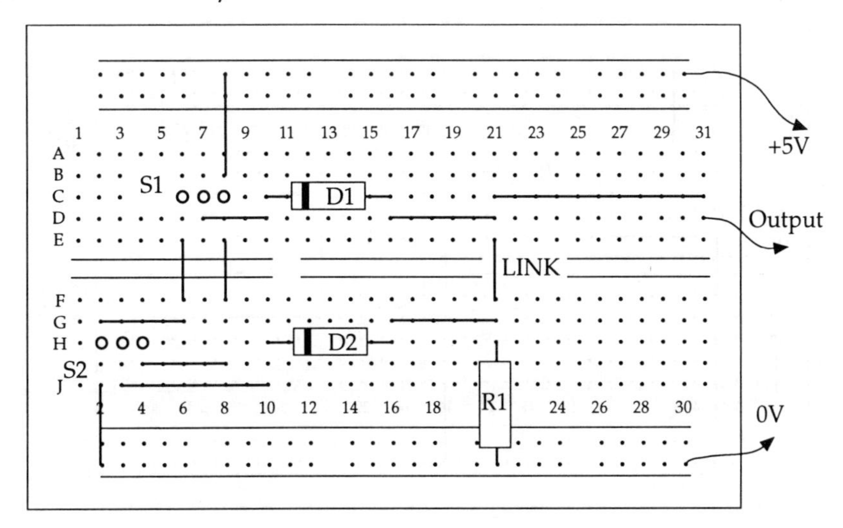

Measure the voltage at the output for the following conditions:

a) S1 and S2 both switched LOW
b) S1 LOW; S2 HIGH
c) S1 HIGH; S2 LOW
d) S1 and S2 both switched HIGH.

Expected Result: You should find that the output will be HIGH only when any input or both inputs are switched HIGH (+5V).

NOT Gate

Connect up the transistor NOT gate as shown below.

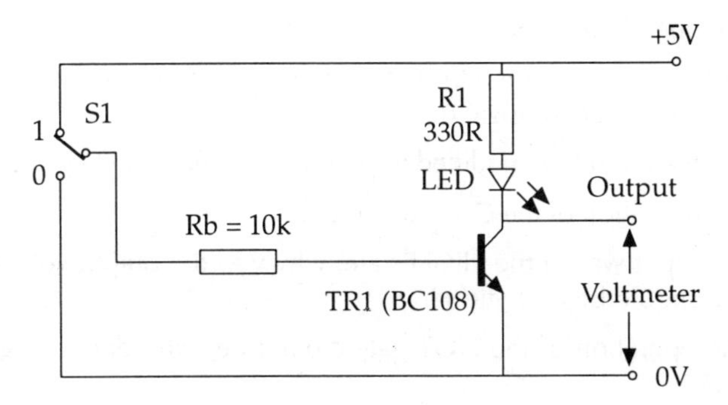

Using a transistor to produce a NOT gate

Suggested breadboard layout

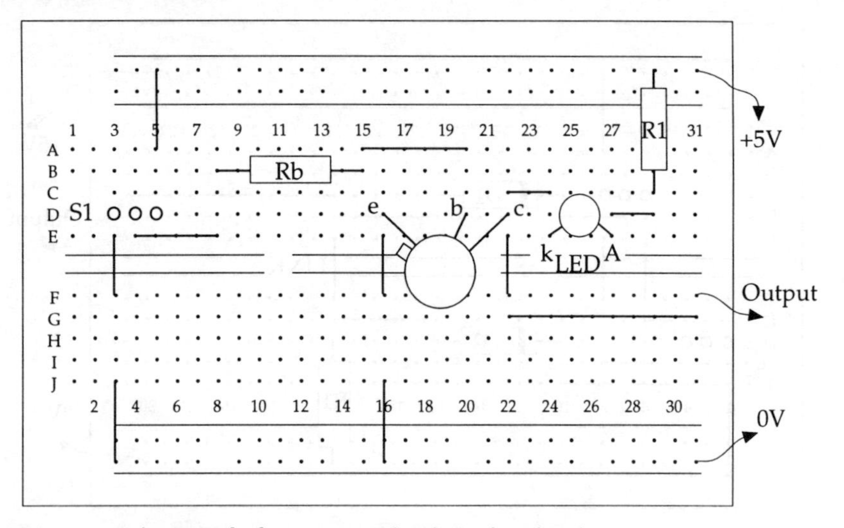

Set the power supply to 5V **before connecting it to the circuit**!

Observe the LED when the base is connected to logic 1. The LED ON indicates that the collector is at a low potential (logic 0). Measure the voltage on the collector (output voltage) when the LED is ON.

Next, observe the state of the LED when the base is connected to logic 0. Once again measure the collector (output) voltgage in this condition.

Self-assessment questions *(answers page 459)*

1) Explain the operation of the AND gate circuit.
2) When the output was in the LOW state, why was the output voltage about 0.7V and not 0V as it should be for logic 0?
3) How could the circuit be modified to produce a 3-input AND gate?
4) Explain the operation of the OR gate circuit.
5) When the output was in the HIGH state, why was the output voltage about 4.3V and not 5V as it should be for logic 1?
6) Explain the operation of the NOT gate circuit: i.e. why does a logic 1 in produce a logic 0 out?

Task b

Investigation of commercial logic gates and their truth tables

Refer to ***Section 2 Information and Skills Bank*** *Truth tables, p.381; Logic families and logic levels, p.389.*

Aims

To investigate the use of the logic probe to test digital logic circuits.
To determine the truth tables of a number of commercial TTL and CMOS logic gate chips using a multimeter and a logic probe.

Equipment

A variable voltage stabilised power supply. A small breadboard and suitable wires. A digital multimeter. A commercial logic probe.

Components
74LS08, 74LS32, TTL logic chips. 4001, 4011 CMOS Logic chips. A 1k potentiometer and a 1k resistor. Two single pole change-over switches (suitable for the breadboard).

Approach

Study the information below and then calibrate your logic probe.

Construct the circuits below on a breadboard one at a time. Use the logic probe to test their operation and produce a truth table.

Logic Probe

This is used to detect logic levels in a circuit under test. There are many manufacturers of logic probes and you must study the probe manual to determine its correct operation and use.

However with most probes, there is an indication of the **High** state, the **Low** state, a **Pulsing** state and a **bad** or **indeterminate** state: i.e. a voltage that is between the logic levels.

A popular probe in common use is the **Global Specialities LP-1**. This probe for example has a display of 3 LEDs which are interpreted as shown in the table.

LED states Hi	Lo	Pulse	*Input signal*	*Comment*
○	●	○	0V	Logic 0 No pulse activity
●	○	○	+V	Logic 1 No pulse activity
○	○	○	?	All LEDs off. Test point open circuit. Indeterminate signal. Probe not powered. Circuit not powered.
●	●	◐		Square wave pulse. Frequency less than 100kHz. Hi-Lo LEDs equally bright.
○	●	◐		Logic 0 with short duration positive going pulses present.
●	○	◐		Logic 1 with short duration negative going pulses present.

Key ● LED on ○ LED off ◐ Blinking LED

Interpreting the LEDs on the Global Specialities LP-1 logic probe

The LP-1's clip leads are used to connect to the power supply of the circuit board under test. The logic family switch is then switched to either TTL or CMOS and the Memory/Pulse switch to the pulse position.

The probe tip is then touched onto the part of the circuit under test and the LEDs will instantly provide an indication of the circuit activity.

Calibrating Check on the Probe

Experience in the use of the logic probe can be obtained by a calibration check. This test is used to detect the range of voltages that will be interpreted as **high**, **low** or **out of tolerance** (indeterminate). The circuit used for the test is shown below.

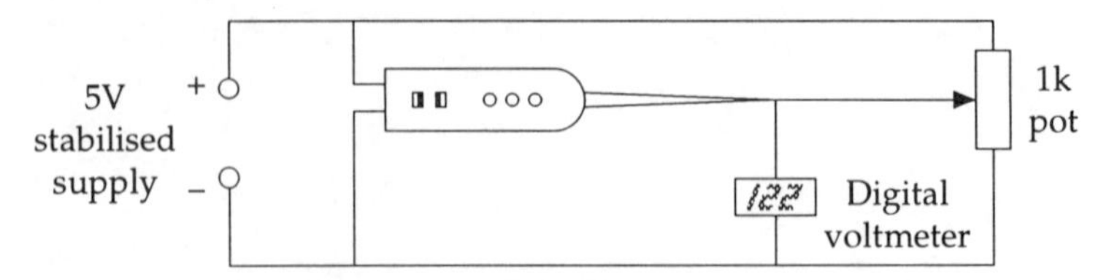

Probe settings for the LP–1:
Pulse/memory switch to *pulse*
TTL/CMOS switch to *TTL*

Test circuit for the calibration check on the probe

The voltmeter should be on no more than a 20V range and the probe switched to *Pulse*. Switch the TTL/CMOS switch to TTL.

Using the digital voltmeter, measure the voltage on the probe at which the LEDs **CHANGE THEIR INDICATION**: i.e. adjust the potentiometer so the voltage at the probe tip increases in APPROXIMATELY 1V steps.

Results Table TTL

Voltage on probe	*State of LEDs*	
	Low LED	*High LED*
0V	ON	OFF

Record the voltage at which the LED's change state in the tube.

All the logic levels in a two state logic system should, under satisfactory operation, fall within a particular voltage range. For a TTL system, the probe will indicate, from the table above:

Logic 0 level is from 0V to _______ V_{max}

Logic 1 level is from _______ V_{min} to _______ V_{max}

The forbidden range of voltages is therefore:

From _______ V_{min} to _______ V_{max}

CMOS Calibration Check

Switch the probe to CMOS and increase the supply voltage to 15V. Repeat the calibration test and complete the table below.

Results Table CMOS

Voltage on probe	*State of LEDs*	
	Low LED	*High LED*
0V	ON	OFF

Record the voltage at which the LED's change state in the tube.

All the logic levels in a two state logic system should, under satisfactory operation, fall within a particular voltage range. For a CMOS system, the probe will indicate, from the table above:

Logic 0 level is from 0V to _______ V_{max}

Logic 1 level is from _______ V_{min} to _______ V_{max}

The forbidden range of voltages is therefore:

From _______ V_{min} to _______ V_{max}

TTL Logic Chips

We will now use the probe to investigate commercial logic chips.

74LS08 Chip

Connect up the 74LS08 and the 74LS32 logic chips as shown in the diagrams below.

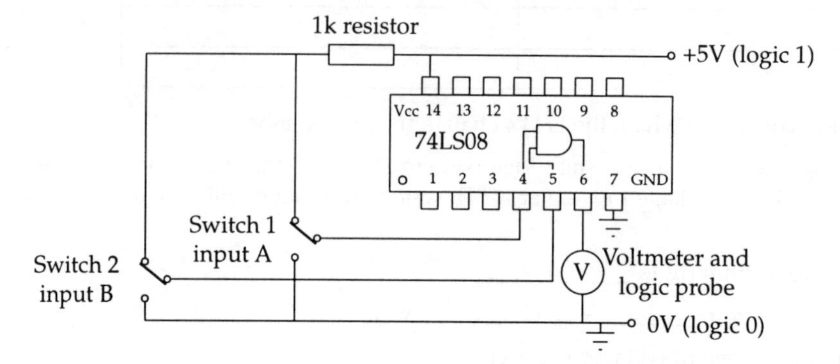

Circuit diagram for 74LS08 (AND gates)

Suggested Breadboard Layout

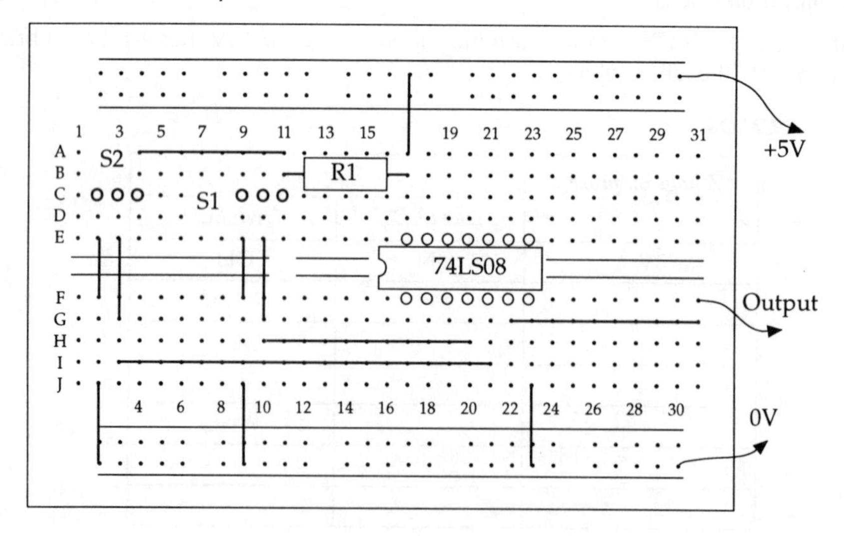

Measure the voltage at the output with the multimeter and the use then probe to indicate the logic level.

a) S1 and S2 both switched LOW
b) S1 LOW; S2 HIGH
c) S1 HIGH; S2 LOW
d) S1 and S2 both switched HIGH.

Construct the truth table for this gate!

74LS32 Chip

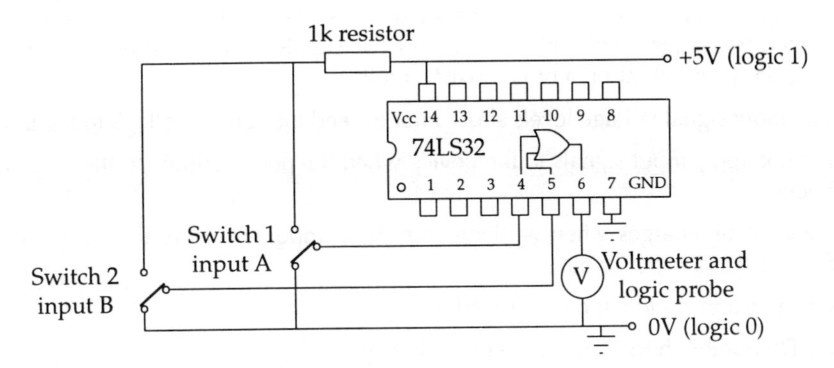

Circuit diagram for 74LS32 (OR gates)

Suggested Board Layout

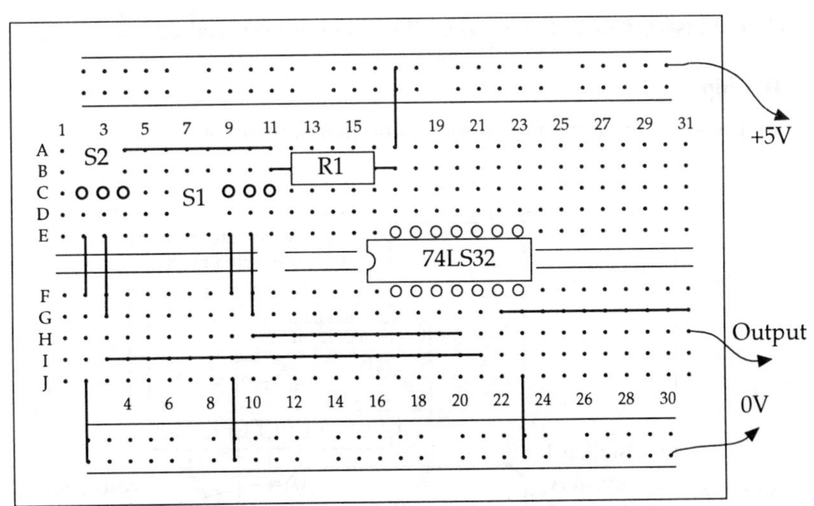

Measure the voltage at the output with the multimeter and the use then probe to indicate the logic level.

a) S1 and S2 both switched LOW
b) S1 LOW; S2 HIGH
c) S1 HIGH; S2 LOW
d) S1 and S2 both switched HIGH.

Construct the truth table for this gate!

CMOS Logic Chips

CMOS chips are sensitive to static. Static electricity may be generated simply by walking across a nylon carpet in slippers. The amount of static generated will depend upon many factors but it is not uncommon to generate voltages of 10kV! When handling CMOS chips certain precautions must be taken.

Handling CMOS Chips

- All inputs on the chip must be connected somewhere! All unused inputs must be connected to a logic level, either ground or the supply, directly or by a resistor. This is especially important on a breadboard circuit.
- The input signal voltage levels should not exceed the supply voltage to the device.
- Do not apply input signals to the device when the power supply to the chip is disconnected.
- Avoid static charges when working with the component. Store the ICs in conductive foam, or on a metallic base.
- Observe anti-static handling procedures:
  - i) Do not touch the pins of the chip directly.
  - ii) Use an anti-static bench mat and wrist earthing straps when inserting/removing ICs.
  - iii) Insert the IC into your circuit last.
  - iv) Do not insert/remove the IC when the circuit is powered up.

4001 NOR Chip

Connect up the 4001 logic chip as shown in the diagrams below.

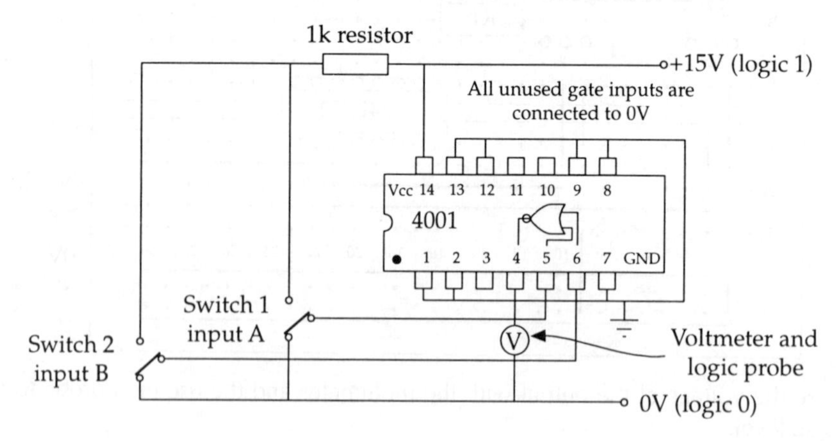

Circuit diagram for 4001 (NOR gates)

Suggested Breadboard Layout

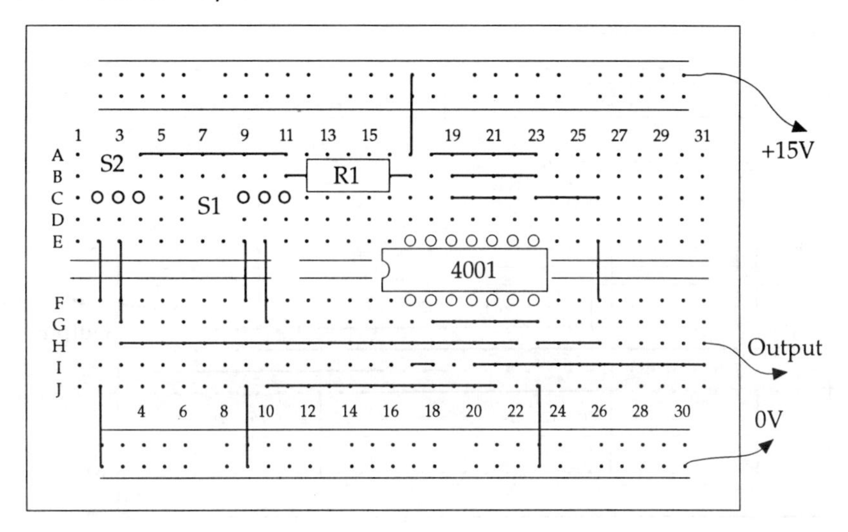

Measure the voltage at the output with the multimeter and the use then probe to indicate the logic level.

a) S1 and S2 both switched LOW
b) S1 LOW; S2 HIGH
c) S1 HIGH; S2 LOW
d) S1 and S2 both switched HIGH.

Construct the truth table for this gate.

4011 NAND Chip

Connect up the 4011 logic chip as shown in the diagrams below.

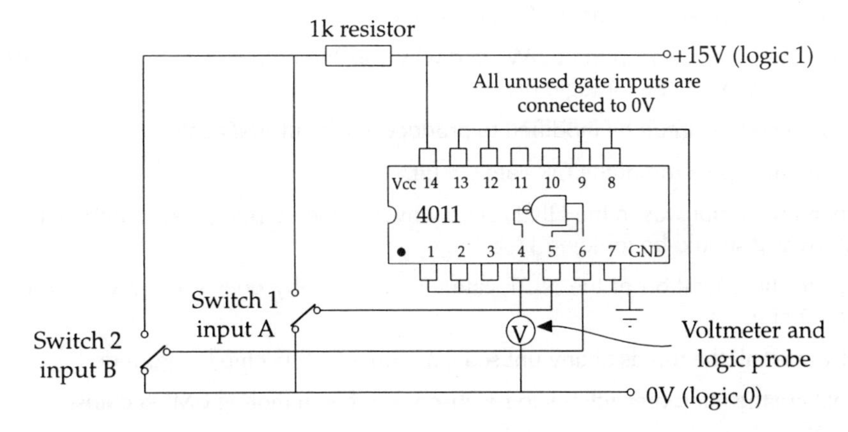

Circuit diagram for 4011 (NAND gates)

Suggested Breadboard Layout

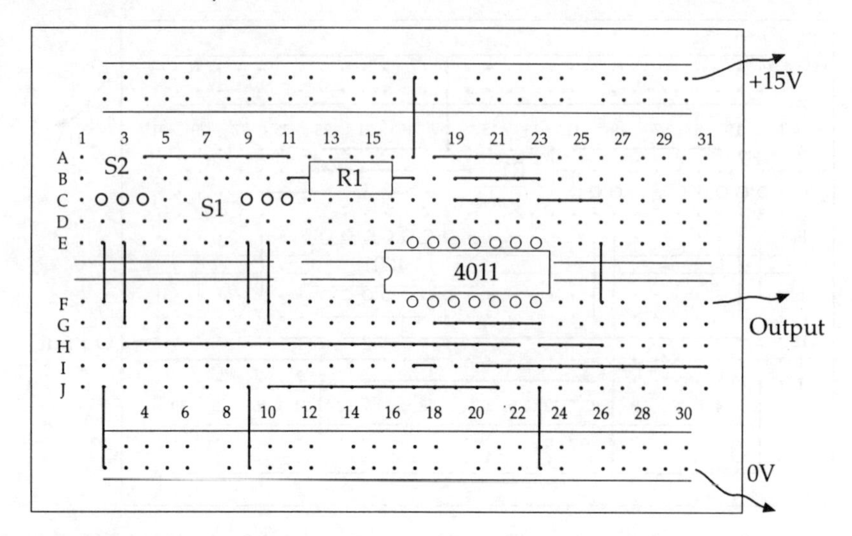

Measure the voltage at the output with the multimeter and the use then probe to indicate the logic level.

a) S1 and S2 both switched LOW
b) S1 LOW; S2 HIGH
c) S1 HIGH; S2 LOW
d) S1 and S2 both switched HIGH.

Construct the truth table for this gate.

Self-assessment questions (answers page 459)

1) Explain the operation of the AND gate circuit.
2) When the output was in the LOW state why was the output voltage about 0.15V and not 0V as it should be for logic 0?
3) How could the circuit be modified to produce a 3-input AND gate?
4) Explain the operation of the OR gate circuit.
5) When the output was in the HIGH state why was the output voltage about 4.19V and not 5V as it should be for logic 1?
6) Explain the operation of the NOT gate circuit, i.e. why does a logic 1 in produce a logic 0 out?
7) Why must all the inputs of any unused gates on a CMOS chip be grounded?
8) What are the main precautions to be observed when handling CMOS chips?

Task c(i)

Investigation of de-Morgan's rules

Task c(ii)

Conversion of logic gate system into all NAND form. Simplification of a system using Boolean algebra

Refer to ***Section 2 Information and Skills Bank*** *Truth tables, p.381; Logic families and logic levels, p.389)*

Aims

To investigate de-Morgan's rules.
Simplify a logic gate diagram using Boolean algebra and convert the diagram into all NAND form.

Equipment

A variable voltage stabilised power supply. A small breadboard and suitable wires. A digital multimeter. A commercial logic probe.

Components

74LS00, 74LS04, TTL logic chips. A 1k resistor. Three single pole switches.

Approach

Task c(i)

Practical verification of de-Morgan's Rules. Construct the circuits given below, one at a time, on a breadboard and use the logic probe to determine the output states.

a) Use de-Morgan's Rules to show that:

$\overline{\overline{A}.\overline{B}} = A + B$

Next construct the circuit below on a breadboard and obtain the truth table to show that it performs the OR function.

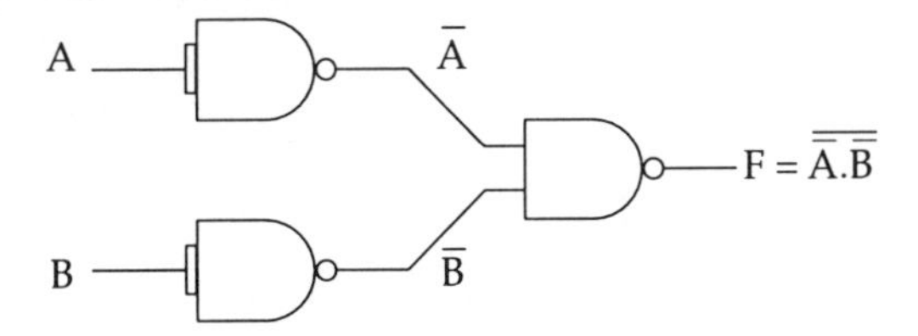

NAND equivalent of the OR function

b) Use de-Morgan's Rules to show that:

$\overline{A}.\overline{B} = \overline{A + B}$

Next construct the circuit below and obtain the truth table to show that it performs the NOR function.

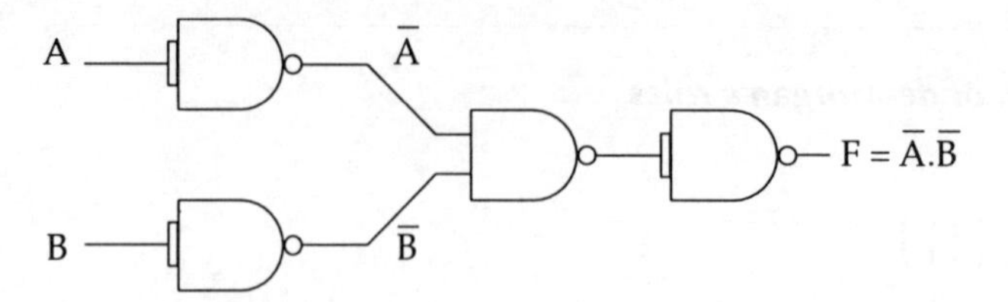

NAND equivalent of the NOR function

c) Use de-Morgan's Rules to show that:

$\overline{\overline{A}+\overline{B}} = A.B$

Next construct the circuit below and obtain the truth table to show that it performs the AND function.

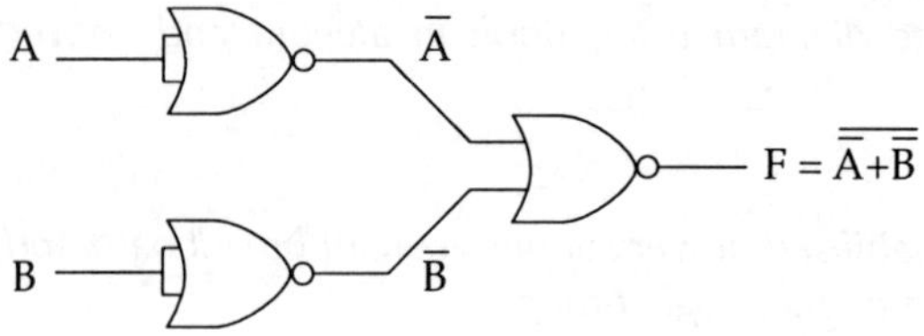

NOR equivalent of the AND function

d) Use de-Morgan's Rules to show that:

$\overline{A}+\overline{B} = \overline{A.B}$

Next construct the circuit below and obtain the truth table to show that it performs the NAND function.

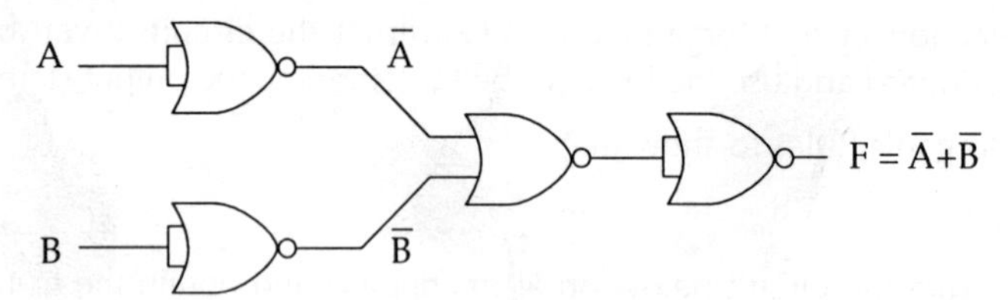

NOR equivalent of the NAND function

Breadboard Layout

Use the breadboard plan below to work out your connections.

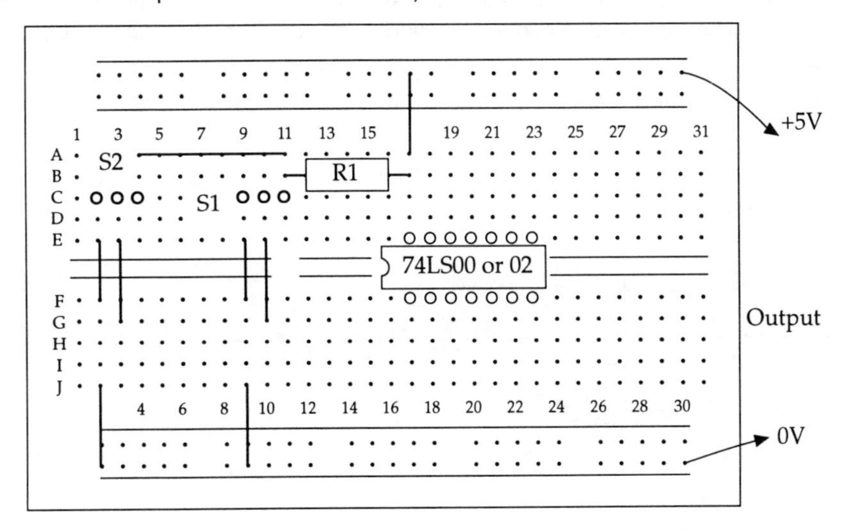

Pin out diagrams

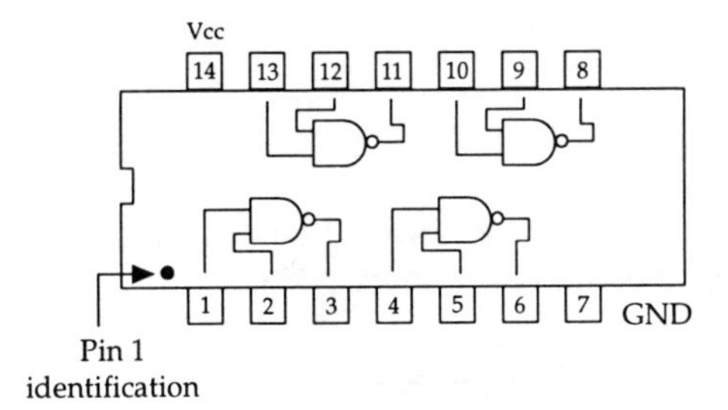

74LS00 pin out diagram

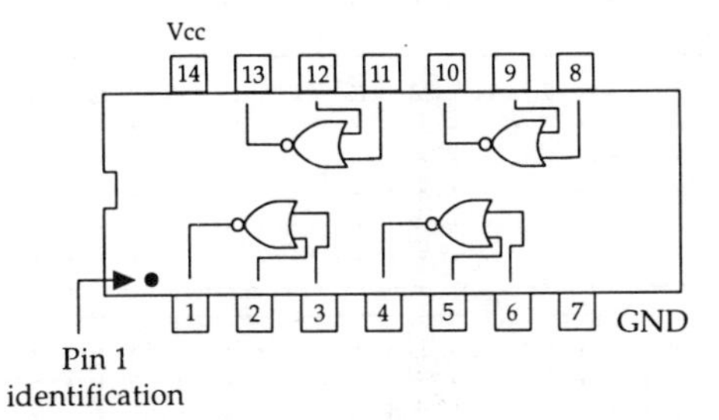

74LS02 pin out diagram

Study Note: If you have difficulty in obtaining the correct function then the most probable cause will be your wiring (or lack of it). Check all connections thoroughly! Remember also that logic 0 is the 0V rail. Do not leave an input floating and assume it is at logic 0! (see **Part 1 Section 2 Information and Skills Bank** How to use breadboards p.96 for breadboard connections.)

Task 2 (Using NAND/NOR Gates)

You have been asked to construct the logic system below in the simplest and cheapest form possible. The logic gate diagram is in pure logic form and therefore will be costly to implement.

In order to simplify and reduce the cost you will need to carry out the following operations:

1) First convert the diagram into all NAND form by replacing gates G1 to G6 with their NAND equivalent.

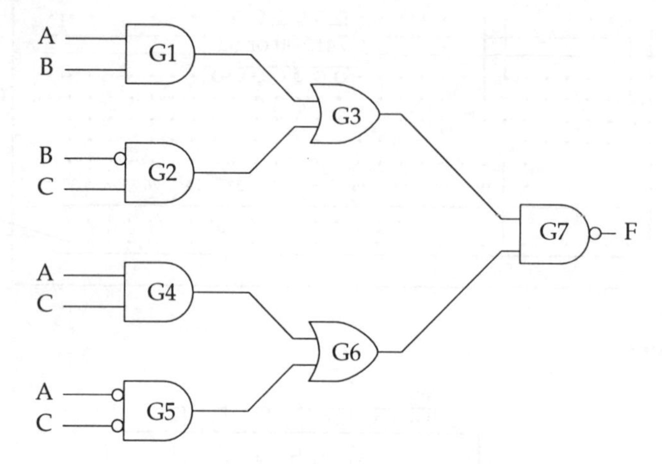

Pure logic gate diagram for task 2

2) Next construct on breadboard the NAND equivalent circuit. (You will need 2 NAND gate Chips and one invertor chip.)

Breadboard diagram for planning your layout

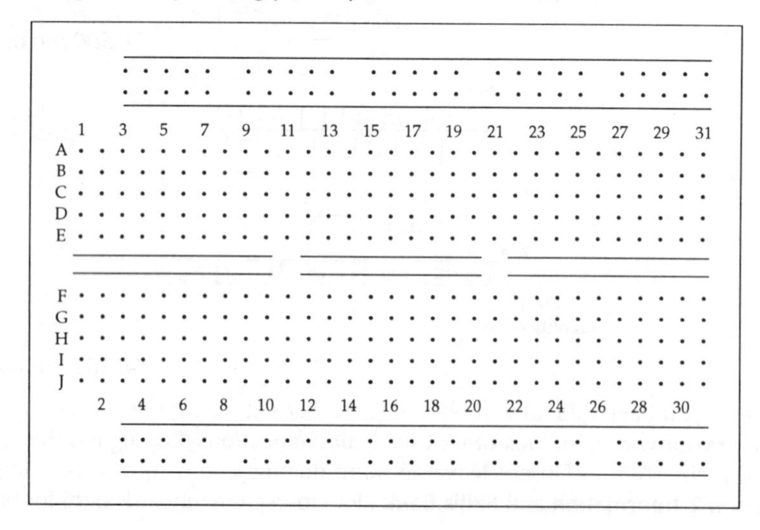

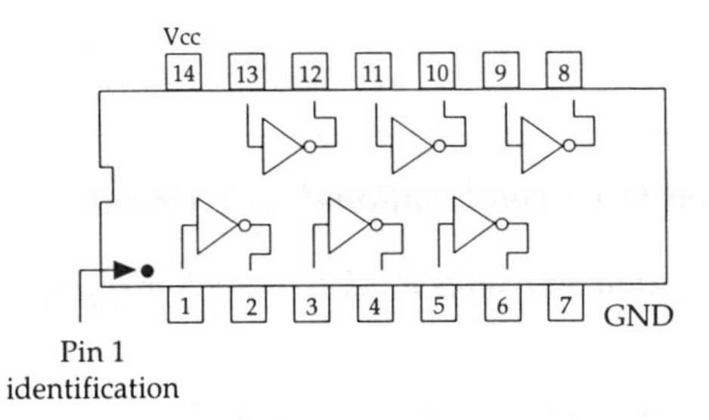

Pin out diagrams for the LS7404 (the LS7400 pin out diagram is shown above in Task 1)

3) Now obtain the truth table from the circuit.

Inputs			Output
C	B	A	F
0	0	0	
0	0	1	
0	1	0	
0	1	1	
1	0	0	
1	0	1	
1	1	0	
1	1	1	

Boolean equation from truth tables:

F =

Truth table for Task 2

4) Next determine the Boolean equation from the truth table.

5) Then simplify the Boolean equation and from the simplified expression produce a simplified gate diagram.

6) Lastly, construct the simplified gate diagram and check its truth table against the original truth table.

7) Finally as a further check convert the original Boolean expression to the simplified expression using de-Morgan's Rules and Boolean algebra simplification methods.

Hint: The output of G3 $= \overline{\overline{(A.B)}.\overline{(\overline{B}.C)}}$

and the output at G6 $= \overline{\overline{(A.C)}.\overline{(\overline{A}.\overline{C})}}$

Convert both these expressions using de-Morgan's Rules. These converted expressions then become the inputs to G7.

Expected Outcome

The final gate diagram should consist of one gate only! This demonstrates the need to simplify before attempting to construct any logic gate system.

Task d

To design, construct and test a combinational logic circuit

Refer to ***Section 2 Information and Skills Bank*** *Karnaugh mapping, p.411.*

Aims

To investigate the operation of seven segment LED displays.
To design, construct and test a seven segment display decoder driver.

Equipment

A variable voltage stabilised power supply. A small breadboard and suitable wires. A digital multimeter.

Components
Logic gate chips. Six 330R resistors. A 7404 Hex invertor chip. Seven segment LED (common anode). Two ultra miniature single pole change-over switches (suitable for a breadboard).

Approach

The first stage is to investigate the construction and operation of a seven segment LED display. We can then determine the requirements of the design of the code converter and driver.

Light Emitting Diode

In a PN junction diode which is **forward biased**, electrons combine with holes. When this happens a small amount of energy is given out. If the semiconductor material is gallium arsenide, the energy is released as infra-red radiation. By mixing other materials with the gallium then visible light is observed.

The diode must be forward biased and passing a few milliamps of current.

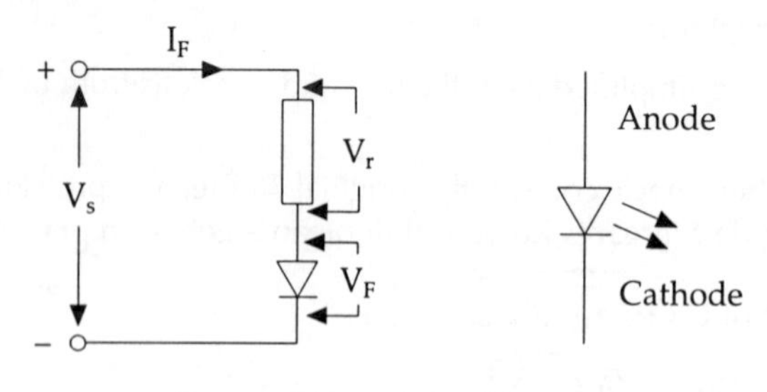

Forward biased diode *Symbol*

Typical LED data

Material	*Colour*	V_F *at*	I_F
Gallium arsenide (GaAsp)	Red	2.0V	10mA
Gallium phosphide (Gap)	Green	2.2V	10mA
Gallium indium phosphide (Galnp)	Yellow	2.4V	10mA
I_{Fmax} is typically 40mA and V_{Fmax} 5V			

Because the diodes have a maximum current, unless this current is limited, the diode will be damaged. A series current limiting resistor is therefore needed.

Determining the Value of the Series Resistor

A (Gap) LED is to be fed from a 5V supply and requires a current of 10mA. Determine the series current limiting resistor required.

$$\text{Voltage across the diode: } V_F = 5V - 2.2V = 2.8V$$

$$\text{Current in the resistor} = 10mA$$

$$\text{Therefore the resistor value will be } \frac{2.8V}{10mA} = 280 \text{ ohms}$$

The nearest preferred value to this will be 270R.

A good bright glow is obtainable with currents of only 10mA (power 20mW). The LED is very efficient because electrical energy is converted directly into light.

Spreading light from the LED can be used to form bars, and these bars form the seven segments of a LED display.

Seven Segment Displays

Individual segments can be lit to represent the decimal numbers 0 to 9 plus a few letters of the alphabet.

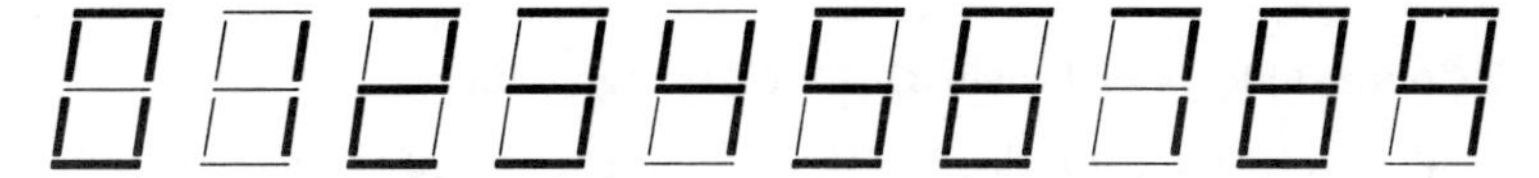

Formation of characters 0 to 9

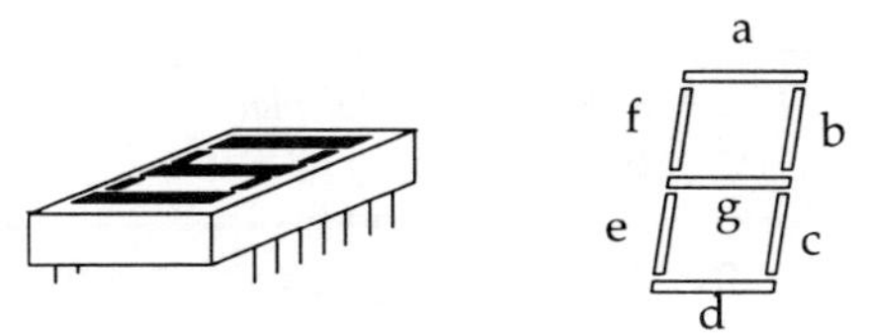

Layout of the seven segments

Common Anode and Common Cathode LED Displays

The diodes used to form the segments may either have all their anodes connected together, or all their cathodes.

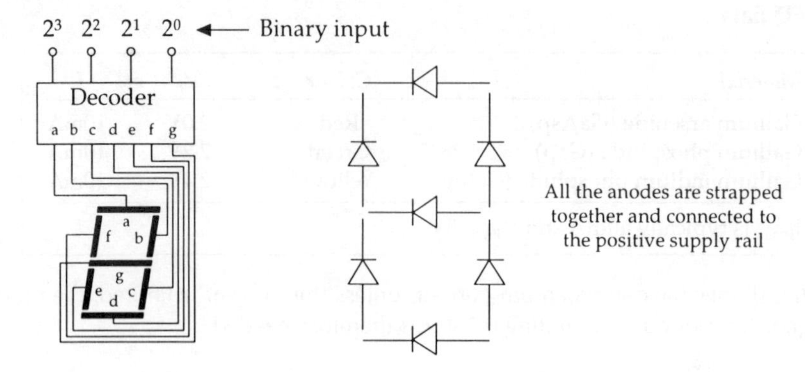

Common anode display

With the common anode display all the anodes are connected to the positive supply rail (+5V). Each individual diode cathode is then switched to ground (0V) to illuminate the diode.

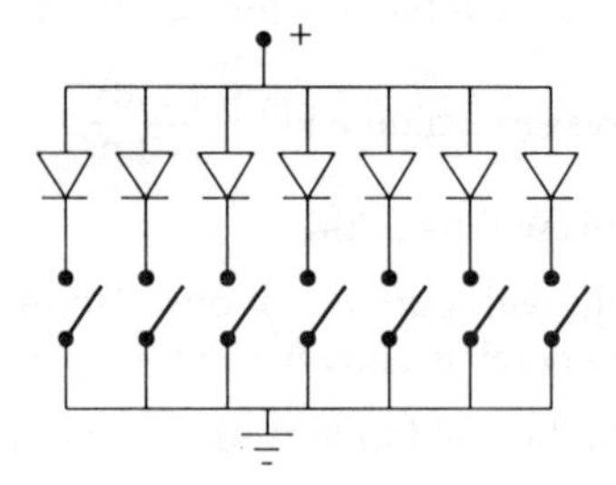

Each cathode is switched to ground to light the LED

Common Cathode Display

With this type of display the cathodes of all the diodes are connected together and the anode of any segment must be switched high (+5V) to illuminate that particular segment.

The Car Park Counter-Decoder-Driver-Display Arrangement

When the count is zero then the display should indicate empty. The binary counter will have a maximum count of 7 and the display should then indicate full. In between these states the display should indicate that there is a space.

Decimal number	*Binary number* C	B	A	*Display*	*Segment code* a	b	c	d	e	f	g
0	0	0	0	E	1	0	0	1	1	1	1
1	0	0	1	S	1	0	1	1	0	1	1
2	0	1	0	S	1	0	1	1	0	1	1
3	0	1	1	S	1	0	1	1	0	1	1
4	1	0	0	S	1	0	1	1	0	1	1
5	1	0	1	S	1	0	1	1	0	1	1
6	1	1	0	S	1	0	1	1	0	1	1
7	1	1	1	F	1	0	0	0	1	1	1

Segment code for a common cathode 7-segment display

Remember that to illuminate a segment then the anode must be switched **HIGH** with the common cathode display (i.e. the segment needs a logic 1).

Block Diagram of the Counter Decoder-Driver LED Display

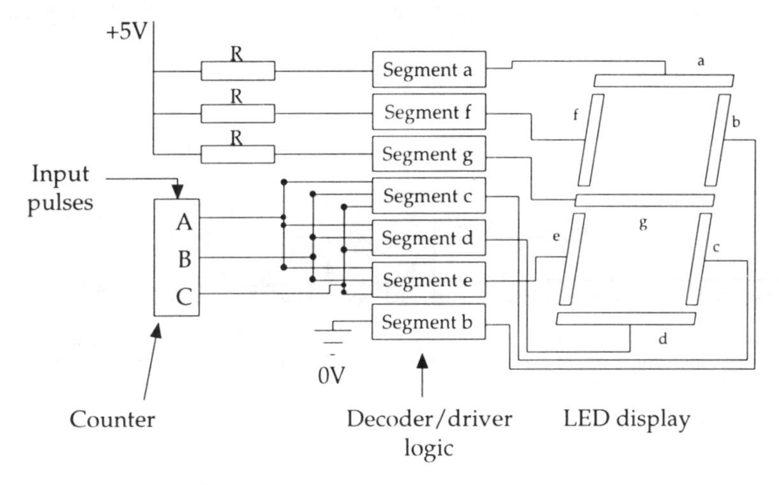

Connections to the decoder-driver

We now have to design the logic circuitry for each segment of the display.

From the table above segments 'a', 'f' and 'g' are always at logic 1. Therefore the anode of these segments can be wired directly via a current limiting resistor to +5V.

Also from the table above segment 'b' is always at logic 0 (0V). Therefore this segment can be wired directly to 0V.

We therefore only need to design the logic circuits for segments 'c', 'd' and 'e'.

The Boolean equations for segments 'c','d' and 'e' can be obtained from the truth table. By using Karnaugh maps these equations can be simplified and a logic gate diagram devised for each of the segments 'c','d' and 'e'.

For example the Boolean equation for segment 'c' in the truth table above is:

$$c = A.\overline{B}.\overline{C} + \overline{A}.B.\overline{C} + A.B.\overline{C} + \overline{A}.\overline{B}.C + A.\overline{B}.C + \overline{A}.B.C$$

Which can be simplified using a Karnaugh map!

The Design Exercise

To ensure that we drive the LED segments sufficiently we are going to use a 7404 Hex invertor chip and a number of series current limiting resistors. The Hex invertor will invert the logic levels. Therefore we will need to change the LED display to a **common anode** type.

Remember that with a common anode display to illuminate a particular segment then that segment's cathode must be switched **LOW**. We are going to use the Hex invertor to switch the cathodes LOW to illuminate a segment.

Step 1. As we are using a common anode display we need to re-construct the segment table above showing the codes required for a common anode display.

Step 2. Produce the Boolean equations for segments 'c','d' and 'e.'

Step 3. Simplify the equations using a Karnaugh map and produce logic gate diagrams for segments 'c','d' and 'e'.

Step 4. Construct the logic gate diagrams and connect each one to the appropriate input of one of the invertors in the 7404. The other three invertors can be employed for segments 'a', 'f' and 'g' as shown below.

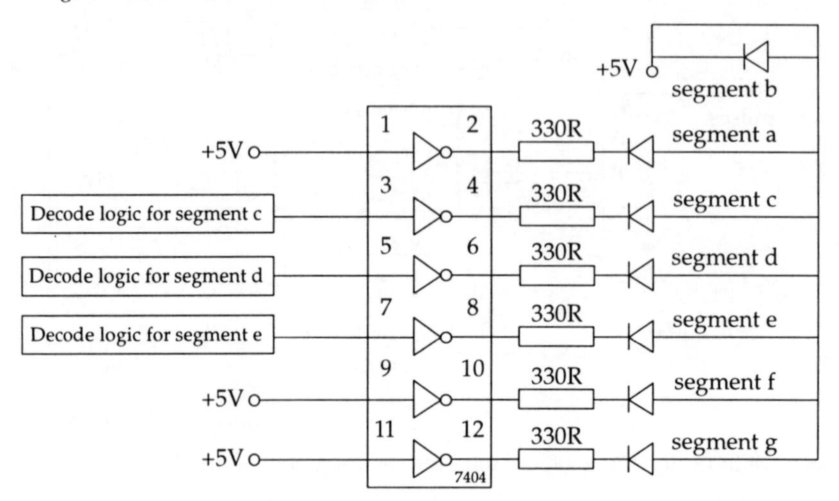

Driving the LED display

Use the breadboard plan below to plan your layout of your circuit. Construct the circuit.

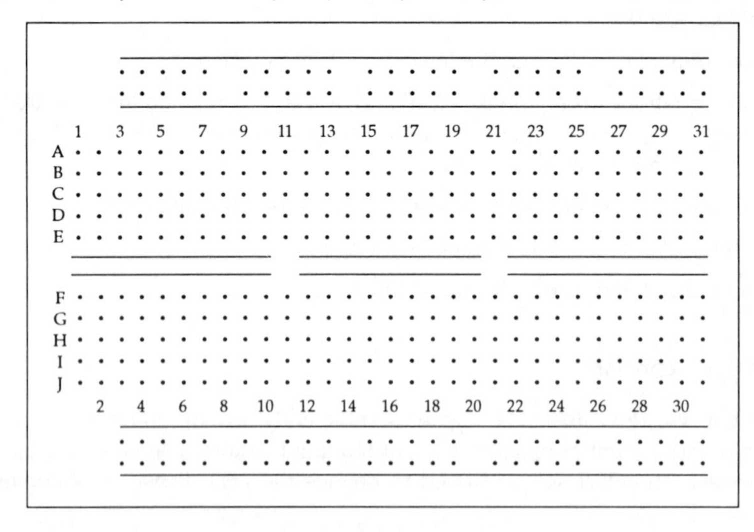

Use three ultra-miniature single pole change-over switches to represent the binary inputs A, B and C (the output from the binary counter).

Remember the most obvious cause if the circuit fails to function correctly will be wiring connections (or lack of them!) For example do not forget to connect the +5V and ground supply rails to each chip!

Check all wiring thoroughly!

Set the output of the power supply to 5V, **switch off and then connect the supply to the breadboard**.

Switch ON the power supply.

Input all 8 combinations of the switches and check that the LEDs give the correct display.

Study Note: Keep the circuit you have just built, you will need it to test the 3-stage Up-Down counter and complete the project for Part 3.

Task e

Construct and test an R.S. bi-stable

*Refer to **Section 2 Information and Skills Bank** Logic families and logic levels, p.389; The bi-stable, p.428.*

Aims
To construct an R.S. bi-stable using discrete components. Use test equipment to investigate the operation of the bi-stable.

Equipment
A variable voltage stabilised power supply. A small breadboard and suitable wires. A digital multimeter. A commercial logic probe.

Components
Two BC108 transistors. Two 3k3 resistors. Four 100k resistors. Two miniature single pole change-over switches (suitable for a breadboard).

Approach

An R.S. bi-stable may be constructed from two transistors as shown below.

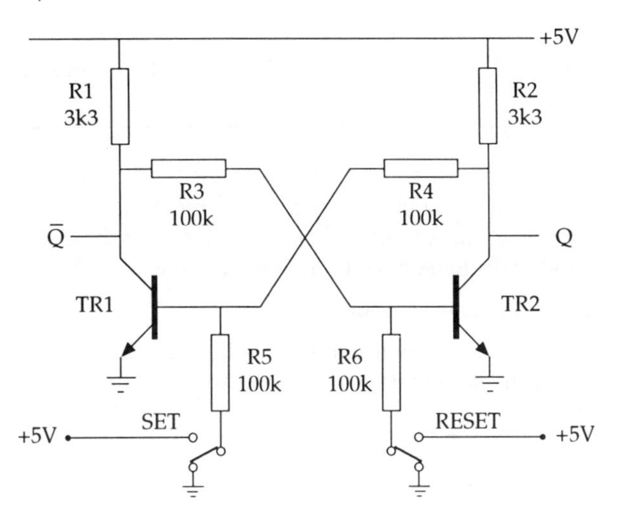

Construct the circuit on a breadboard. A suggested layout is shown below.

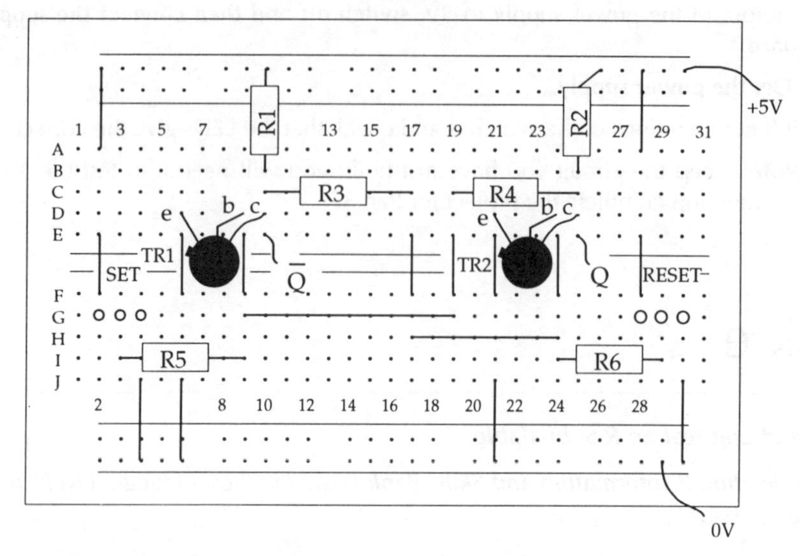

Set the variable power supply to +5V, switch OFF and then connect it to the circuit.

Switch ON the power supply. Operate the switches to SET, or RESET the bi-stable. Use a digital multimeter and record the voltages at the following points in the circuit for both the SET and RESET conditions. Use a logic probe to determine the logic level.

SET Condition (Q =1)

Voltage	TR1	TR2	Logic level (probe)
Collector			
Base			

RESET Condition (Q = 0)

Voltage	TR1	TR2	Logic level (probe)
Collector			
Base			

Study Note: If you have difficulty in obtaining the correct function then the most probable cause will be your wiring (or lack of it). Check all connections thoroughly! (See **Part 1 Section 2 Information and Skills Bank** How to use breadboards, p.96.)

Finally, construct a truth table for your bi-stable.

In the comments column state which conditon is the SET, RESET, and No change. Also state which condition is indeterminate and why it is so.

Inputs		*Outputs*		*Comment*
S	R	Q	$\overline{Q}$	
0	0			
0	1			
1	0			
1	1			

Truth table for the R.S. bi-stable

Task f

Construct and test R.S. and J.K bi-stables

Refer to ***Section 2 Information and Skills Bank*** *R.S. and J.K. bi-stables, p.434.*

Aims

To construct a LED display unit for checking the logic levels at the output of the bi-stables.
To construct and test cross-coupled NOR and NAND bi-stables.
Construct and test a D-type bi-stable.
To investigate the operation of the J, K, Clock, Preset and Clear inputs of a J.K. bi-stable and produce its truth table.

Equipment

A variable voltage stabilised power supply. Two small breadboards and suitable wires. A digital multimeter.

Components

One 74LS02 (NOR gate) chip. One 74LS00 (NAND gate) chip. One 74LS76 J.K. Bi-stables chips. One 74LS04 invertor chip. Four LEDs. Four 330R resistors. Five ultra-miniature single pole change-over switches (suitable for a breadboard).

Approach

Construct the LED display unit.

Connect up the cross-coupled NAND and NOR gate circuits below one at a time and investigate their operation. Connect an invertor between the R and the S inputs to create a D-type bi-stable and investigate its operation.

Finally connect up the 74LS76 J.K. bi-stable and investigate the operation of the asynchronous and synchronous inputs to produce the truth table.

LED Display Unit

Construct and test the LED display unit below.

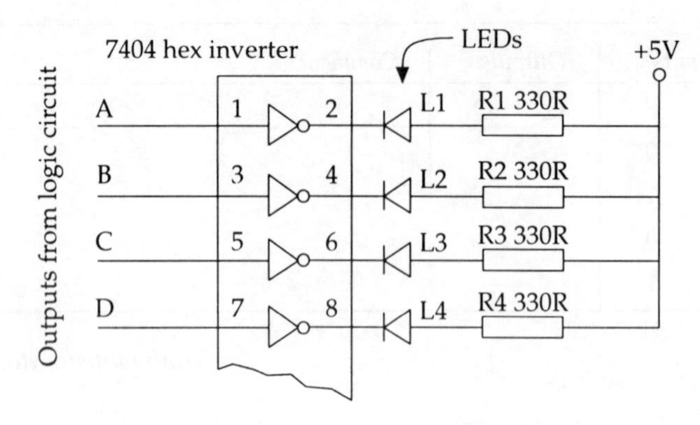

LED display for indicating a logic 1 level

Suggested Breadboard Layout

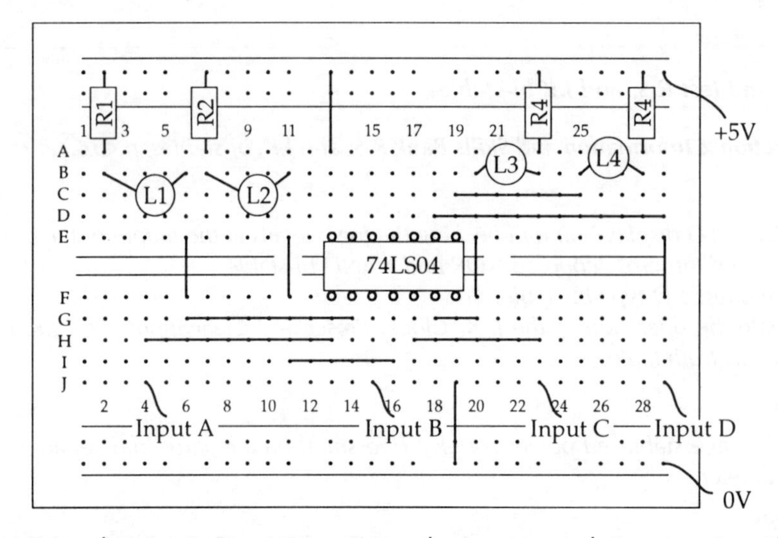

Connect +5V to the circuit. By putting +5V on the input to each invertor via a 1k resistor, check that the associated LED is illuminated.

Study Note: Keep the display unit wired up after you have completed this task. It will be useful for checking the output states of counters and shift registers in later tasks.

The NOR gate R.S. Bi-stable.

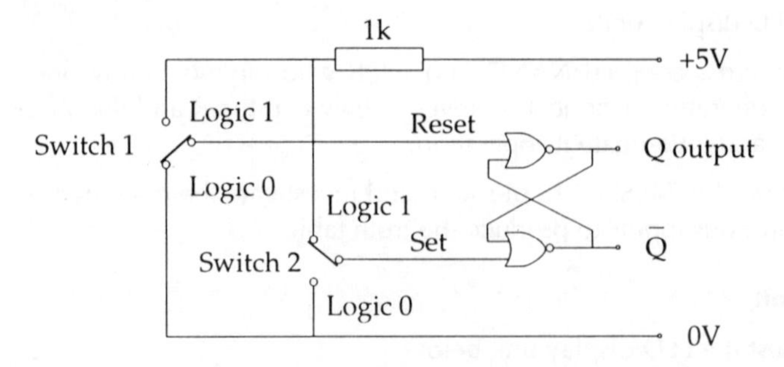

Circuit diagram of a cross-coupled NOR bi-stable

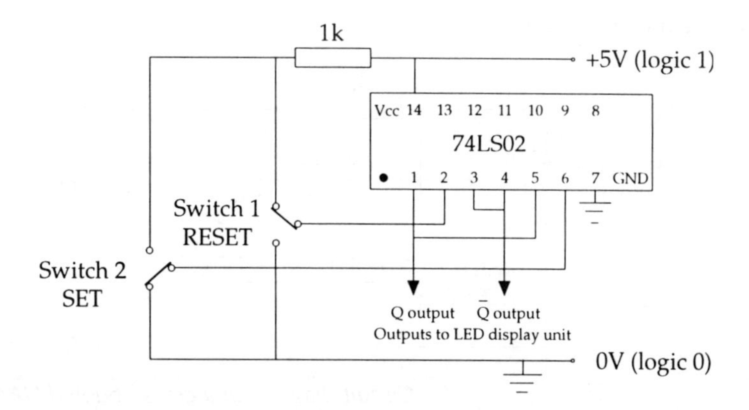

RS bi-stable using a 74LS02 (X-coupled NOR gates)

Suggested Breadboard Layout

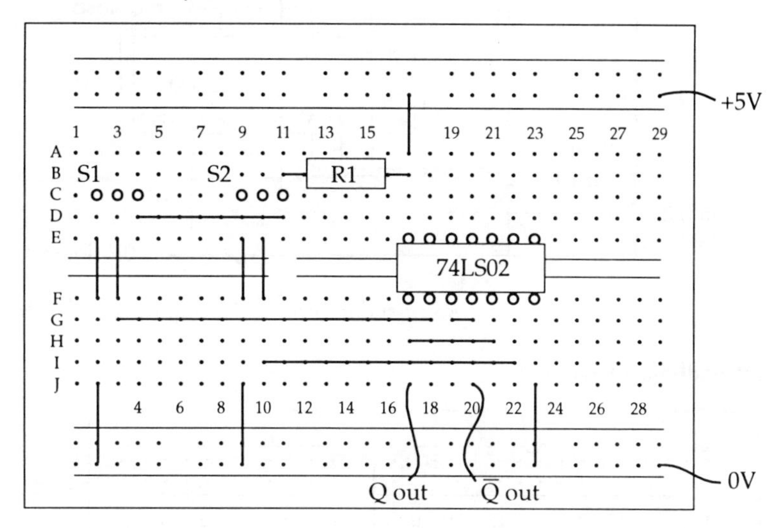

Connect up the circuit and obtain the truth table!

Inputs		*Outputs*		*Comment*
S	R	Q	$\overline{Q}$	
0	0			
0	1			
1	0			
1	1			

The 4-NAND gate R.S. bi-stable

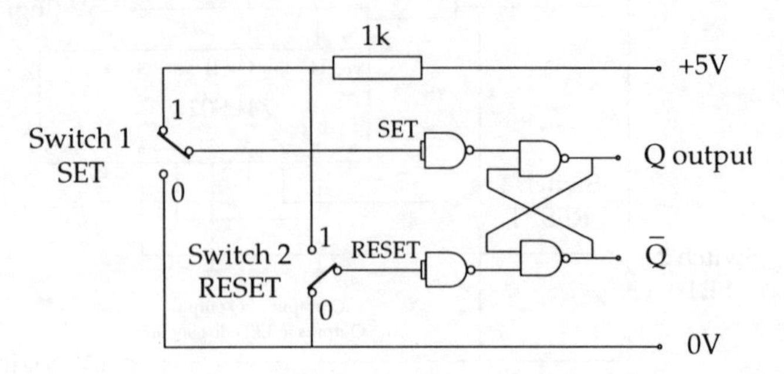

Circuit diagram of a cross-coupled NAND bi-stable

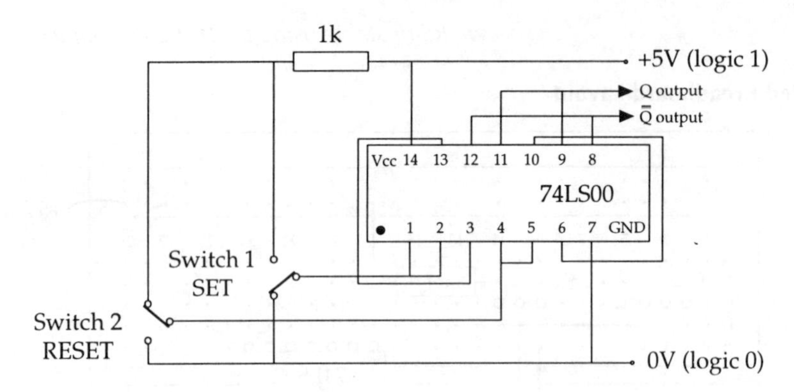

RS bi-stable using a 74LS00 (X-coupled NAND gates)

Suggested Breadboard Layout

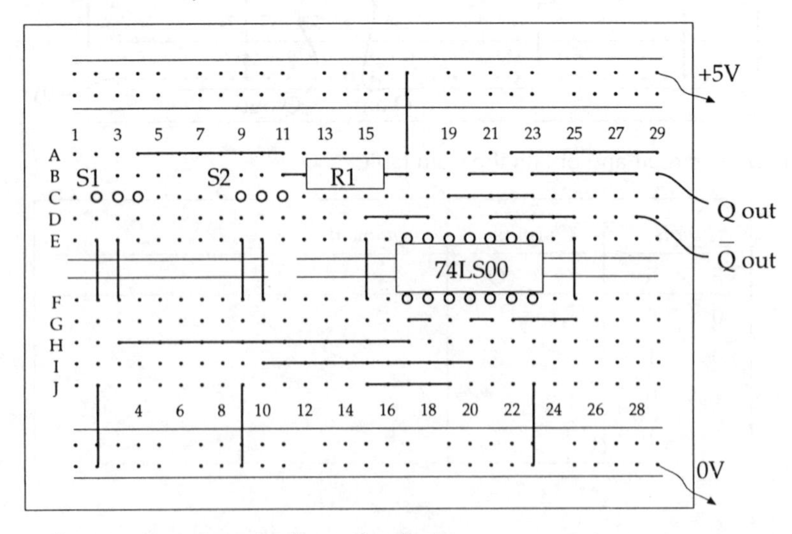

Connect up the circuit and obtain the truth table!

Inputs		*Outputs*		*Comment*
S	R	Q	$\overline{Q}$	
0	0			
0	1			
1	0			
1	1			

Clocked R.S. Bi-stable

A clocked R.S. bi-stable can easily be obtained from the 4-NAND cross-coupled bi-stable circuit above.

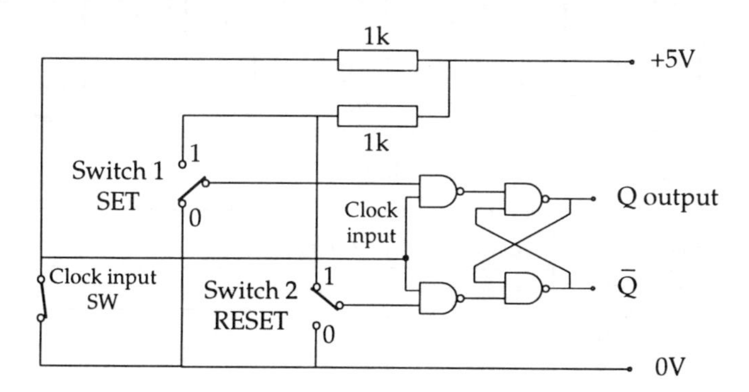

Clocked RS bi-stable formed from NAND gates

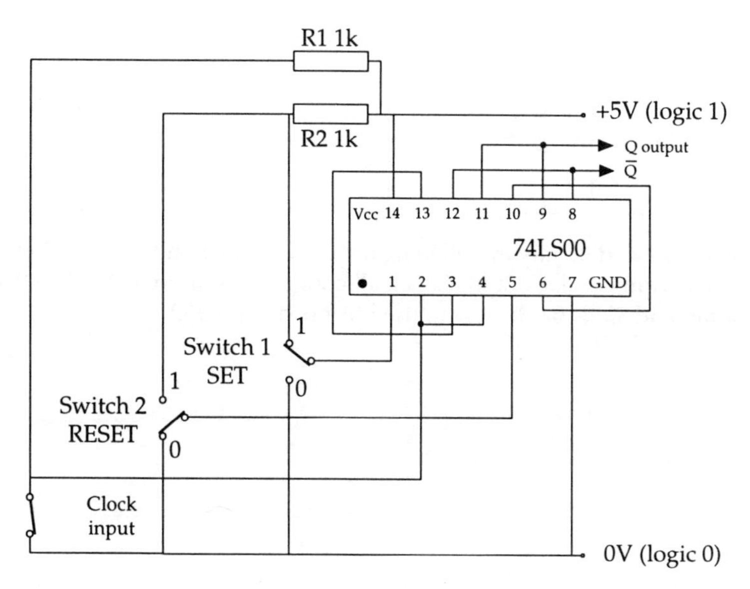

Clocked RS bi-stable using a 74LS00 (NAND gates)

Suggested Breadboard Layout

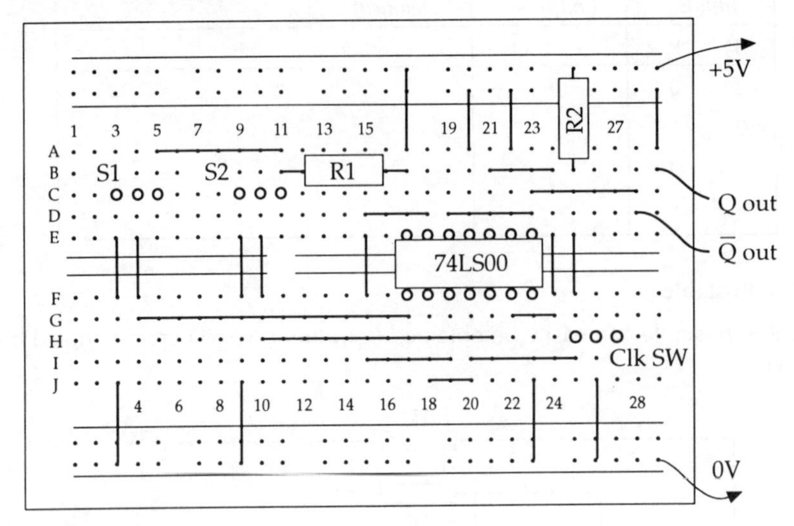

Connect up the circuit and obtain the truth table!

Inputs			*Outputs*		*Comment*
S	R	Clk	Q	$\overline{Q}$	
0	0	0			
0	1	0			
1	0	0			
1	1	0			
0	0	1			
0	1	1			
1	0	1			
1	1	1			

Study Note: If you have difficulty in obtaining the correct function then the most probable cause will be your wiring (or lack of it). Check all connections thoroughly! (See **Part 1 Section 2 Information and Skills Bank** How to use breadboards, p.96.)

The Dual J.K. Bi-stable (74LS76)

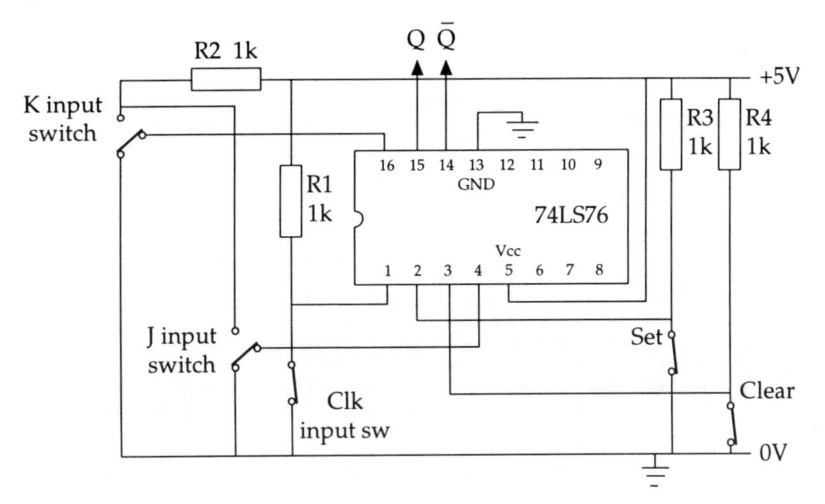

Test circuit for the dual J.K. bi-stable 74LS76

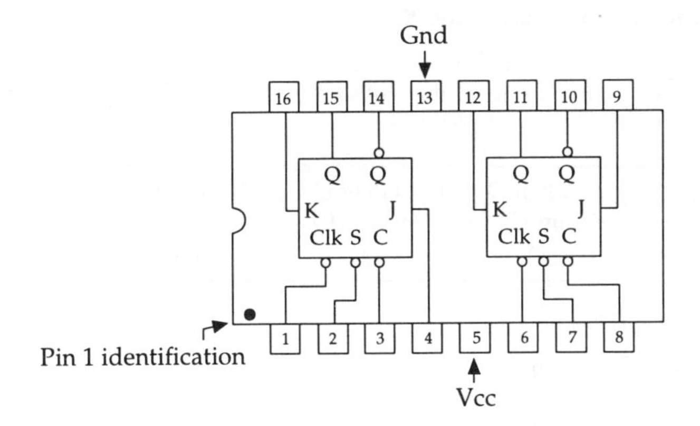

74LS76 (dual J.K. bi-stable) pin out diagram

Suggested Breadboard Layout

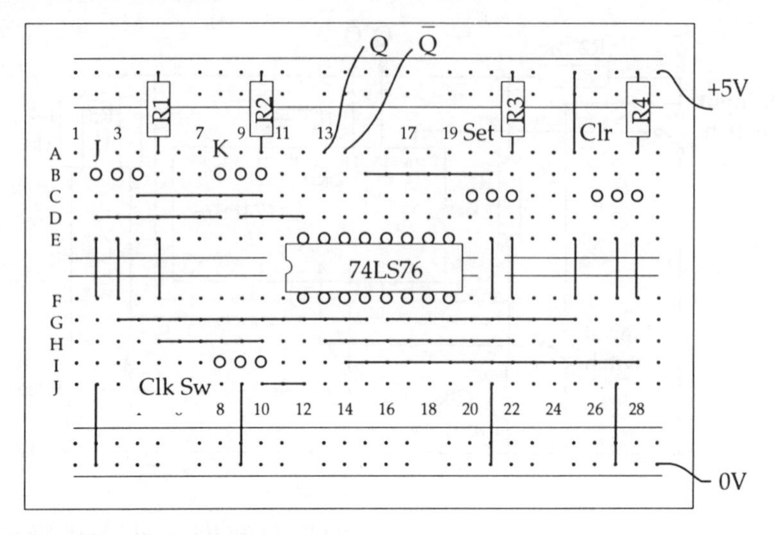

Truth Table (Synchronous Inputs J.and K.)

Connect up the circuit; switch the SET and CLEAR inputs to logic 1. Set the J and K switches as required in each line of the truth table. Input a clock pulse and complete the truth table below!

Inputs J	K	*Output Q (before C.P.)* Q	*Output Q (after C.P.)* Q+	*Comments*
0	0	0		
0	0	1		
0	1	0		
0	1	1		
1	0	0		
1	0	1		
1	1	0		
1	1	1		

Note: C.P. = clock pulse

J.K. truth table (synchronous inputs)

Truth Table (Asynchronous inputs SET and CLEAR)

Switch the SET input to logic 0 and the CLEAR input to logic 1. Switch the J and K inputs to logic 1. Input a number of clock pulses. What is the logic state at Q and does the Q output change at all?

Switch the CLEAR input to logic 0 and the SET input to logic 1. Switch the J and K inputs to logic 1. Input a number of clock pulses. What is the logic state at Q and does the Q output change at all?

Finally, using only the SET and CLEAR switches, complete the truth table for the asynchronous inputs.

Inputs Set	Clear	*Output* Q	*Comments*
1	1		
0	1		
1	0		
0	0	?	Indeterminate state

Truth table for the asynchronous inputs

The DATA or D-Type bi-stable

A data latch or D-type bi-stable can be made by inserting an invertor between the J and the K inputs of a J.K. bi-stable. This simply means that the inputs J and K are always complementary (when J=1, K=0 and vice versa) and that now there are only two inputs; the D input and the clock input. (The asynchronous inputs are not used in this configuration.)

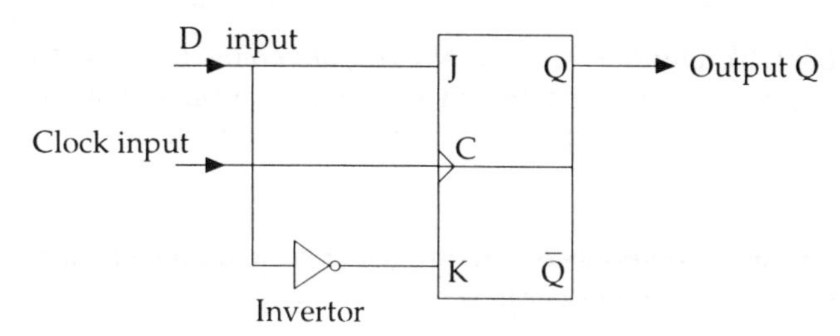

Converting the J.K. bi-stable into a D-type

The operation of the circuit is simply:

The logic state at output Q follows the logic state at input D, after a clock pulse.

Connect an invertor between one set of J and K inputs on the 74LS76. Use either a NAND chip 74LS00 with one of the gates connected as a NOT gate or an 74LS04 Hex invertor chip.

Devise your own breadboard circuit!

With the SET and CLEAR inputs set to logic 1, change the logic level at the D input, apply a clock pulse and determine the output logic state. Does output Q follow input D after a clock pulse?

Task g

Tests on a three-stage binary counter

Refer to ***Part 3 Section 2 Information and Skills Bank*** *Binary counters, p.447.*

Aim

To construct an asynchronous and a synchronous three stage binary Up-Down counter from J.K. bi-stables and test the operation of the circuits. To construct a de-bounced switch to provide the clock pulse input for the counters.

Equipment

A variable voltage stabilised power supply. A small breadboard and suitable wires. A digital multimeter. The LED display unit constructed on a breadboard for Task f.

Components

Two 7476 J.K bi-stable chips. Two 7400 NAND gate chips. One 7404 Hex invertor chip. Three ultra miniature single pole change-over switches (suitable for a breadboard).

Approach

First construct the switch de-bounce unit on a separate small breadboard. (This is required for this Task and for Task h Shift Registers).

Study the explanation of an Up-Down counter given below. Then construct and test the asynchronous and synchronous counters. Use the switch de-bounce unit for the clock pulse and the LED display unit to test the counters and determine the logic state of each output.

Finally connect one of the counters to the decoder driver circuit you constructed in Task d, and test the operation of the complete car park monitoring system.

Switch De-bounce Unit

A normal switch suffers from contact bounce and if used as the input to a counter then it often gives rise to erratic operation.

A pair of cross coupled NANDs may be connected as an R.S. latch and used to de-bounce the contacts. Alternatively a pair of invertors may be employed as shown below.

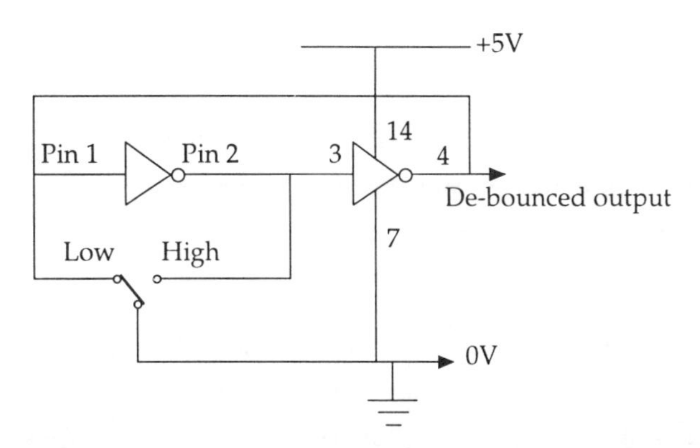

A switch de-bounce unit (1/3 of a 7404 hex invertor chip)

Construct the circuit on a separate small breadboard and test its operation.

When the switch is in the LOW position the output should be approximately 0V. When in the HIGH position the output should be approximately +5V.

This unit will now be used to provide the input clock pulses for the Up-Down counters.

An Up-Down Counter

A three stage UP counter can be constructed from three J.K. bi-stables as shown.

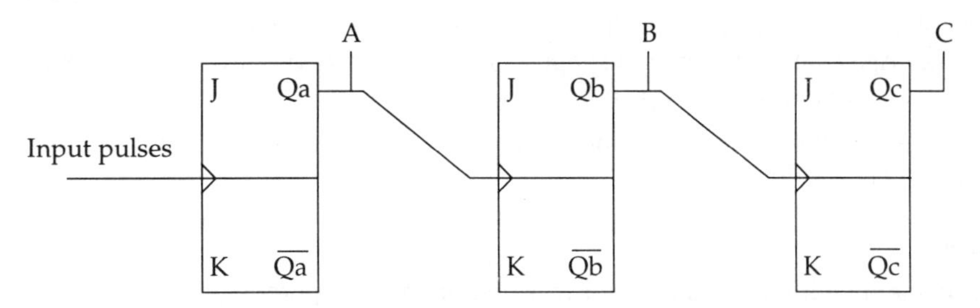

An asynchronous UP counter

A three stage DOWN counter can also be constructed from three J.K. bi-stables as shown below.

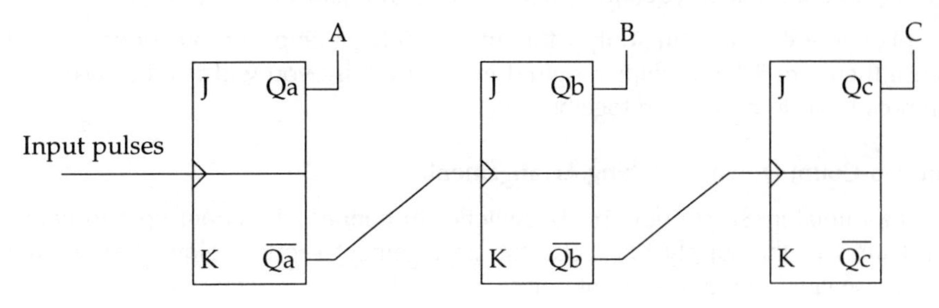

An a synchronous DOWN counter

For an UP counter the following bi-stable is clocked from the Q output of the preceding bi-stable. Whereas for a down counter, the following bi-stable is clocked from $\overline{Q}$.

By employing suitable logic and an Up-Down control line the circuit can be turned into an Up-Down counter.

Up-Down Counter

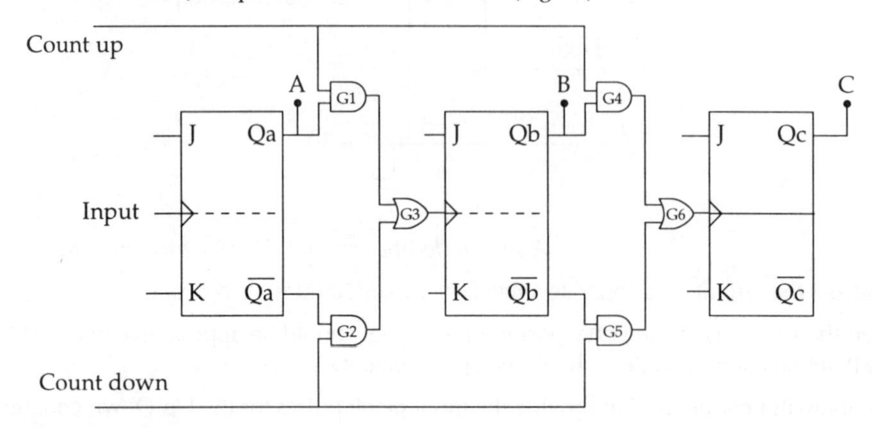

An asynchronous up-down counter

Circuit Operation

With a logic 1 on the count up line, gates G1 and G4 are enabled and G2 and G5 disabled (an AND gate with one of its inputs at logic 0 can never have a logic 1 output). The circuit therefore becomes an up counter.

Similarly when the count down line has a logic 1, gates G2 and G5 are enabled and the circuit becomes a down counter.

Task g

The problem with the circuit as shown is that it employs pure logic (AND and OR gates) which makes it more difficult to construct.

The first step is to convert the logic gates into an all NAND form: i.e. replace each gate in the diagram with its NAND equivalent (see **Part 3 Section 2 Information and Skills Bank** Simplification of Boolean Equations, p.397 and de-Morgan's Rules, p.405).

Next, plan your circuit, using only 7400 and 7476 logic chips, on breadboards. (Note that because there are 5 logic chips required in this exercise you will need to use two of the small breadboards, connected together.)

Count Up-Count Down Switching Arrangement

Use ultra-miniature single pole p.c.b. switches to connect the count up and count down control lines to the supply rails. Use the arrangement shown below. Use another ultra miniature switch to input the count pulses.

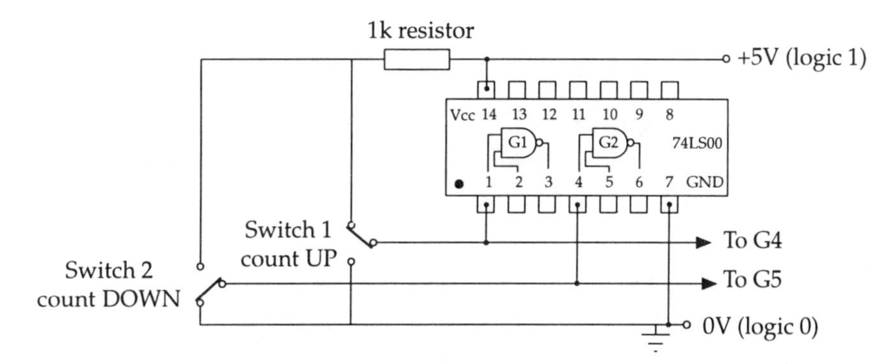

Count UP – count DOWN switching arrangements

Now construct the circuit and test its operation.

Remember the most obvious cause if the circuit fails to function correctly will be wiring connections (or lack of them!) For example do not forget to connect the +5V and ground supply rails to each chip!

Check all wiring thoroughly!

Set the output of the power supply to 5V, **switch off and then connect the supply to the breadboard.**

Switch ON the power supply.

Testing the Circuit

To test your circuit, connect the outputs A,B and C to the LED display unit you constructed in Task f.

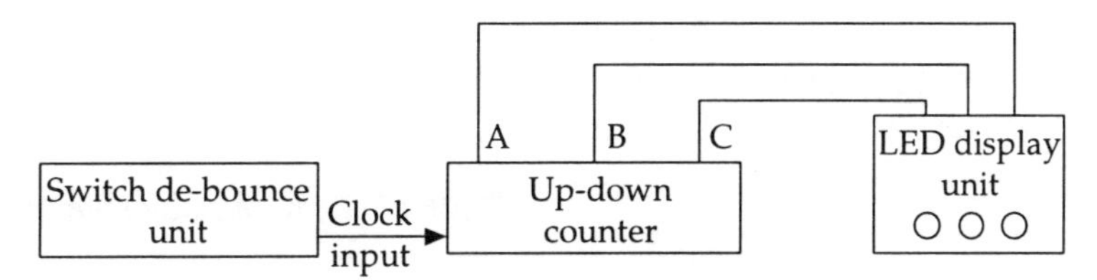

Block diagram of de-bounce unit, counter and display

Connect the de-bounce and display units to the counter.

By operating the input switch and setting count up line to logic 1, make the counter increment up to the count of 7.

Observe the LED display and check its operation. LED A will be the least significant digit and LED C the most significant. Now switch the circuit to act as a down counter and decrement the count back to zero.

A Synchronous Counter

A synchronous 3 stage counter circuit is very similar and only requires a few modifications to your existing circuit.

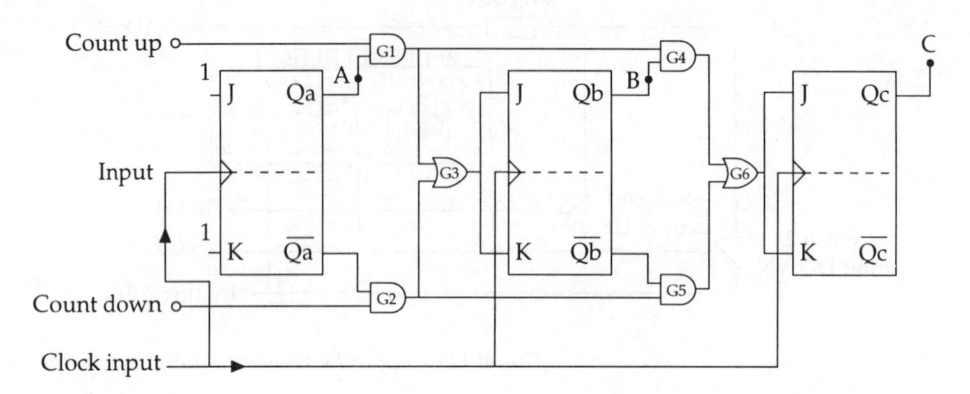

An asynchronous UP-DOWN counter

Modify your existing circuit to the one above and test its operation as before.

7476 Pin Out Diagram

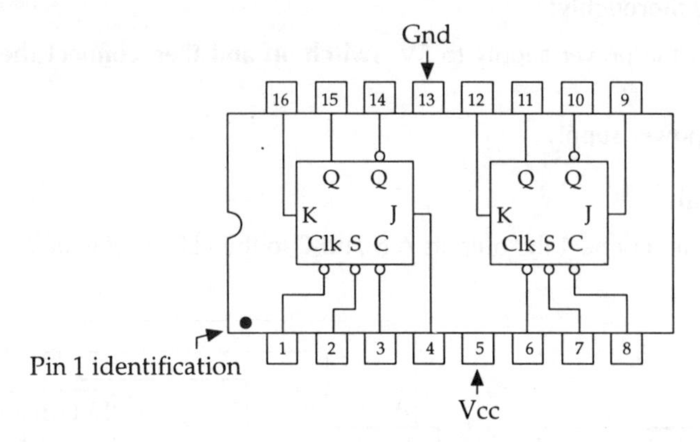

7476 (dual J.K. bi-stable) pin out diagram

Task h

Construct and test a shift register and ring counter

*Refer to **Section 2 Information and Skills Bank** Shift registers, p.454.*

Aims

To investigate the use of J.K. bi-stables to form shift registers and ring counters.

Equipment

A variable voltage stabilised power supply. A small breadboard and suitable wires. A digital multimeter.

Components

Two 7476 J.K. Bi-stables chips. LED display unit constructed in Task f. De-bounced switch unit constructed in Task g. Two ultra-miniature single pole p.c.b. change-over switches (suitable for a breadboard).

Approach

A four-stage Shift register may be constructed from two 7476 J.K. bi-stable chips as shown below.

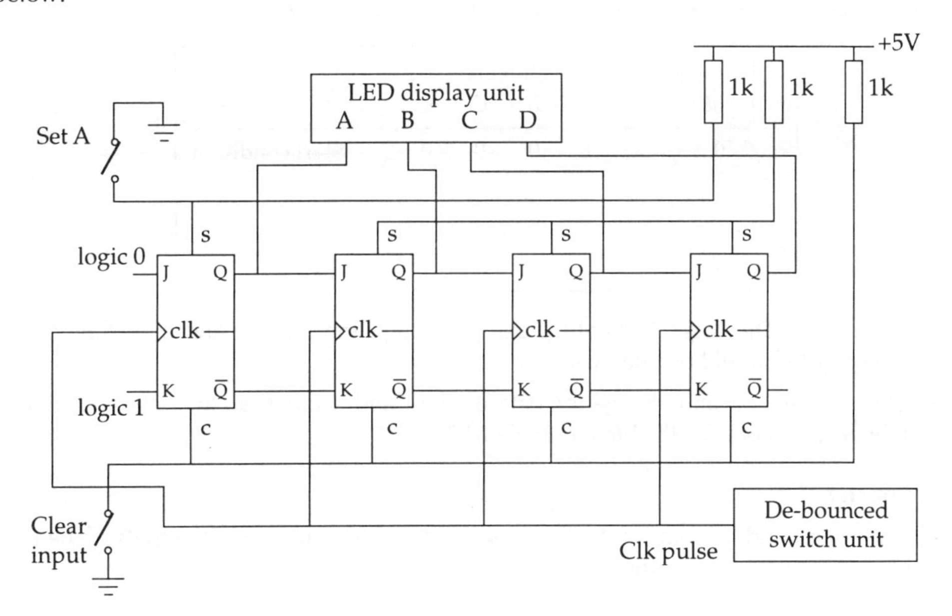

Four-stage shift register

Shift Register

Connect up the circuit as shown. Momentarily operate the clear switch. This will set all the outputs A,B,C and D to logic 0.

Momentarily operate the SET switch to set output A to logic 1. LED A should now be ON and all the other LED's should be OFF.

Input a series of four clock pulses and record the result in the table below.

Clock pulse number	*A*	*B*	*C*	*D*	
0	1	0	0	0	Start condition
1					
2					
3					
4					

Expected Result: The logic 1 should progress through the register and after 4 clock pulses the register should be filled with 0s.

Now reverse the connections to the J. and K. inputs of the first bi-stable: i.e, make J = logic 1; K = logic 0.

Momentarily operate the clear switch. This will set all the outputs A,B,C and D to logic 0.

Momentarily operate the SET switch to set output A to logic 1. LED A should now be ON and all the other LEDs should be OFF.

Input a series of four clock pulses and record the result in the table below.

Clock pulse number	*Outputs* A	B	C	D	
0	1	0	0	0	Start condition
1					
2					
3					
4					

Expected Result: The logic 1 should progress through the register and after four clock pulses the register should be filled with 1s.

Now clear the register and change the data on the J and K inputs so that after four clock pulses the register contains the binary number 1010.

Ring Counter

Re-connect the J and K inputs of the first bi-stable to the Q and NOT Q outputs of the last bi-stable as shown to form a ring counter.

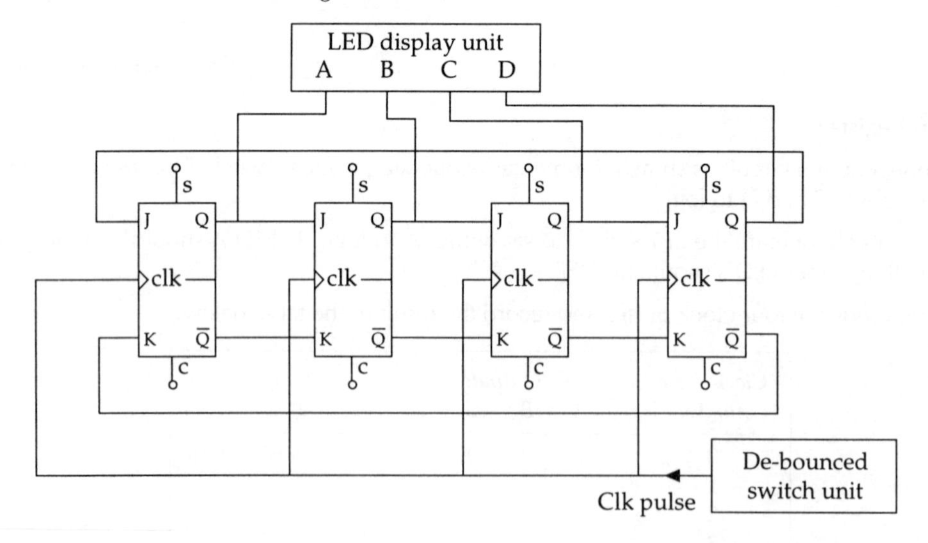

Four-stage ring counter

Set output A to logic 1 and outputs B,C, and D to logic 0 using the SET and clear inputs as before.

Input five clock pulses and record the result in the table below.

Clock pulse number	*Outputs A*	*B*	*C*	*D*	
0	1	0	0	0	Start condition
1					
2					
3					
4					
5					

Expected Result: The logic 1 should progress through the register and after four clock pulses it should be back at output A: i.e. LED A illuminated.

Twisted Ring Counter

Rearrange the J.K. inputs of the first bi-stable by connecting J to the NOT Q output and K to the Q output as shown in the diagram below to form a twisted ring counter.

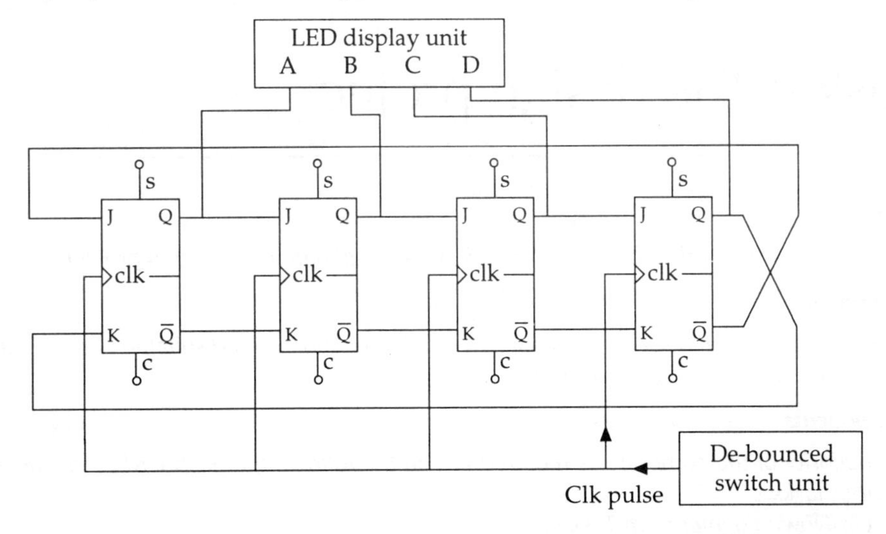

Twisted ring counter

SET all the outputs to logic 0 by using the CLEAR inputs.

Input nine clock pulses and record the result in the table below.

Clock pulse number	Outputs A	B	C	D	
0	1	0	0	0	Start condition
1					
2					
3					
4					
5					
6					
7					
8					
9					

Study Note: If you have difficulty in obtaining the correct function then the most probable cause will be your wiring (or lack of it). Check all connections thoroughly! (See **Part 1 Section 2 Information and Skills Bank** How to use breadboards, p.96.)

Task i: The design project

Aims

To construct and test the complete car park system outlined in the design project.

Equipment

A variable voltage stabilised power supply. Small breadboards and suitable wires. A digital multimeter. A commercial logic probe.

Components

The majority of the components required are on the following breadboards constructed in previous tasks.
The Up-Down counter from Task g.
The decoder/driver and LED unit from Task d.
The switch de-bounce unit from Task g.

In addition:
Two ultra miniature single pole p.c.b. change-over switches (suitable for a breadboard). One 74LS04 Hex invertor chip. One 74LS02 NOR gate chip.

Approach

The basic system of the car park is shown below. Connect the Up-Down counter to the decoder/driver and the display units as shown below.

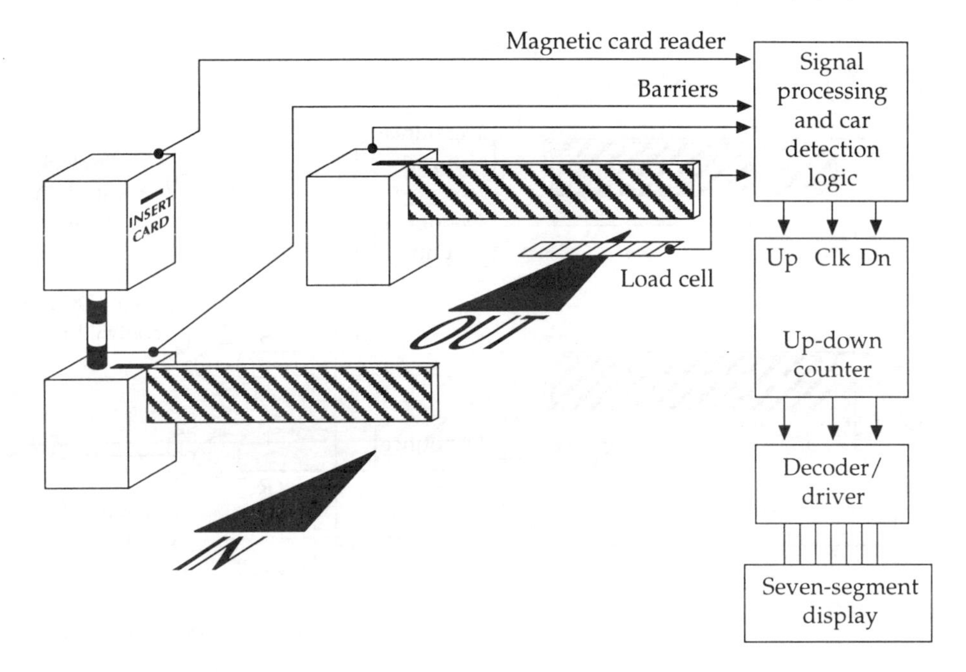

The car-park system

We now need to sort out the signal processing and car detection logic circuitry; i.e. the signals from the barriers and sensors which drive the Up-Down counter.

Up-Down Control signals

These signals can be obtained from two NOR gates connected as an R.S. bi-stable. The outputs of the bi-stable are complementary and can drive the Up-Down control lines directly.

Using a switch to represent the magnetic card reader, feed the output from the switch via a switch debounce unit to the SET input of an R.S. bi-stable constructed from the two cross coupled NOR gates. Connect the Q output of the bi-stable to the UP control line on the Up-Down counter.

Use another de-bounced switch from the load cell in the road surface as a RESET input to the R.S. bi-stable. Connect the NOT Q output to the DOWN control line of the Up-Down counter.

Barrier Signals

Use two more switches to represent the car park barriers. Feed the output from each switch via switch de-bounce units to a two input OR gate (two NOR gates connected to give the OR function). The output from the OR gate then becomes the clock input to the counter.

Input clock pulses from the de-bounce unit (Task g) and check the correct operation of the circuit and display.

The arrangement is shown below.

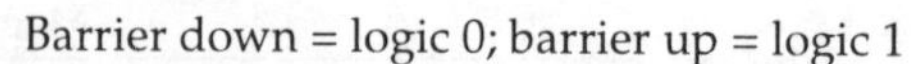

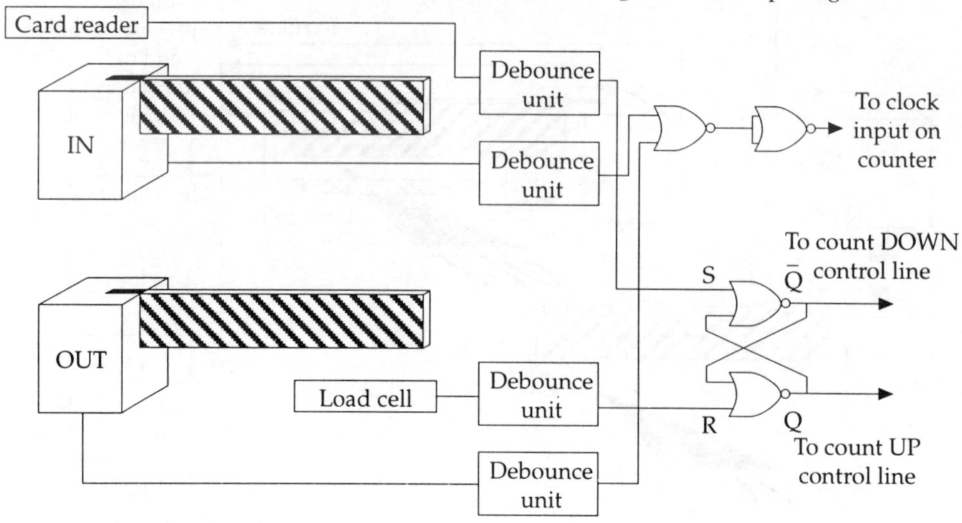

Barrier and sensor logic system

Circuit Operation

When the magnetic card reader is activated then it SETs the R.S. bi-stable which puts a logic 1 on the UP control line. When the IN barrier rises it will input a logic 1 on the clock pulse input of the counter via the OR gate. When the barrier is lowered the logic 1 goes to logic 0 which increments the counter.

When the load cell is activated then it RESETs the bi-stable which puts a logic 1 on the DOWN control line. When the OUT barrier rises it will input a logic 1 on the clock pulse input of the counter via the OR gate. When the barrier is lowered the logic 1 goes to logic 0 which decrements the counter.

De-bounce Units

In Task g you should have constructed a switch de-bounce unit from two invertors. You will need to construct three more similar units. Use two of the spare invertors on the existing switch de-bounce unit constructed in Task g. Create another two de-bounce units with another 74LS04 chip.

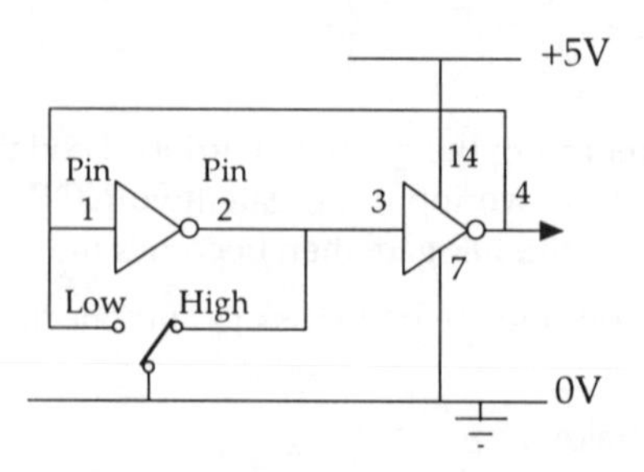

A switch de-bounce unit (1/3 of a hex invertor chip)

Next use the NOR gates in the 74LS02 chip to make a cross-coupled R.S. bi-stable as shown below.

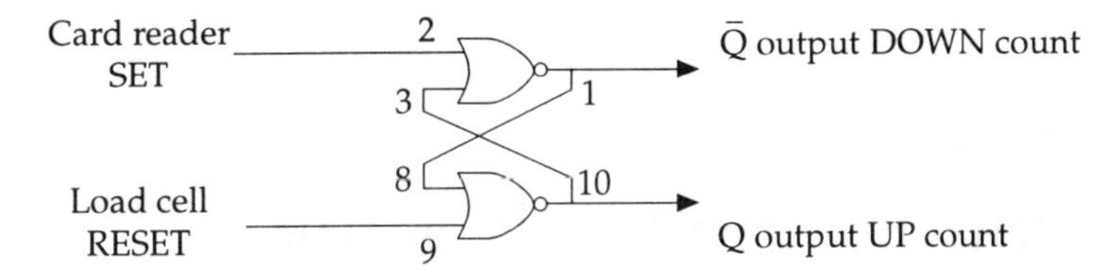

Cross-coupled NOR bi-stable (7402 pin connections)

Connect the R.S bi-stable outputs to the Up-Down counter circuit of your counter-decoder-display system. Connect the barrier outputs to the clock input of your counter-decoder-display system.

Operate the card reader switch (ON-OFF). Operate the IN barrier switch several times and check that the system is counting UP and gives the correct display.

Now operate the load cell switch, (ON-OFF) Operate the OUT barrier switch several times and check that the system is counting DOWN and gives the correct display.

Next, operate all the switches to simulate the following and check the operation of the system:

Step 1 3 cars enter the car park.
Step 2 2 cars leave the car park.
Step 3 6 cars enter the car park.
Step 4 4 cars leave the car park.
Step 5 1 car enters.
Step 6 4 cars leave.

Finally, you will need to consider a method of preventing the counter advancing beyond the count of 7 should another car attempt to gain access to the car park; i.e. if another UP clock pulse is received on the count of 7, the display will change from **F**-ull to **E**-mpty!

Part 3: Digital Electronics

Section 2
Information and Skills Bank

Contents

Introduction to digital control

Digital Circuits and Logic Control

Logic circuits are increasingly used in industry and in everyday life to control machines and processes. Some typical applications are: sequencing of a multiple operation machine; lift control; automatic warehouse control; checkweigher operation; welding sequence control; automatic car park barrier operation.

Before we can control any process we need signals from input transducers (sensors) that represent a physical quantity or condition (e.g. temperature measurement or a limit switch closed).

Physical quantities such as temperature, pressure, flow, humidity, light or sound level etc, are all examples of analogue quantities.

Analogue Signals

In an ANALOGUE System, the input transducers (sensors) convert the physical quantities into electrical voltages or currents.

The physical quantity can vary continuously by very small or large amounts. Therefore the electrical voltage or current will vary in a similar way.

An a.c. sinewave is a good example of an analogue signal, as it varies continuously and can have any value.

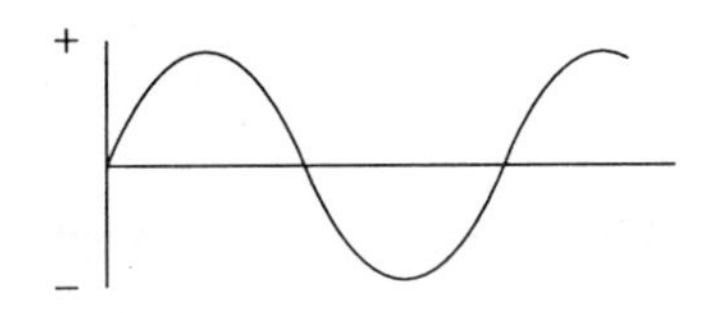

An analogue signal can vary continuously and have any value

Digital Signals

A digital signal can only have discrete values just like a box of ball bearings. You can have one ball bearing or two ball bearings etc, but not half a ball bearing or one and three quarters of a ball bearing.

In a DIGITAL system, therefore, the physical quantity we measure must first be converted into a digital form: either into a multi-level digital signal (a number of ball bearings) or a two-state (binary) signal (there is either one ball bearing or no ball bearing).

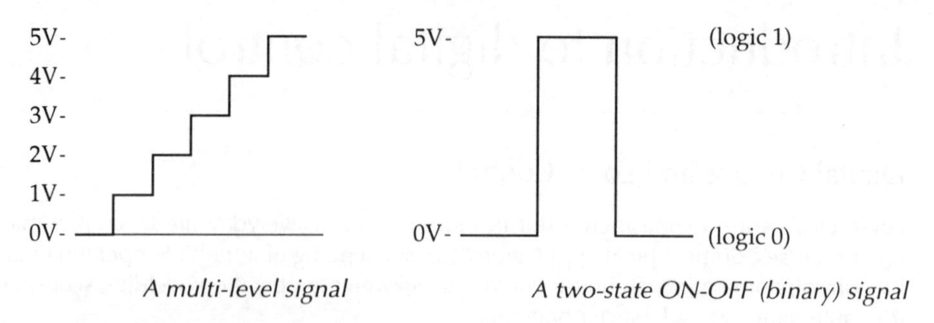

A multi-level signal *A two-state ON-OFF (binary) signal*

Example of Analogue and Digital Signals

If, for example, we have a checkweigher which can weigh up to 10kg, in an analogue system the output could be a voltage which can vary from zero to 10v, the voltage being directly proportional to the weight in the pan. This voltage could be measured and indicated by a pointer type (analogue) voltmeter.

Again taking a 10kg checkweigher as our example, the output could be a series of digits, say 000 to 10-0, to represent weights from zero to 10kg in discrete steps of 0.01kg (a multi-level digital signal).

The diagrams below show the difference between an analogue reading and a digital reading for the checkweigher.

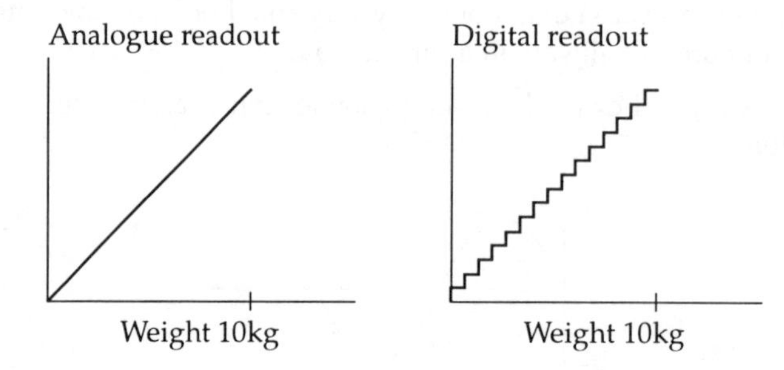

The difference between an analogue and a digital readout of the checkweigher

Note that the ANALOGUE systems output can vary smoothly and continuously. By contrast, the output of the DIGITAL system increases in steps; each step being a unit increase in the least significant digit of the readout.

For example, the weight could either be 9.99kg or 10.0kg in a three digit readout; the difference 0.01kg would be the RESOLUTION of the system.

Resolution

The resolution of a system is the smallest change that can be detected by the system: i.e. in the above digital system a change of less than 0.01kg would not be detected.

On-Off Digital Control.

Many digital control systems do not depend upon a measurement of a physical quantity such as temperature or light level but upon On-Off (two state) signals.

On-Off signals are received from limit switches, push buttons, photoelectric heads, magnetic censors, force sensors and similar devices.

A switch is either closed or open. A push button contact is made or not; a photo cell is illuminated or it is not; there is no ambiguity in such a system!

Two-state (binary) digital control has the advantage over a multi-level digital system in that it uses ON-OFF signals. If however, you wish for close control of the temperature of a furnace for example, then a measurement of the actual temperature is required, before conversion into a digital form.

At the output of the control system, some action must result; for instance, a contactor or solenoid valve must operate, a motor must be started up, or a ticket is to be issued.

A Typical Control System

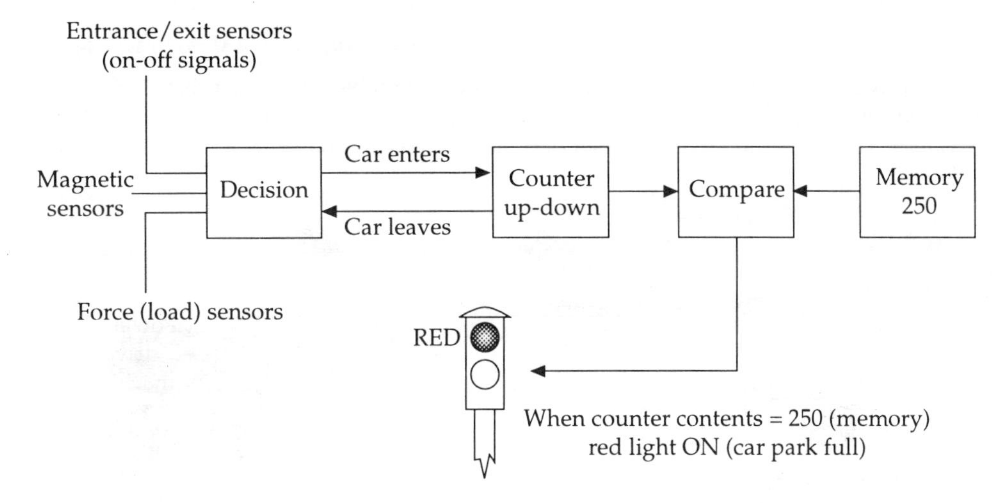

Automatic car park barrier

In the system above, the entrance and exit of the car park have a number of sensors which provide on-off signals. A decision is then made as to whether a car has entered or left the car park.

The counter is then incremented or decremented accordingly. The contents of the counter are continually compared with the contents of the system's memory. When the two are equal, no more cars can be parked and the red stop light at the entrance is illuminated.

Binary Digital Logic Control

Digital logic control which employs a two-state method uses ON-OFF or binary control signals. Whenever we have to store or process numbers in a digital system, we use the binary number system. This system has only two digits: 1 and 0.

There are many electrical devices which have two distinct states, which can be used to store and represent binary digits.

For Example:

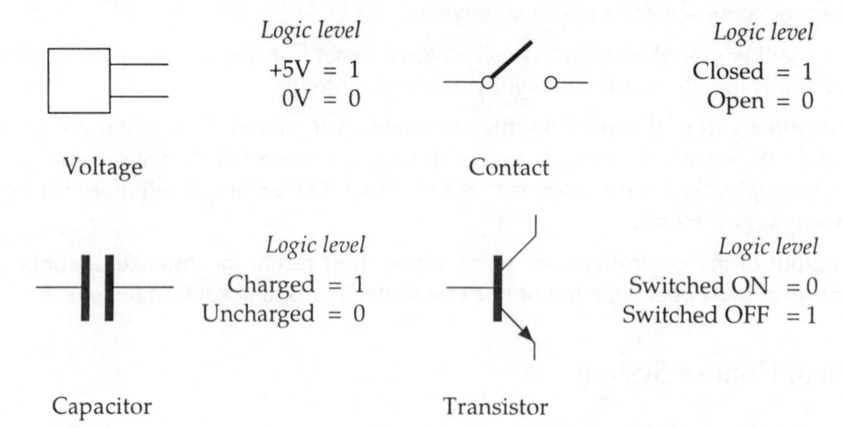

Electrical devices that can produce or store binary signals

A typical digital control system will therefore have a number of devices that can provide an on-off signal. These signals are then processed and a relay or contactor connects electrical power to the machine that is being controlled. A typical arrangement is shown below.

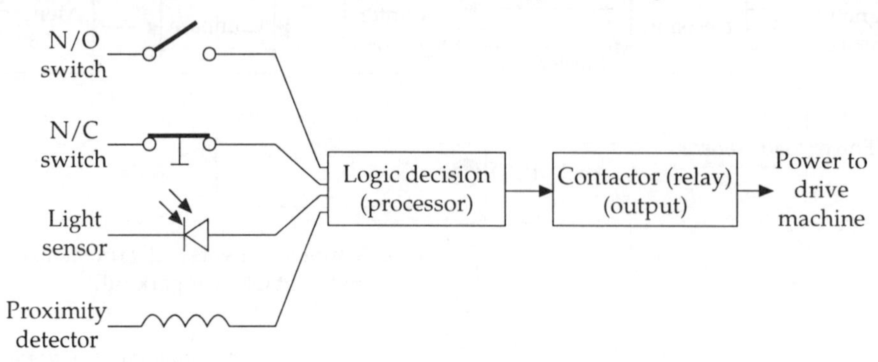

Digital control

The Processor

Between input and output of the system, the information must be 'processed': logical decisions must be made so that an input is given only when the correct combination of input signals has been received, or after the correct time interval, or after a preset number of operations has been counted.

To make these control decisions we use solid state (semiconductor) circuits which act as **gates**, **counters**, **memories** and **timers**. The heart of the control system is the **Processor**.

Mechanical switches such as relays and uni-selectors have formerly been used to do these jobs. These switches suffer from contact wear, bounce, sticking and oxidation. Solid state switches (transistors operated as a switch) do not have these disadvantages and can operate much faster.

Binary Numbers

In a digital system all quantities can be represented as binary numbers, and these numbers are able to be counted, manipulated and stored in memories.

The use of binary numbers result, in digital systems which are fast, reliable, compact and versatile. The binary scale is clear and distinct (ON-OFF).

However, binary numbers are not easily recognised, and there must be conversion from the ordinary decimal numbers of everyday life to binary numbers at the INPUT and OUTPUT ends of a digital system.

In practice, the conversions from decimal to binary at the input of a system and from binary to decimal at the output end are done automatically by electronic logic circuits.

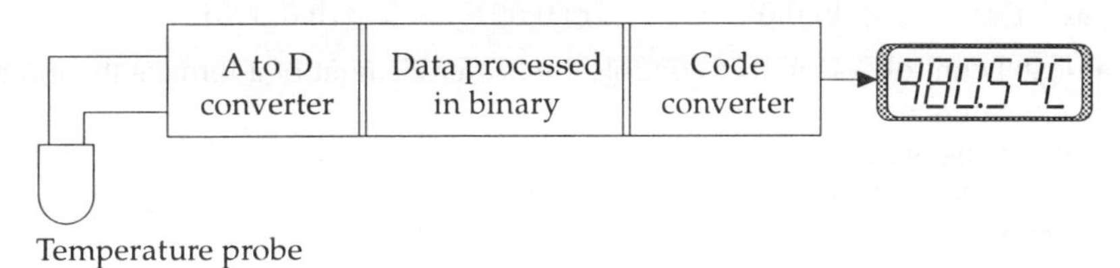

A digital thermometer

In the example of a digital thermometer below the temperature is sensed and then converted into a coded binary signal by an analogue to digital converter. At the output of the system the binary signal is converted by a code converter into decimal form to provide a decimal readout.

Summary

- ❒ An analogue signal is a continuously varying signal and can have any value: e.g. a sine wave.
- ❒ A digital signal can only have discrete values like a box of ball bearings. A binary digital signal only has two states: e.g. +5V (logic 1) and 0V (logic O).
- ❒ The resolution of a system is the smallest change that can be detected by the system.
- ❒ Digital logic systems use ON-OFF or binary control.
- ❒ Many electrical devices have two distinct states which may be used to represent and store binary numbers.
- ❒ In digital systems all quantities can be represented as binary numbers.

Self-assessment questions *(answers page 460)*

1) A digital signal is:
   a) a multi level signal
   b) a two state signal
   c) a mixture of a) and b)
   d) either a) or b)

2) An analogue signal is:
   a) a continuously variable signal.
   b) a fixed value of signal
   c) a variable signal that can have any value
   d) a signal with discreet values

3) Which of the following are analogue devices:
   a) power switch b) lamp dimmer
   c) railway signal d) mercury in glass thermometer
   e) boiler thermostat f) volume control

4) A voltmeter has a four digit readout 0000 to 9999. The resolution of the meter if the full scale reading is 10V is:
   a) 0.1 b) 0.01 c) 0.001 d) 0.0001

5) In a digital control system that controls the heater element of a furnace the input from the temperature probe will be:
   a) an analogue signal
   b) a multi-level digital signal
   c) a multi-level analogue signal
   d) a binary digital signal

6) In the digital control system in question 5 the signal from the temperature probe must be:
   a) converted by a digital to analogue converter
   b) converted by an analogue to digital converter
   c) converted to an on-off signal
   d) amplified but not converted

7) The heater element of the furnace in the digital control system of question 5 is switched in by a contactor. The contactor is brought in by:
   a) a binary signal
   b) a multi-level digital signal
   c) a multi-level analogue signal
   d) an analogue signal.

8) Which of the following is **NOT an advantage** of the binary signal over the multilevel signal:
   a) the binary signal has two well defined voltage levels whereas the multilevel signal has many voltage levels.
   b) with the binary signal there is either a voltage or no voltage present.
   c) attenuation (reduction) of the binary signal is not such a problem since the receiver only has to detect a voltage or no voltage.
   d) a distorted binary signal (pulse of voltage) may easily be regenerated.
   e) most physical quantities are analogue in nature and the multilevel signal is closer in form than the binary signal to an analogue signal.

Number systems

Introduction

Digital computers, numerically controlled machine tools, electronic counters and instruments with a digital readout all use BINARY because of the ease of storing binary numbers, and the speed of processing numbers in binary. Therefore, whenever we have to store or process numbers in a digital system, we use the binary number system. This system has only two digits: 1 and 0, whereas the decimal or denary system has 10 digits (0 to 9).

We will now briefly review three number systems to show the differences and similarities between them.

Decimal (base 10)

This **common**, everyday system has ten digits:

0, 1, 2, 3, 4, 5, 6, 7, 8, 9

After 9 we write 10 which means 1 ten and 0 units. Notice that the PLACE of a digit determines its VALUE in a number.

Example

4563 = 4 thousand + 5 hundreds + 6 tens + 3 units

To show that this is base 10 the number would be written as 4563_{10}; the subscript 10 indicates the base of the number.

Place Values in the Decimal System

Thousands	*Hundreds*	*Tens*	*Units*		*Tenths*	*Hundredths*
1,000	100	10	1	•	$\frac{1}{10}$	$\frac{1}{100}$
10^3	10^2	10^1	10^0	•	10^{-1}	10^{-2}

Every place value is a **power of 10** and ten is the **base** of the decimal system.

Binary (base 2)

The binary number system has only two digits, 0 and 1. After 1 we write 10 which means 1 two and 0 units.

Place Values in the Binary System

As with the denary or decimal system we can represent each place value as a power of two. Remember that with binary the base of the system is TWO.

16	8	4	2	1	•	$\frac{1}{2}$	$\frac{1}{4}$	$\frac{1}{8}$	$\frac{1}{16}$
2^4	2^3	2^2	2^1	2^0	•	2^{-1}	2^{-2}	2^{-3}	2^{-4}

Note that every place is a power of 2, and that 2 is now the base of the system. Note also that moving one place value to the right divides by 2, and moving one place value to the left multiplies by 2.

As with the decimal system, to represent larger numbers we place digits to the left of the binary point.

To represent binary fractions we place digits to the right of the **binary point.**

Example

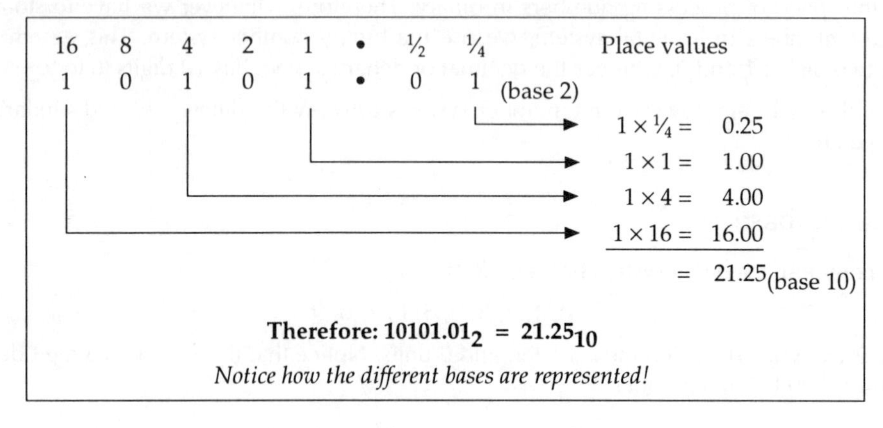

The place values are 'summed' to obtain the decimal equivalent

Counting in binary we have:

Binary	*represents*	*Decimal*
0	"	0
1	"	1
10	"	2
11	"	3
100	"	4
101	"	5
110	"	6
111	"	7
1000	"	8
1001		9

The binary equivalent of the decimal numbers 0 to 9

Binary Digits

In the binary system there can only be 0 or 1. Therefore with one binary digit there is only two possible alternatives 0 or 1. A **B**inary Dig**IT** is called a **BIT**. With two bits there are four possible combinations of 0 and 1

00, 01, 10, 11

With 3 Bits there will be 8 unique combinations:

000, 001, 010, 011, 100, 101, 110, 111

The number of combinations can be determined from:

No of combinations = 2^n where n = number of bits.

For example:

2 bits	$= 2^2$	= 4 combinations
3 bits	$= 2^3$	= 8 combinations
4 bits	$= 2^4$	= 16 combinations
6 bits	$= 2^6$	= 64 combinations
10 bits	$= 2^{10}$	= 1024 combinations

If these combinations are arranged in terms of the place values for the binary system, then the natural binary code is produced.

Consider an arrangement with 4 bits, A, B, C, and D. Then by convention:

A is the least significant digit L.S.D.(place value 1).
B is the next more significant digit (place value 2).
C is the next more significant digit (place value 4).
D is the most significant digit M.S.D. (place value 8).

Natural Binary Code

	Binary				
Bits →	*D*	*C*	*B*	*A*	*Decimal*
	0	0	0	0	0
	0	0	0	1	1
	0	0	1	0	2
	0	0	1	1	3
	0	1	0	0	4
	0	1	0	1	5
	0	1	1	0	6
	0	1	1	1	7
	1	0	0	0	8
	1	0	0	1	9
	1	0	1	0	10
	1	0	1	1	11
	1	1	0	0	12
	1	1	0	1	13
	1	1	1	0	14
	1	1	1	1	15
Place value →	**8**	**4**	**2**	**1**	

A 4 bit natural binary code

You do not have to memorise the natural 4 bit code above (although eventually it would be an advantage): just remember the pattern!

With 4 bits there will be 16 combinations:

The least significant bit pattern A, **going down** the column, is 0 1 0 1 0 1 0 1 etc.
The next more significant bit pattern B is 00 11 00 11 etc.
The next more significant bit pattern C is 0000 1111 etc.
The most significant bit pattern D is 00000000 11111111.

If there were 5 bits then the most significant bit E would have 16 0s and 16 1s going down the column.

Hexadecimal (Base 16).

The natural binary code shown above requires at least 4 bits (binary digits) to represent the decimal numbers 0–9. (3 bits will only represent up to decimal 7.)

With 4 bits there are 16 different combinations ($2^4=16$). Therefore with 4 bits there is a **base of 16**. Each **different combination** must be represented by its **own unique character**: i.e the decimal characters 0–9 (**DECIMAL**) are used as normal, but to represent the other 6 (**HEX**) combinations, the first 6 letters of the alphabet are employed.

	Binary					
Bits →	*D*	*C*	*B*	*A*	*Decimal*	*Hex*
	0	0	0	0	0	0
	0	0	0	1	1	1
	0	0	1	0	2	2
	0	0	1	1	3	3
	0	1	0	0	4	4
	0	1	0	1	5	5
	0	1	1	0	6	6
	0	1	1	1	7	7
	1	0	0	0	8	8
	1	0	0	1	9	9
	1	0	1	0	10	**A**
	1	0	1	1	11	**B**
	1	1	0	0	12	**C**
	1	1	0	1	13	**D**
	1	1	1	0	14	**E**
	1	1	1	1	15	**F**
Place value →	**8**	**4**	**2**	**1**		

The Hexadecimal Code

Note:

1) Do not confuse the letter A representing bit A (L.S.D) in the binary system with the letter A in the Hex code which represents decimal 10.

2) There are 16 combinations but the highest decimal number you can represent is 15 (F). The first combination represents 0!

Using the Hex Code

Computers employ large strings of binary digits; 8, 16, 32 or 64 bits. With 8 bits for example there will be: $2^8 = 256$ combinations.

Using Hexadecimal each of the 256 combinations can be represented by a 2-digit hex code. Therefore if an 8-bit binary number were as follows:

00101010

then this would be represented as two 4 bit numbers: 0010 and 1010.

Each 4-bit binary pattern is now represented by its Hexadecimal equivalent:

0010 = 2 and 1010 = A

So the binary pattern 00101010 is now represented as 2A!

This makes programming computers easier.

Decimal, binary and hex codes

Decimal	*Binary*		*Hex*
0	0000	0000	00
1	0000	0001	01
2	0000	0010	02
3	0000	0011	03
4	0000	0100	04
5	0000	0101	05
6	0000	0110	06
7	0000	0111	07
8	0000	1000	08
9	0000	1001	09
10	0000	1010	0A
11	0000	1011	0B
12	0000	1100	0C
13	0000	1101	0D
14	0000	1110	0E
15	0000	1111	0F
16	0001	0000	10
17	0001	0001	11
Similarly			
28	0001	1100	1C
36	0010	0100	24
42	0010	1010	2A
And so on, up to			
255	1111	1111	FF

Examples of decimal, binary and hex codes

Self-assessment questions (1) (answers page 461)

1) Give the decimal equivalent of the following binary numbers: 1011_2, 1001.11_2, 11111.01_2

2) A binary digit is called a BIT. How many bits are required to represent any number from zero to one thousand?
   a) 8 b) 9 c) 10 d) 11

3) Check that the following is true, and write in the bases that makes the equation true:

100,000,100 = 260

4) The contents of a register in a digital computer are represented by a row of lamps.

XXXX, XXXX, XXXX

i) If Lamp On = 1; and Lamp OFF = 0, what is the largest decimal number that can be shown by this register?
a) 1023 b) 2047 c) 4095 d) 8191

ii) How many lamps would be needed by a similar register if it is to display any number from zero to 4000?
a) 10 b) 11 c) 12 d) 13

5) What is the largest number that can be represented by ten BITS:

(i) If the binary point is in the middle of the ten bits.
a) 31.96875 b) 32 c) 63.9375 d) 64

(ii) If it is at the right-hand end of the ten bits.
a) 511 b) 512 c) 1023 d) 1024

6) If $22_{10} = 10110_2 = 1 \times 2^4 + 0 \times 2^3 + 1 \times 2^2 + 1 \times 2^1 + 0 \times 2^0$; write down the binary equivalent of 55_{10} in the same way.

7) The base of a number is also called the RADIX. What is the radix in:
(a) decimal (b) hex (c) binary d) octal.

8) The binary number 1010, 1110, 1011, 1001 in hexadecimal would be:
a) ADB9 b) AFB9 c) AEC9 d) AEB9

9) Convert the following hex codes into binary:
a) 3C b) 1D c) 59 d) 72

10) Convert the following binary patterns into the equivalent hex code:
a) 0000 1111 b) 1010 1100 c) 0011 1011 d) 0111 1110

Binary Arithmetic

We often need to convert between the number systems. Binary to hex; decimal to binary etc. Many modern calculators will now perform this conversion for you but if you do not have a suitable calculator a few rules are given below.

Binary Decimal Conversion

Rule: sum the place values, beginning at the binary point.

Example

$$1010111_2 = 64 + 0 + 16 + 0 + 4 + 2 + 1 = 87_{10}$$

$$0.1101_2 = \tfrac{1}{2} + \tfrac{1}{4} + 0 + \tfrac{1}{16} = 0.8125_{10}$$

$$1101.101_2 = 8 + 4 + 0 + 1 + \tfrac{1}{2} + 0 + \tfrac{1}{8} = 13.625_{10}$$

Decimal Binary Conversion

Whole Numbers Rule: Divide the decimal number successively by 2, and record the remainders, which form the digits of the binary number.

Example

Convert 43_{10} to a binary number.

Divisor	*Dividend*	*Quotient*	*Remainder*	
2	43	21	**1**	LSB (least significant bit)
2	21	10	**1**	
2	10	5	**0**	
2	5	2	**1**	
2	2	1	**0**	
2	1	0	**1**	MSB (most significant bit)

Therefore $43_{10} = 101011_2$.

Note that the last remainder is always the most significant bit. It must always be 1.

Fractions Rule: Multiply the fraction successively by 2 and save the whole-number parts, which must be 1 or 0 at each step.

Example

Convert 0.812510 to a binary number.

		.8125	× 2
MSB	**1**	.6250	× 2
	1	.2500	× 2
	0	.5000	× 2
LSB	**1**	.0000	

Note: only the decimal part of any number after the decimal point is multiplied by 2.

Therefore $0.8125_{10} = 0.1101_2$.

Binary Hexadecimal Conversion

To convert from binary to Hex all we need to remember are the 16 binary patterns that represent the Hex digits: i.e.

Binary		*Hex*	*Binary*		*Hex*
0000	=	0	1000	=	8
0001	=	1	1001	=	9
0010	=	2	1010	=	A
0011	=	3	1011	=	B
0100	=	4	1100	=	C
0101	=	5	1101	=	D
0110	=	6	1110	=	E
0111	=	7	1111	=	F

Simply group the binary number into 4s and write down the Hex equivalent.

Example

$$1110101100010111_2$$

Grouping into 4s gives:

1110,1011,0001,0111

Writing the hex equivalent:

1110 1011 0001 0111
E B 1 7

Therefore $1110,1011,0001,0111_2 = EB17_H$.

Note that the subscript H is often used to denote base 16!

Hexadecimal Binary Conversion

Simply reverse the above process.

Decimal to Hexadecimal Conversion

Convert the decimal number to binary using the method outlined above, then convert the binary to hex.

Self-assessment questions (2) *(answers page 461)*

1) Convert the following decimal numbers to binary:
   a) 299 b) 51.375 c) 11 5/64 d) 15/32 e) 50.625

2) Convert the following binary numbers to decimal:
   a) 10111 b) 10110 c) 110001 d) 11110 e) 101010

3) Convert to following to Hexadecimal:
   a) 10110011_2 b) 10100101_2 c) 235_{10} d) 162_{10} e) 2482_{10}

4) Convert the following Hex numbers to decimal:
   a) FF b) C8 c) 12 d) A9

Binary Arithmetic

In digital electronic equipment, any arithmetic which must be done is performed on binary numbers automatically by the hardware of the system. It is only necessary for the technician to understand enough binary arithmetic to allow the appreciation of how the hardware operates.

Addition

0 + 0 = 0
0 + 1 = 1
1 + 0 = 1
1 + 1 = 10 (0 and carry 1)

Example:

```
  11101      i.e. 29 + 26 = 55
  11010 +
 ------
 110111
```

Subtraction

0 – 0 = 0
1 – 0 = 1
0 – 1 = 1 (if we borrow 1)
1 – 1 = 0

Example:

```
  11101      i.e. 29 –18 = 11
  10010 –
  -----
   1011
```

Multiplication

$0 \times 0 = 0$
$0 \times 1 = 0$
$1 \times 0 = 0$
$1 \times 1 = 1$

Example:

```
     110011       i.e. 51 × 13 = 663
       1101  x
   ----------
     110011       Multiply by 1
   110011         Multiply by 100
  110011          Multiply by 1000
  ----------
  1010010111
```

In binary, multiplication is simple because either the number is multiplied by 1, in which case it is 'added-in' without alteration, or it is multiplied by 0, giving zero to add on. In practice multiplication can be performed by a shift-and-add process.

Remember multiplication is repeated addition!

Division

Normal long division procedure is used. Enter 1 in quotient if the divisor can be subtracted from the dividend: if the divisor cannot be subtracted, enter a 0 and 'bring down' the next digit.

Example 1

```
Divisor Dividend Quotient
101) 1111101 )11001
     101
      101
      101
      . . . 101
            101
            000    remainder
```

Example 2

```
100) 1010.0)10.1          40)10 = 2.5
     100
       100
       100
       000
```

Example 3

1.11)11.1 i.e. 11.1 divided by 1.11

this is the same as:

```
111) 1110                 (both numbers x 100)
111) 1110)10              i.e. 1.75)3.5 = 2
     111
     0000
```

Remember: division is repeated subtraction!

Division can be achieved by subtraction: consider 27 divided by 9.

If we take 9 from 27 and add 1 to a counter each time, then at the end of the process the counter contents contain the answer!

		Counter contents
27		
9	–	1
18	remainder	
9	–	2
9	remainder	
9	–	3
0	remainder	

i.e. 9 can be taken from 27 three times, and electronic circuits can take advantage of this process to perform division.

Self-assessment questions (3) *(answers page 461)*

1) Carry out the operations indicated. In each case convert the binary numbers to decimal afterwards to check your answers.

a) 10111	b) 10110	c) 10110	d) 110001
11111 +	1010 +	1010 –	11110 –

2) Convert the following decimal numbers to binary:
   a) 297 b) 511 c) 63 d) 255

3) A binary digit is called a bit. How many bits are required to represent any number from 0 to 99?

4) What are the advantages of using the binary number scale?

   List five electrical devices that can be used to represent binary digits, and say which state corresponds to logic 1 and which to logic 0.

5) Convert the following decimal numbers to binary. Do the operations indicated in binary, and check the answers by converting them back to decimal numbers.

a) 325	b) 401	c) 27	d) 91
88 +	513 +	14 –	55 –

6) What is the largest number that can be represented using 8 binary digits if:
   a) the binary point is at the right-hand end?
   b) the binary point is in the middle?

7) Convert the following decimal numbers to binary:
   a) 300 b) 0.375 c) 260.0625

8) Write out the table in full (as far as 256):

Decimal	*Power of 2*	*Binary*
1	2^0	1
2	2^1	10
4	2^2	100
8	2^3	1000
256	—	—

Logic gates

Introduction

Early control systems employed switches, relays and mechanical timers to process the on-off (digital) input signals.

Electronic logic gates which can perform switching, timing and decision-making functions are now the building blocks of control systems, computers and digital instruments. They are smaller, cheaper, more reliable, able to switch at very high frequencies, and consume less energy.

Many thousands of these logic gates can be built on a small slice of silicon to form a Digital Logic Integrated Circuit (I.C.)

Any operation that can be specified logically in words can be performed by a suitable combination of logic gates.

Example 1

A lift motor is to drive forward if the lift doors are closed **AND** the up button has been pressed on a floor above the lift, but **NOT** if the emergency button has been pressed.

Example 2

An automatic machine is to start if there is material in the feed hopper, **AND** the moulding tool is at the correct temperature, but **NOT** if the last piece was not ejected **OR** if more than 10 seconds have elapsed since the last material was used.

Descriptions of systems such as these contain the words **AND**, **OR**, **NOT**. These are logical connectives and these functions can be carried out by logic gates.

Combinational Logic

A gate is an electronic circuit which gives an output when the appropriate combination of inputs is applied to it. Combinational logic is the use of logic gates to produce an output for a specific and unique combination of inputs.

The operation of logic gates or switching circuits is quickly and compactly described by Boolean algebra. (George Boole a Mathematician, and Claude Shannon a Telephone Engineer, were pioneers in this field.)

The use of Boolean algebra can be illustrated by writing equations for switching circuits, and for logic gates which perform the same functions.

The AND Function

Switching Circuit

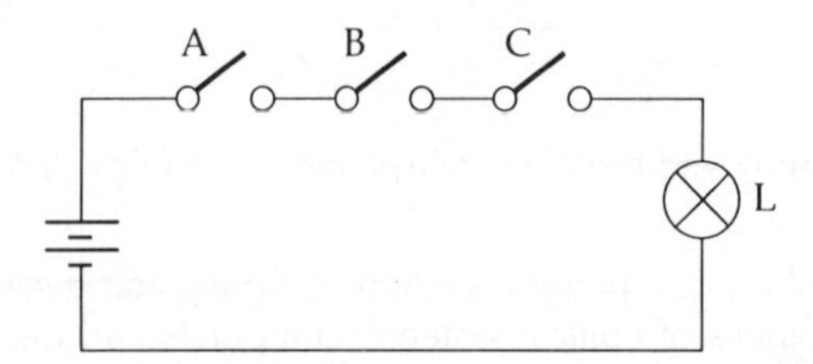

The circuit is complete, and the lamp L lights provided that contacts A AND B AND C are closed.

The inputs, with the switching circuit, are the mechanical operation of the switches A, B and C. The output is the illumination of the lamp L.

The same switching (logic) function can be performed by logic gates, but the inputs and output are now electrical voltages.

The AND Logic Gate

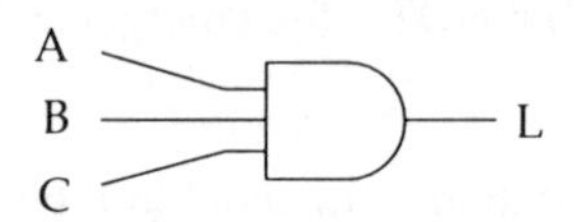

There is an output voltage on wire L provided that there are input voltages on wires A AND B AND C

The inputs to A,B and C are now the presence or absence of a small voltage. For example: +5V can represent logic 1; 0V represents logic 0

The output from the gate will also be either +5V (logic 1) or 0V (logic 0).

Only when all the inputs are present together will there be an output with the AND gate.

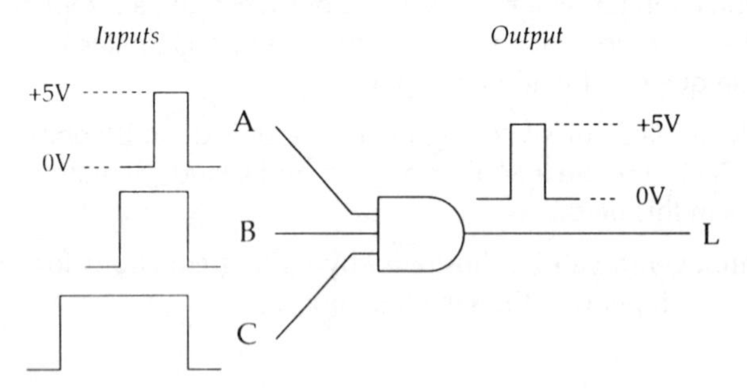

Only for the time that A AND B AND C,.are present together will there be an output on wire L

The AND logic function can be represented by a Boolean equation.

Boolean Equation. (AND Function)

The Boolean equation for both the switching circuit and the logic gate would be:

$$L = A.B.C$$

This means **AND**

The logic gate performs the same function as the switching circuit but is not simply switches in series; the logic gate employs transistors to do the switching and is capable of switching many thousands of times a second.

Also unlike switching circuits the logic gates need well defined voltage levels to drive and operate them.

The OR Function

Switching Circuit

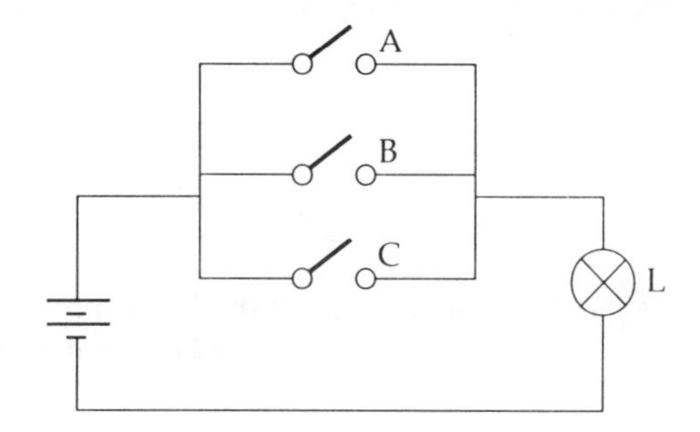

The circuit is complete, and the lamp L lights if contact A OR B OR C is closed

OR Logic Gate

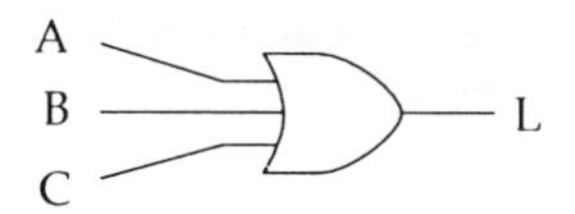

The gate will give a logic 1 output on wire L if there is a logic 1 input on wire A OR B OR C

Boolean Equation

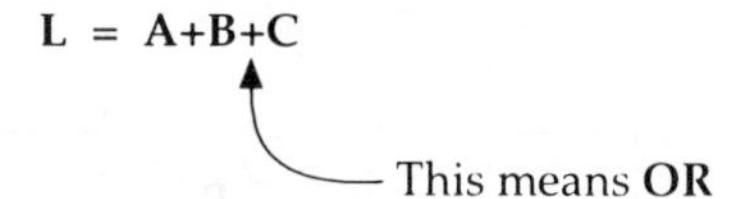

The circuits and logic gates shown so far all have 3 switches or inputs. In practice there are 2, 3, 4, or 8 input gates available commercially.

Remember that with the switch circuits an INPUT is the operation of a switch: i.e. an input is pressing a switch.

With the logic gates an input is a voltage: i.e. a voltage is applied to an input terminal.

A transistor may be used as a switch. If it is used in logic circuits it has the effect of inverting the logic level: i.e. no input gives an output and vice versa. An invertor in logic terms is called a NOT gate.

The NOT Gate

Switching Circuit

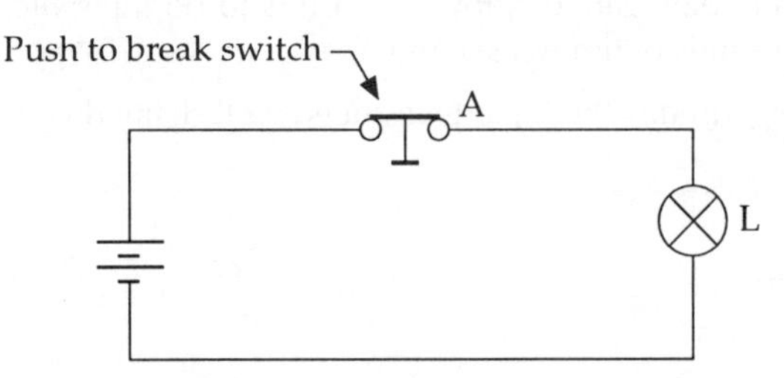

The circuit is complete, and lamp L lights provided that contact A is NOT operated. A is a normally closed switch

The NOT Logic Gate

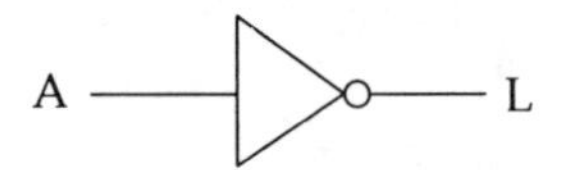

There is a logic 1 output on wire L provided that there is NOT a logic 1 at A: i.e. a logic 0 at A produces a logic 1 on L

Boolean Equation

$$L = \overline{A}$$

The bar over the A means **NOT**

The NOT gate may be used after the AND and OR gates to produce two other logic functions; the **NAND** and **NOR** logic functions.

The NAND Function

Switching Circuit

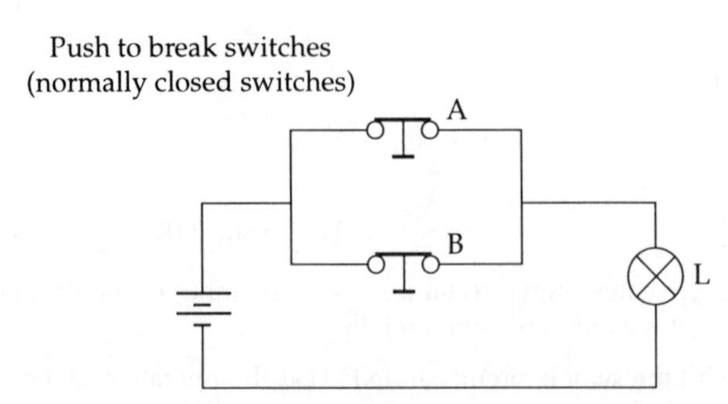

The circuit is complete, and the lamp L lights provided that contacts A AND B are NOT operated

The NAND Logic Gate

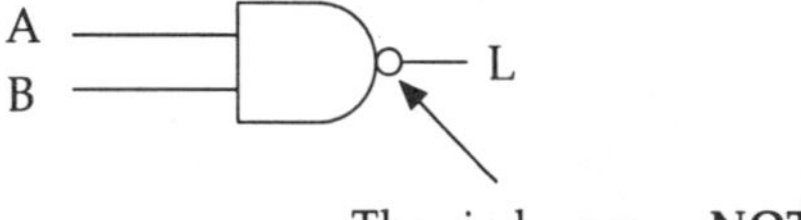

There is a logic 1 output on wire L provided that there is a logic 0 input on either wire A OR B.

Boolean Equation

$$L = \overline{A.B}$$

With a NAND gate there will always be an output provided that one of the inputs is at logic 0. i.e. if any input is low (logic 0) then the output is high (logic 1).

The NOR Function

Switching Circuit

Push to break switches
(normally closed switches)

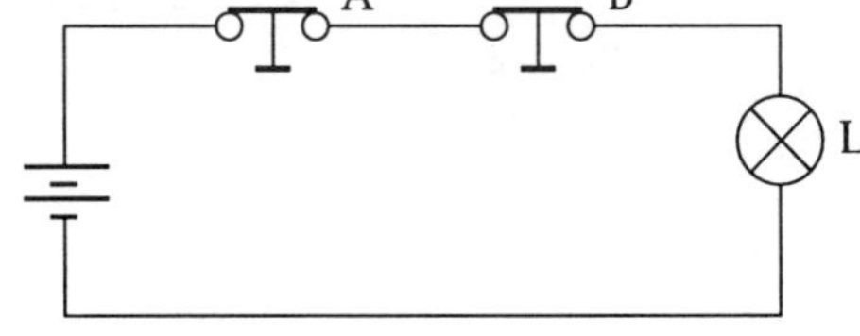

The circuit is complete, and the lamp L lights if contact A and B are NOT operated

The NOR logic gate

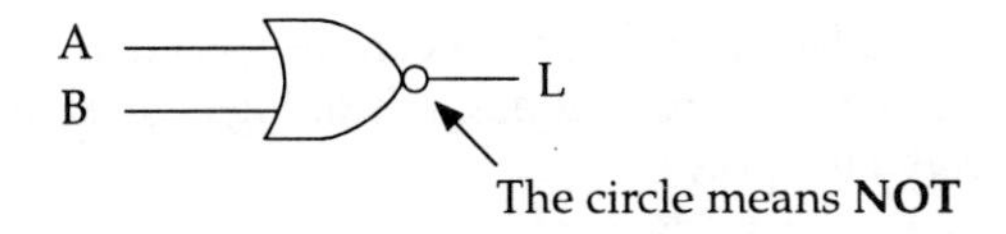

The gate will give a logic 1 output on wire L only when there is a logic 0 on both inputs A AND B

Boolean Equation

$$L = \overline{A+B}$$

With the NOR gate, if any input is high (logic 1), then the output will be low. Only when both inputs are low (logic 0) will the output be high.

The logic function of any switching circuit can be represented and implemented using logic gates.

Example

Consider the switching circuit shown below.

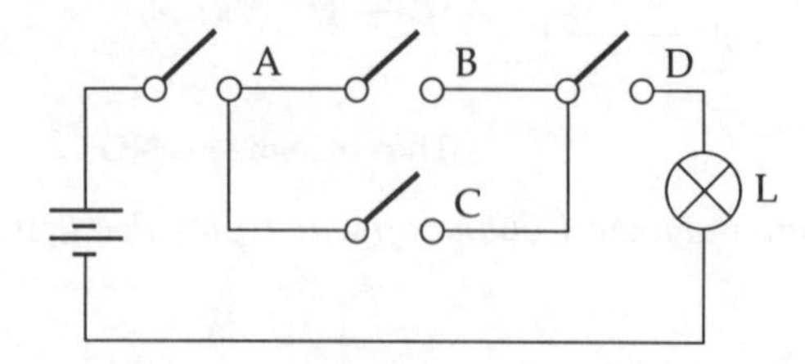

Lamp L lights if contacts A AND D AND either B OR C are closed

Boolean Equation

$$L = A.D.(C+B)$$

Notice that switch A is in series with switch D (AND function). Switch C is in parallel with Switch B (OR function).

Logic Gate Diagram

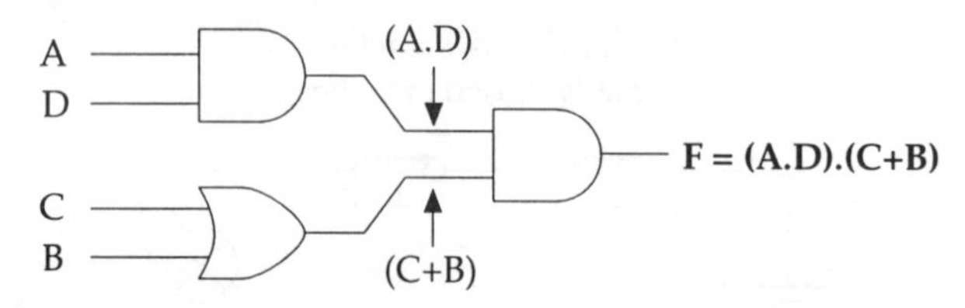

Notice that instead of using 'L', we often use the letter 'F' (for Function) at the output of a logic gate.

Summary

- Switching circuits that perform logic operations can be replaced by logic gates.
- Logic gates have the advantage that they can be switched at very high frequencies (MHz); also I.C.s can contain many thousands of logic gates in a very small package, which consumes very little power.
- The input to switching circuits is the mechanical operation of a switch. With logic gates the input is a small voltage: e.g. +5V represents logic 1; 0V represents logic 0.
- +5v (logic 1) is also called **HIGH** logic level, whereas 0V (logic 0) is called a **LOW** logic level.
- With an AND gate, the output is at logic 1 only when all the inputs are at logic 1.

  Boolean equation $F = A.B$
- With an OR gate, there is a logic 1 at the output when any one or more inputs are at logic 1.

  Boolean equation $F = A + B$
- With the NOT gate the output is at logic 1 when the input is at logic 0 and vice versa.

  Boolean equation $F = \overline{A}$

- With a NAND gate, if any input is low (logic 0), the output is high (logic 1). With all inputs at logic 1 the output is low.

  Boolean equation $F = \overline{A.B}$

- With a NOR gate, if any input is high (logic 1), the output is low (logic 0). With all inputs at logic 0 the output is high.

  Boolean equation $F = \overline{A + B}$

Self-assessment questions *(answers page 462)*

Question 1

Derive the Boolean equations for the switching circuits shown.

(a)

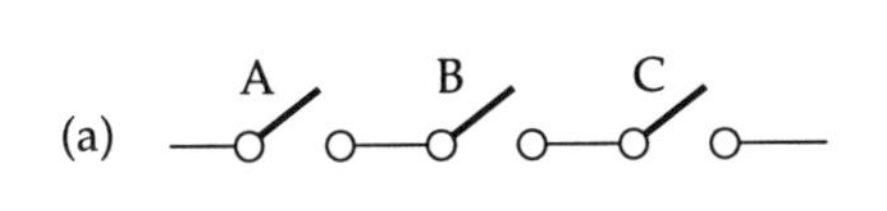

F = ? ____________

(b)

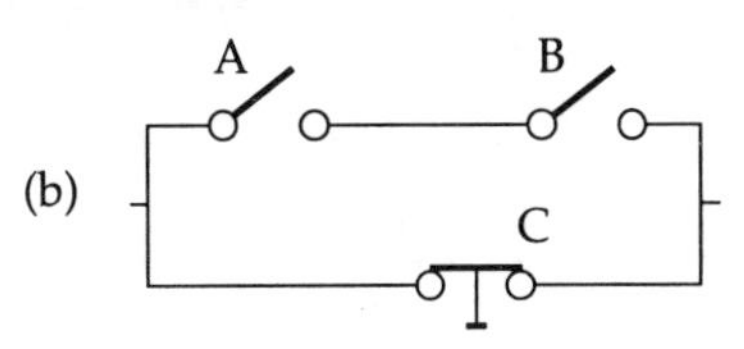

F = ? ____________

(c)

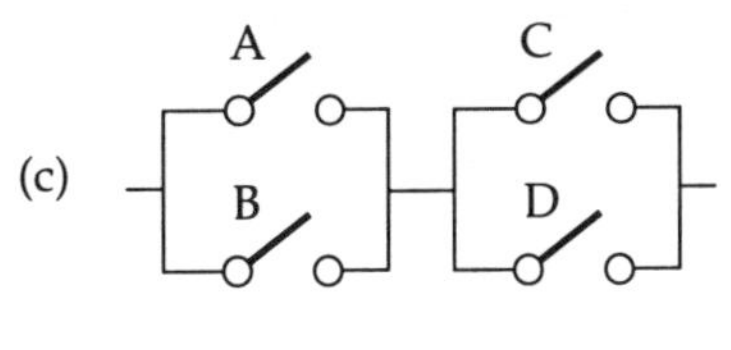

F = ? ____________

(d)

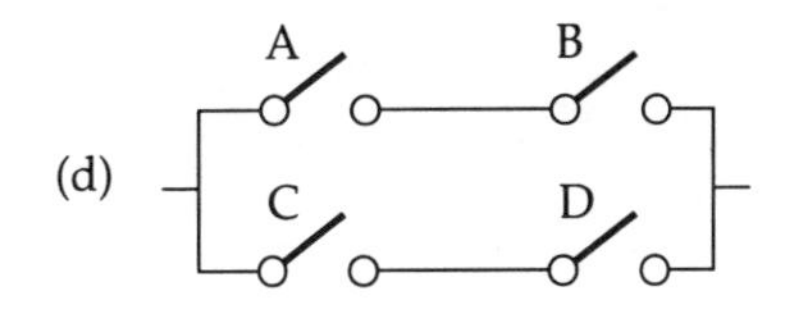

F = ? ____________

Question 2

Write down the Boolean equations for the logic gates shown.

(a)

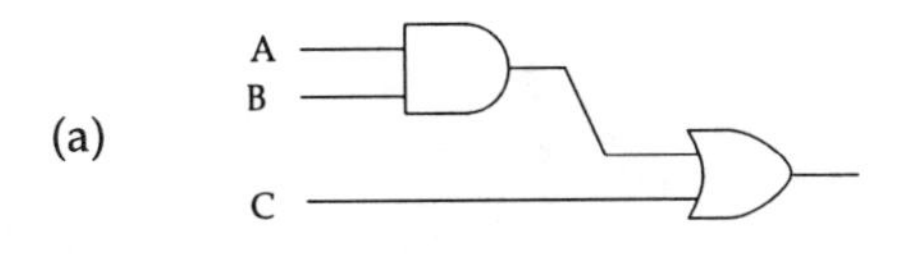

F = ? ____________

(b)

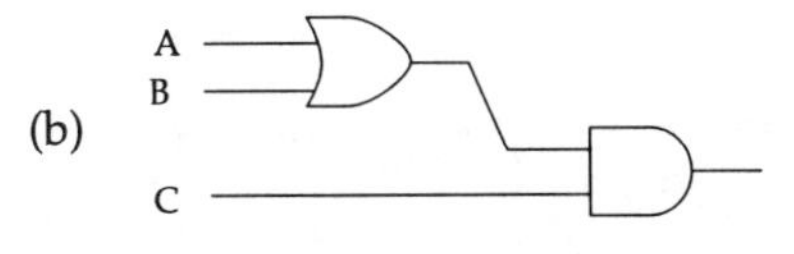

F = ? ____________

(c)

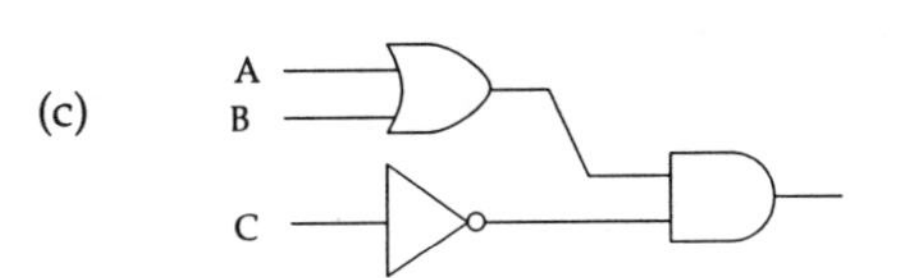

F = ? ____________

(d)

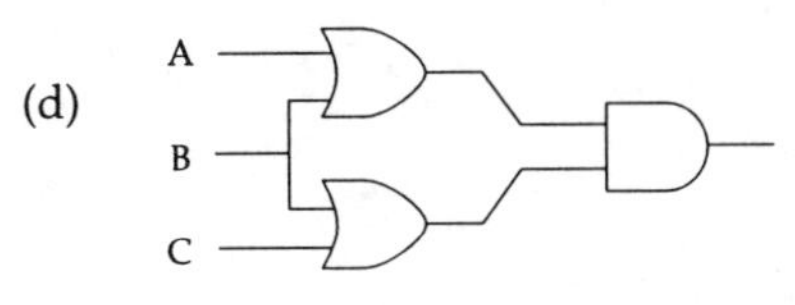

F = ? ____________

Question 3

Draw the logic gate diagrams for the Boolean equations given below:

(a) $A.B + C.\overline{B}$

(b) $A.\overline{B} + B.\overline{A}$

(c) $(A + B + C).\overline{D}$

(d) $A.B.C + A.B.D + \overline{D}$

(e) $(A + B).(C + D)$

(f) $A.B + \overline{A}.\overline{B}$

(g) $\overline{A.B} + \overline{(C + B)}$

(h) $\overline{A.B + C.D}$

Truth tables

Introduction

A truth table is a table of all the possible combinations of the inputs to a gate and the resulting output. By using truth tables we can check the operation of a logic gate system and derive the Boolean equations. The truth table is also used to compare one Boolean equation against another: i.e. if two Boolean equations produce the same truth table they are equivalent to each other.

The 'AND' Operator

To obtain an output (F) from a 2 input AND gate there must be two inputs (A & B) present.

F will only be logic 1 when both A and B are at logic 1.

Since there are two variables, there will be $2^2 = 4$ combinations of these variables. If there were 3 inputs (A.B.C), then there would be $2^3 = 8$ combinations of these variables.

A table is constructed showing all the possible combinations of the inputs, together with the corresponding outputs.

The AND gate truth table

Inputs		*Output*	*AND operator*
A	*B*	*F*	
0	0	0	0 AND 0 = 0
0	1	0	0 AND 1 = 0
1	0	0	1 AND 0 = 0
1	1	1	1 AND 1 = 1

The AND operator is identical to BINARY Multiplication!

i.e. $0 \times 0 = 0$
$0 \times 1 = 0$
$1 \times 0 = 0$
$1 \times 1 = 1$

Therefore, A AND B is represented as A.B.

The 'OR' Operator

To obtain an output at F there must be an input at either A OR B, or inputs at both A and B.

The OR Gate Truth Table

Inputs		*Output*	*OR operator*
A	*B*	*F*	
0	0	0	0 OR 0 = 0
0	1	1	0 OR 1 = 1
1	0	1	1 OR 0 = 1
1	1	1	1 OR 1 = 1

The OR gate is similar to binary addition!

It is not identical however; the only difference is that 1 OR 1 = 1; not 0 carry 1 as in binary addition.

i.e. binary addition

0 + 0 = 0
0 + 1 = 1
1 + 0 = 1
1 + 1 = 0 (carry 1)

Therefore A OR B is represented as A + B.

The NOT Gate

The inverse of a LOGIC LEVEL is very important in logic circuits; an invertor or NOT gate is employed. The invertor will change logic 1 to logic 0, and logic 0 to logic 1.

The NOT Gate Truth Table

A	$\overline{A}$
0	1
1	0

The NAND Gate

The NAND gate operates as if it were made from a AND gate followed by a NOT gate.

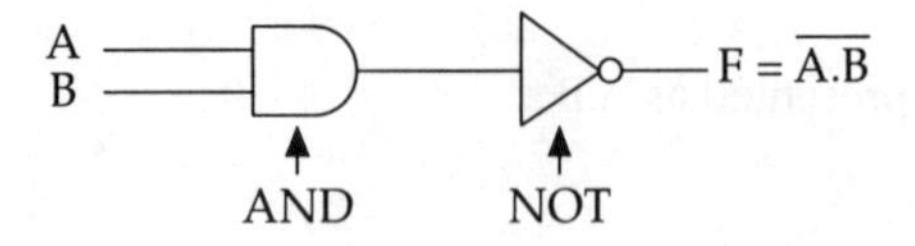

The NAND gate is an AND-NOT

The NAND Gate Truth Table

Inputs		*Output*	
		AND	*NAND*
A	*B*	*A.B*	$\overline{A.B}$
0	0	0	1
1	0	0	1
0	1	0	1
1	1	1	0

The NAND gate truth table is the AND gate table with all the 0s changed to 1s and the 1 to a 0

The NOR Gate

The NOR gate operates as if it were made from an OR gate followed by a NOT gate.

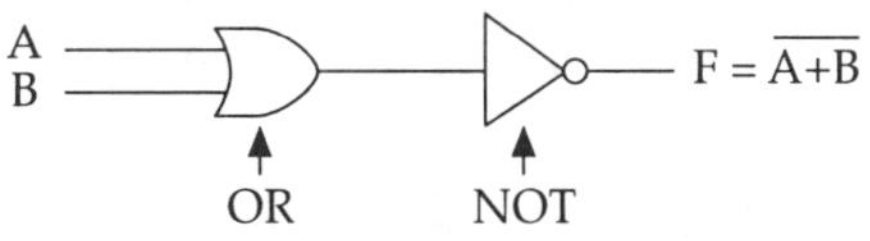

A
B
$F = \overline{A+B}$
OR
NOT

The NOR gate is an OR-NOT

The NOR Gate Truth Table.

Inputs		*Output*	
		OR	*NOR*
A	*B*	*A+B*	$\overline{A+B}$
0	0	0	1
1	0	1	0
0	1	1	0
1	1	1	0

The NOR gate truth table is the OR gate table with all the 1s changed to 0s and the 0 to a 1

Constructing Truth Tables

Example 1

Consider the Boolean Expression $F = \overline{A.B + C}$

Step 1 There are 3 variables so there will be: $2^3 = 8$ combinations of these 3 variables.

Step 2 Construct the table of inputs. Fill in the column A (by convention A is designated as being the least significant digit) with alternate 0s and 1s: i.e. Column B with 2 0s, 2 1s: Column C with 4 0s, 4 1s etc.

Inputs		
C	B	A
0	0	0
0	0	1
0	1	0
0	1	1
1	0	0
1	0	1
1	1	0
1	1	1

3 inputs will produce 8 combinations shown here as the natural binary code

Step 3 Next logically AND column A with B and enter the results into a column (A.B) to the right as shown below.

C	**B**	**A**	*A.B*
0	**0**	**0**	0
0	**0**	**1**	0
0	**1**	**0**	0
0	**1**	**1**	1
1	**0**	**0**	0
1	**0**	**1**	0
1	**1**	**0**	0
1	**1**	**1**	1

A ANDed with B

Step 4 Next produce another column to the right and logically OR the A.B column with C and enter the results as shown below.

C	*B*	*A*	***A.B***	***A.B + C***
0	0	0	**0**	0
0	0	1	**0**	0
0	1	0	**0**	0
0	1	1	**1**	1
1	0	0	**0**	1
1	0	1	**0**	1
1	1	0	**0**	1
1	1	1	**1**	1

Column A.B ORed with column C

To obtain the final expression we now have to invert the last column.

C	B	A	$A.B$	$A.B + C$	$\overline{A.B + C}$
0	0	0	0	0	1
0	0	1	0	0	1
0	1	0	0	0	1
0	1	1	1	1	0
1	0	0	0	1	0
1	0	1	0	1	0
1	1	0	0	1	0
1	1	1	1	1	0

This now gives the final truth table for the function:

$$F = \overline{A.B + C}$$

The last column gives the output conditions for all the possible combinations of the inputs.

Example 2

It is often necessary to draw up truth tables to verify that one Boolean equation is equivalent to another. For example; use truth tables to verify that:

$$\overline{A + B} = \bar{A}.\bar{B}$$

In order to show that these two equations are the same we need to draw up the truth tables of;

$F = \overline{A + B}$ and $\bar{A}.\bar{B}$ and compare them with each other.

Truth tables of $F = \overline{A+B}$ and $\bar{A}.\bar{B}$

B	A	$\overline{A+B}$	$\bar{A}$	$\bar{B}$	$\bar{A}.\bar{B}$
0	0	1	1	1	1
0	1	0	0	1	0
1	0	0	1	0	0
1	1	0	0	0	0

These two columns are the same therefore the two equations are equivalent

Truth tables may also be used in the design of digital logic systems.

Designing Combinational Logic Circuits

Consider a two-way lighting circuit. This is the lighting circuit required on stairways. When both switches are in the ON or OFF position the light is OFF. Only when the two switches are different should the light come ON.

The 2-way lighting circuit.

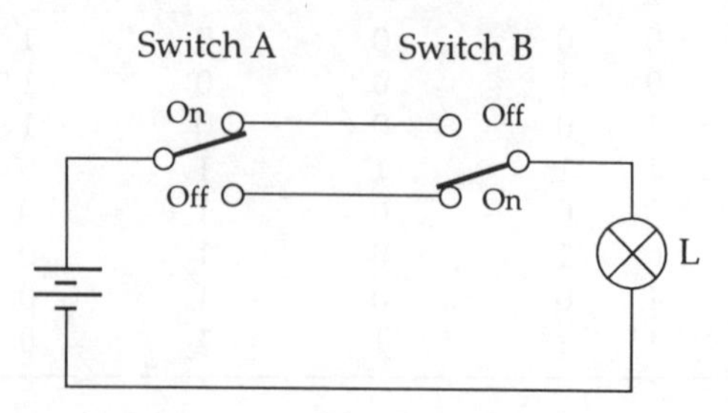

The lamp is only ON when one switch is in the ON position and the other is in the OFF position

Truth tables can be used to design a logic gate circuit to perform the same function as this switching circuit.

To design the logic gate diagram carry out the following:

Step 1 Construct a truth table from the specification required for the circuit.

Step 2 Obtain the Boolean equation from the truth table.

Step 3 Simplify the Boolean equation if possible.

Step 4 Draw the gate diagram from the simplified Boolean equation.

Step 1: The truth table.

There are two inputs (switches) A and B. There must be an output (logic 1) whenever there is a logic 1 on A or B; but NOT if there is a 1 on both A and B at the same time.

Inputs		*Output*		
B	*A*	*F*		
0	0	0		
0	1	1	→ $F = \overline{B}.A$	Therefore $F = A.\overline{B} + B.\overline{A}$
1	0	1	→ $F = B.\overline{A}$	
1	1	0		

Obtaining the Boolean equation from the truth table

Step 2

To obtain the Boolean equation: write down the input conditions that give a 1 in the output (F) column.

There are two conditions that produce a 1 in the output (F) column:

when B=1; A=0 and when B=0; A= 1

Therefore we OR these two conditions together to produce the full Boolean equation:

$$F = A.\overline{B} + B.\overline{A}$$

Step 3

This Boolean equation cannot be simplified.

This is a NOT equivalent circuit; there is an output only when the inputs are different. It is more commonly known as an Exclusive OR (**Ex–OR**) logic gate.

Step 4: Draw the logic gate diagram.

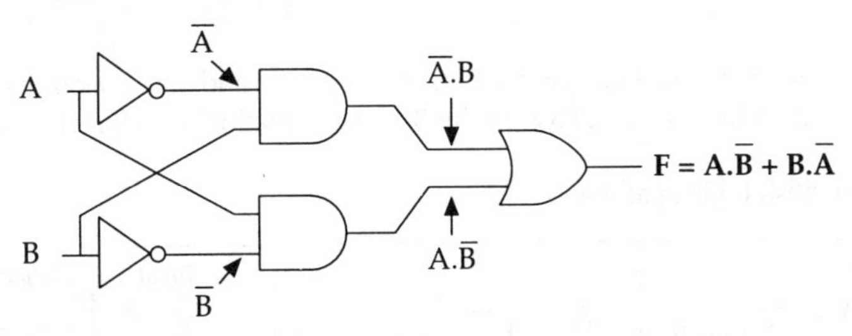

Summary

- Tables of truth list all the possible input combinations together with the outputs for those input combinations.
- Truth tables may be used to verify Boolean equations and in the design of logic gate systems.
- Overleaf is a summary of logic gate symbols and their truth tables.

Self-assessment questions (answers page 463)

Question 1

Obtain the Boolean expressions from the following truth tables:

a)

B	*A*	*F*
0	0	1
0	1	0
1	0	0
1	1	0

b)

B	*A*	*F*
0	0	1
0	1	0
1	0	1
1	1	0

c)

B	*A*	*F*
0	0	1
0	1	1
1	0	0
1	1	1

d)

C	*B*	*A*	*F*
0	0	0	0
0	0	1	1
0	1	0	1
0	1	1	0
1	0	0	0
1	0	1	0
1	1	0	0
1	1	1	0

e)

C	*B*	*A*	*F*
0	0	0	1
0	0	1	0
0	1	0	0
0	1	1	0
1	0	0	0
1	0	1	0
1	1	0	1
1	1	1	0

Question 2

Verify by the use of truth tables the following Boolean expressions:

a) B + B.A = B   b) A + A.$\overline{B}$ = A

c) A.(A + B) = A   d) $\overline{A}$ + $\overline{A}$.B = $\overline{A}$

e) $A.B + A.\overline{B} = A$ f) $\overline{A.B} = \overline{A} + \overline{B}$

g) $A.B = \overline{\overline{A} + \overline{B}}$ h) $\overline{A+B} = \overline{A}.\overline{B}$

Question 3

Design an equivalent circuit that has two inputs A and B and has an output whenever the inputs are at the same logic level. Develop the Boolean equation and draw the gate diagram.

American and British logic symbols

British Standard logic gate symbols: BS3939	Military and American Standards: MS806B	Truth tables for logic gates
A, B — [&] — F	A, B — $F = A.B$ **AND gate**	*Inputs A B / Output F* 0 0 / 0 0 1 / 0 1 0 / 0 1 1 / 1
A, B — [≥] — F	A, B — $F = A+B$ **OR gate**	*Inputs A B / Output F* 0 0 / 0 0 1 / 1 1 0 / 1 1 1 / 1
A — [1]o — F	A — $F = \overline{A}$ **NOT or INVERTER**	*Input A / Output F* 0 / 1 1 / 0
A, B — [&]o — F	A, B — $F = \overline{A.B}$ **NAND gate**	*Inputs A B / Output F* 0 0 / 1 0 1 / 1 1 0 / 1 1 1 / 0
A, B — [≥]o — F	A, B — $F = \overline{A+B}$ **NOR gate**	*Inputs A B / Output F* 0 0 / 1 0 1 / 0 1 0 / 0 1 1 / 0
A, B — [=1] — F	A, B — $F = A \oplus B$* **EXCLUSIVE-OR gate** *$\oplus$ symbol means EXOR. $A \oplus B$ is equivalent to $F = A.\overline{B} + B.\overline{A}$	*Inputs A B / Output F* 0 0 / 0 0 1 / 1 1 0 / 1 1 1 / 0

Logic families and logic levels

Introduction

Digital logic integrated circuits may contain many transistors, diodes and resistors, together with all the interconnections to form a complete circuit or logic function. The integrated circuit is formed on a substrate and the transistors may be formed using either uni-polar (Field Effect Transistor) or bi-polar devices (NPN or PNP transistors).

This gives rise to two main logic families; **transistor-transistor logic** (ttl) and **complementary metal oxide semiconductors** (cmos).

Both these logic families may be used in the following types of integrated circuit (I.C.);

Small-scale integration	(SSI)	1 to 10 gates
Medium-scale integration	(MSI)	10 to 100 gates
Large-scale integration	(LSI)	100 to 1000 gates
Very-large-scale integration	(VLSI)	1000 to 10,000 gates
Super-large-scale integration	(SLSI)	10,000 to 100,000 gates per I.C.

TTL Devices

The most popular range is the 74 series, which will operate over the temperature range 0°C to 70°C. The supply voltage required for these gates is:

Maximum:	+5.25V
Typical:	+5.00V
Minimum:	+4.75V

A pair of digits are used to code the device and these distinguish the logic function of the chip.

For example:

7400 is a standard Quad 2 INPUT NAND gate
7402 is a standard Quad 2 INPUT NOR gate.
7408 is a standard Quad 2 INPUT AND gate
7432 is a standard Quad 2 INPUT OR gate.

Sub-families of the 74 series

There are many sub families of this group which are distinguished by infix letters in the device coding: e.g.

74 LS 00 device code

'LS' are the infix letters

The sub-families are compared by their switching speeds and power consumption as shown in the table below.

	Type of device	*Switching speed*	*Power consumption*
No infix letter	Standard gate	10ns	10mW
L	Low power	33ns	1mW
LS	Low power Schottky series	10ns	2mW
AS	Advanced Schottky	1.5ns	22mW
ALS	Advanced LS Schottky	4ns	1mW
H	High speed	6ns	22mW
S	Schottky	3ns	20mW

Sub-families of the 74 series of TTL gates

CMOS Devices

The most popular range is the 4000 series which will operate over the temperature range –40° C to +85°C.

For example:

4011 is a Quad 2 INPUT NAND gate
4001 is a Quad 2 INPUT NOR gate.
4081 is a Quad 2 INPUT AND gate
4071 is a Quad 2 INPUT OR gate.

The supply voltage required for these gates is from +3V to + 15V.

CMOS Versions of TTL Devices

The main advantage of TTL devices is their speed, whereas the advantage with CMOS is its low-power consumption. There are currently high-speed versions of CMOS which have equivalent functions and Pin-outs to TTL devices.

These CMOS I.C.s are therefore a direct replacement for TTL chips and have comparable speed with lower power consumption.

These devices are also coded with infix letters; e.g.

74HCXX = high-speed CMOS version of TTL with CMOS compatible inputs.
74HCTXX = high-speed CMOS version of TTL with TTL compatible inputs.

PIN-Out Diagrams

The most popular method of packaging digital logic I.C.s is Dual-in-Line (DIL). The number of logic gates that can be built onto a chip is however limited mainly by the number of connections to the chip.

With a 14-pin DIL package then it is only possible to have 4 2-Input gates. Each gate will require 3 pins (2 pins for its input and 1 pin for its output), so 4 gates will need 12 pin connections. Two further pins are required for the supply voltage making 14 pins in total. Only three 3-input gates can be packaged in a 14-pin DIL I.C. and only two 4-input gates.

The pin-outs of the basic TTL gates are shown below.

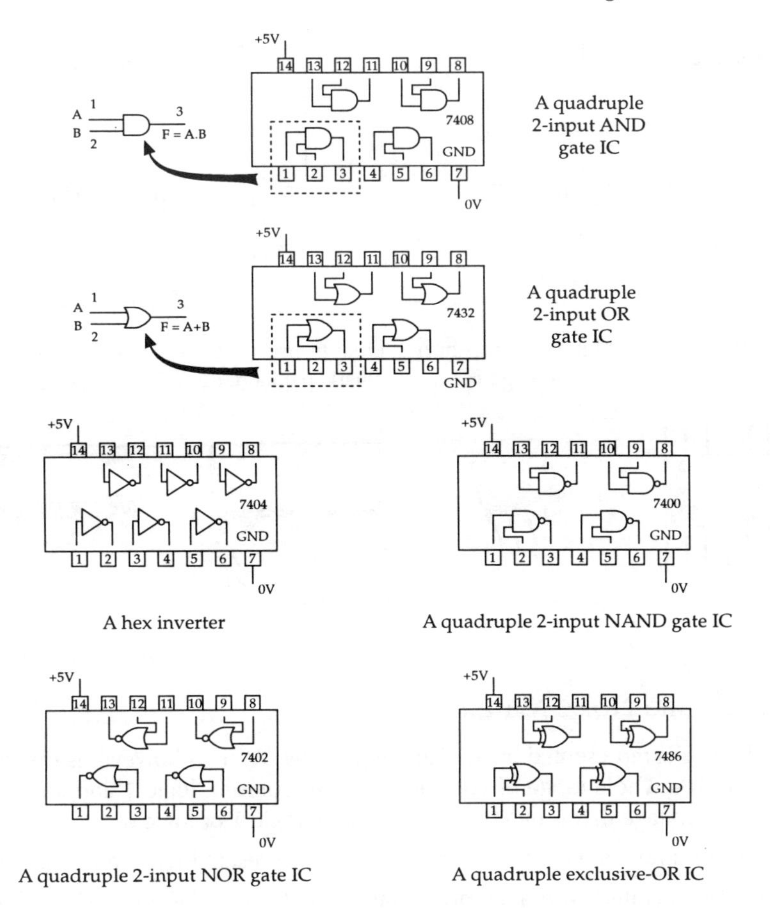

74 Series Basic Logic Gates and Pin-Out Diagrams

DIL devices come in a variety of sizes from 8-pin to 40-pin packages.

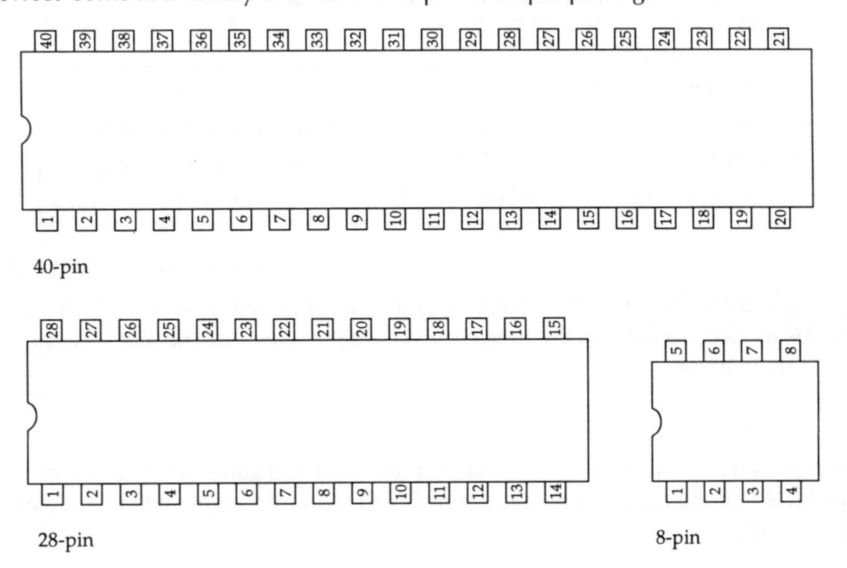

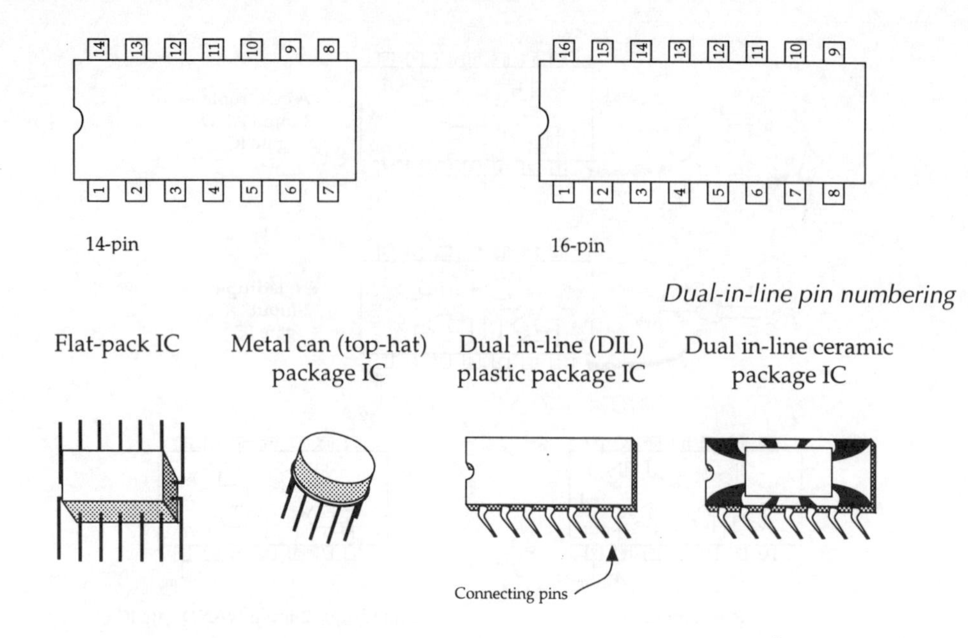

Dual-in-line pin numbering

Integrated circuit packages

Logic levels and Noise Margins

With TTL logic 1 is represented by +5V and logic 0 by 0V. This however is the ideal and in practice there is a whole range of voltages that can represent logic 1 and logic 0. There is also a range of voltage that is indeterminate, neither logic 1 or logic 0.

Logic levels are simply the range of voltages used to represent logic 1 and logic 0.

CMOS may be operated with a supply voltage of between +3V and +15V. Therefore the logic levels with CMOS are very different to TTL.

The logic levels with CMOS depend upon the supply voltage used whereas with TTL they are firmly fixed.

Logic state	*TTL*	*CMOS*
Logic 1	2V to 5V	⅔ of the supply voltage
Logic 0	0V to 0.8V	⅓ of the supply or less
Indeterminate	0.8V to 2V	⅓ to ⅔ of the supply voltage

Logic levels with TTL and CMOS

Voltages present at the inputs and output of a gate must be maintained at the levels shown and not be allowed to fall into the indeterminate range; otherwise the gate will not give the correct logic function.

Noise

Noise is any small unwanted signal voltage at the input to a gate. Noise can force the input voltage level into the indeterminate range and the output of the gate will then be an unreliable logic state.

For example, the output of a NOT gate is logic 1 with its input at 0.8V (just in the logic 0 state). If now a number of gates in the system switch (outputs change state), then this may cause a voltage spike on the supply rail of 0.5V. This voltage spike or noise may then raise the input voltage to the NOT gate which causes the gate output to switch to logic 0.

Noise Margin

Logic gates therefore need some ability to reject noise, a small margin of voltage which when added to the existing input voltage will not cause the output to change state. The ability of a logic device to reject noise is measured in terms of its noise margin.

TTL Noise Margins

High State

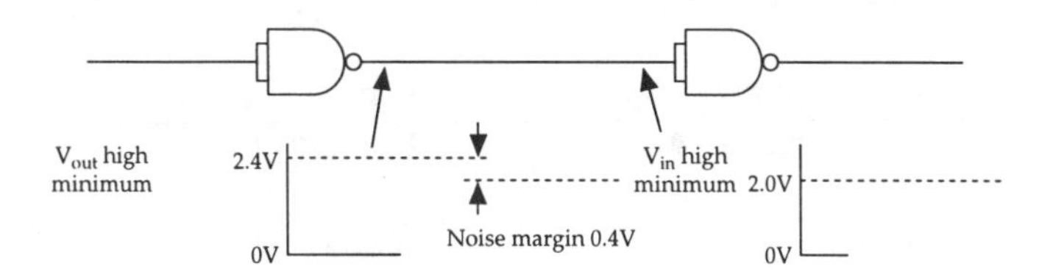

Output of gate guaranteed not to fall below 2.4V in the HIGH state

Minimum voltage accepted as a logic 1 input is 2.0V

TTL noise margin in the HIGH state

Low State

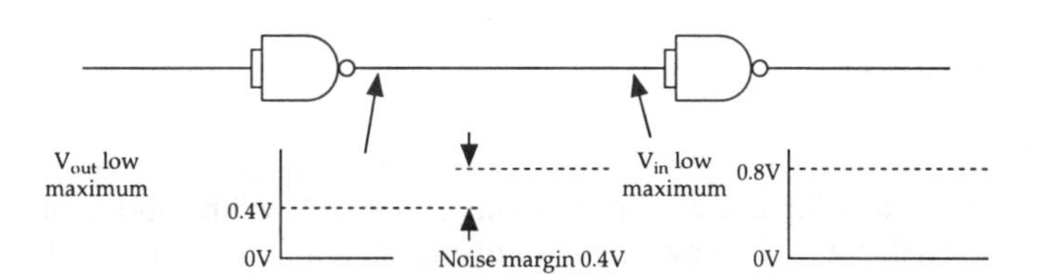

Output of gate guaranteed not to rise above 0.4V in the LOW state

Maximum voltage accepted as a logic 0 input is 0.8V

TTL Noise margin in the LOW state

CMOS Noise Margins

These are much better than TTL but the actual margin will depend upon the supply voltage. If we assume a supply voltgage of 9V then the noise margin would be $^1/_3$ of the supply, 3V.

High State

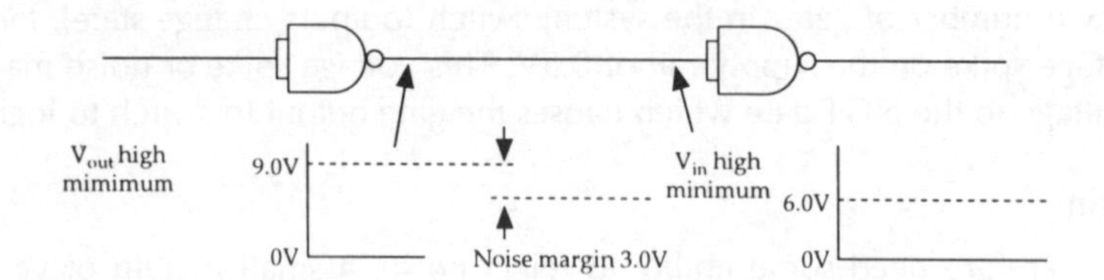

Output of gate will not fall much below 9V in the HIGH state

Minimum voltage accepted as a logic 1 input is 6.0V

CMOS noise margin in the HIGH state (supply voltage 9V)

Low State

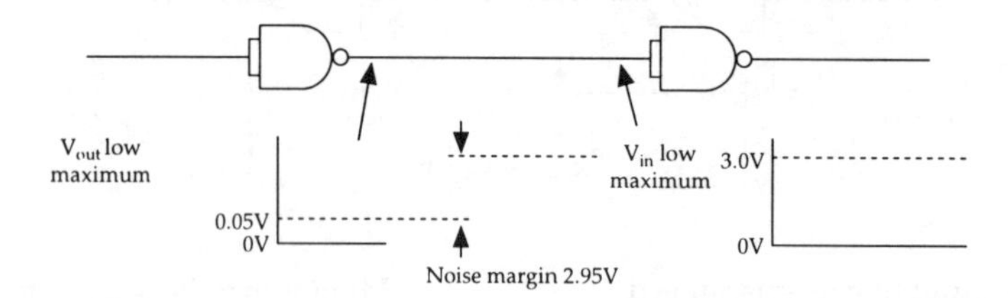

Output of gate guaranteed not to be above 0.05V in the LOW state

Maximum voltage accepted as a logic 0 input is 3V

CMOS noise margin in the LOW state (supply voltage 9V)

Testing Logic Chips

Logic Probes

These are used to detect logic levels in a circuit under test. With most probes, there is an indication of the **High** state, the **Low** state, a **Pulsing** state and a voltage that is between the logic levels; i.e. a **bad** or **indeterminate** logic level.

The probe will also have to distinguish between TTL and CMOS devices operating from differing supply voltages. This is achieved because the probe supply comes from the circuit under test and in the case of CMOS, if the supply voltage is known, the indeterminate logic levels can be computed by the probe.

Logic Pulser

The logic pulser is used to inject a pulse into a logic circuit. The injected pulse has sufficient current to overide the existing output at a gate. For example if the TTL gate output is logic 0 (0V) the pulser will force the output up to logic 1 (+5V) momentarily, changing the logic state. To avoid damage to the chip, the duration of the pulse is restricted to about 1.5 microseconds.

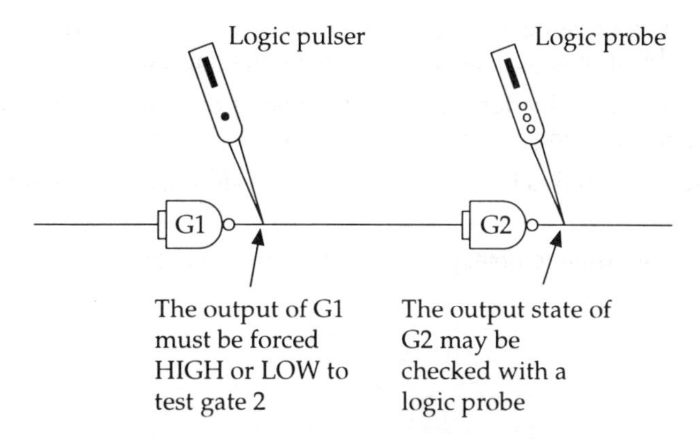

A logic pulser and probe being used to check gate G2

A commercial pulser must be capable of sinking or sourcing a current of at least 100mA.

When the output of a gate is high and has to be forced low then the pulser must sink (lose) current. When the output of the gate is low and must be forced high then the pulser must source (provide) current.

The pulser is normally powered from the circuit (logic chip) under test.

In between pulses the pulser has a high resistance of around 300k ohms. This will allow all logic families to function normally even though the pulser probe tip is still in contact with the circuit under test.

The operation of the pulser can be either a single shot, one pulse only, or a continuous mode, which provides a train of pulses.

The logic pulser may be used in conjunction with the logic probe. The pulser is used to produce a change in the logic state at the input of a gate, and the probe used to detect the change at the output.

Summary

- There are two main logic families: TTL produced from bi-polar devices and CMOS produced from uni-polar devices.
- The TTL family is the 74 series of logic chips.

  The CMOS family is the 4000 series of logic chips.
- There are several sub-families of the 74 series which are distinguished by an infix code e.g. 74LSXX is the low power Schottky range of digital chips.
- There are pin-compatible CMOS versions of TTL chips and these are coded with the letters HCT: e.g. 74HCT00 is a CMOS quad 2-input NAND chip with TTL-compatible inputs.
- There are a range of voltages that are considered to be logic 1 and logic 0. In between this range are voltages that will give an indeterminate state (neither logic 1 or logic 0).
- The noise margin is the ability of the gate to reject noise.

  The noise margin is the difference between:

a) the minimum value for logic 1 at the output of a gate and the minimum value acceptable as a logic 1 on the input to a following gate.

b) the maximum value for logic 0 at the output of a gate and the maximum value acceptable as a logic 0 on the input to a following gate.

- Logic pulsers are used to inject pulses into logic gate systems and override the existing logic states.
- Logic probes are used to detect the logic states in digital logic gate systems.

Simplification of Boolean equations

Introduction

When we design and then build practical logic gate circuits, we require the logic gate diagram to be in its simplest form. By simplifying the Boolean equations we can often reduce the number of logic gates needed in the logic gate diagram.

Below is a list of logic rules that can be used for simplification of Boolean equations. These rules are illustrated using the equivalent switching circuits. In order to fully understand these rules then:

a) All switches with the same label will operate together.

   For example: if one switch is normally open (N.O.) and the other is normally closed (N.C) then if they both have the same label A, the N.O. will close and the N.C. will open when switch A is operated.

b) The **switch circuits** are only intended to **illustrate the operation**, and the logic function would normally be implemented with logic gates.

c) The meanings of the abbreviations used are:
   S/C = short circuit. (very low resistance)
   O/C = open circuit. (very high resistance)
   N.O. = normally open (push to make switch)
   N.C. = normally closed (push to break switch)

1) $A+A=A$

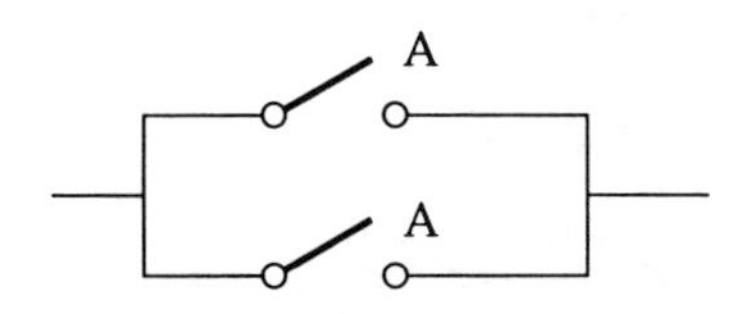

Two switches in parallel, one switch is redundant.

2) $A.A=A$

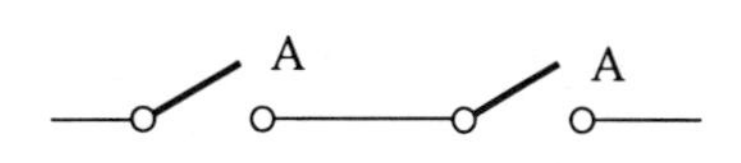

Two switches in series, one switch is redundant.

3) $A+\overline{A}=1$

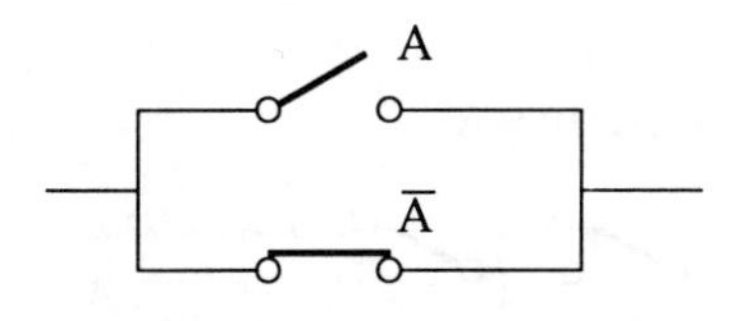

A N/O switch in parallel with a N/C switch will always give an output.

4) $A.\overline{A}=0$

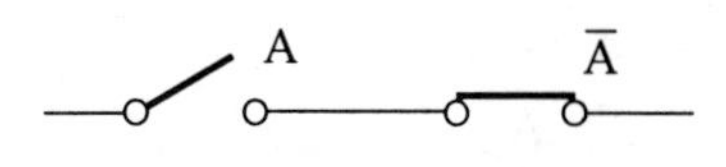

A N/O switch in series with an N/C switch will never give an output.

5) A+0 = A

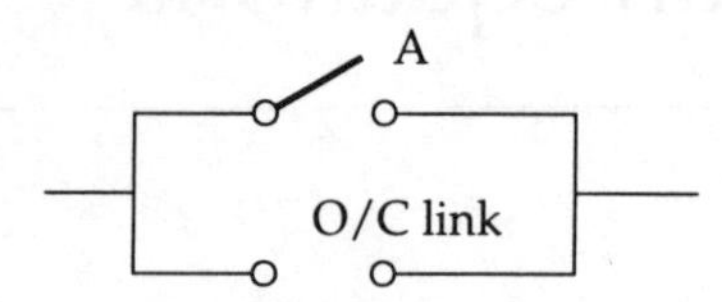

Switch A in parallel with an O/C link is simply switch A.

6) A+1 = 1

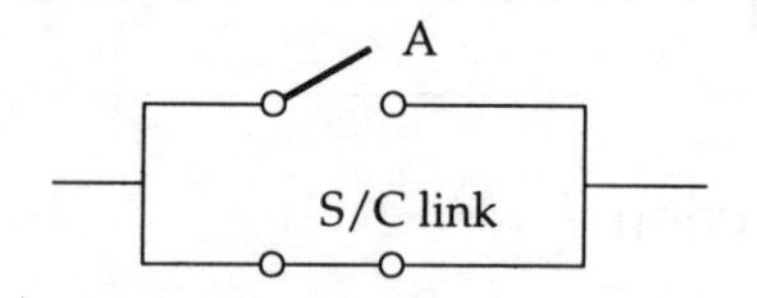

Switch A in parallel with a S/C link is shorted out.

7) A.1 = A

Switch A in series with a S/C link is simply switch A.

8) A.0 = 0

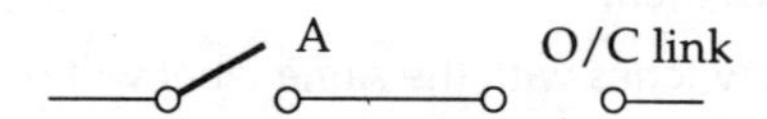

Switch A in series with an O/C link can never give an output.

9) A.B = B.A

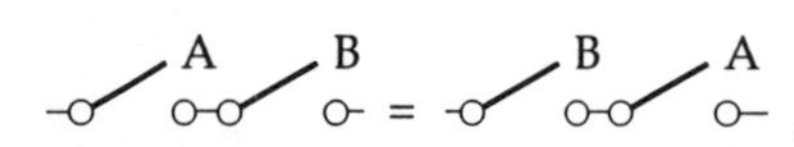

The circuit is the same either way round.

10) A.(B+C) = A.B + A.C

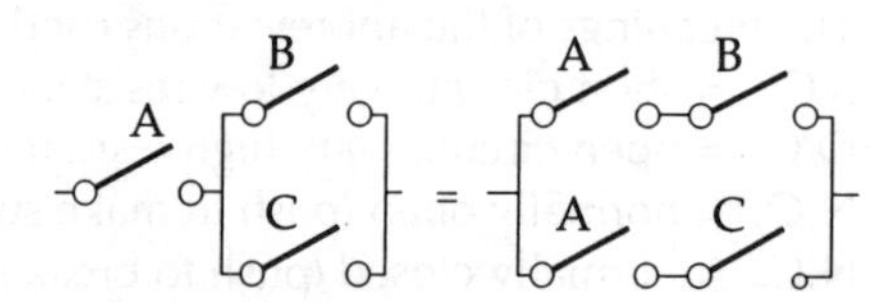

Both circuits perform the same function.

11) A + B.C = A.A + B.C

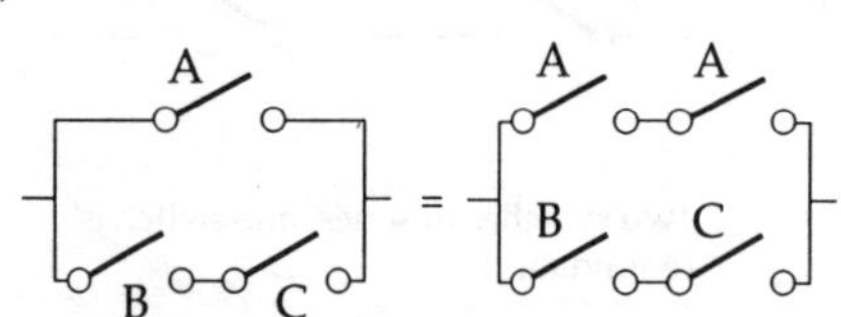

Both circuits perform the same function.

12) A.(B+$\overline{B}$) = A

B and $\overline{B}$ effectively form a S/C link.

13) A + A.B = A

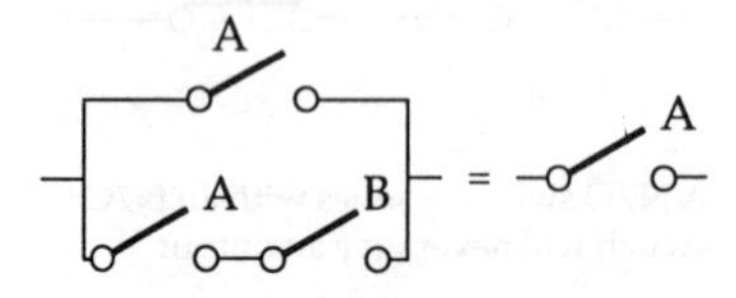

An output is obtained if A alone is closed. Switch B operated alone will not give an output, only if operated with A. B is therefore redundant!

14) A.(A+B) = A

Output obtained if A is operated. No output obtained if B is operated alone, only if operated with A. B is therefore redundant.

15) $A + \overline{A}.B = A+B$

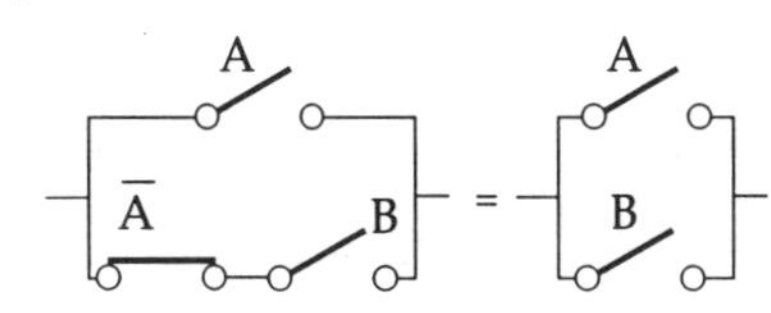

An output is obtained if A is operated alone or if B is operated alone. Switch NOT A is redundant.

16) $B.(A+\overline{B}) = A+B$

B A $\overline{B}$ = A B

Output occurs only when A AND B are operated together. NOT B is redundant.

The basic rules above may be used to simplify Boolean equations.

Study Note: The normal algebraic rules of factorisation and multiplying out can be employed. This is illustrated by rule 10.

Example

Simplify the Boolean equation $F = A.B + A.\overline{C} + B.C + A.C$

Solution

$$\mathbf{F = A.B + A.\overline{C} + B.C + A.C}$$

$= A.B + B.C + A(\overline{C} + C)$ (from rule 10)

$= A.B + B.C + A.1$ (since $\overline{C} + C = 1$: rule 3)

$= A.B + B.C + A$ (since $A.1 = A$: rule 7)

$= A(1 + B) + B.C$ (since $A(1+B) = A + A.B$)

$= A.1 + B.C$ (since $1 + B = 1$ rule 6)

Therefore:

$$\mathbf{F = A + B.C}$$ (since $A.1 = A$ rule 7)

To build the logic gate diagram for the Boolean equation:

$$\mathbf{F = A.B + A.\overline{C} + B.C + A.C}$$

requires four 2-input AND gates and a 4-input OR gate.

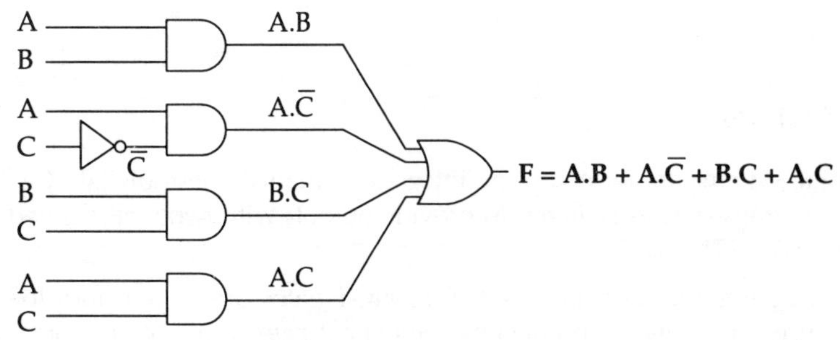

Logic gate diagram for the equation $F = A.B+\overline{A.C}+B.C+A.C$

Study Note: Notice that in the above logic gate diagram inputs A, B and C appear more that once. All this means is that in practice (when you actually build the gate diagram) all the input wires labelled A must be connected together. Similarly so must all the input wires labelled B, etc. Labelling the inputs on the gate diagram in this way makes the diagram look less complicated.

The simplified Boolean equation only requires one 2-input OR gate and one 2-input AND gate, which is far less complicated to build.

Logic gate diagram for the Boolean equation F = A+B.C

Summary of Logic Rules

1) $A + A = A$
2) $A.A = A$
3) $A + \bar{A} = 1$
4) $\bar{A}.A = 0$
5) $A + 0 = A$
6) $A + 1 = 1$
7) $A.1 = A$
8) $A.0 = 0$
9) $A.B = B.A$
10) $A(B+C) = A.B + A.C$
11) $A + B.C = A.A + B.C$
12) $A(\bar{B}+B) = A$
13) $A + A.B = A$
14) $A(A+B) = A + B$
15) $A + \bar{A}.B = A + B$
16) $B(A+\bar{B}) = A + B$

Self-assessment questions (1) (answers page 465)

1) Use the rules listed in the summary table above to simplify the following Boolean equations.

    a) $A(A + B)$

    b) $A.B + B + B$

    c) $B.(\bar{B} + C)$

    d) $A.\bar{B} + A.B$

    e) $A.(1 + B + C) + B$

    f) $\bar{A}.(A + B) + A.B$

Using NAND-NOR Logic

Practical logic systems are formed using integrated circuits. These digital I.C.s have output stages that use transistors as switches to provide outputs with two well defined logic levels (+5V and 0V with TTL chips).

Logic gate diagrams that contain AND, OR, NOT gates are called pure logic gate diagrams. Because each logic chip contains 4 identical gates it is not very practical to have gate diagrams in pure logic form.

With **NAND** and **NOR** gates **all the other logic functions can be obtained**. Therefore it is more economical practically to have all the gates in the system in either all NAND or all NOR form, rather than in pure logic form. Logic functions are therefore implemented using all NAND or all NOR gates.

Obtaining Other Logic Functions From NAND Gates

AND Function

Two NAND gates may be used to give the AND function. The output from the first gate feeds a second gate in which the inputs are strapped together. This second gate then behaves simply as a NOT gate or invertor.

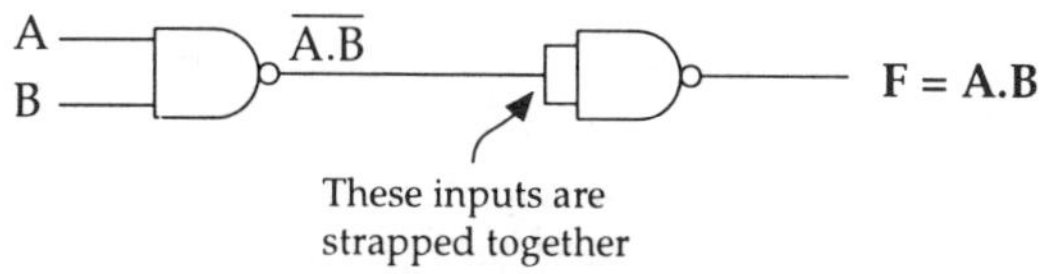

The AND function using two NAND gates

The logic level is inverted at the output of the first NAND gate and it is further inverted by the second NAND gate connected as an invertor. This double inversion means that the logic level is unaffected: i.e. if you invert logic 1 twice, you still have logic 1!

B	A	$\overline{A.B}$	$\overline{\overline{A.B}}$	
0	0	1	0	
0	1	1	0	
1	0	1	0	
1	1	0	1	**The AND function**

Truth Table for the AND function using NAND gates

OR Function

Two levels of NAND logic may be used to achieve the OR function. The NAND gates in the first stage or level are connected as invertors.

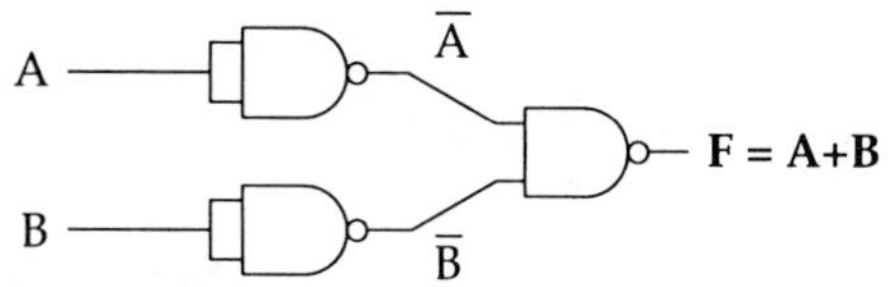

The OR function using three NAND gates

B	A	$\overline{A}$	$\overline{B}$	$\overline{A}.\overline{B}$	$\overline{\overline{A}.\overline{B}}$	
0	0	1	1	1	0	
0	1	0	1	0	1	
1	0	1	0	0	1	
1	1	0	0	0	1	**The OR function**

Truth Table for the OR function using NAND gates

Obtaining Other Logic Functions From NOR Gates

OR Function

Two NOR gates may be used to give the OR function.

The output from the first gate feeds a second gate in which the inputs are strapped together. This second gate then behaves simply as a NOT gate or invertor.

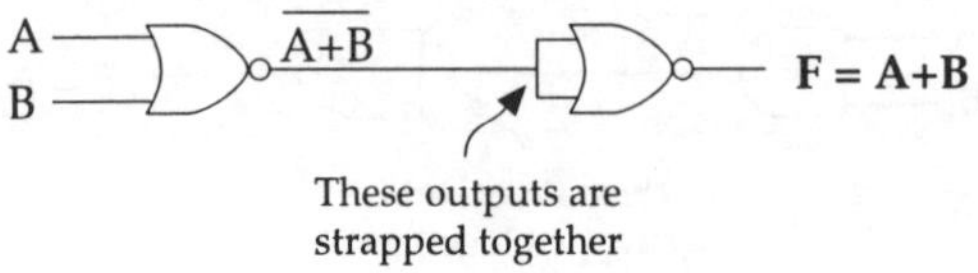

A, B, $\overline{A+B}$, F = A+B

These outputs are strapped together

The OR function using two NOR gates

B	A	$\overline{A+B}$	$\overline{\overline{A+B}}$
0	0	1	0
0	1	0	1
1	0	0	1
1	1	0	1

The OR function

Truth Table for the OR function using NOR gates

AND Function

Two levels of NOR logic may be used to achieve the AND function. The NOR gates in the first stage or level are connected as invertors.

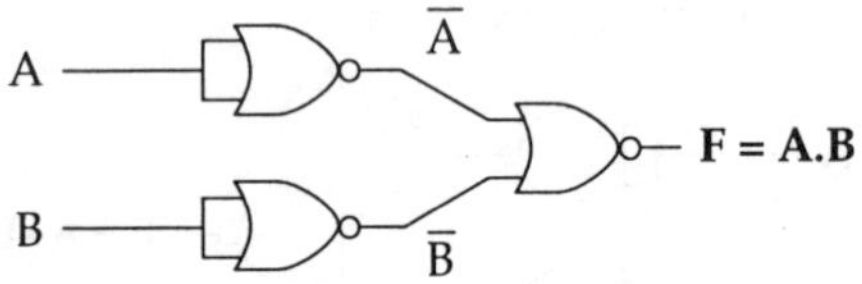

A, $\overline{A}$, B, $\overline{B}$, F = A.B

The AND function using three NOR gates

B	A	$\overline{B}$	$\overline{A}$	$\overline{A}+\overline{B}$	$\overline{\overline{A}+\overline{B}}$
0	0	1	1	1	0
0	1	1	0	1	0
1	0	0	1	1	0
1	1	0	0	0	1

The AND function

Truth Table for the AND function using NOR gates

Logic Gate Diagrams in all NAND/NOR Form

Because NAND (or NOR) logic gates can be used to perform other logic functions and in a logic chip there is more than one gate, it makes construction of logic gate circuits more simple if we use all NAND or all NOR chips.

To achieve this we simply replace any AND or OR functions with its NAND equivalent to create an all NAND form of the circuit.

Example

Devise a NAND gate logic circuit for the function F = A.B + C.

Pure Logic Gate Diagram

We now replace the AND and the OR gate with its equivalent NAND form.

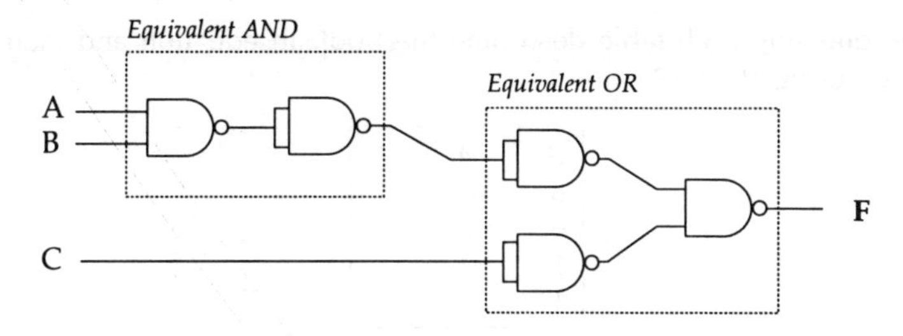

The AND and OR gates are replaced with equivalent NANDS

We now remove any double inversions. Remember a double inversion means that the logic level is unaffected: i.e. if you invert logic 1 twice, you still have logic 1!

Logic gate diagram after double inversions are removed

Summary

- To achieve simple practical logic gate circuits requires the Boolean equations to be in their simplest form.
- Basic logic chips have a number of similar gates and if only one gate is used within a chip then up to 75% of that chip can be redundant.
- NAND and NOR gates can be configured to perform other logic functions. To make construction of logic gate circuits more simple and utilise more of the available gates per chip, we can convert the logic gate diagram into an all NAND (or NOR) form.
- The AND or OR gate is replaced with its equivalent NAND or NOR form and any double inversions removed.
- To obtain these other logic functions, NAND or NOR gates within a chip, have to be connected as invertors. To do this in practice requires the input pins on the chip for a particular gate, to be physically wired together.

Self-assessment questions (2) (answers page 465)

1) Draw the gate diagram for a NAND gate using all NOR gates.
2) Draw the gate diagram for a NOR gate using all NAND gates.
3) From the following truth table determine the Boolean equation and then implement the function using all NAND gates.

B	*A*	*F*
0	0	0
0	1	1
1	0	1
1	1	0

4) From the following truth table determine the Boolean equation and then implement the function using all NOR gates.

B	*A*	*F*
0	0	1
0	1	0
1	0	0
1	1	1

5) Devise a NOR gate circuit for the function:

   $F = (A + B).C$

6) Devise a NOR gate circuit for the function:

   $F = \overline{A}.B.C + A.\overline{B}.C$

de-Morgan's rules

Introduction

To build simple practical logic gate circuits, the Boolean equation needs to be in its simplest form. But also the equation often needs to be in all NAND or all NOR form.

In **Part 3 Section 2 Information and Skills Bank** Simplification of Boolean equations, p.397, we replaced AND or OR functions with their equivalent NAND or NOR circuits. de-Morgan's rules are a technique for writing the AND equivalent of an OR gate, or the OR equivalent of an AND gate, directly.

Consider the NAND Gate Truth Table

B	A	$A.B$	$\overline{A.B}$
0	0	0	1
0	1	0	1
1	0	0	1
1	1	1	0

The NAND function is an AND gate followed by a NOT gate. Therefore the Boolean equation is:

$$F = \overline{A.B}$$

Also from the truth table, when any input is at logic 0, the output is at logic 1.

The switching circuit for the NAND gate is:

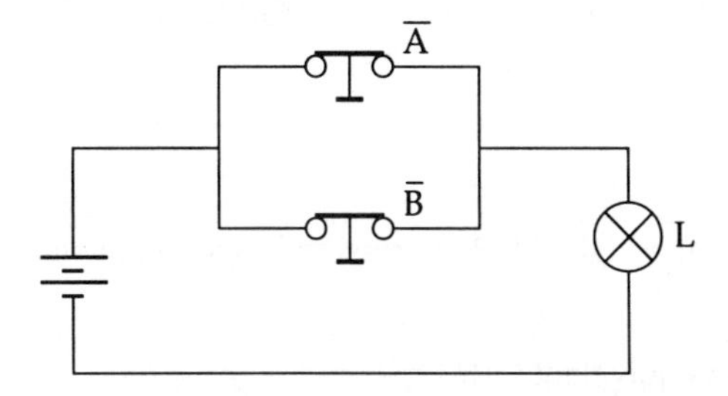

The function of the circuit is $\mathbf{F = \overline{A} + \overline{B}}$. The lamp is ON if A is not operated. The lamp is ON if B is not operated, or if both A and B are not operated. The lamp is OFF if both A and B are operated.

The expressions $F = \overline{A} + \overline{B}$ and $F = \overline{A.B}$ both describe the same truth table. Therefore these two expressions must be equivalent to each other!

Consider the NOR Gate Truth Table

B	A	$A+B$	$\overline{A+B}$
0	0	0	1
0	1	1	0
1	0	1	0
1	1	1	0

The NOR function is an OR gate followed by a NOT gate. Therefore the Boolean equation is:

$F = \overline{A+B}$

Also from the truth table, when any input is at logic 1, the output is at logic 0.

The switching circuit for the NOR gate is:

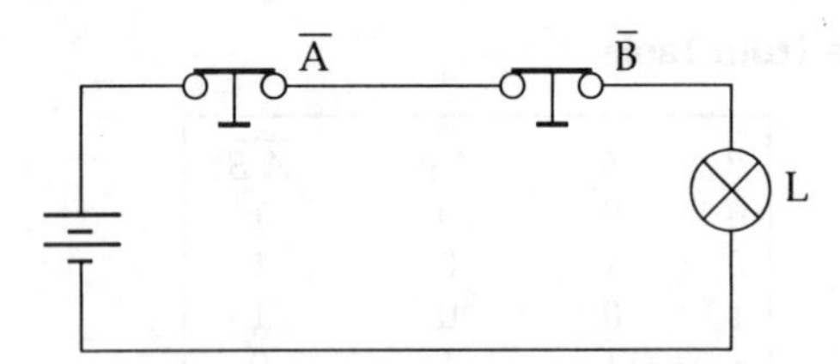

The function of the circuit is $\mathbf{F = \overline{A}.\overline{B}}$. The lamp is OFF if A is operated. The lamp is OFF if B is operated, or if both A and B are operated. The lamp is ON if both A and B are not operated.

The expressions $F = \overline{A}.\overline{B}$ and $F = \overline{A+B}$ both describe the same truth table. Therefore these two expressions must also be equivalent to each other!

These relationships show that there is an equivalent OR for an AND circuit and vice versa. This is the basis of de-Morgan's Rules.

de-Morgan's Rules

These, put simply, state that a logical OR expression has an equivalent AND;

e.g. $\overline{A + B} = \overline{A}.\overline{B}$

and a logical AND has an equivalent OR;

e.g. $\overline{A.B} = \overline{A} + \overline{B}$

Converting to one form or the other may be achieved by employing 3 simple rules.

1) Bar all unbarred terms.
2) Un-bar all barred terms.
3) Change all (+) to (.) and all (.) to (+).

When using these rules remember that:

- The rules can be applied in any order.
- It may not be possible to apply all the rules.
- You must not undo using one rule what was previously done using another rule.

- Each individual term in the expression may be barred as well as the whole expression.

The whole expression is barred!

$$\overline{\overline{A} + \overline{B}}$$

Individual terms are barred!

In logic gate terms this means:

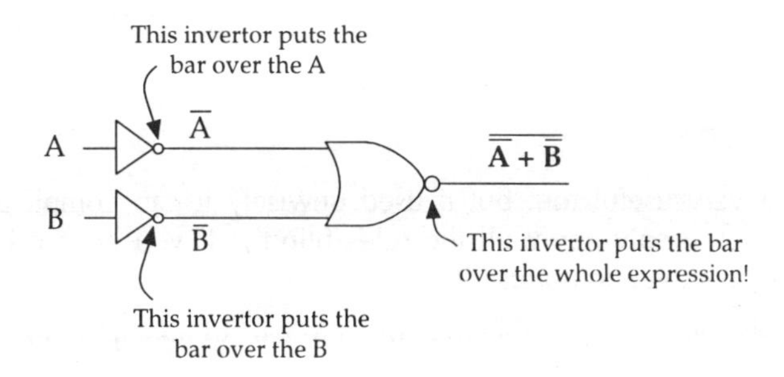

Example 1 $F = \overline{A.B}$

$F = \overline{\overline{A}.\overline{B}}$	[rule 1 barring all un-barred terms]
$F = \overline{A}.\overline{B}$	[rule 2 de-barring the whole expression] (Notice that you do not undo the terms you barred using rule 1.)
$F = \overline{A}+\overline{B}$	[rule 3 changing the (.) for a (+)]

Therefore: $\overline{A.B}$ is equivalent to $\overline{A} + \overline{B}$

Example 2 $F = \overline{A + B}$

$F = A + B$	[rule 2 de-barring all barred terms]
$F = \overline{A} + \overline{B}$	[rule 1 barring all unbarred terms]
$F = \overline{A}.\overline{B}$	[rule 3 changing the (+) for a (.)]

Therefore: $\overline{A + B}$ is equivalent to $\overline{A}.\overline{B}$.

de-Morgan's is very useful in determining the function of logic gates. The logic gate symbol used on a drawing does not always coincide with the logic gate chip used to implement that function!

Also to simplify drawings, NOT gates (invertors) at the input to a gate are represented as circles: e.g.

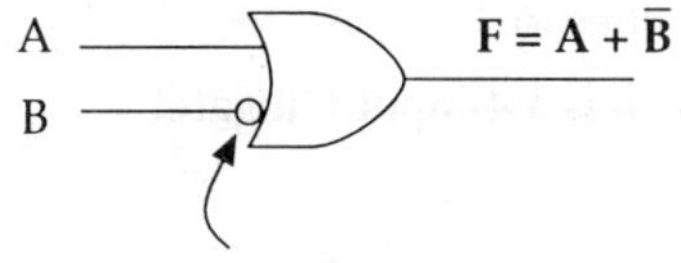

This circle means B is inverted before it enters the OR gate

Self-assessment questions (1) *(answers page 467)*

Use de-Morgan's to determine the functions of the following circuits:

1) 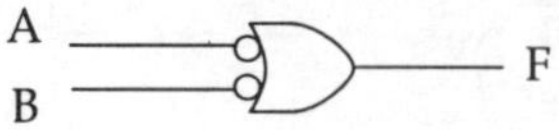 A B F

2) 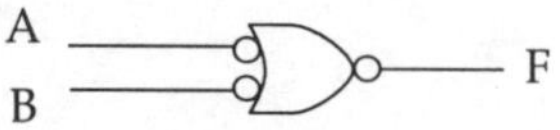 A B F

3) 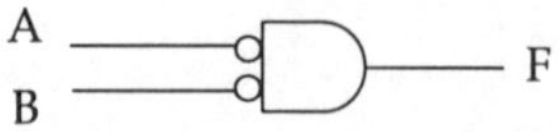 A B F

4) 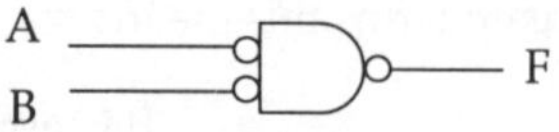 A B F

Using de-Morgan's

de-Morgan's is a very useful tool but if used unwisely it can complicate your Boolean expressions. Do not simply apply all the rules blindly! If you do not have a reason for applying de-Morgan's then do not use it!

When using de-Morgan's it is possible to apply it to **part of an expression** only!

Example 1: To remove an OR term

Consider the equation $F = \overline{\overline{(A.B)}.(\overline{C} + \overline{D})}$

In order to use de-Morgan's on complicated terms it is often useful to simplify the terms before applying de-Morgan's.

The expression above can be put in the form:

$F = \overline{X.Y}$ where $X = \overline{A.B}$ and $Y = (\overline{C} + \overline{D})$

We now perform de-Morgan's **on Y only!**

and $Y = \overline{C.D}$ [by de-Morgan's]

We now substitute back into the original expression to obtain:

$F = \overline{\overline{(A.B)}.\overline{(C.D)}}$ **which is now in all NAND form!**

Example 2: To remove an AND term

Consider the equation: $F = \overline{\overline{(A + B)}.\overline{(C + D)}}$

This again can be put in the form: $F = \overline{\overline{X}.\overline{Y}}$

where $X = A + B$ and $Y = C + D$

We now do de-Morgan's on the whole expression (Not to X and Not to Y individually).

$F = X + Y$ [by de-Morgan's]

which gives: $F = (A + B) + (C + D)$ substituting for X and Y

The expression therefore simplifies to:

$F = A + B + C + D$ **which is a 4-input OR gate!**

Example 3: To remove bars (invertors)

Consider the equation: $F = \overline{\overline{(A.\overline{A.B})}.\overline{(B.\overline{A.B})}}$

This can be put in the form: $F = \overline{\overline{X}.\overline{Y}}$

where $X = A.\overline{A.B}$ and $Y = B.\overline{A.B}$.

Doing de-Morgan's on the whole expression gives:

$$
\begin{aligned}
F = \overline{\overline{X}.\overline{Y}} &= X + Y && \text{[by de-Morgan's]} \\
F &= A.\overline{A.B} + B.\overline{A.B} && \text{substituting back X and Y} \\
&= (A + B).(\overline{A.B}) && \text{[factorizing]} \\
&= (A + B).(\overline{A} + \overline{B}) && \text{[de-Morgan's on } \overline{A.B}\text{]} \\
&= A.\overline{A} + A.\overline{B} + B.\overline{A} + B.\overline{B} && \text{[multiplying out]} \\
&= A.\overline{B} + B.\overline{A} && \text{[since } A.\overline{A} \text{ \& } B.\overline{B} = 0\text{]}
\end{aligned}
$$

Therefore:

$F = A.\overline{B} = B.\overline{A}$ which is the exclusive OR!

Example 4: To get an equation in all NAND or NOR form

Consider the equation: $F = P.Q + A.B.C$

We can put this into the form: $F = X + Y$

where $X = P.Q$ and $Y = A.B.C$.

Using de-Morgan's gives: $F = \overline{\overline{X}.\overline{Y}}$.

Substituting back for X and Y gives:

$F = \overline{\overline{P.Q}\ .\ \overline{A.B.C}}$

which is in all NAND form!

Summary

- A logical OR function has a equivalent AND and a logical AND function has an equivalent OR. This is the basis of de-Morgan's rules.
- de-Morgan's should only be used for a specific reason.
- de-Morgan's may be used to:

  a) get rid of an AND or OR term in an expression.

  b) to remove bars (invertors) in an expression.

  c) to convert the expression to all NOR or NAND form.

  d) to determine the function of a series of logic gates.

Self-assessment questions (2) *(answers page 467)*

1) Use de-Morgan's to show that:

   a) $\overline{A.B + C} = (\overline{A}+\overline{B}).\overline{C}$

   b) $\overline{(A + B).(C + D)} = (\overline{A.B}) + (\overline{C.D})$

c) $\overline{(A+\overline{B}+\overline{C})} + \overline{(\overline{A}+\overline{C}+B)} = \overline{A}.B.C + A.\overline{B}.C$

2) Show by truth table and de-Morgan's that:

$$\overline{\overline{(\overline{A}.B)}.\overline{(\overline{B}.A)}} = A.\overline{B} + B.\overline{A}$$

3) For the two gate diagrams given, derive the output expression and simplify using de-Morgans.

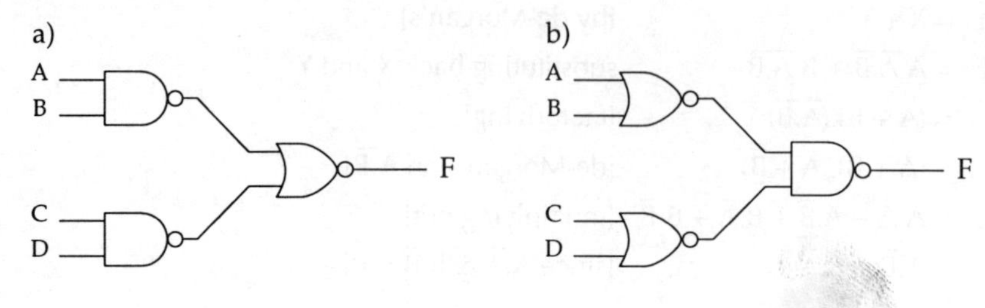

Karnaugh mapping

Introduction

Karnaugh mapping is a graphical method of representing the truth table of a particular logic function. It is a powerful method of obtaining a simplified Boolean equation easily and producing logic gate diagrams free from static hazards.

The map is a rectangular diagram divided into cells where the number of cells in the map = 2^n and *n* is the number of variables A,B,C etc. Therefore with 2 variables there will be 4 cells; with 3 variables 8 cells and so on.

Each variable in the Karnaugh map is **represented by half the total area** and its **complement by the other half**.

One-variable Map (2-cells)

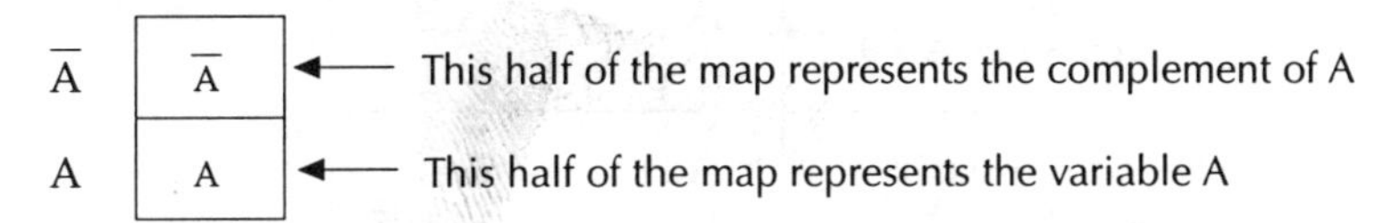

Two-variable Map (4-cells)

The map now has to represent the two variables A and B and their complements.

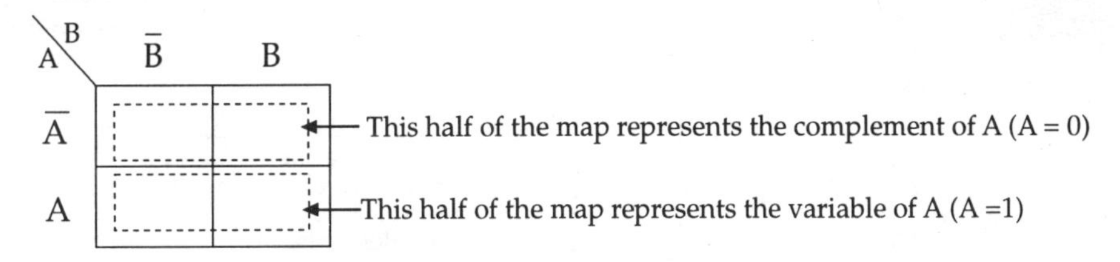

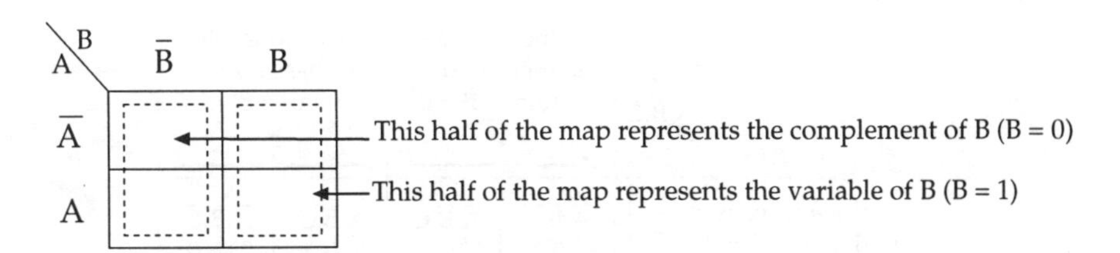

Each cell on the map now represents both the variables A and B or their complements.

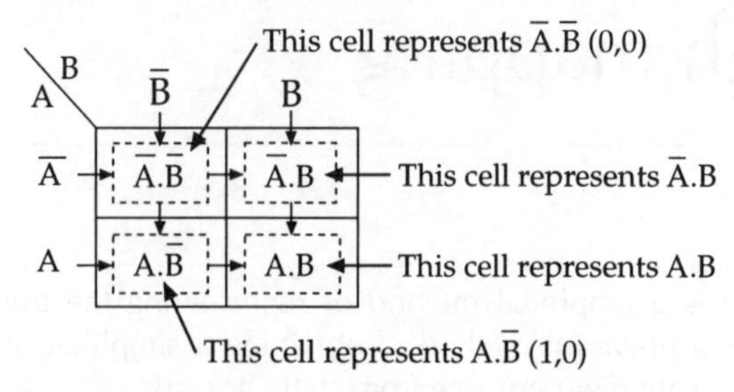

Because this is a graphical representation of the truth table the letters in the map may be replaced with logic 1s and 0s.

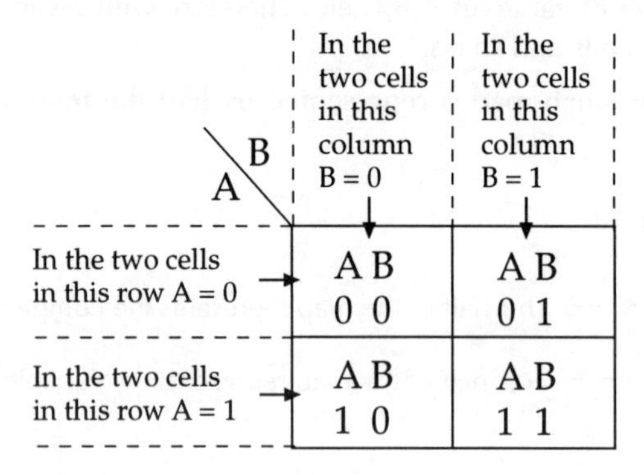

The Karnaugh map is a graphical representation of the truth table.

A two-variable map is therefore drawn as shown below.

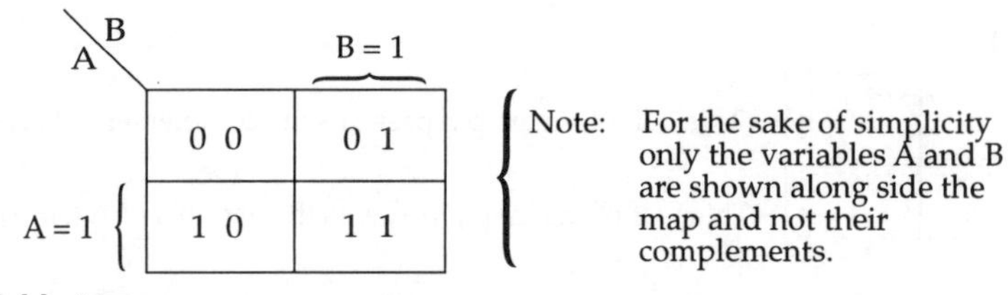

Three-variable Map

In the four cells in these two columns B = 0 ($\overline{B}$); in the four cells in these two columns B = 1 (B).

A \ B.C	$\overline{B}$		B	
In the four cells in this row A = 0 — $\overline{A}$	$\overline{A}.\overline{B}.\overline{C}$ 0 0 0	$\overline{A}.\overline{B}$.C 0 0 1	$\overline{A}$.B.C 0 1 1	$\overline{A}$.B.$\overline{C}$ 0 1 0
In the four cells in this row A = 1 — A	A.$\overline{B}.\overline{C}$ 1 0 0	A.$\overline{B}$.C 1 0 1	A.B.C 1 1 1	A.B.$\overline{C}$ 1 1 0
	$\overline{C}$	C	C	$\overline{C}$

$\overline{C}$: C = 0 in the two cells in this column. C: C = 1 in the four cells in these two columns. $\overline{C}$: C = 0 in the two cells in this column.

A Four-variable Map

	$\overline{C}$		C		
$\overline{A}$	$\overline{A}.\overline{B}.\overline{C}.\overline{D}$	$\overline{A}.\overline{B}.\overline{C}.D$	$\overline{A}.\overline{B}.C.D$	$\overline{A}.\overline{B}.C.\overline{D}$	$\overline{B}$
$\overline{A}$	$\overline{A}.B.\overline{C}.\overline{D}$	$\overline{A}.B.\overline{C}.D$	$\overline{A}.B.C.D$	$\overline{A}.B.C.\overline{D}$	B
A	$A.B.\overline{C}.\overline{D}$	$A.B.\overline{C}.D$	$A.B.C.D$	$A.B.C.\overline{D}$	B
A	$A.\overline{B}.\overline{C}.\overline{D}$	$A.\overline{B}.\overline{C}.D$	$A.\overline{B}.C.D$	$A.\overline{B}.C.\overline{D}$	$\overline{B}$
	$\overline{D}$	D	D	$\overline{D}$	

As before, the letters in the map may be replaced with logic 1s and 0s.

AB \ CD	00	01	11 (C=1)	10 (C=1)	
00	0000	0001	0011	0010	
01	0100	0101	0111	0110	B=1
11 (A=1)	1100	1101	1111	1110	B=1
10 (A=1)	1000	1001	1011	1010	
		D=1	D=1		

Notice in the map that the digits in the cells do not follow the natural binary code as they do in a truth table, but are arranged so that as you go from one cell to another, only one digit changes. Also as you go from top to bottom or side to side only one digit changes.

The positions marked A, B, C and D are the cells in the map in which that respective variable has a value of 1. The complements (the other half of the map) are not normally shown.

Using the Map

We can now use the map and plot a 1 in a cell to correspond with a logic 1 in the output column of a truth table.

For example, if in our truth table $\overline{A}.\overline{B}.\overline{C}.\overline{D} = 1$, then we place a 1 in the top left-hand cell.

Similarly, if in our truth table $A.\overline{B}.C.\overline{D} = 1$, we place a 1 in the bottom right-hand cell.

When we have plotted all the 1s in the output column of the truth table on the map, we can loop cells together and obtain a simplified Boolean expression.

Example

Consider the Boolean equation: $F = A.B + \overline{A}.C + A.C$

Draw up the truth table.

A	B	C	$A.B$	$B.C$	$A.C$	$A.\overline{C}$	$F = A.B + A.\overline{C} + B.C + A.C$
0	0	0	0	0	0	0	0
1	0	0	0	0	0	1	1 → $F = A.\overline{B}.\overline{C}$
0	1	0	0	0	0	0	0
1	1	0	1	0	0	1	1 → $F = A.B.\overline{C}$
0	0	1	0	0	0	0	0
1	0	1	0	0	1	1	1 → $F = A.\overline{B}.C$
0	1	1	0	1	0	0	1 → $F = \overline{A}.B.C$
1	1	1	1	1	1	1	1 → $F = A.B.C$

We can now draw up the Karnaugh map from the truth table. 3 variables gives 8 cells.

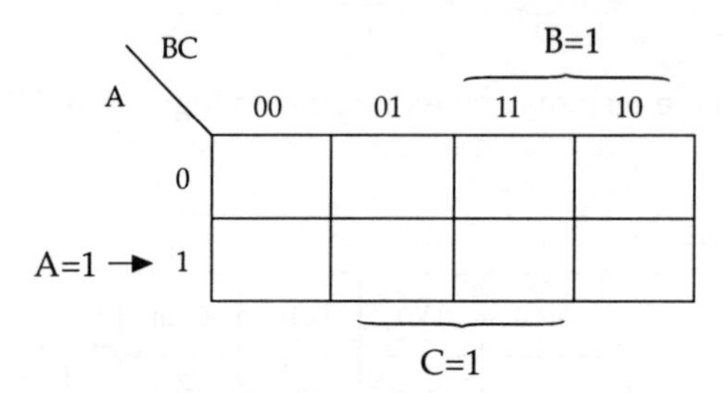

Step 1 Look across the truth table and where there is a 1 in the output column place a 1 in the cell corresponding to the 3 inputs (A.B.C) that produced the 1:

e.g. A=1; B=0; C=0; gives a 1 in the cell on the map where A = 1 and B and C both equal 0, as shown

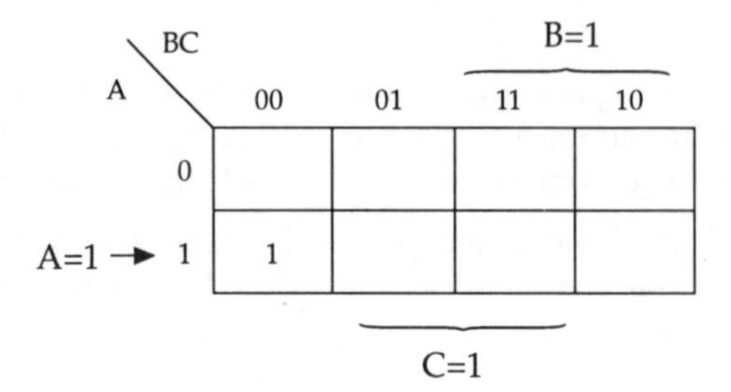

Now continue down the truth table and place 1s in the cells for all the other 1s in the output column of the truth table. The map should then be as follows:

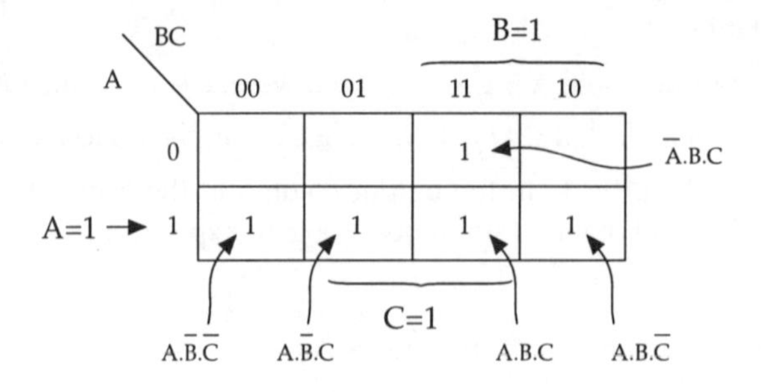

Step 2 When we have a 1 in each cell that corresponds to each and every logic 1 in the truth table, we can group the cells by drawing a loop around groups of 2, 4, or 8 cells.

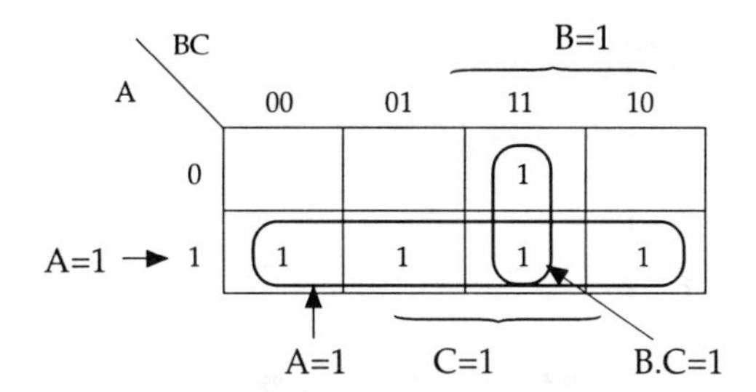

Step 3 From the looped cells the simplified equation can be obtained.

For example, there is a 1 in each cell that corresponds to when A = 1.

Therefore F = A

Also there is a 1 in each cell that corresponds to B = 1 AND C = 1.

Therefore F = B.C

The simplified equation becomes:

F = A + B.C

Check using Boolean algebra

$F = A.B + A.\overline{C} + B.C + A.C$

$= A.B + B.C + A.(\overline{C}+C)$

$= A.(1 + B) + B.C$

$= A + B.C$

Cell Looping

In the example of the 3-variable map, if 4 cells were looped then it gave rise to 1 term only (F = A). When only two cells were looped it gave rise to two terms (F = B.C). If cells cannot be looped then it will give rise to 3 terms.

Examples

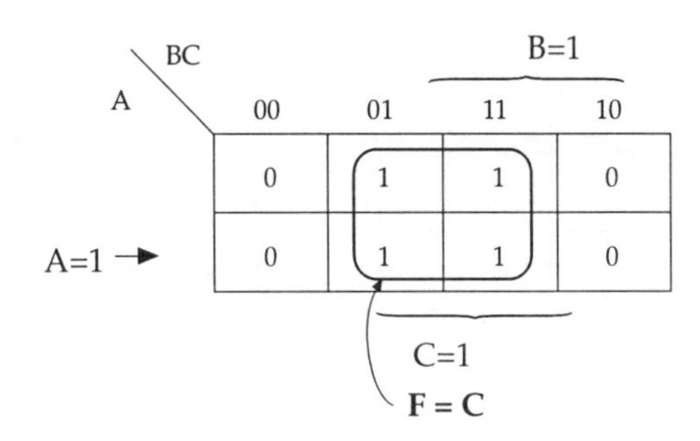

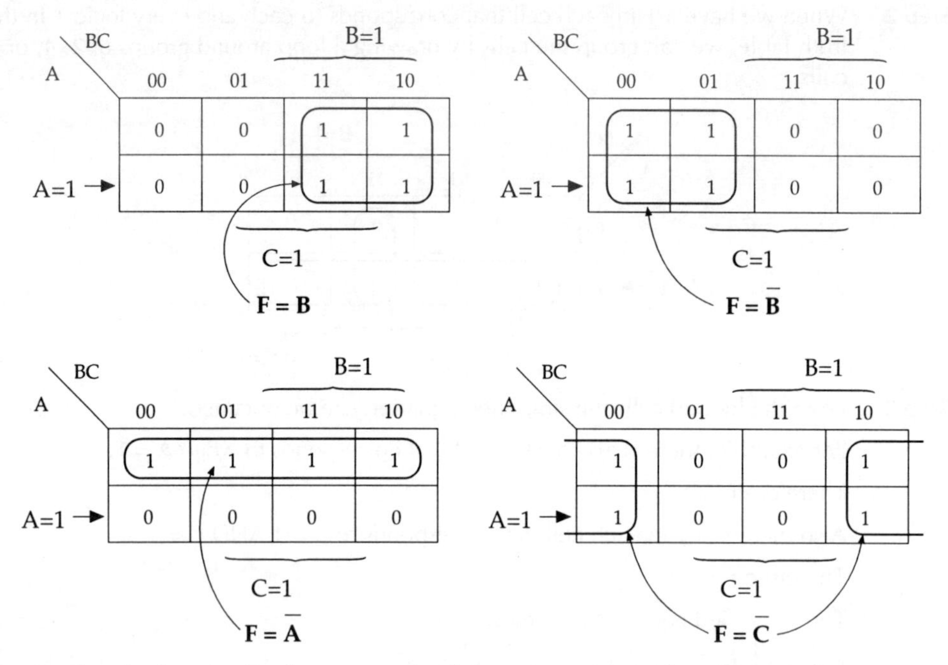

Notice that the looping has been extended from side to side in the last example.

Self Assessment Questions 1 *(answers page 468)*

Derive the Boolean Equations from the following Karnaugh Maps.

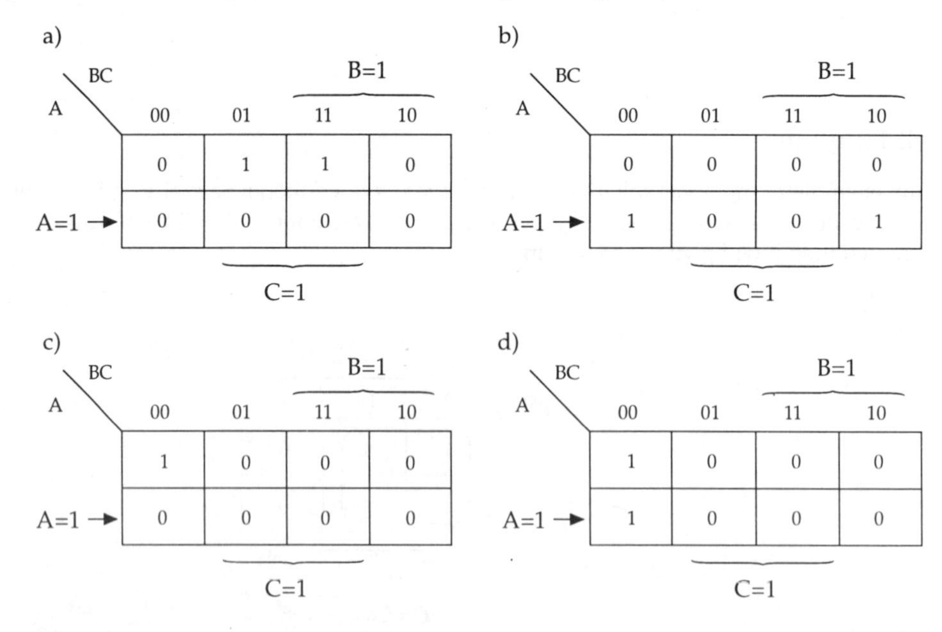

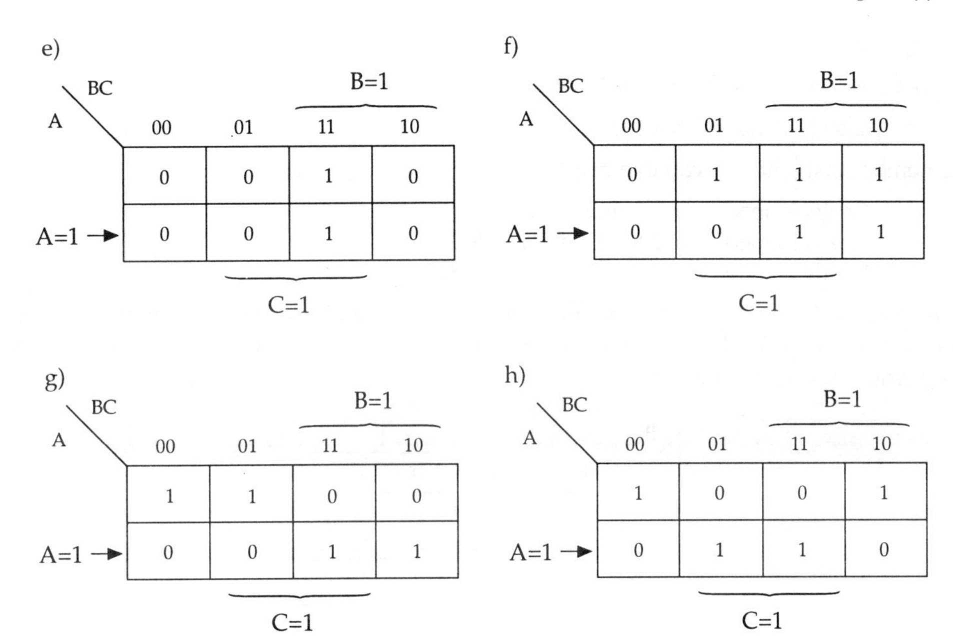

Direct Karnaugh mapping

It is not necessary to draw up the truth table in order to plot the Karnaugh map. The mapping may be done directly from the Boolean equation.

Example 1

Simplify the equation $F = A.B.C + \overline{A}.\overline{B}.\overline{C} + A.B.\overline{C} + \overline{A}.B.\overline{C}$

by Karnaugh map and check using Boolean algebra.

Karnaugh map

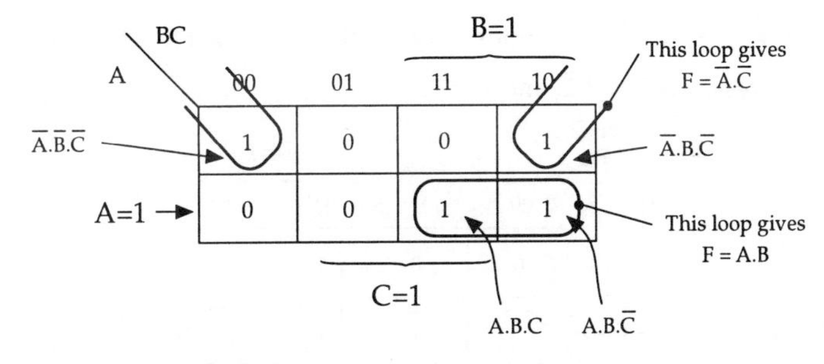

Looping gives $F = \overline{A}.\overline{C} + A.B$

Boolean

$$F = A.B.C + \overline{A}.\overline{B}.\overline{C} + A.B.\overline{C} + \overline{A}.B.\overline{C}$$

$$F = A.B\,(C+\overline{C}) + \overline{A}.\overline{C}(\overline{B}+B) \qquad \text{[factorizing]}$$

$$F = A.B + \overline{A}.\overline{C} \qquad \text{[since } B+\overline{B} = 1;\ C+\overline{C} = 1\text{]}$$

Example 2

Consider the equation $F = A.B + A.C + B.C + A.\overline{C}$

This may also be mapped directly.

Remember that with a 3-variable map:

4 cells looped	= 1 term (e.g. F = A)
2 cells looped	= 2 terms (e.g. F = A.B)
1 cell	= 3 terms (e.g. F = A.B.C)

Therefore in the example each product term (e.g. A.B) will give rise to putting 1s in two cells on the map. Look at the first term in the expression and put a 1 in the cells on the map where A = 1; AND B = 1.

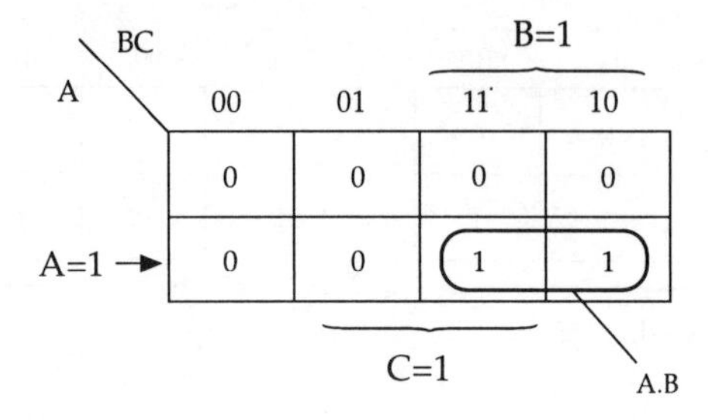

Now plot the next term F = A.C.

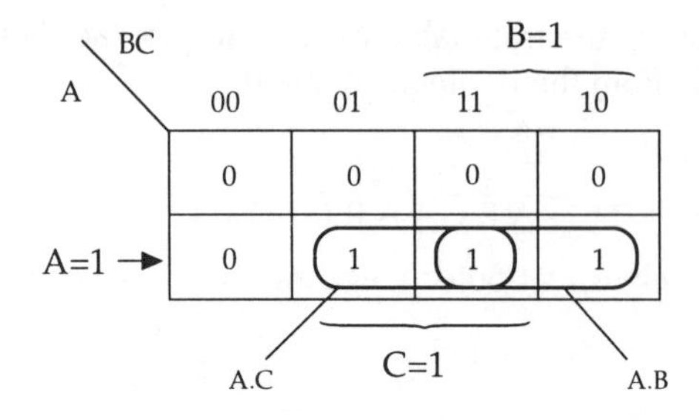

Notice that where two expressions refer to one cell you only put a 1 in that cell once!

Now plot the expressions $F = B.C$ and $F = A.\overline{C}$.

A \ BC	00	01	11	10
0	0	0	1	0
A=1	1	1	1	1

$A.\overline{C}$ C=1   B.C   B=1

When all the 1s have been placed in the cells, other larger loops may be drawn to simplify the equation.

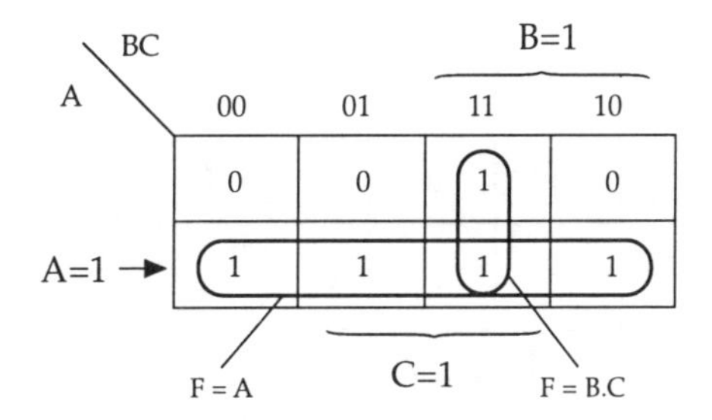

The simplified expression therefore is F = A + B.C.

A Four-Variable Karnaugh Map

With the four-variable map there are 16 cells. Therefore when looping:

looping 8 cells	= 1 term (e.g. F = A)
looping 4 cells	= 2 terms (e.g. F = A.B.)
looping 2 cells	= 3 terms (e.g. F = A.B.C)
1 cell	= 4 terms (e.g. F = A.B.C.D)

Consider the Equation:

$F = \overline{A}.\overline{B}.C.\overline{D} + A.\overline{B}.C.\overline{D} + \overline{A}.B.C.\overline{D} + A.B.C.\overline{D} + \overline{A}.\overline{B}.\overline{C}.D + A.\overline{B}.\overline{C}.D$

Mapping gives:

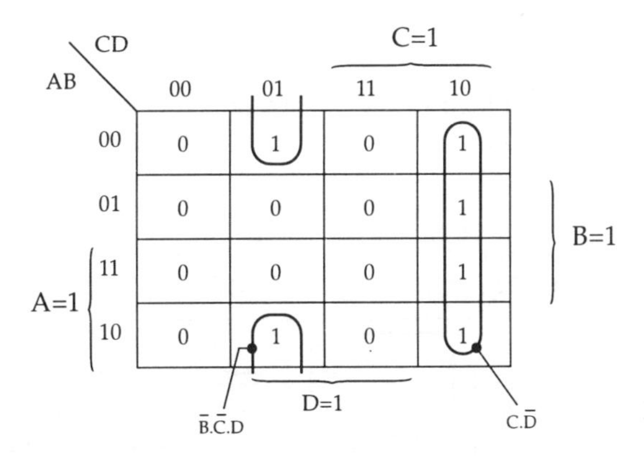

From which $F = C.\overline{D} + \overline{B}.\overline{C}.D$

Inverse Function

In some cases it is more economical in terms of logic gates to implement the inverse (the complement) of the function. To do this we simply loop the 0s rather than the 1s.

For example, consider the equation:

$F= A.B.\overline{C} + \overline{A}.B.C + \overline{A}.C.\overline{D} + A.\overline{C}.\overline{D}$

Mapping gives:

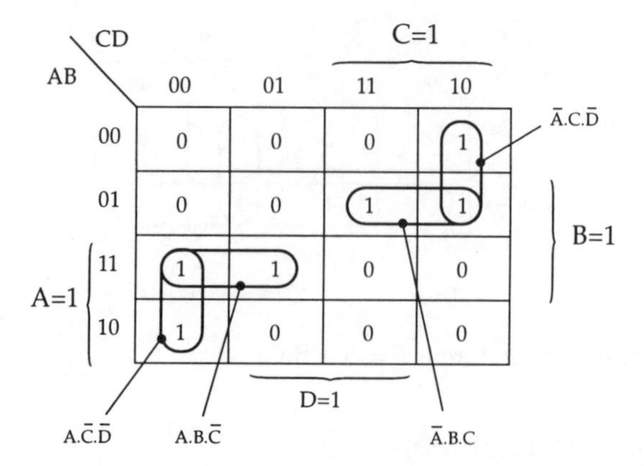

There is no further simplification possible and the logic gate diagram would be as shown below.

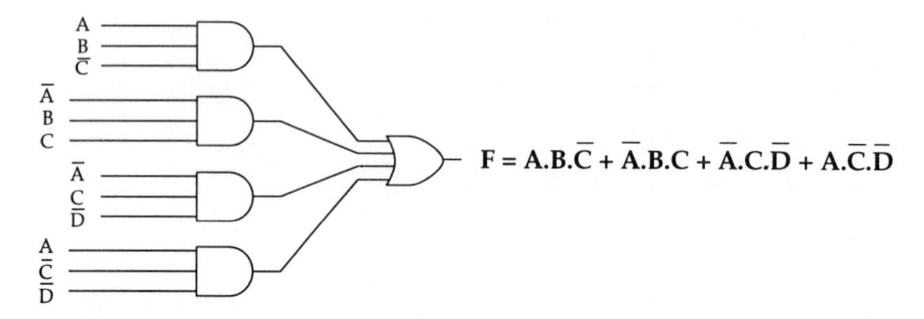

If we now loop all the 0s instead of the 1s, we are solving for the complement of the function.

Mapping gives:

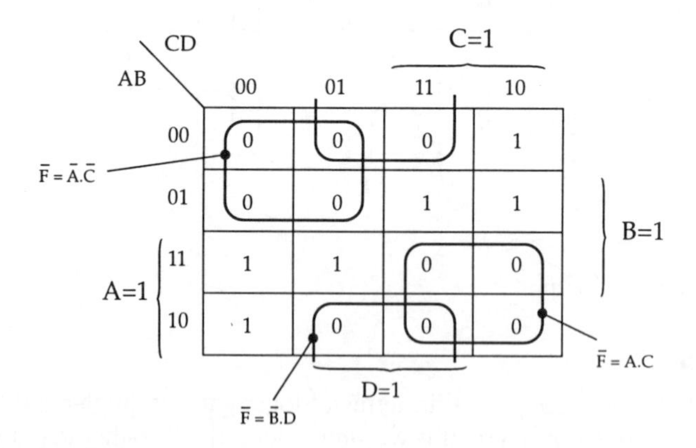

From which the complement:

$$\overline{F} = \overline{A}.\overline{C} + A.C + \overline{B}.D$$

To find the function F we simply invert both sides of the equation!

$$F = \overline{\overline{A}.\overline{C} + A.C + \overline{B}.D}$$

This will produce a much simpler logic gate diagram!

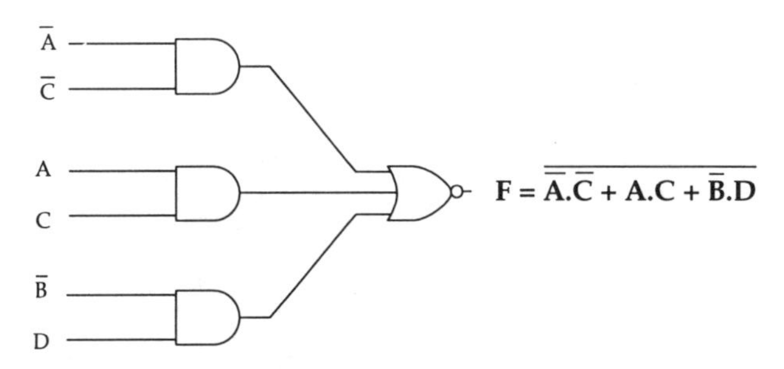

The logic gate diagram is now much easier to construct.

It uses only three 2-input AND gates instead of four 3-input gates, and one 3-input NOR gate instead of a 4-input OR gate!

Use of NAND and NOR Gates

Because the NAND or NOR gate can be used to perform other logic functions, it is often necessary to obtain the Boolean equation in NAND or NOR form from the Karnaugh map. This can often be achieved by complementing the function F.

Example 1

Consider the Karnaugh map:

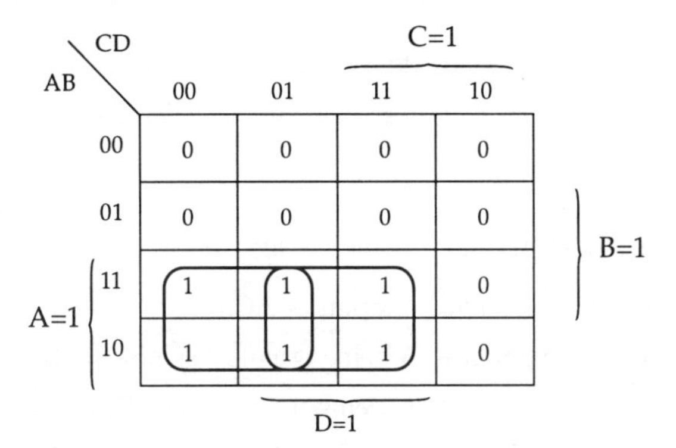

AB \ CD	00	01	11	10
00	0	0	0	0
01	0	0	0	0
11	1	1	1	0
10	1	1	1	0

From the two loops shown:

$$F = A.\bar{C} + A.D$$

which can be implemented using pure logic gates. However by complementing the function F we obtain:

$$\bar{F} = \overline{A.\bar{C} + A.D}$$

$$\bar{F} = (\overline{A.\bar{C}}) . (\overline{A.D}) \qquad \text{by De-Morgan's}$$

$$F = \overline{(\overline{A.\bar{C}}) . (\overline{A.D})} \qquad \text{by re-complementing}$$

The equation now contains NAND terms only and may be constructed using only NAND gates!

Example 2

Consider the same Karnaugh map:

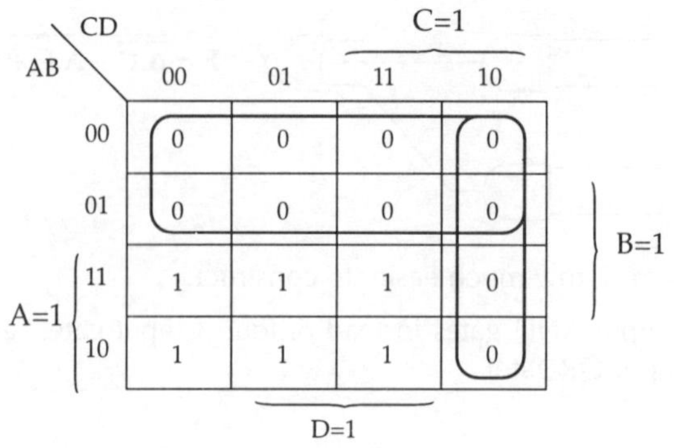

AB \ CD	00	01	11	10
00	0	0	0	0
01	0	0	0	0
11	1	1	1	0
10	1	1	1	0

C=1 (columns 11, 10); D=1 (columns 01, 11); A=1 (rows 11, 10); B=1 (rows 01, 11)

This time we loop the cells containing 0s and obtain the complement of the function. From the two loops shown:

$$\overline{F} = \overline{A} + C.\overline{D}$$

By complementing the function F we obtain:

$$F = \overline{\overline{A} + C.\overline{D}}$$

$$F = \overline{\overline{A} + (\overline{\overline{C}+D})} \qquad \text{by de-Morgan's on (C.D).}$$

The equation now contains NOR terms only and may be constructed using only NOR gates!

Self-assessment questions (2) *(answers page 468)*

A logic gate system is required having 4 inputs A.B.C.D

A and B represent a binary number X (A being the most significant bit).

C and D represent a binary number Y (C being the most significant bit).

The output of the system (F) is logic 1 only when X is greater than Y.

Draw up the truth table, simplify by using Karnaugh mapping and produce the logic gate diagram using all NAND gates.

Don't Care Terms

In practice it may not matter whether a cell, or group of cells, contains either a logic 1 or a logic 0. These are known as **don't care terms**. In such cases an asterisk (or a letter d) can be written in the cell so that when the loops are being drawn these cells can either be included or excluded as required. By including some of these 'don't care' terms the Boolean equation can often be simplified further.

Example

Consider a binary to decimal convertor. This is a logic system with 10 outputs and 4 inputs (ABCD) which represent a 4-bit binary number (A being the most significant bit).

If the binary input is less than or equal to decimal 9 then the appropriate decimal output is at logic 1.

The cells on the map where the binary input is greater than decimal 9 may be used as 'don't care' cells and used for simplification purposes.

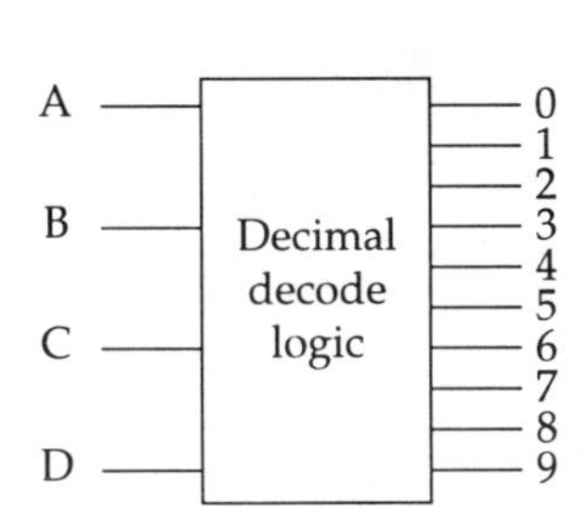

Decimal	*A*	*B*	*C*	*D*
0	0	0	0	0
1	0	0	0	1
2	0	0	1	0
3	0	0	1	1
4	0	1	0	0
5	0	1	0	1
6	0	1	1	0
7	0	1	1	1
8	1	0	0	0
9	1	0	0	1
10	1	0	1	0
11	1	0	1	1
12	1	1	0	0
13	1	1	0	1
14	1	1	1	0
15	1	1	1	1
Weighting	**8**	**4**	**2**	**1**

A binary-decimal convertor

The 6 states 1010 to 1111 can't happen but in practice they may be used for simplification purposes.

Remember, in this example we are not simplifying for one single output but for 10 outputs! For this reason the mapping is made easier if we fill each cell on the map with the decimal number it represents. The decimal numbers 10 to 15 are represented with an asterisk.

Where there is a number, it corresponds to logic 1 and an asterisk is a 'don't care' term.

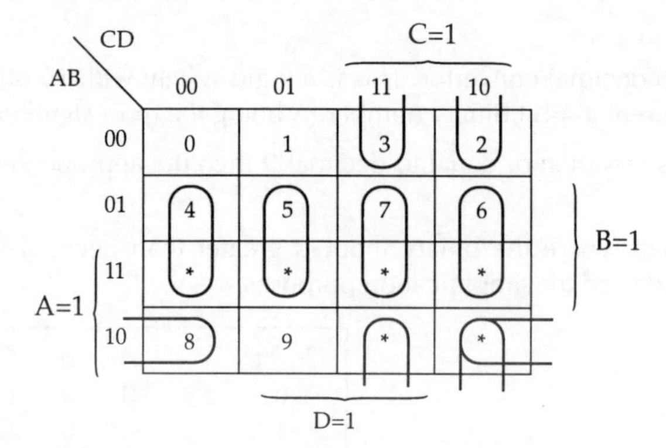

Looping the cells as shown gives:

Cell 0 = $\bar{A}.\bar{B}.\bar{C}.\bar{D}$ and cannot be looped with any * cells

Cell 1 = $\bar{A}.\bar{B}.\bar{C}.D$ and cannot be looped with any * cells

Cell 2 = $\bar{B}.C.\bar{D}$ and can be looped with * cells

Cell 3 = $\bar{B}.C.D$ and can be looped with * cells

Cell 4 = $B.\bar{C}.\bar{D}$ and can be looped with * cells

Cell 5 = $B.\bar{C}.D$ and can be looped with * cells

Cell 6 = $B.C.\bar{D}$ and can be looped with * cells

Cell 7 = $B.C.D$ and can be looped with * cells

Cell 8 = $A.\bar{D}$ and can be looped with * cells

Cell 9 = $A.D$ and can be looped with * cells

The logic gate diagram may now be drawn.

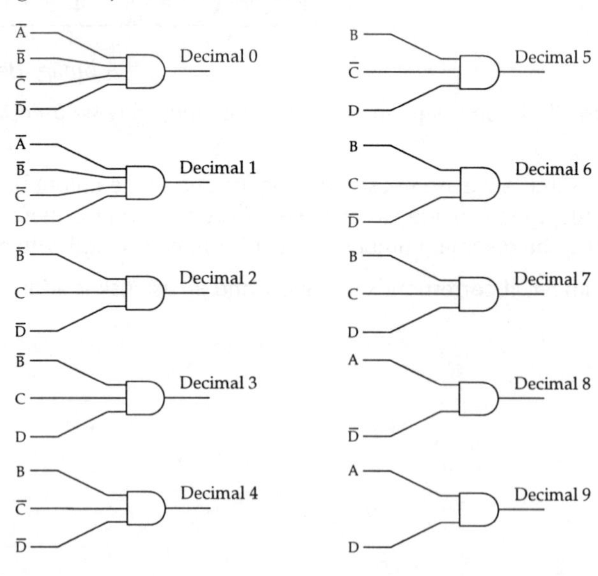

Gate Diagram; binary to decimal convertor

Static Hazards

All logic gates have a **propagation delay**. That is, the logic level at the output of the gate will change a few micro seconds after the input to the gate has been changed.

Consider the logic gate diagram shown below.

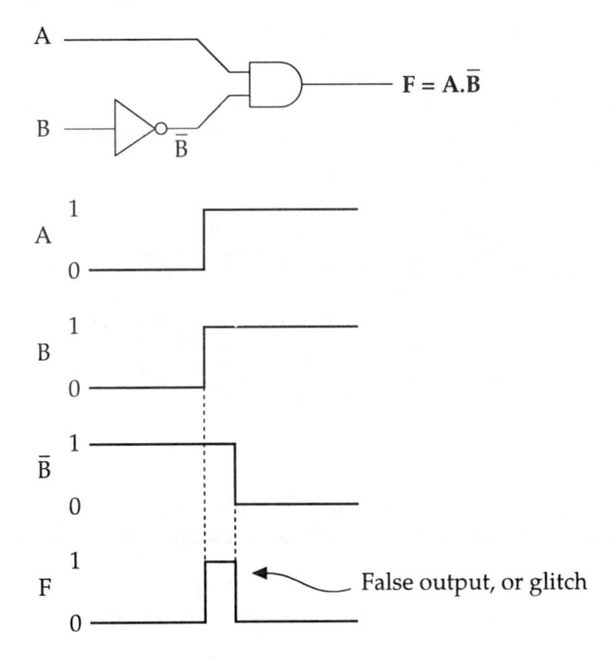

Logic gate diagram with static hazard

When inputs A and B are both applied at the same moment in time then input A will be applied to the AND gate directly but input B is inverted by the invertor and will suffer a delay. This means that a logic 1 will appear at the output of the AND gate for a few micro seconds.

A short duration pulse or **glitch** will appear at the output of the AND gate which may cause other gates in the logic system to switch. This is called a **static hazard**.

This hazard can be eliminated by ensuring that all **adjacent 1 squares** in the Karnaugh map **are looped together**. This will mean that the simplest solution cannot always be used if static hazards are to be avoided!

Example

Consider the logic gate diagram shown below.

A 1
C
B 1
C $\bar{C}$

F = A.C + B.$\bar{C}$

Logic gate diagram with a static hazard

When A and B are at logic 1 then the function of the circuit is:

$F = 1.C + 1.\overline{C}$

$F = C + \overline{C} = 1$ [since C OR $\overline{C}$ = 1]

i.e. there is an output of logic 1 at F.

Now if C changes from 1 to 0; $\overline{C}$ should change, in the same time, from 0 to 1 keeping the output F at logic 1.

However due to the propagation delay of the invertor, $\overline{C}$ will only change to 1 after a short delay and for this period the output will be at logic 0, producing a glitch.

Mapping the equation $F = A.C = B.\overline{C}$:

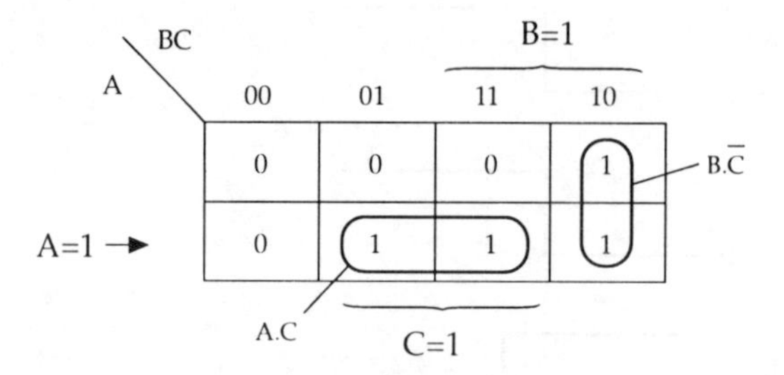

This produces two separate loops. To eliminate static hazards the adjacent squares containing a logic 1 must be linked so that all cells with a logic 1 are within a loop.

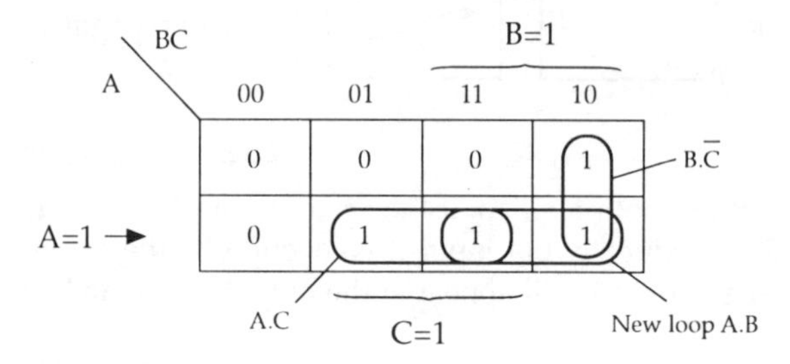

Adding the new loop gives the equation:

$F = A.C + B.\overline{C} + A.B$

Now when A and B are logic 1 the output F is at logic 1 because of the A.B term, irrespective of the state of C and $\overline{C}$.

The glitch has been eliminated!

Self-assessment exercise *(answers page 470)*

Design a combinational logic circuit that has ten inputs and one output. The output is logic 1 when any one, or more, of the following inputs are at logic 1: 2,5,6 and 7.

The circuit should be free from static hazards and employ NAND gates only.

Summary

- A Karnaugh map is an alternative way of representing a truth table.
- The map is divided into cells such that for each variable is represented by half of the map and its complement by the other half.
- By looping cells in the map a simplified Boolean expression can be obtained.
- With a 3-variable map there will be 8 cells. Cells are looped in 2s or 4s.
  A loop of 4 cells is described by one variable
  A loop of 2 cells is described by two variables
  A single cell is described by 3 variables.
- With a 4-variable map there will be 16 cells. Cells are looped in 2s, 4s or 8s.
  A loop of 8 cells is described by one variable
  A loop of 4 cells is described by two variables
  A loop of 2 cells is described by three variables
  A single cell is described by 4 variables.
- The cells with logic 0 in them may be looped to find the inverse or complement of the function.
- Certain cells may contain either a logic 1 or a logic 0. These are called 'don't care' terms but may still be included in loops to simplify the Boolean equation.
- By ensuring that all adjacent cells with a logic 1 are included in a loop, static hazards can be prevented.

The bi-stable

Study Note: Before attempting this Information and Skills Bank reference you will need to be familiar with Bi-polar transistor, p.209, and Field effect transistor,p.255.

Introduction

If the output of a logic gate system depends solely upon a combination of inputs, then this is a **combinational** logic system. If the output depends upon both the present inputs and the previous logical state of the output, then this is an example of a **sequential** logic circuit.

A **bi-stable** or **flip-flop** is an example of sequential logic.

Bi-stable

The bi-stable is a circuit which has two stable states and will remain in one of the states until it is forced into the other state. It is used as a memory or latch.

The bi-stable depends for its operation on a very fast switching action. Logic gates, bi-stables and almost all digital logic integrated circuits employ transistors as saturated switches. This gives two great advantages. Firstly the power dissipated within the transistor is lower and secondly, it provides two well-defined voltage levels which can represent the binary states 0 and 1.

The Transistor as a Switch

To operate the transistor as a switch then we bias the device to be fully ON or OFF and ensure that it is not biased anywhere in between these two conditions.

The bi-polar transistor is OFF without any base bias current to start the transistor action. It is fully ON (saturated) with sufficient bias current flowing into the base.

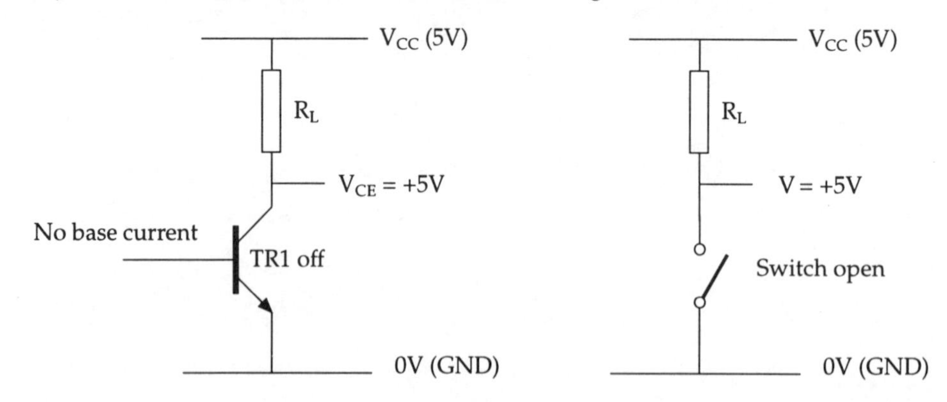

No Bias, TR1 is OFF *The equivalent switch circuit*

The bi-polar transistor is fully ON (saturated) with sufficient bias current flowing into the base. The base is connected to +5V via the base resistor R_b. The value of R_b must be chosen to fully saturate TR1.

With a large base current flowing, the transistor is fully saturated and the large collector current produces a large voltage drop across the load resistor. The collector voltage therefore falls to a very low saturation value $V_{CE_{sat}}$ (approximately 0.2V). For practical purposes this can be considered as 0V.

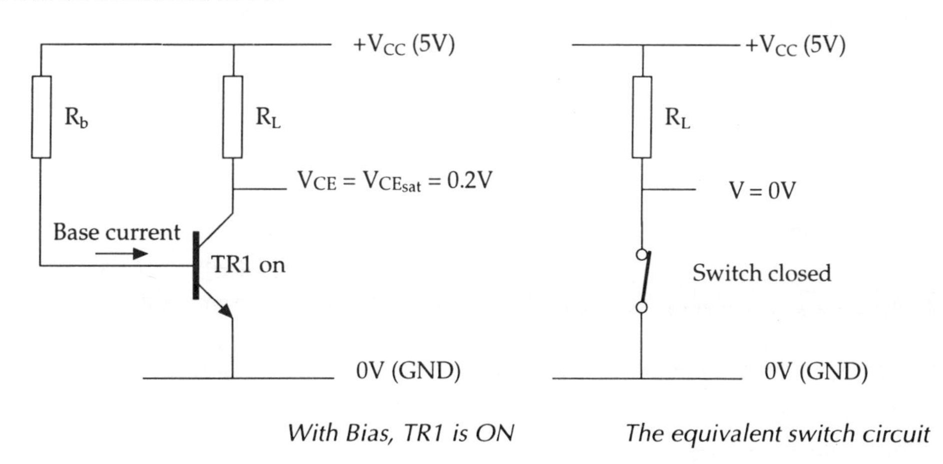

With Bias, TR1 is ON *The equivalent switch circuit*

Power Dissipation in the Transistor

The power dissipated in the transistor P_C is at all times equal to $I_C \times V_{CE}$. This power is wasted and only heats up the transistor. In digital logic circuits, the transistor power dissipation is small because when the output is LOW then the collector voltage V_{CE} is almost zero (0.2V).

When the output is HIGH (+5V) then the collector current I_C is almost zero. Provided that the switching between these two states is fast, the collector power dissipation will always be negligible.

Logic Levels

The output at the collector therefore provides well-defined logic levels: +5V for logic 1 and 0.2V for logic 0. Notice however that an input of +5V (logic 1) applied to the base via Rb produces an output of 0.2V (logic 0) at the collector. The transistor acts as an invertor or NOT gate!

The transistor may also be used to simply switch a relay, filament lamp or an LED.

Using a Transistor to Switch a Load

A 12V, 1W lamp is to be switched by a silicon transistor having an $h_{fe_{min}}$ of 50. We are now going to calculate the value of the base resistor R_b and state the voltage, current and power rating of the transistor.

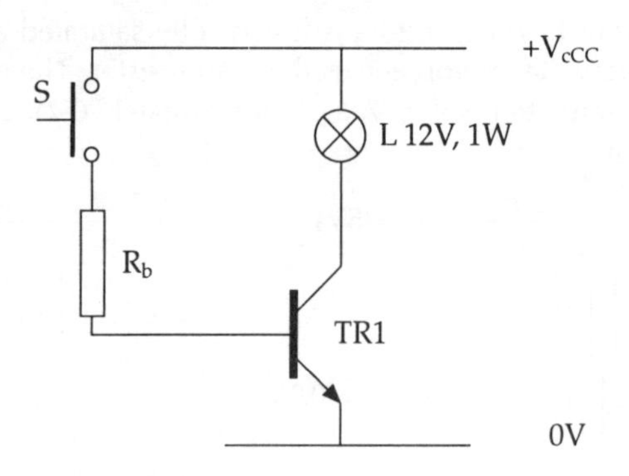

Using a transistor to switch a lamp

As the transistor, when on, will only have a small voltage across it (0.2V) the circuit supply can be 12V. The lamp being 1W operated from 12V will require a current of $^1/_{12}$ A or 83.33mA. Since the minimum h_{fe} is 50 the base current required for saturation will be:

$$I_B = I_C/h_{fe} = 83.33\text{mA}/50 = 166.66\mu\text{A}.$$

With a silicon transistor, if we assume that V_{BE} = 0.7V then the voltage across the base resistor will be:

$$12 - 0.7 = 11.3\text{V}$$

Therefore the value of the base resistor R_b will be:

$$R_b = 11.3/166.66 \times 10^{-6} \text{ ohms} = 67.8 \text{ k ohms}$$

The nearest preferred value is 68k but since we need to insure that the transistor is truly saturated, (otherwise the power dissipation increases), then we should go for the next value down in the range. In the E24 series this will be 62k.

The transistor will therefore be required to have a V_{CEmax} of more than 12V; a current rating I_{Cmax} of 100mA or more; and a power dissipation of:

$$100\text{mA} \times 0.2\text{V} = 20\text{mW}$$

If however the transistor is required to cope with being switched only half ON then I_C will be 50mA and V_{CE} will be 6V. This will require a power rating of:

$$50\text{mA} \times 6\text{V} = 300\text{mW}$$

The FET as a Switch

Similarly the FET may also be used as a switch. The N-channel FET will be fully ON, with no gate bias voltage. The saturation voltage across the FET (V_{DSsat}) is in the range of 0.2V to 1V. The ON resistance is between 30 to 200 ohms.

To turn the FET OFF a sufficiently negative voltage is applied to the gate. The drain source voltage will rise to the supply (V_{DD}) and the drain leakage current will be in the order of nano amps.

The FET consumes less power than the bipolar transistor but is slower in switching due to capacitive effects. Between the gate and channel there is capacitance and time is needed to charge and discharge this gate to channel capacitance. This will increase the turn ON and turn OFF time.

The Transistor bi-stable circuit

A bi-stable may be constructed from two transistors as shown below.

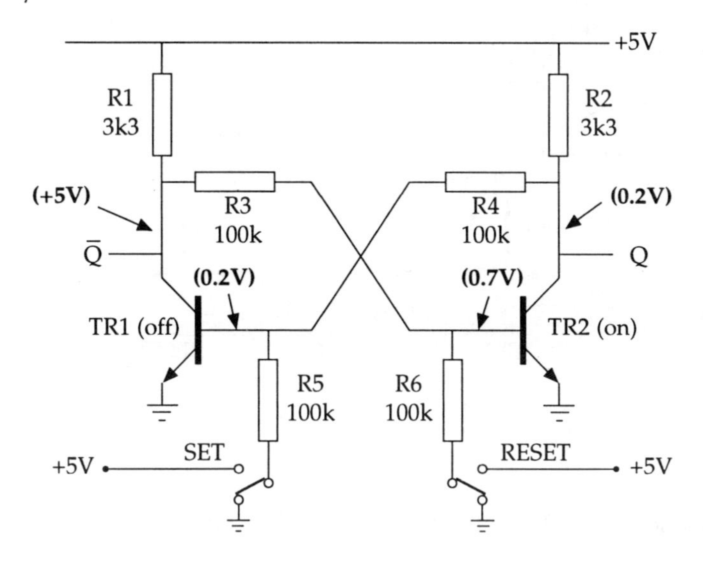

Transistor Bi-stable Circuit

Initially one of the transistors will conduct a little earlier than the other and this transistor will switch fully ON and the other will be fully OFF. In the diagram above TR2 is ON and TR1 is OFF.

With TR2 conducting its collector voltage falls to $V_{CE_{sat}}$ (0.2V) and this becomes the output Q (logic 0). This voltage is coupled to the base of TR1 via R4 and is too small to switch TR1 on.

With TR1 OFF its collector is at +5V and this is the output voltage at $\overline{Q}$ (logic 1). This voltage is coupled to the base of TR2 via R3 and is large enough to maintain TR2 in the ON state.

The bi-stable will remain in this state until a SET input is applied to the base of TR1 to turn it ON.

When the SET switch is operated, TR1 is biased into conduction, and its collector voltage falls, lowering the voltage on the base of TR2.

TR2 conducts less, and its collector voltage rises, increasing the voltage on the base of TR1. This makes TR1 conduct more, lowering its collector voltage and the base voltage of TR2 still further.

This action is a very fast regenerative switching action which quickly switches TR1 ON and TR2 OFF.

The output Q is now at Logic 1 and $\overline{Q}$ at logic 0. The bi-stable has been SET.

The conditions in the circuit are now as shown below.

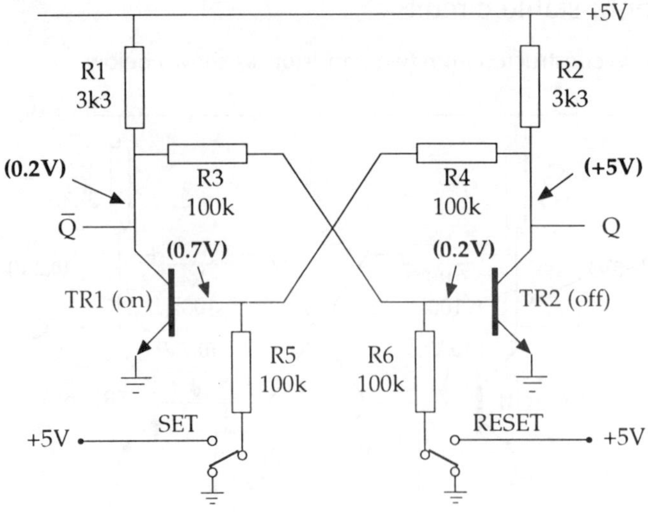

Transistor Bi-stable Circuit

With TR1 conducting and TR2 off, then Q is at logic 1 and $\overline{Q}$ at logic 0. The bistable has been SET.

If now the RESET switch is momentarily active, then TR2 will be made to conduct, lowering its collector voltage and the voltage on the base of TR1 and turning TR1 off. The rising voltage on the collector of TR1 will turn on TR2 which will turn TR1 off completely.

We are now back in the original state with TR1 OFF ($\overline{Q}$ at logic 1) and TR2 ON (Q at logic 0). The bi-stable has been RESET.

Summary

- Bi-polar and Field Effect Transistors may be operated as switches.
- When operated as switches the transistors dissipate very little power and may be switched at very high frequencies. The bi-polar is faster in operation than the FET but consumes more power.
- When transistors are used as switches in logic circuits they give very well-defined logic levels. The logic level is however inverted because the switching transistor behaves as a NOT gate.
- A bi-stable may be formed from two transistors with cross coupling between the collectors and bases. This provides fast regenerative switching. The bi-stable has two complementary outputs Q and $\overline{Q}$. Q may be SET to logic 1 or RESET to logic 0.

Self-assessment question *(answers page 471)*

A transistor is employed as a saturated switch, controlling current through a lamp. The lamp is required to light if any input in the circuit below is switched to the 1 position.

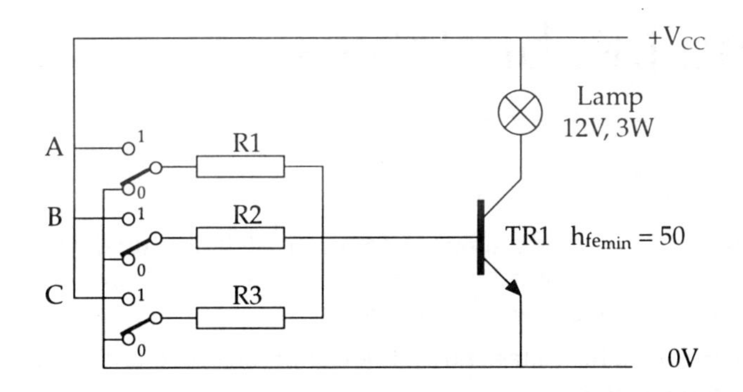

Circuit for the self assessment question

a) Calculate the values of R1,R2 and R3, and give a suitable preferred value.

b) Calculate the voltage, current and power ratings of the silicon transistor TR1.

c) State what logic function this circuit performs if, when the lamp is lit, this is taken to indicate logic 1.

R.S. and J.K. bi-stables

Introduction

The bi-stable is mainly used as a simple memory or latch. It is used whenever it is necessary to retain or remember some logic state in equipment. Its output state therefore often depends upon some previous signal or logic level.

The bi-stable has two stable states and will remain in one state or the other until changed or the power is turned off.

The bi-stable has two outputs which are complementary to each other: i.e. when one output is logic 1 the other is always logic 0.

The RESET – SET (R.S.) bi-stable has an input to SET output Q to 1 and a RESET input which sets Q to 0.

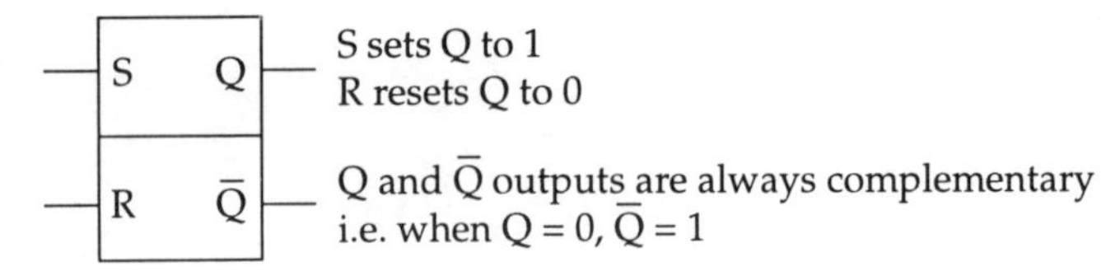

Bi-stable Symbol: 2 boxes to represent the two states

R.S. Bi-stable Operation

A momentary signal applied to S sets Q to 1 and $\overline{Q}$ to 0.

The bi-stable will remain in this state, after the set input is removed, until the reset input is activated.

A momentary signal applied to R resets Q to 0 and $\overline{Q}$ to 1.

The bi-stable will remain in this state, after the reset input is removed, until the set input is activated.

An R.S. bi-stable can be made from two transistors with **cross coupling** from the collectors to the bases (see The bi-stable, p.428). The R.S. bi-stable can also be made from a pair of NAND or NOR gates. The output of each gate is **cross-coupled** back to one of the inputs of the other gate.

The Cross-Coupled NOR Gate Bi-stable

A pair of NOR gates are used to form a bi-stable.

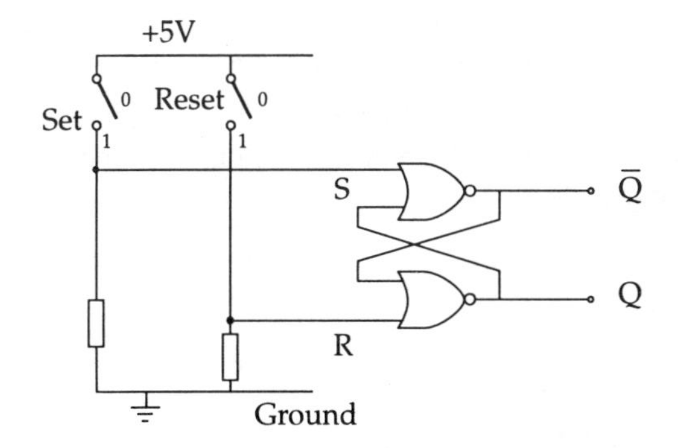

X-coupled NOR gate bi-stable

With the switches open, the two inputs are held at ground potential (logic 0) by the two resistors. Closing a switch will put +5V (logic 1) on the associated input. Remember however that you cannot SET and RESET the bi-stable (operate both switches) at the same moment in time. This is like attempting to switch the electric light ON and OFF at the same time!

Circuit Operation

The circuit normally rests with both S and R held low (logic 0). In the RESET condition Q = 0; and $\overline{Q}$ = 1. We will assume that the circuit is in this reset state. The logic levels on the inputs and outputs of the bi-stable are therefore as shown below.

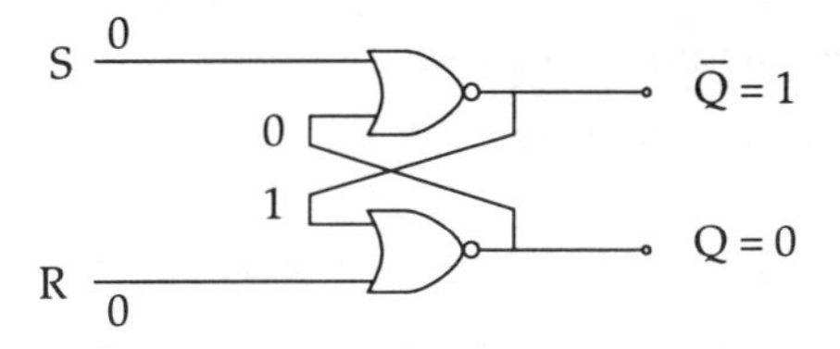

Logic levels in the reset state

The output at $\overline{Q}$ is fed back to the lower NOR gate and holds this input at logic 1. The output at Q is fed back to the upper NOR gate and holds this input at logic 0.

With these input conditions we can see from the NOR gate truth table that the outputs will remain in the states shown: i.e. two logic 0's on the inputs to the upper NOR gate will sustain an output of logic 1.

The NOR gate truth table

Inputs		*Outputs*	
		OR	NOR
A	*B*	*A+B*	$\overline{A+B}$
0	0	0	1
1	0	1	0
0	1	1	0
1	1	1	0

With both inputs at 0 the output is 1. When any input is logic 1, the output is logic 0

The bi-stable will remain in this state until a SET input is applied. If a RESET input is applied then the circuit is already in the reset state and no change will take place!

If now a momentary logic 1 (+5V) is applied to the SET input then the output of the upper gate will change from 1 to 0. This will change the input conditions on the lower gate and cause its output to change state.

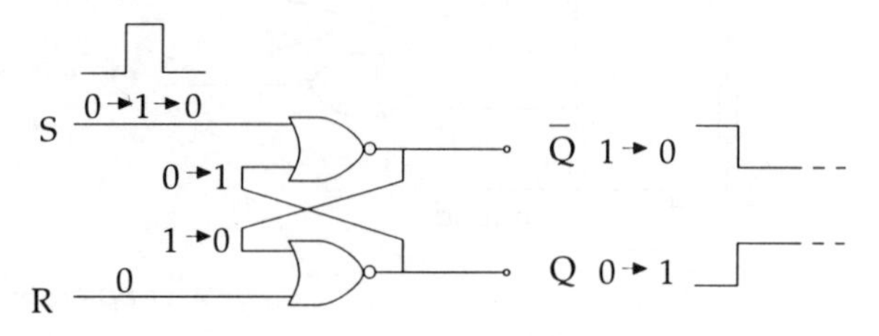

Applying a momentary 1 to the SET input

Circuit Action

A momentary change at the S input from 0 to 1 will cause a change in the upper gate output from 1 to 0. This in turn alters the input conditions on the lower gate from 1 to 0. This will alter the output of the lower gate from 0 to 1.

Although the original input at S has now disappeared the circuit remains in its changed-over state. This is because the upper gate, even after the S input has disappeared, still has (due to the cross coupling) a logic 1 and 0 on its inputs, which from the NOR gate truth table means that the output must be at logic 0.

This means that the bi-stable **memorised the last instruction** even though it was only momentary!

Now it is in the SET state, the bi-stable will not respond to another 1 on the SET input.

The circuit can be reset to its original state by applying a momentary 1 to the RESET input.

The X-Coupled NOR Bi-stable Truth Table

Inputs		*Outputs*		*Comment*
S	*R*	*Q*	*$\overline{Q}$*	
0	0	Q	$\overline{Q}$	No change in state
0	1	0	1	Reset Q = 0
1	0	1	0	Set Q = 1
1	1	?	?	Indeterminate state

Notice that the last state is forbidden because it is not possible to set and reset the bi-stable at the same moment in time: i.e. S and R must not be logic 1 at the same time!

It is also possible to create a bi-stable using a pair of cross-coupled NAND gates.

The NAND Gate Bi-stable

Using a pair of cross-coupled NANDs in the same circuit will also produce a bi-stable. The only difference will be that to get a logic 0 output from one of the gates will require both inputs of that gate to be at logic 1.

This means that in the rest state both S and R inputs must rest at logic 1 and not logic 0 as before. The S and R inputs are now taken up to the +5V supply via resistors.

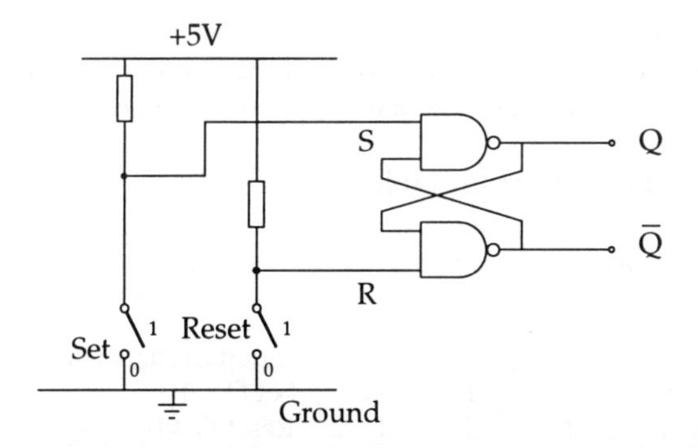

X-Coupled NAND bi-stable

With the switches open then +5V (logic 1) is applied to the R and S inputs of the bi-stable. When a switch is closed then that input is taken low and the resistor ensures that a short circuit is not placed across the +5V and 0V supply rails.

Resistors used in this way are often called 'pull-up' resistors. They pull the inputs up towards +5V when the switches are open.

The NAND gate truth table shows that both inputs to a gate now have to be logic 1 to give a logic 0 output.

The NAND Gate Truth Table

Inputs		*Outputs* AND	NAND
A	*B*	*A.B*	$\overline{A.B}$
0	0	0	1
1	0	0	1
0	1	0	1
1	1	1	0

When any input is at logic 0 the output is logic 1. Both inputs at logic 1, the output is 0

Circuit Action

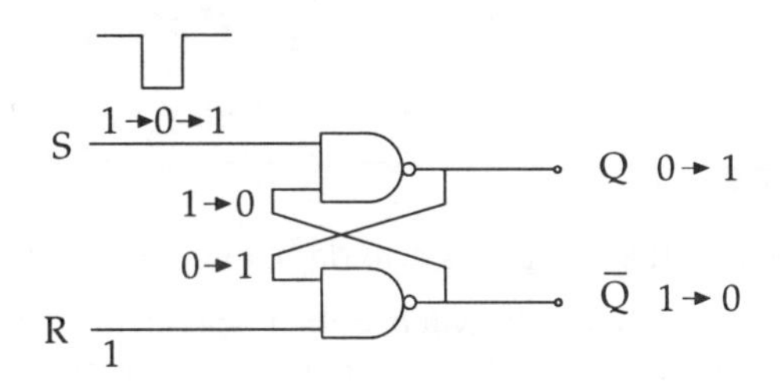

Applying a momentary 0 to the SET input

Applying a momentary 0 at the S input will change the output of the upper gate from 0 to 1. This in turn will change the conditions on the input of the lower gate from 0 to 1, changing the output of this gate from 1 to 0.

Although the original input S has now disappeared the circuit remains in the changed-over state. This is because the upper gate now has a logic 0 on one of its inputs.

The X-Coupled NAND Bi-stable Truth Table

Inputs		*Outputs*		*Comment*
S	*R*	*Q*	$\overline{Q}$	
0	0	?	?	Indeterminate state
0	1	1	0	Set Q = 1
1	0	0	1	Reset Q = 0
1	1	Q	$\overline{Q}$	No change in state

Notice that the first state is forbidden because it is not possible to set and reset the bi-stable at the same moment in time: i.e. S and R must not be logic 0 at the same time!

Summary of Cross-Coupled NANDS

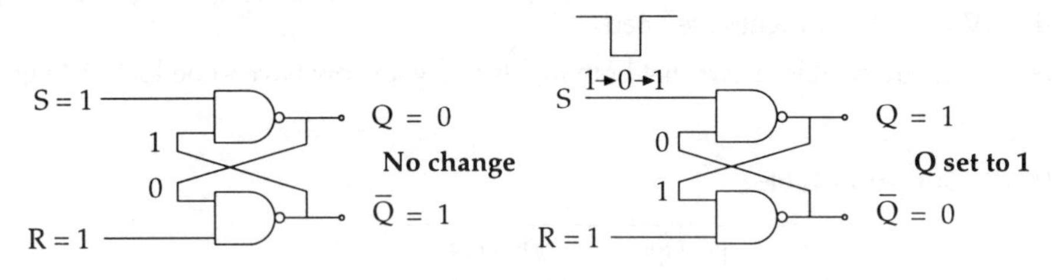

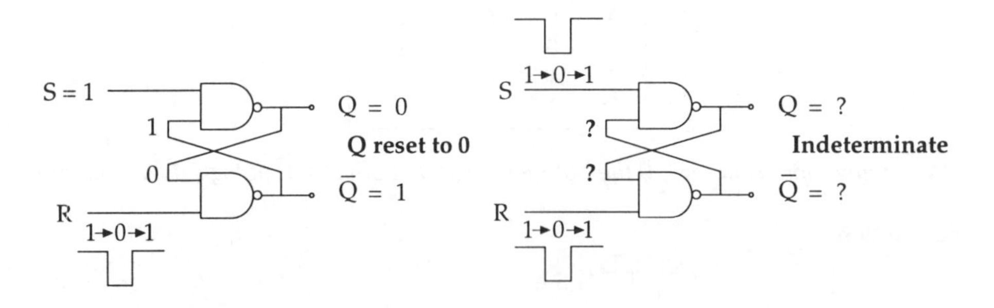

Switch De-bouncing

One practical application of R.S. bi-stables is to de-bounce switches.

The spring contacts of a mechanical switch, when operated, can bounce together and spring apart for up to 10ms. If the switch is the input to a counter then the counter can register several inputs for just one operation of the switch!

The R.S. latch is used to de-bounce the switch. The latch, which is very fast acting, will change state for the first closing and will not respond to any additional making and breaking of the contacts.

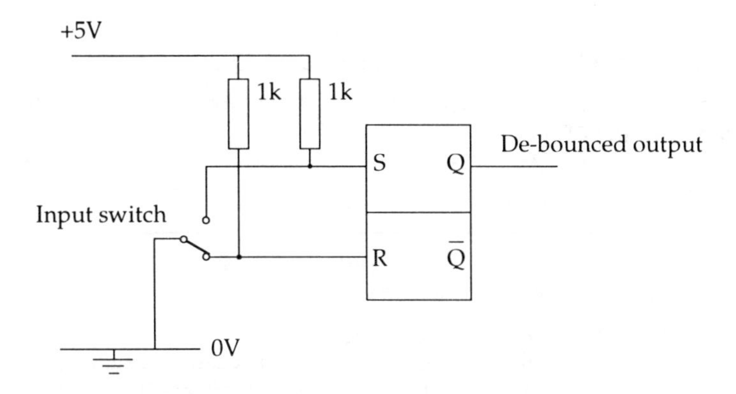

Using an R.S. latch to de-bounce a switch

Clocked R.S. bi-stable

The bi-stables discussed so far will change state every time the appropriate logic levels on the R and S inputs are present. If we have a number of bi-stables we often require them to change state all at the same time. To achieve this we need a synchronizing or timing pulse to be applied to all the bi-stables to tell them when to change state. This timing pulse is often called a clock pulse.

Our bi-stables therefore need a third input, called the clock input, to create a clocked R.S. bi-stable.

Data can only enter this type of bi-stable when the logic level on the clock input is high. This allows the entering of data to be synchronized by the timing or **clock pulses** that control and time the operations within a digital logic system.

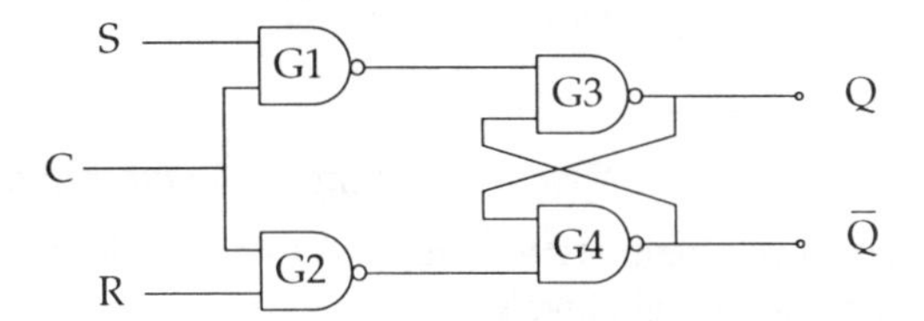

A Clocked R.S. bi-stable using NAND gates

Circuit Action

Remember that if any input to a NAND gate is low then the output is high. Therefore if the clock input is held low in this circuit the outputs of G1 and G2 will be high. This is the normal rest state for the cross-coupled NAND bi-stable.

This means that unless the clock pulse goes high the outputs of G1 or G2 can never go low to set or reset the bi-stable.

With input C low, the outputs of G1 and G2 are high and the bi-stable is in the rest state.

With C high and S high, the output of G1 will go low and the bi-stable will be SET (assuming it is not already set state); i.e. output Q will go to 1.

With C high and R high, the output of G2 will go low and the bi-stable will be RESET (assuming it is not already in the reset state); i.e. output Q will go to 0.

The usual procedure is to make either R or S HIGH (not both) and then apply a clock pulse to input C to SET or RESET.

Clocked R.S. Bi-stable Truth Table

Inputs before a clock pulse		*Outputs after a clock pulse*		*Comment*
S	*R*	*Q*	$\overline{Q}$	
0	0	Q	$\overline{Q}$	No change of state
0	1	0	1	Q is always 0
1	0	1	0	Q is always 1
1	1	?	?	Indeterminate state

There is still an indeterminate state which can be avoided if the R and S inputs receive complementary logic levels. This is easily done by placing an invertor between the S and R inputs and forming what is now called a Data or D-type latch.

Data or D-Type Bi-stable

This bi-stable only has two inputs; a data input and a clock input.

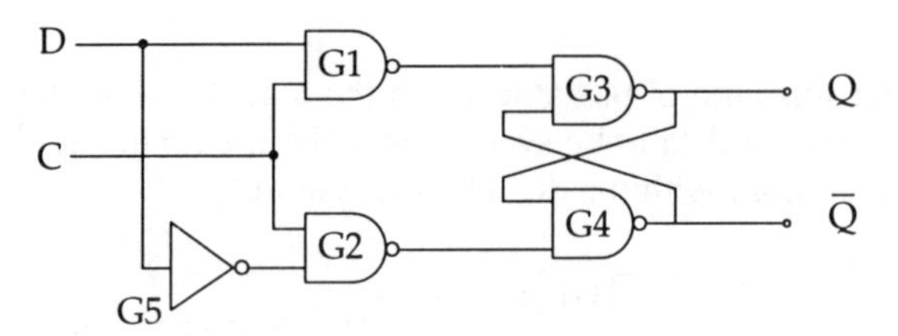

Using an invertor to form a D-type Bi-stable

The invertor inverts whatever logic state is applied to the D input, so that S and R now always receive complementary logic levels.

As before, gates G1 and G2 will both give a high output when the clock pulse is low and there can be no change in the state of Q without a clock pulse.

The logic state at Q will follow the logic state at D after a clock pulse!

A number of D-type bi-stables are often used as a temporary store for a word of data. They are often used in shift registers and they have the advantage of having no indeterminate state.

To produce a counter we require a bi-stable in which the output Q will change its state every time a clock pulse is applied. This is called toggling. By making a modification to the clocked R.S. bi-stable we can produce a bi-stable that will toggle.

The J.K. Bi-stable

This bi-stable is formed by making the gates G1 and G2 three input gates and feeding back the cross-coupled output logic levels to these third inputs.

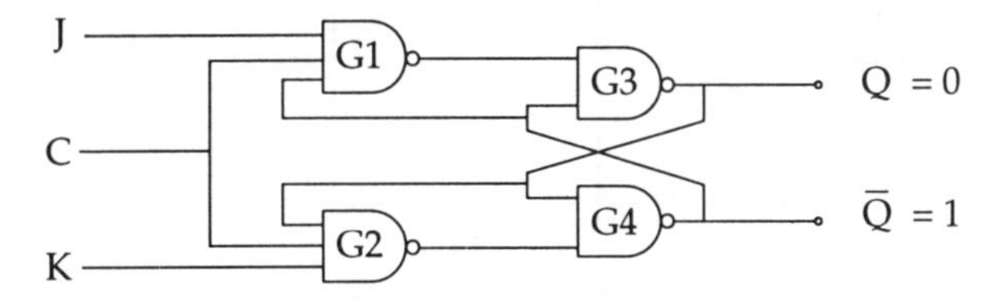

A J.K. bi-stable formed from 4 NAND gates

Now for the bi-stable to change state, all three inputs of either G1 or G2 must be HIGH together.

For example: if $\overline{Q}$ = 1 and J = 1 then when the clock pulse goes HIGH,the output of G1 will go LOW and Q will be SET to 1.

On the next clock pulse, if K = 1 and Q has just been SET to 1, then the output of G2 will go LOW and RESET Q to 0.

Therefore if we keep the inputs J and K at logic 1, then the output will change state with each clock pulse: i.e. the bi-stable toggles.

Truth Table for a J.K. bi-stable

Inputs			*Outputs after CP*		
J	*K*	*Clock*	*Q*	$\overline{Q}$	*Comment*
0	0	⎍	Q	$\overline{Q}$	No change
0	1	⎍	0	1	Reset
1	0	⎍	1	0	Set
1	1	⎍	$\overline{Q}$	Q	Toggles

There is however one major disadvantage of this circuit. The clock pulse must be of a shorter duration than the switching time of the bi-stable, otherwise the bi-stable will continue to toggle for the duration of the clock pulse. This can be achieved by using a bi-stable with edge triggering.

Edge Triggering

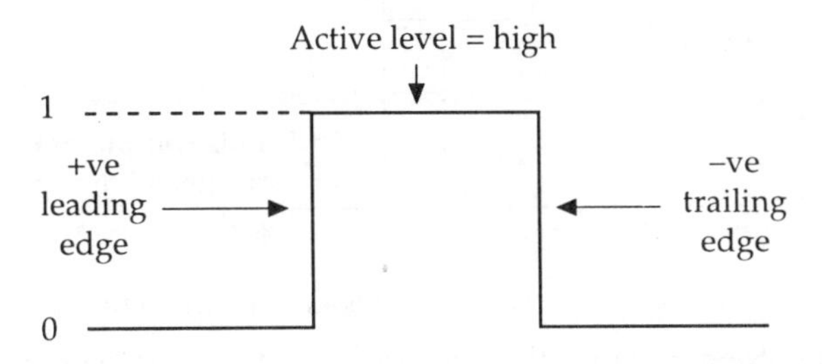

Positive going clock pulse

The clock pulse above has a positive leading edge (0V to +5V) and a negative trailing edge (+5V to 0V).

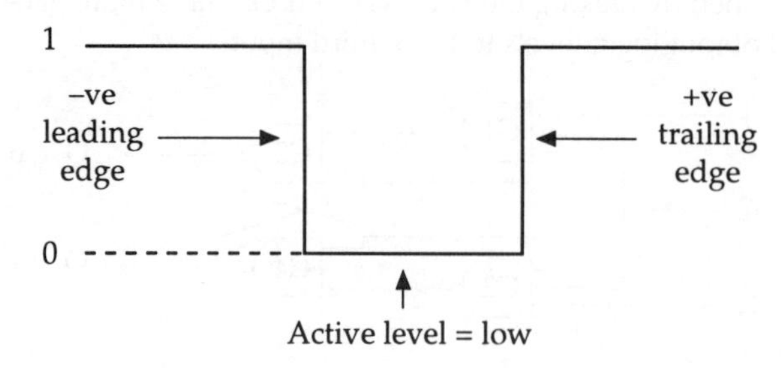

Negative going clock pulse

This clock pulse has a positive trailing edge (0V to +5V) and a negative leading edge (+5V to 0V).

The bi-stable may be clocked by a positive or negative edge. Many bi-stables however are negative edge triggered: i.e. on the transition from logic 1 to logic 0.

A bi-stable may be made edge-sensitive by using the propagation delay of a gate.

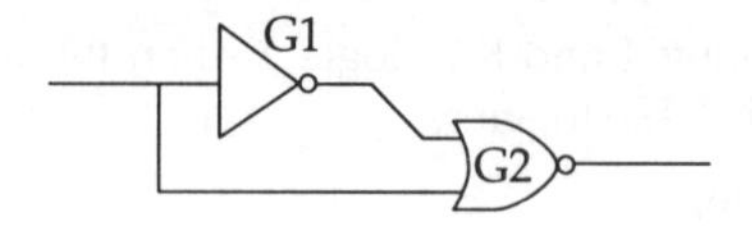

Circuit for making a bi-stable edge-sensitive

G2 gets one of its inputs direct and the other after the propagation delay of G1. This ensures that the output of G2 will be a very short duration pulse that occurs on the trailing edge of the clock pulse.

Remember that with a NOR gate only when both inputs are LOW will there be a logic 1 at the output!

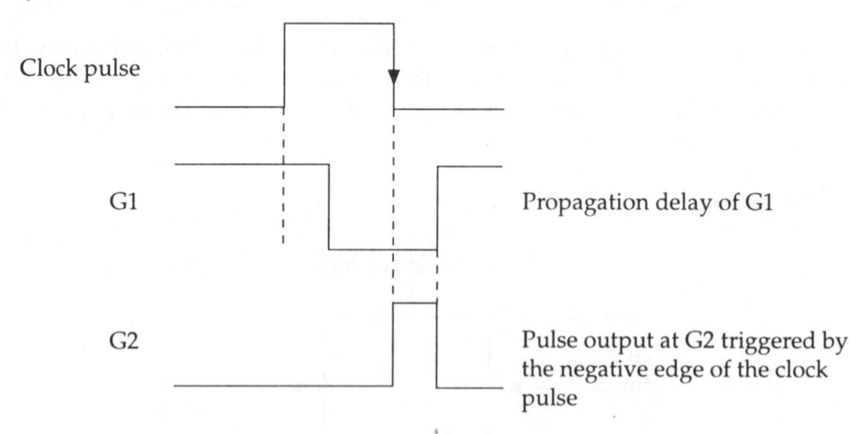

Using propagation delay to produce a short clock pulse

When the clock pulse goes LOW the output of G1 remains LOW due to the propagation delay of the gate. This means that for a short period both inputs to G2 are LOW. This produces a short duration output pulse to trigger the bi-stable on the negative edge of the clock pulse.

Commercial J.K. Bi-stables (74LS76)

Commercial J.K. bi-stables are edge-triggered, toggle, have direct PRESET and CLEAR inputs and also allow new data to be presented to the inputs at the same time as previous data is being transferred to the outputs.

They are formed from two R.S. bi-stables with a primary bi-stable controlling a secondary bi-stable. This is done to control the movement of data through the bi-stable and to make it edge-triggered.

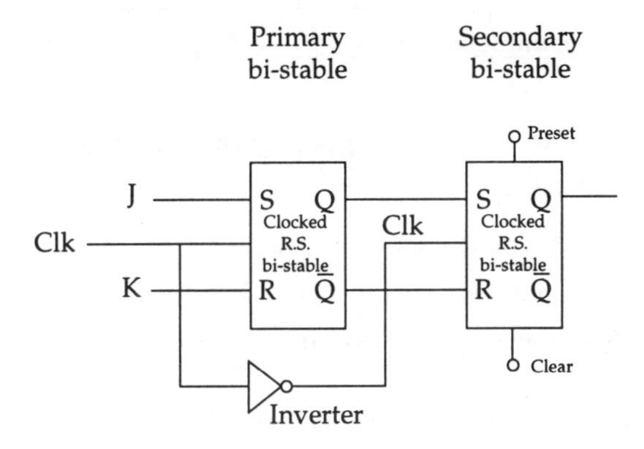

A J.K. bi-stable formed from two R.S. bi-stables

Operation

Data to be input is connected to J and K. When the clock pulse is HIGH the data can enter the PRIMARY bi-stable. When the clock pulse returns to LOW the output of the inverter is HIGH and data is transferred from the PRIMARY to the SECONDARY bi-stable. With the clock now LOW, data is present at the output Q and fresh data may be presented at the inputs J and K without it affecting the output at Q.

PRESET and CLEAR inputs

There are also direct connections to the X-coupled NANDs of the secondary R.S. bi-stable, which allows the J.K. to be CLEARed or PRESET. The J.K. PRESET input is used to set the output Q directly to 1 i.e. when this input is activated it will set Q to 1 irrespective of the logic levels on the inputs J, K and Clk.

The J.K. CLEAR input will also override the J, K and Clk inputs and set Q to 0.

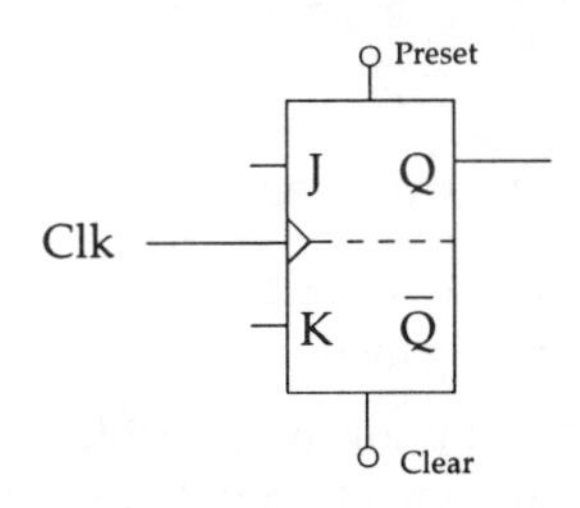

The J.K. bi-stable circuit symbol

The commercial J.K. (74LS76) contains two J.K. bi-stables in the same package, as shown below.

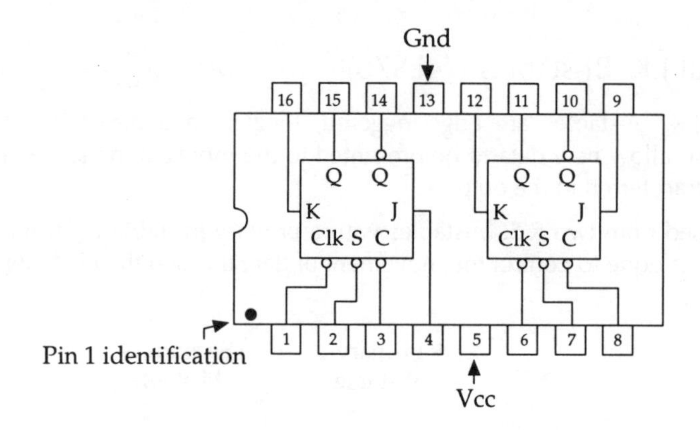

The 74LS76 (dual J.K. bi-stable) pin-out diagram

Summary

- Bi-stables are used as a simple memory or latch.
- The bi-stable has two complementary outputs Q and $\overline{Q}$
- A simple R.S. bi-stable may be formed from a pair of cross-coupled NAND or NOR gates.
- With the R.S. bi-stable there is a forbidden condition. It is not possible to both SET and RESET the bi-stable at the same moment in time.
- An R.S. Bi-stable may be used to de-bounce switches on keypads.
- When a number of bi-stables are used in a logic circuit it is often necessary that they should all change state together. To achieve this a clocked R.S. bi-stable may be obtained using two more NAND gates.
- By putting an invertor between the R and S inputs of a clocked R.S. bi-stable, a D-type bi-stable is formed. This circuit has the advantage that there is no indeterminate state. The operation of the D-type is that Q follows D after a clock pulse.
- For counters a toggling or J.K. bi-stable is needed. The output Q changes state with each applied clock pulse.
- To prevent erratic operation of bi-stables they are made edge-triggered e.g. the bi-stable only changes state on the negative edge of the clock pulse.
- A commercial J.K. bi-stable employs two R.S. bi-stables and has a an additional PRESET and a CLEAR input.

Self-assessment questions *(answers page 472)*

1) Sketch the circuit to de-bounce a switch using cross-coupled NOR gates. Show clearly the input switching arrangement.

2) Below is the truth table showing all the possible inputs to a clocked R.S. bi-stable together with the outputs before the clock pulse. Enter into the output column the output states after the clock pulse.

Inputs		*Output before clock pulse*		*Outputs after clock pulse*	
S	*R*	*Q*	$\overline{Q}$	*Q*	$\overline{Q}$
0	0	0	1		
0	0	1	0		
0	1	0	1		
0	1	1	0		
1	0	0	1		
1	0	1	0		
1	1	0	1		
1	1	1	0		

Truth table for clocked R.S. bi-stable

3) In the diagram below data is entered into the bi-stables from a keypad one key at a time. Data in all the bi-stables can then be read out simultaneously and sent down the data highway or BUS. Study the diagram and decide on the following:
   a) Which buttons must be pressed to enter data into the bi-stables?
   b) Which buttons must be pressed in order to read data out of the bi-stables?
   c) What is the largest number that can be stored in the register?
   d) Which button is pressed to clear the register?
   e) Is the output from the register taken in serial or parallel form?

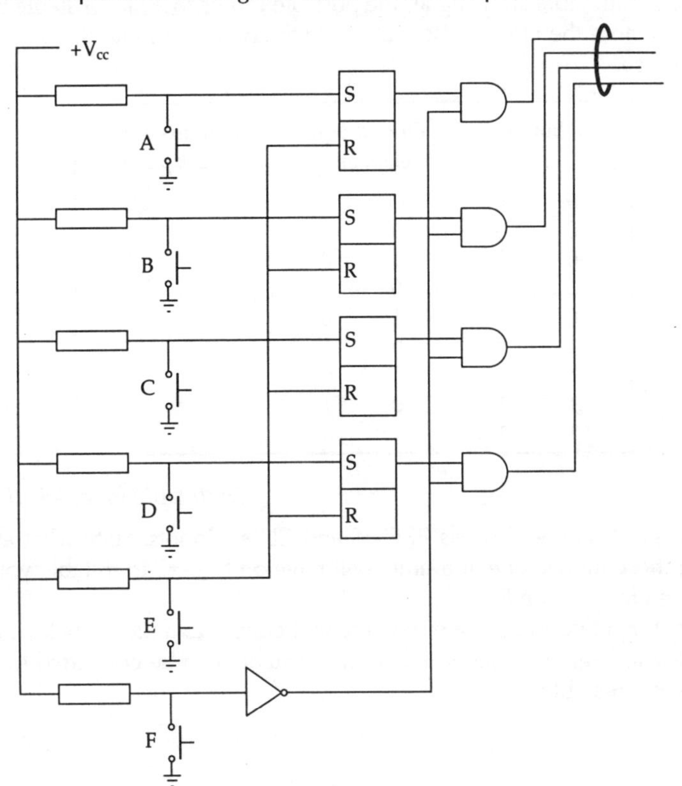

Strobed Keypad

4) Below is a diagram for a sequence start-up for a machine. The push-buttons have to be pressed in a particular order to energise the relay and start the machine. Study the diagram and answer the following:

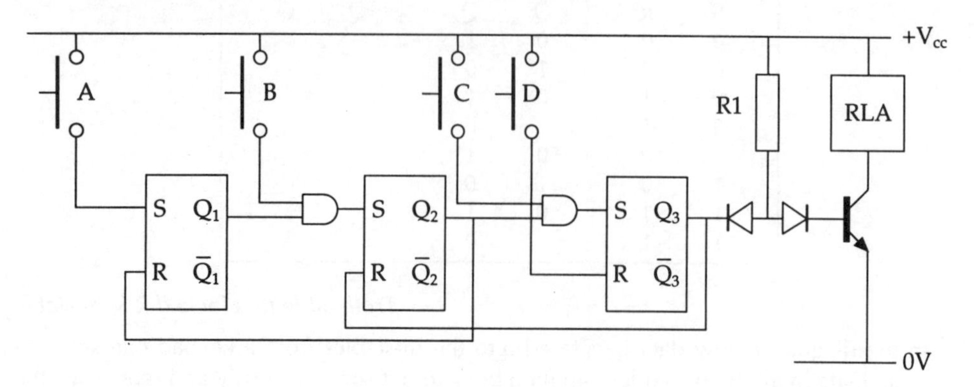

Start-up sequence

a) What buttons must be pressed, and in what order, to energise the relay?
b) How is the relay made to release? Explain fully how the transistor is made to switch ON or OFF and the purpose of the two diodes.
c) Are these bi-stables made from X-coupled NAND or NOR gates?

5) Below is the truth table showing all the possible inputs to a J.K. bi-stable together with the outputs before the clock pulse. Enter into the output column the output states after the clock pulse.

Inputs		*Output before clock pulse*		*Outputs after clock pulse*	
S	*R*	*Q*	$\bar{Q}$	*Q*	$\bar{Q}$
0	0	0	1		
0	0	1	0		
0	1	0	1		
0	1	1	0		
1	0	0	1		
1	0	1	0		
1	1	0	1		
1	1	1	0		

Truth table for clocked J.K. bi-stable

6 a) The J.K. bistable also has PRESET and CLEAR inputs. State what effect activating these inputs, one at a time, will have on the J.K. truth table you have completed in question 5.
b) What input conditions are required at the input to a J.K. to make it toggle?
c) Draw a diagram to show how a J.K. bi-stable can be converted to operate as a D-type bi-stable.

Binary counters

Introduction

J.K. bi-stables have a toggle facility and this is necessary to form a binary counter.

When the J.K bi-stable has both its J and K inputs HIGH (+5V), then its output at Q will change state every time a pulse is applied to the clock pulse input.

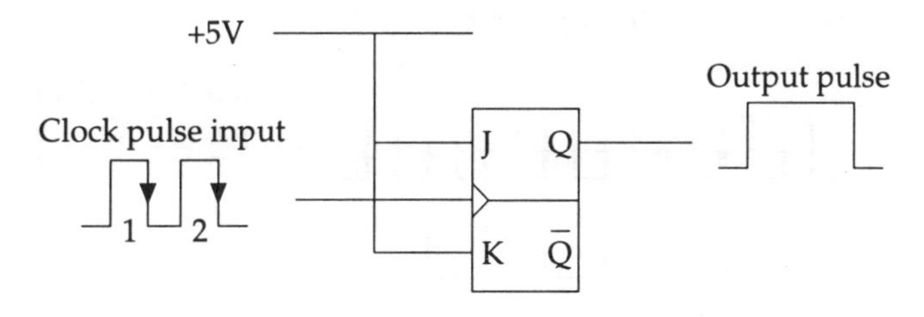

Two clock pulses in produce one pulse out

The negative edge of the first clock pulse sends the output at Q HIGH; the negative edge of the next pulse makes the output go LOW. Two pulses at the input produce one pulse at the output.

The bi-stable therefore counts in two's. Also if we apply a repetitive pulse at the input, the output frequency will be half the input frequency. The bi-stable may therefore be used as a frequency divider.

By cascading two bistables, we connect the Q output of the first bi-stable to the clock input of the second bi-stable.

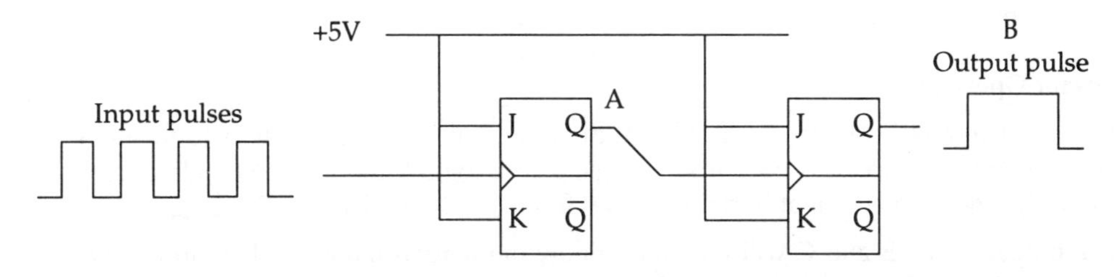

Two bistables will count in 4s

Four pulses applied to the input of the first bi-stable will give two pulses at its output A. These two pulses form the input to the second bi-stable, so the output at B is a single pulse.

The two bi-stables will therefore provide one output pulse at B for every four input pulses to the first bi-stable.

Alternatively the output frequency at B will be one quarter of the input frequency: i.e. divide by 4.

Similarly three cascaded bi-stables will count in eights and four cascaded bi-stables will count in sixteens.

The counters may be configured to count up or count down.

Serial UP-Counter

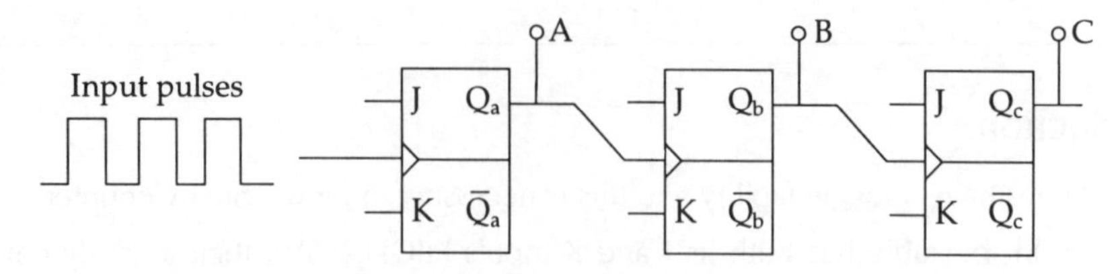

The J.K. bi-stable will change state on the **negative edge** of the input or clock pulse.

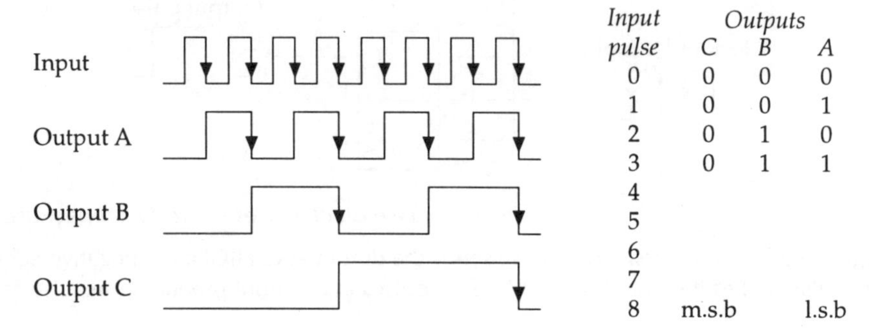

Input pulse	*Outputs* C	B	A
0	0	0	0
1	0	0	1
2	0	1	0
3	0	1	1
4			
5			
6			
7			
8	m.s.b		l.s.b

3 Stage Serial UP-Counter

Study the waveforms in the diagram above and complete the truth table!

Study Note: all the J and K inputs, in the diagram above, are not shown connected to +5V (logic 1), to simplify the diagram. Whenever a J or K input is not shown connected in this way then assume they are connected to +5V. In practical circuits you must connect the J and K inputs to obtain the correct circuit operation.

Circuit Operation

The negative edge of each clock pulse at the input will cause bi-stable A to toggle. Every time the output pulse at A goes from a logic 1 to a logic 0 then it will toggle bi-stable B. Similarly when the output at bi-stable B goes from 1 to 0 the bi-stable C is toggled.

The outputs at A, B and C will therefore follow the natural binary code with A being the least significant digit and C the most significant.

Note

a) After eight pulses the counter has returned to its original condition. In this example 8 is called the modulo of the count.

b) The maximum count is 7; one less than the modulo.

c) The bi-stables change state on the negative edge of the clock pulse.

d) On the 4th and 8th pulses, all the bi-stables change state. There is a **ripple through** which is a disadvantage of this type of counter. i.e. it increases the count time and reduces the maximum rate of counting.

e) This type of counter is therefore called a **ripple through** or **asynchronous** counter. Each bi-stable operates in turn: that is, the first bi-stable must change state before the second bi-stable can change its state.

Cascading bi-stables will form counters of any modulo that is a power of 2.

No of Bi-stables	3	4	5	6
Power of 2	2^3	2^4	2^5	2^6
Modulo	8	16	32	64
Maximum Count	7	15	31	63

By rearranging the connections a down counter can be produced.

Serial Down Counter

With this counter, when all the inputs are reset to zero, the first input pulse sets all the outputs A,B,C to logic 1 (maximum count).

The next input pulse will reduce the count by one. For each input pulse the count is reduced by one; the down counter therefore counts down from its maximum count to zero. To produce a down counter the output is taken from the Q output of each bi-stable as before, but the next bi-stable in the chain gets its clock input from the NOT Q or complementary output.

A Three-Stage Down Counter

Note: all J.K. inputs are connected to +5V! The outputs are taken from A.B.and C.

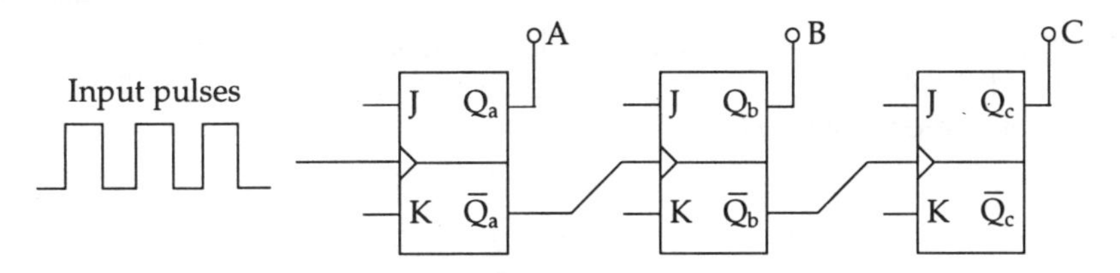

The next bi-stable is triggered from the Q output.

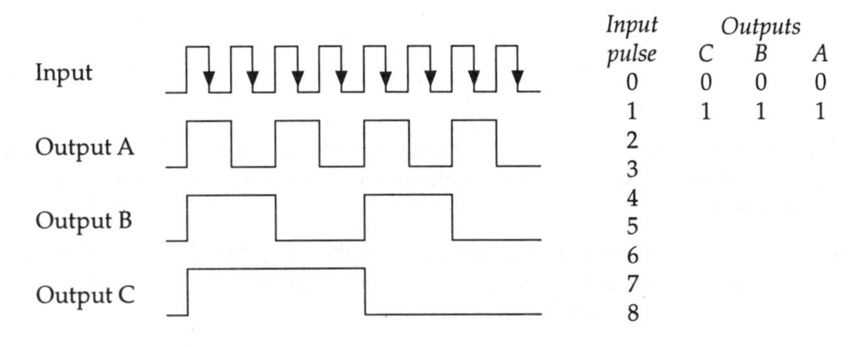

Input pulse	*Outputs* C	B	A
0	0	0	0
1	1	1	1
2			
3			
4			
5			
6			
7			
8			

Complete the truth table! Remember that the bi-stable change state on the negative edge of the clock pulse.

Asynchronous Down Counter

With three stages the maximum count will be 7 (Modulo 8). If it is required to have a count of less than 7 then the counter needs to be modified.

Reducing the Count

To reduce the count some of the counter's 8 states must be missed out; i.e. if a maximum count of 5 is required (modulo 6) then two of the eight states must be omitted.

One of the ways to reduce the number of states is to reset the counter when the desired modulo has been reached. This may be achieved by using the reset inputs on the J.K. bi-stables.

For example, if a maximum count of 5 is required, then when the count is 6, all the reset inputs are activated on the bi-stables, which sets all the outputs to zero.

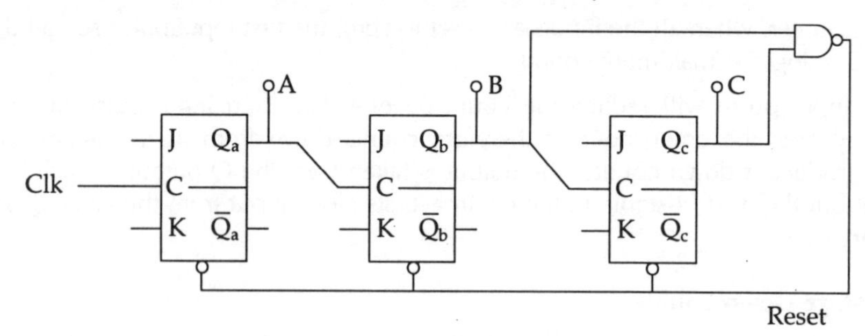

Modulo 6 Counter

The count proceeds normally as before until the count of 6 is reached. At this point the outputs B and C will both be at logic 1. The output of the NAND gate will go LOW resetting all the bi-stable outputs to 0.

States	*C*	*B*	*A*
1	0	0	0
2	0	0	1
3	0	1	0
4	0	1	1
5	1	0	0
6	1	0	1
7	1	1	0
8	1	1	1

When the seventh state (count of 6) is reached, the bi-stables are reset and the outputs A, B and C return to state 1 (counter reset to 0).

The state of 6 is transitory and can only exist for the amount of time it takes the NAND gate to change its state (a few ns). Therefore the maximum count is 5, the 6th pulse resetting the count to zero.

The NAND gate is used because with many J.K. bi-stables the PRESET and CLEAR inputs are active low; that is, a logic 0 is required on the clear input to clear the bi-stable.

Summary

- Bi-stables with a toggle facility are needed to make a binary counter.
- Each bi-stable will count or divide by 2.

- Bi-stables may be cascaded to increase the count.
- The number of unique states the counter has is called the modulo; the maximum count is always one less than the modulo.
- Bi-stables connected in cascade form serial or asynchronous counters. They are also called ripple counters.
- Serial counters may be configured to count up or down.

Self-assessment questions (1) *(answers page 473)*

1) What type of bi-stable is needed to form a counter?
2) What is meant by the term ripple counter?
3) How does the ripple-through-effect determine the maximum rate of counting?
4) How many J.K. bi-stables are needed to form a decade counter?
5) Explain what is meant by the term 'sequential circuit'.
6) Design a counter that has a modulo of 7 using J.K. bi-stables.
7) The circuit below is a three stage asynchronous counter using D-type bi-stables. Explain the operation of the circuit, and determine its modulo and maximum count by drawing up a truth table.

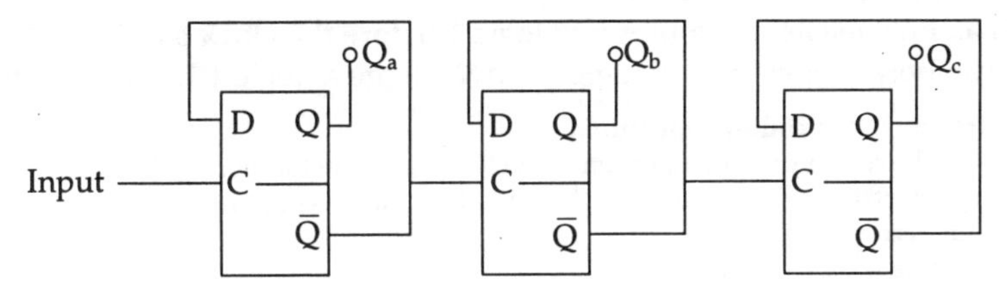

Circuit for Question 7

8) Design a circuit that will produce an output frequency of 200Hz when the input frequency to the circuit is a 1kHz squarewave.

Parallel or Synchronous Counters

With parallel counters all the bi-stables are clocked simultaneously. This means that all the bi-stables will change state at the same time.

The connections of the J and K inputs determine which bi-stable will change state. This gives two advantages:

1) All the bi-stables change state at the same moment in time under the control of the input clock or synchronizing pulse.
2) This will produce a faster operation than serial counters because the ripple-through-effect is eliminated.

This increases the maximum rate of counting.

In order to determine the operation of synchronous counters the following points must be understood:

1) With synchronous counters all bi-stables change state at the same time.
2) The logic levels at each J and K input for each bi-stable **before a clock pulse** input will determine which bi-stables will change state.
3) When deciding which bi-stables will change state you must look at the **logic level** on **each bi-stable before the clock pulse input**.

A Three Stage Parallel Counter

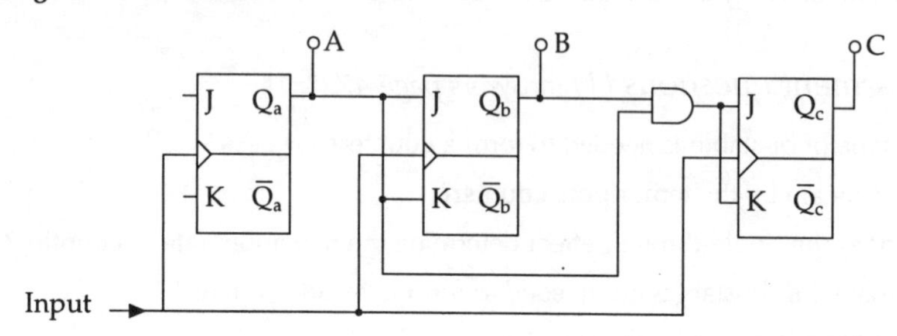

Synchronous or parallel counter

Circuit Operation

A logic 1 on J and K on the first bi-stable (bi-stable A) means that this bi-stable will toggle (change state) with each input clock pulse.(CP)
Bi-stable B can only toggle if output A is at logic 1 **before the clock pulse.**
Bi-stable C can only change state if outputs A **AND** B are at logic 1 **before the clock pulse.**

To determine the counter operation:
Look at the conditions of J and K on each bi-stable before the clock pulse and decide which bi-stable will toggle when the clock pulse is present.
Remember:
- ❐ Bi-stable A toggles with each CP
- ❐ Bi-stable B will toggle only when output A = 1
- ❐ Bi-stable C will toggle only when outputs A AND B = 1.

Input pulse	*Outputs* A	B	C	*Comments*
0	**0**	**0**	**0**	No change: reset state.
1	1	0	0	Before CP: A = 0; A AND B = 0. A toggles; no change on B and C.
2	0	1	0	Before CP: A = 1; A AND B = 0. A toggles; B toggles; no change on C.
3	1	1	0	Before CP: A = 0; A AND B = 0. A toggles; no change on B and C.
4	0	0	1	Before CP: A = 1; A AND B = 1. A toggles; B toggles and C toggles. All change state.
5	1	0	1	Before CP: A = 0; A AND B = 0. A toggles; no change on B and C.
6	0	1	1	Before CP: A = 1; A AND B = 0. A toggles; B toggles; C does not change.
7	1	1	1	Before CP: A = 0; A AND B = 0. A toggles; B and C do not change.
8	0	0	0	Before CP: A = 1; A AND B = 1. A, B and C toggle. All change.

Summary

- ❐ With the parallel or synchronous counter all the bi-stables change state at the same time under the control of the clock pulse.
- ❐ They are faster than serial counters because the ripple-through-effect is eliminated. The maximum rate of counting is increased.
- ❐ To determine the operation of a parallel counter you must look at the logic levels on the J and K inputs **before a clock pulse arrives** in order to determine which bi-stables will change state.

Self-assessment questions (2) (answers page 474)

1) Produce the truth table for the counter shown below.

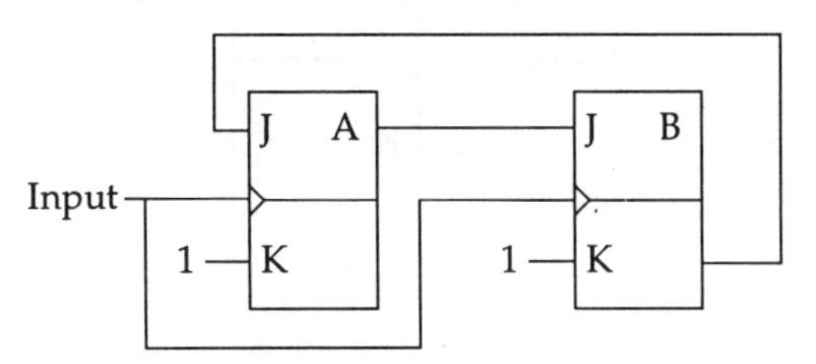

Diagram for Question 1

2) Produce the truth table for the counter shown below.

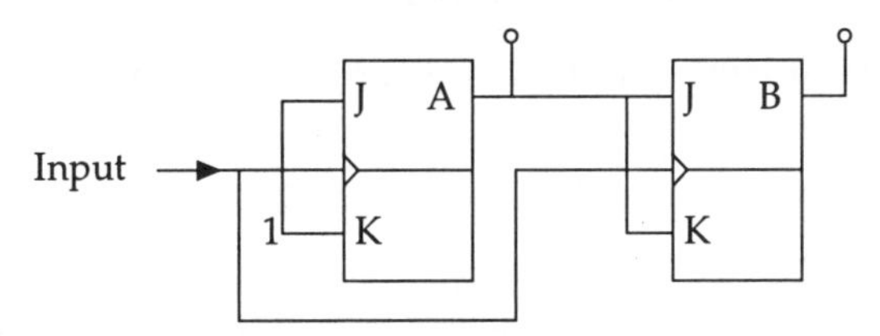

Diagram for Question 2

3) Produce the truth table for the counter shown below.

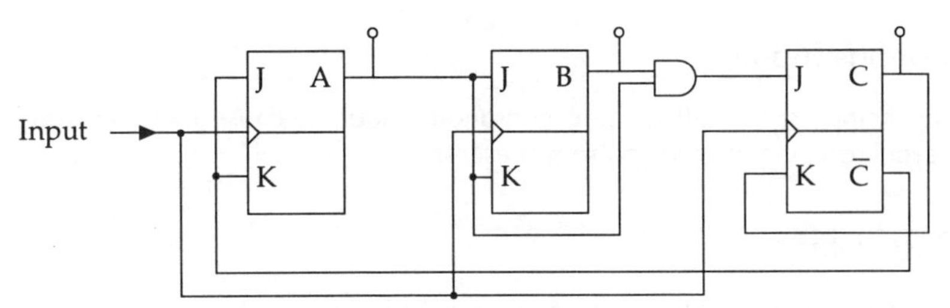

Diagram for Question 3

Shift registers

Introduction

A shift register is a device used for the temporary storage of data. This data can then be shifted out and new data shifted in. The shift register is formed from J.K. bi-stables. Before we investigate shift registers we will revise the operation of the J.K. bi-stable.

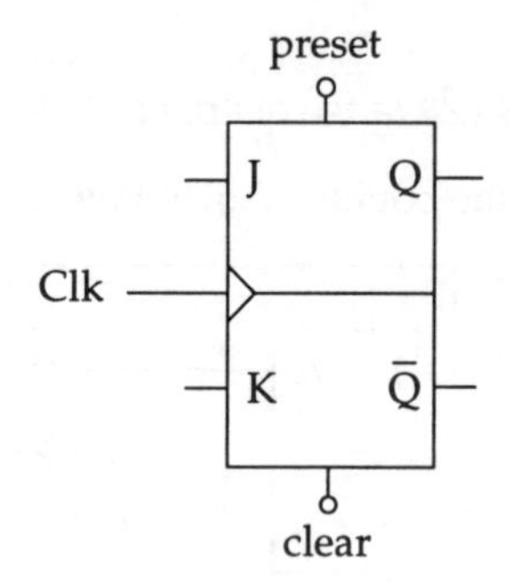

J.K. bi-stable

The J.K. has a preset input which may be used to set the output Q to 1 directly: i.e. when this input is activated it will set Q irrespective of the logic levels on the inputs J, K, and Clk.

The J.K. also has a clear input. This input will also override the J,K, and Clk inputs and set Q to 0.

Asynchronous Inputs

These two inputs, preset and clear, are called the asynchronous inputs because they may be activated at any moment in time and do not depend upon a timing or synchronizing pulse.

Synchronous Inputs

The J and K inputs are called the synchronous inputs because they can only be active when a synchronizing or clock pulse is present.

A Shift Register

Four J.K.s may be connected together to form a 4-bit shift register.

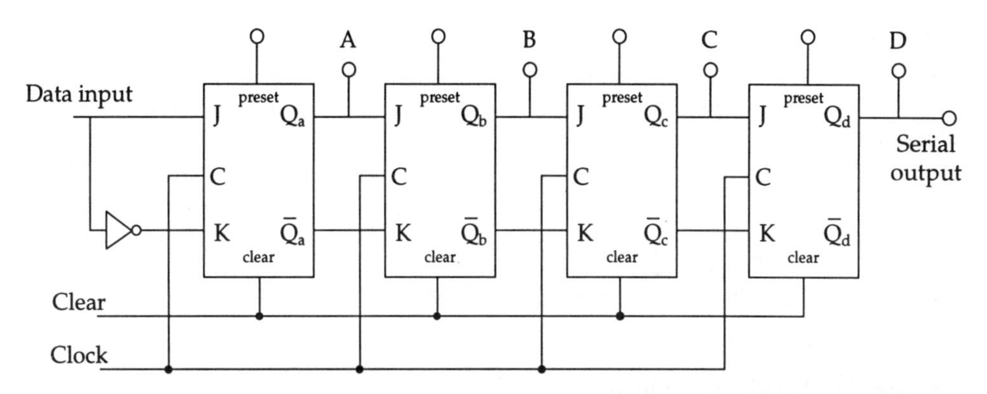

4-bit Shift Register

The Four Methods of Operation

The shift register can be operated in one of four ways.

1) Serial in – Parallel out (SIPO)

The 4-bit data word is fed serially to the data input together with 4 clock pulses. This shifts the 4-bit word into the register. The data word may now be read from the 4 Q outputs of the bi-stables. This can be used as a *serial to parallel converter.*

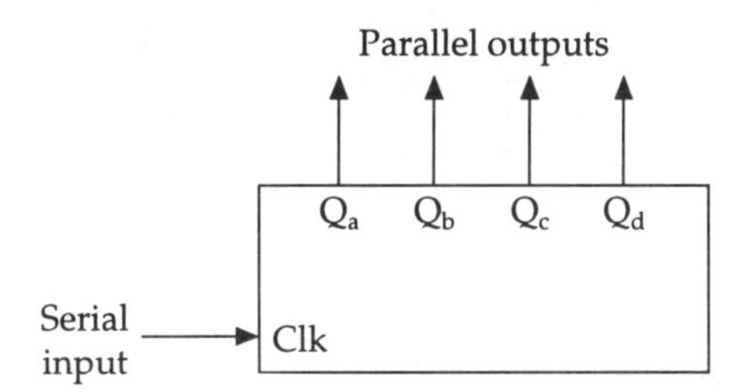

Serial in – parallel out (SIPO)

2) Parallel in – Serial out (PISO)

By using the preset inputs a 4-bit data word may be fed in parallel. Then by applying 4 clock pulses the 4-bit word will be shifted out serially. This may be used as a *parallel to serial converter.*

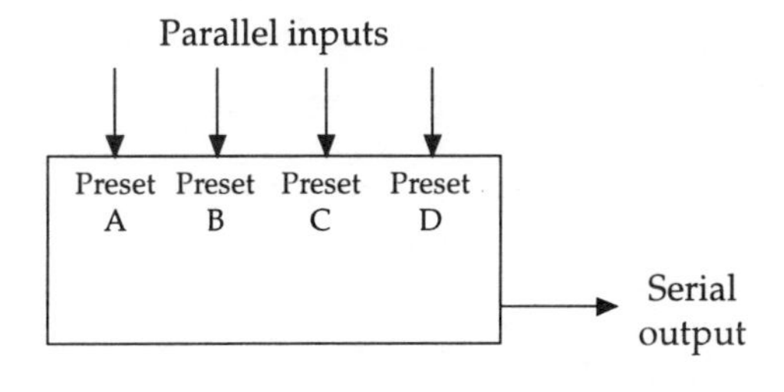

Parallel in – serial out (PISO)

3) Serial in – Serial out (SISO)

To shift in: the 4-bit data word is applied to the serial input together with 4 clock pulses. To shift out: 4 more clock pulses are applied and the data is shifted out serially at the output terminal.

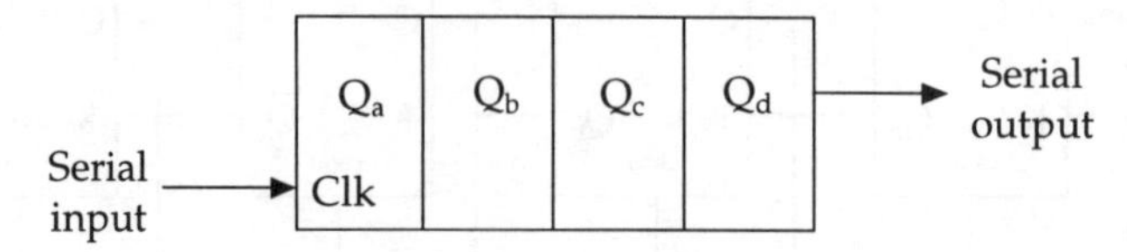

Serial in – serial out (SISO)

4) Parallel in – Parallel out (PIPO)

The preset inputs are used to input data directly. The output is then obtained from the normal Q outputs of the bi-stables.

This register does not require any clock pulses. It is used for temporary storage and is much faster in operation than the other 3 methods.

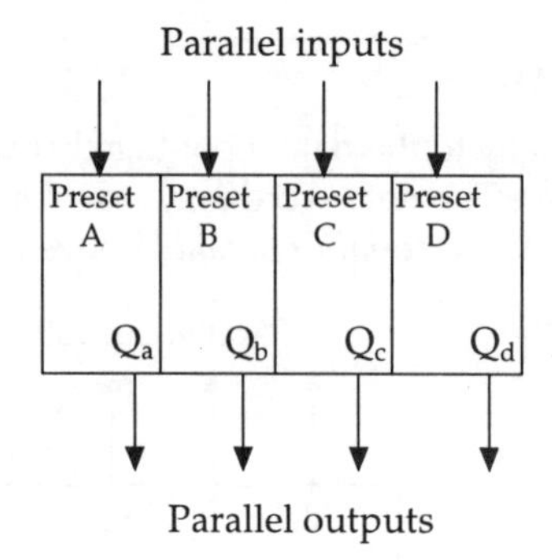

Parallel in – parallel out (PIPO)

Shift Register Operation

The first J.K. bi-stable has an invertor between the J and K inputs. It is therefore connected as a D type.

Because the outputs of the J.K. bi-stable are always complementary, the remaining bi-stables in effect are operating as D-types.

The operation of a D-type is: the output Q *follows* D *after a clock pulse*!

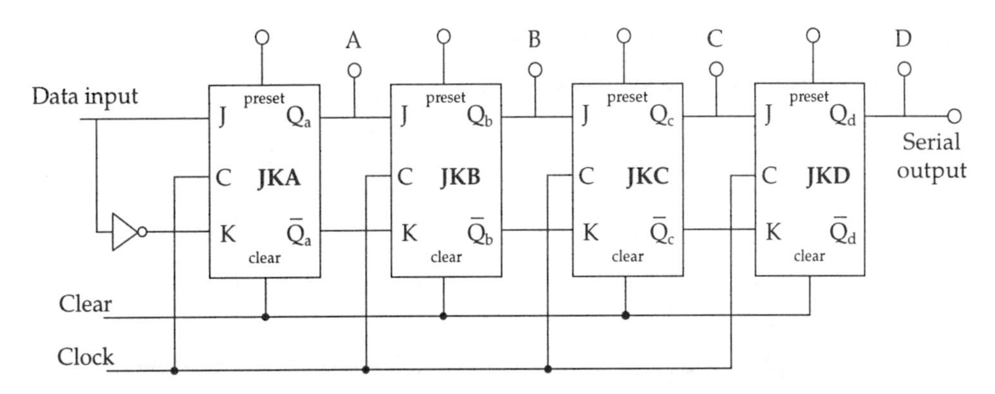

4-bit SISO shift register

Serial Operation

1) The register is first cleared by applying a pulse to the reset inputs. All the Q outputs are now zero.

2) The first bit (1) of the data input (say 1001) is applied to the data input together with a clock pulse.

   Bi-stable JKA now has its Q output set to 1.

   The input to bi-stable JKB is now 1 on J and 0 on K (remember the outputs of bi-stable JKA must be opposite).

3) The second bit (0) of the data word is applied together with another clock pulse to the data input. The Q output of bi-stable JKA will follow the data input (0) and will be unchanged. The output of bi-stable JKB however will follow its J input (1).

   Therefore the logic 1 at the output of JKA will shift to the output of JKB.

4) The third bit (0) together with a clock pulse will shift the logic 1 from JKB to JKC.

5) On the fourth clock pulse the logic 1 will be shifted from JKC to JKD and a logic 1 will be shifted in at JKA.

   The register is now storing the 4-bit word.

6) If now the data input is held at logic 0 and four clock pulses applied, then the data word will be shifted out serially and all the Q outputs will be at logic 0.

   Note: If the data input is held at logic 1 and four clock pulses applied, then the Q outputs would all be at logic 1.

Universal Shift Register

These can be used in any of the 4 modes and are also capable of shifting data either to the left or right.

Typical chips: 74194: 4-bit universal
74299: 8-bit universal
4035: 4-bit universal

Self-assessment questions (answers page 475)

All questions refer to the diagram below.

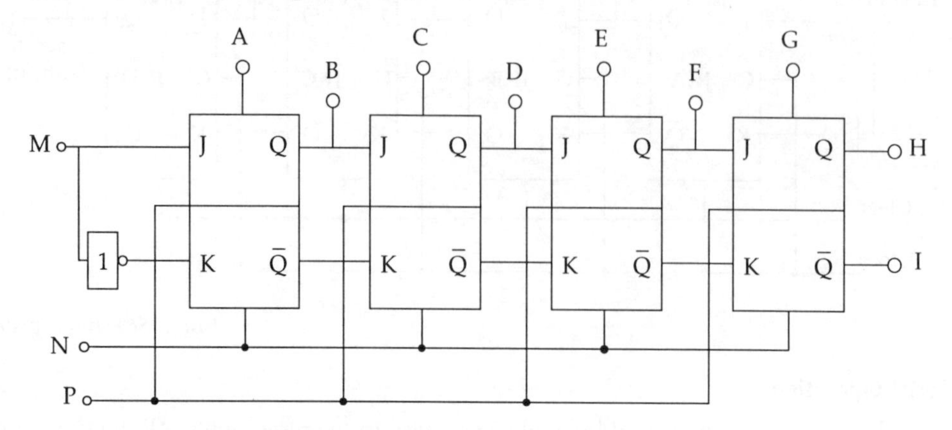

1) To which terminal would you connect the clock pulse?
2) Which terminal would you use to reset the shift register?
3) Explain how you would connect the register for serial input, parallel output, SIPO.
4) Explain how you would connect the register for serial input, serial output, SISO.
5) Explain how you would connect the register for parallel input, parallel output, PIPO.
6) Explain how you would connect the register for parallel input, serial output, PISO.
7) Explain the procedure to enter the data 1011 serially into the register.
8) Explain the procedure to enter the data 1011 in parallel form and shift it out in serial form.
9) Why is there an inverter between the J and K inputs of the first bi-stable?
10) Why are there no inverters needed between the J and K inputs of the other bi-stables?

Answers to questions in Part 3: Digital Electronics

Section 1

Task a Investigation of logic gates

1) When both diode 1 AND diode 2 are at logic 1 (+5V) then the diodes are non-conducting and the output at F is HIGH (+5V).

   I f either diodes cathode is switched to ground then that diode will conduct bringing the output F down to about 0.7V (LOW).

2) The volt drop across a conducting silicon diode is approximately 0.7V. This voltage is fairly constant and does not depend very much on the current flowing.

3) Simply add another diode and switch. The anode of the diode connected to R1.

4) When both inputs are at logic 0 (0V) then the diodes are non-conducting and there is no output voltage. If either input is connected to logic 1 (+5V) then that diode will conduct and the current that flows will produce an output voltage across R1.

5) Because of the volt drop across the diode the maximum output voltage will be

$$5V - 0.7V = 4.3V.$$

6) When the input is switched to logic 0 (0V) then the transistor has no input current to start the transistor action. Therefore no collector current flows and the collector voltage is HIGH (logic 1). When the input is switched to logic 1 (+5V) a large base current flows which saturates the transistor. The collector voltage 'bottoms' (all the voltage is dropped across R1) and the collector voltage falls to a low value Vcesat = 0.2V.

Task b Investigation of logic gates and truth tables

1) Only when both inputs A AND B are present will there be an output.

2) The logic gate uses transistors as saturated switches. The transistor is switched fully ON to give a LOW (logic 0) output. Under these conditions the collector voltage is VCE_{sat} about 0.15V and not 0V.

3) Two AND gates may be employed as shown below.

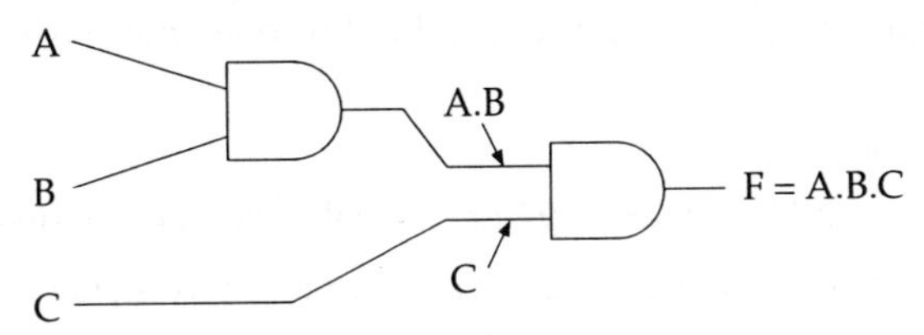

4) There will be an output (logic 1) from the gate when input A OR input B OR both inputs are present.

5) The output stages of logic gates employ transistors as switches to switch the output up to +5V or down to 0V. Because these transistors are not perfect switches (they do not have zero resistance in the ON state) a small voltage will be lost across the ON resistance of the switching transistor, lowering the voltage at the output of the gate.

6) The NOT gate is a transistor operated as a switch. With a base voltage applied (logic 1) the transistor switches fully ON and its collector voltage (output) falls to almost 0V (logic 0).

7) Unused CMOS inputs if left floating (unconnected) may pick up stray signals (noise) which causes the output stage of the gate to be partially switched ON. Under these conditions the logic gate chip dissipates more power, which can cause the gate to overheat. The heat may be sufficient to cause the failure of the whole chip.

8) The main procedure when handling MOS chips are:
   - All inputs on the chip must be connected somewhere! All unused inputs must be connected to a logic level, either ground or the supply, directly or by a resistor. This is especially important on a breadboard circuit.
   - The input signal voltage levels should not exceed the supply voltage to the device.
   - Do not apply input signals to the device when the power supply to the chip is disconnected.
   - Avoid static charges when working with the component. Store the ICs in conductive foam, or on a metallic base.
   - Observe anti-static handling procedures.i.e:
     i) Do not touch the pins of the chip directly.
     ii) Use an anti-static bench mat and wrist earthing straps when inserting/removing ICs.
     iii) Insert the IC into your circuit last.
     iv) Do not insert/remove the IC when the circuit is powered up.

Section 2 Information and Skills Bank

Introduction to digital control

1) d) a digital signal may be a binary signal or a multilevel signal.

2) c) The analogue signal can have any value but does not have to be continously variable.

3) b), d) and f)

4) c) The reading could be either 9.999 or 10.00 the difference being 0.001V.

5) d) the output from the probe is converted into a binary signal before being fed to the input of the digital control system.

6) b) the output from the probe is an analogue signal and must be converted into a digital form.

7) a) an ON-OFF control signal is a binary signal.

8) e) a mutilevel signal is closer to an analogue signal than binary.

Number Systems

SAQ 1

1) 11, 9.75, 31.25

2) 10, $2^{10} = 1024$

3) $100{,}000{,}100_2 = 260_{10}$

4) i) (b) $2^{12} = 4096$ binary patterns – less the 1st pattern (0) gives the maximum number of 4095.

   ii) (c) ; 11 lamps would only contain a maximum number of 2047.

5) i) (a) ii) (c)

6) $55_{10} = 110111_2 = 1x2^5 + 1x2^4 + 0x2^3 + 1x2^2 + 1x2^1 + 1x2^0$

7) a) 10 b) 16 c) 2 d) 8

8) (d)

9) a) 0011 1100 b)0001 1101 c)0101 1001 d) 0111 0010

10) 1) 0F 2) AC 3) 3B 4) 7E

SAQ 2

1) a) 100101011 b) 110011.011 c) 1011.000101

   d) 0.1111 e) 110010.101

2) a) 23 b) 22 c) 49 d) 30 e) 42

3) a) B3 b) A5 c) EB d) A2 e) 9B2

4) a) 255 b) 200 c) 18 d) 169

SAQ 3

1) a) 110110; b) 100000 c) 1100; d) 10011.

2) a) 100101001 b) 1,1111,1111

   c) 111111 d) 1111,1111

3) 7 bits

4) Well defined logic levels.

   Many electrical devices can represent binary numbers:

	Logic 1	**Logic 0**
Transistor	Off	On
Capacitor	charged	uncharged
Switch	closed	open

voltage source	voltage	no voltage
current source	current	no current
magnetic ferrite ring	magnetised	de-magnetised

5) a)
```
 101000101
   1011000 +
 110011101
```
b)
```
 110010001
1000000001 +
1110010010
```
c)
```
11011
 1110
 1101
```
d)
```
1011011
 110111
 100100
```

6) a) 255 b) 15.9375

7) a) 100101100 b) 0.011 c) 100000100.101

8)

Decimal	Power of 2	Binary
1	2^0	1
2	2^1	10
4	2^2	100
8	2^3	1000
16	2^4	10000
32	2^5	100000
64	2^6	1000000
128	2^7	10000000
256	2^8	100000000

Logic Gates

1) a) F = A.B.C b) F = A.B + $\overline{C}$ c) (A+B).(C+D) d) A.B + C.D

2) a) F = A.B + C b) F = (A+B).C

c) F = (A+B).$\overline{C}$ d) F = (A+B).(B+C)

3) a)

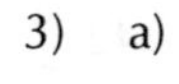

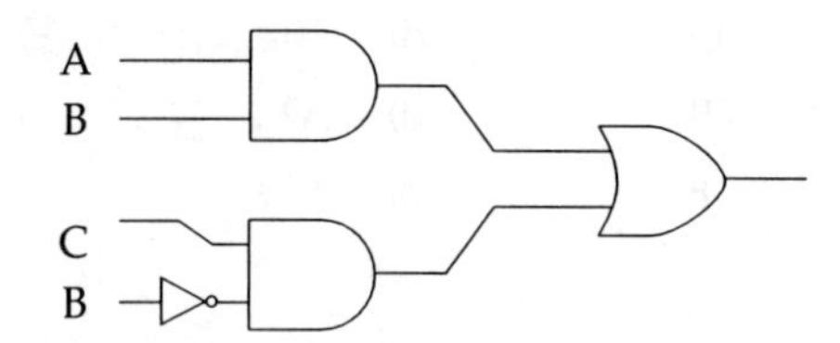

b)

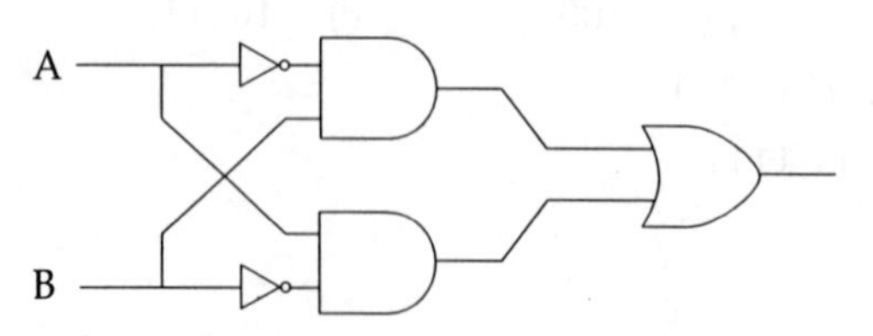

c)

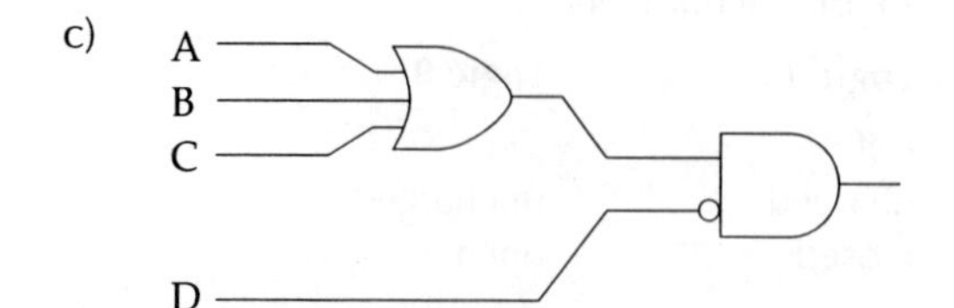

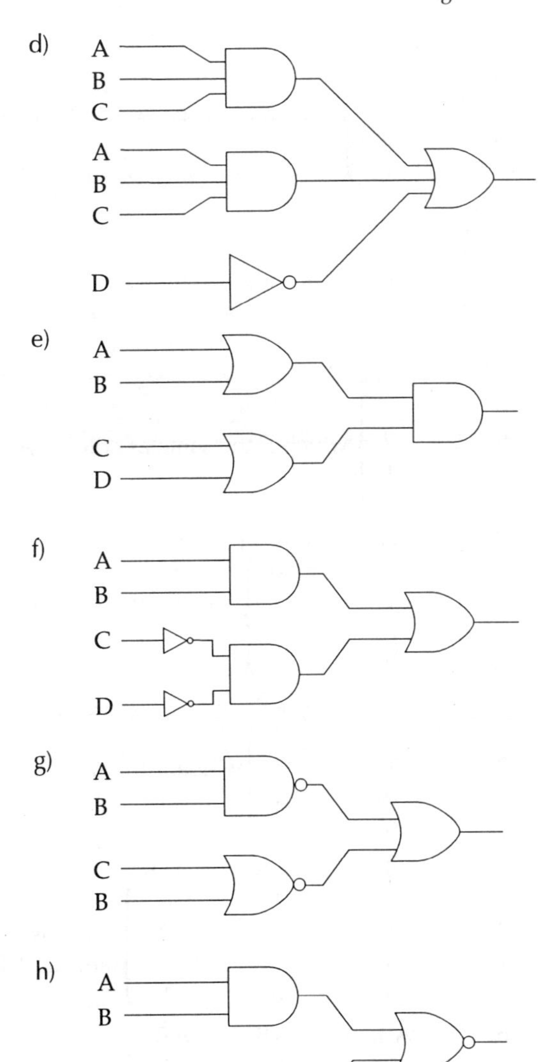

Truth Tables

1) a) $F = \overline{A}.\overline{B}$

b) $F = \overline{A}.\overline{B} + B.\overline{A}$

c) $F = \overline{A}.B + A.\overline{B} + A.B$

d) $F = A.\overline{B}.\overline{C} + \overline{A}.B.\overline{C}$

e) $F = \overline{A}.\overline{B}.\overline{C} + \overline{A}.B.C$

2. a)

B	A	A.B	A.B + B
0	0	0	0
0	1	0	0
1	0	0	1
1	1	1	1

which is the same as column B

b)

B	A	$\overline{B}$	$A.\overline{B}$	$A.\overline{B} + A$
0	0	1	0	0
0	1	1	1	1
1	0	0	0	0
1	1	0	0	1

which is the same as column A

c)

B	A	A+B	A.(A+B)
0	0	0	0
0	1	1	1
1	0	1	0
1	1	1	1

which is the same as column A

d)

B	A	$\overline{A}$	$\overline{A}.B$	$\overline{A}+\overline{A}.B$
0	0	1	0	1
0	1	0	0	0
1	0	1	1	1
1	1	0	0	0

which is the same as column $\overline{A}$

e)

B	A	A.B	$\overline{B}$	$A.\overline{B}$	$A.\overline{B}+A.B$
0	0	0	1	0	0
0	1	0	1	1	1
1	0	0	0	0	0
1	1	1	0	0	1

which is the same as A

f)

B	A	$\overline{A.B}$	$\overline{A}$	$\overline{B}$	$\overline{A} + \overline{B}$
0	0	1	1	1	1
0	1	1	0	1	1
1	0	1	1	0	1
1	1	0	0	0	0

which is the same as $\overline{A.B}$

g)

B	A	A.B	$\overline{A}$	$\overline{B}$	$\overline{A}+\overline{B}$	$\overline{A+B}$
0	0	0	1	1	1	0
0	1	0	0	1	1	0
1	0	0	1	0	1	0
1	1	1	0	0	0	1

which is the same as A.B

h)

B	A	$\overline{A}$	$\overline{B}$	$\overline{A}.\overline{B}$	A+B	$\overline{A.B}$
0	0	1	1	1	0	1
0	1	0	1	0	1	0
1	0	1	0	0	1	0
1	1	0	0	0	1	0

which is the same as $\overline{A}.\overline{B}$

3) Truth table.

B	A	F	
0	0	1	$F = \overline{A}.\overline{B}$
0	1	0	
1	0	0	
1	1	1	$F = A.B$

Boolean equation $F = \overline{A}.\overline{B} + A.B$

Gate diagram

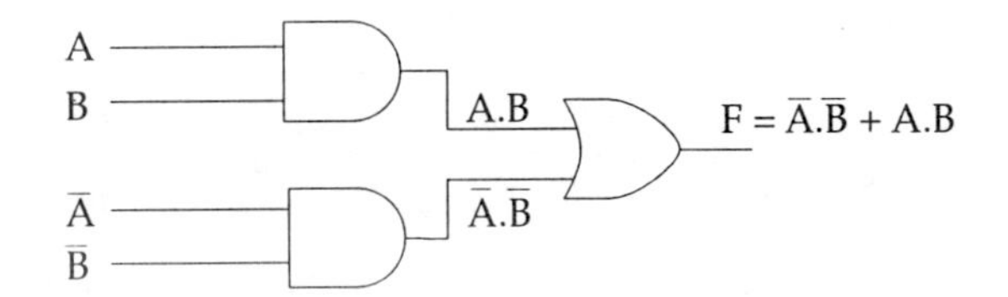

Note: This circuit is called an exclusive NOR and its logic function is the inverse of the exclusive OR.

Simplification of Boolean equations

SAQ 1

1) a) $F = A.A + A.B$ — mutiplying out.
$= A + A.B$
$= A$ — rule 13

b) $F = A.B + B$ — since B+B = B — rule 1
$= B$ — rule 13

c) $F = B.\overline{B} + B.C$ — multiplying out
$= B.C$ — since $B.\overline{B} = 0$ — rule 4

d) $F = A(B + \overline{B})$ — factorizing
$= A.1$ — since $B + \overline{B} = 1$ — rule 3
$= A$ — since A.1 = A — rule 7

e) $F = A + B$ — since 1+B+C = 1 — rule 6

f) $F = \overline{A}.A + \overline{A}.B + A.B$ — multiplying out
$= 0 + \overline{A}.B + A.B$ — since $\overline{A}.A = 0$ — rule 4
$= B(\overline{A} + A)$ — factorizing
$= B.1$ — since $\overline{A} + A = 1$ — rule 3
$= B$ — since B.1 = B — rule 7

SAQ 2

1)

2)

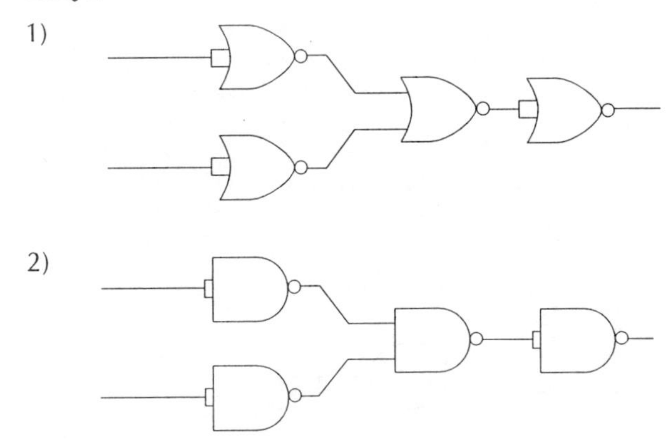

3) $F = A.\overline{B} + B.\overline{A}$

Pure Logic diagram.

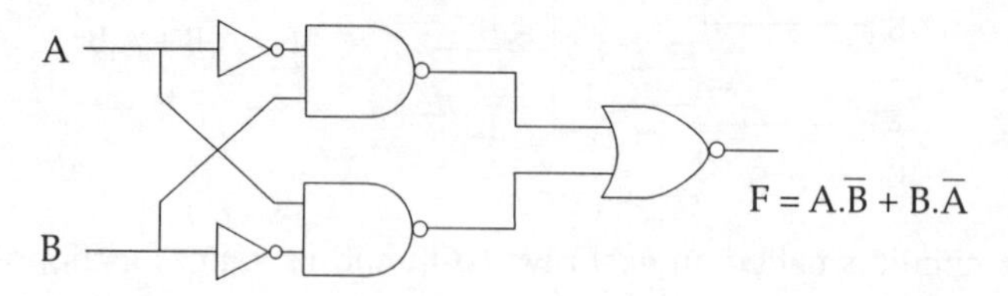

NAND gate diagram.

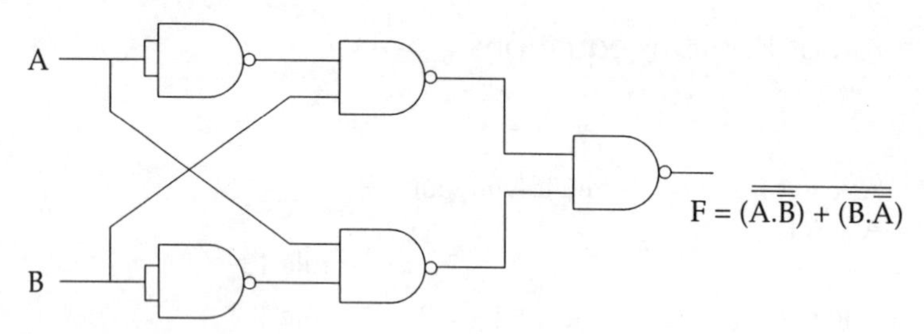

4) $F = \overline{A}.\overline{B} + B.A$

Pure Logic diagram.

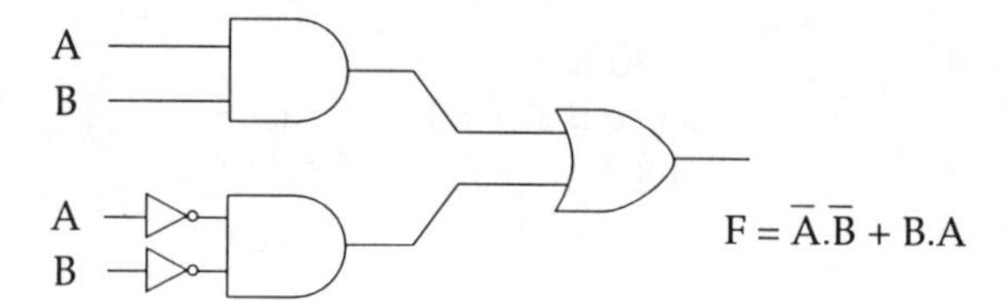

NOR gate diagram.

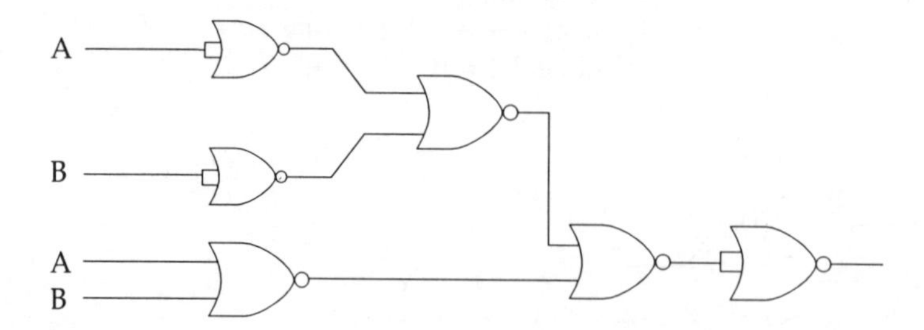

5) $F = (A+B).C$

Pure logic gate diagram.

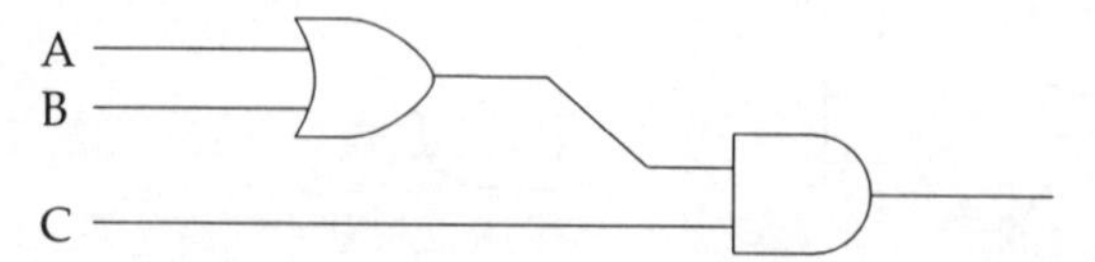

NOR Gate diagram.

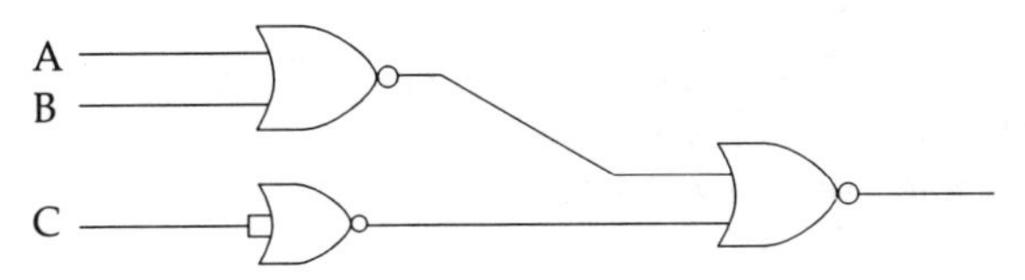

6) $F = \overline{A}.B.C + A.\overline{B}.C$

Pure logic.

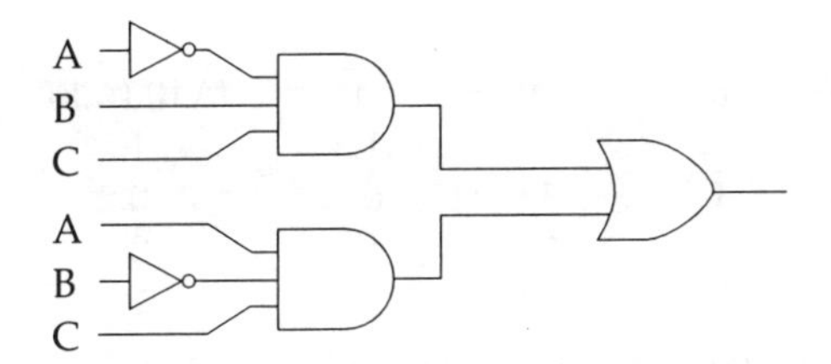

NOR gate diagram.

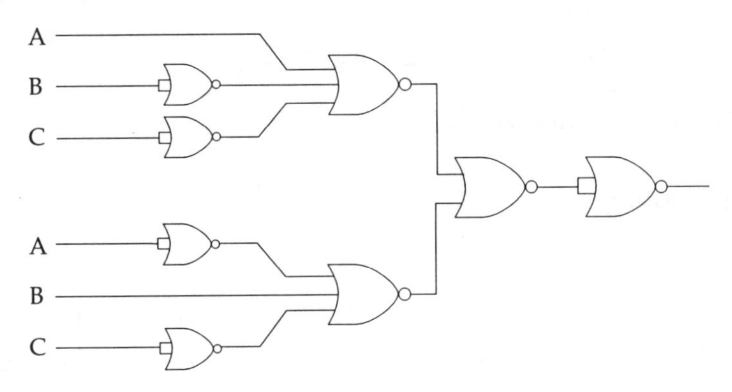

de-Morgan's Rules

SAQ 1

1) $F = \overline{A.B}$ NAND
2) $F = A.B$ AND
3) $F = \overline{A+B}$ NOR
4) $F = A+B$ OR

SAQ 2

1) a) $F = \overline{A.B + C}$ which can be written as $F = \overline{X + C}$
By de-Morgan's $F = \overline{X}.\overline{C}$
Also de-Morgan's on X gives $\overline{A} + \overline{B}$
Therefore $F = (\overline{A}+\overline{B}).C$ which equals $(\overline{A}+\overline{B}).\overline{C}$

b) $F = \overline{(A+B).(C+D)}$ which can be written as $F = \overline{X.Y}$

By de-Morgan's $F = \bar{X} + \bar{Y}$
Therefore $F = (\overline{A+B}) + (\overline{C+D})$

c) $F = \overline{A+\bar{B}+\bar{C}} + \overline{\bar{A}+\bar{C}+B}$ which can be written as $\bar{X} + \bar{Y}$
By de-Morgan's on X; $X = \bar{A}.B.C$
By de-Morgan's on Y; $Y = A.\bar{B}.C$
Therefore $F = \bar{A}.B.C + A.\bar{B}.C$

2) $F = \overline{\overline{\bar{A}.B} \;.\; \overline{A.\bar{B}}}$ which can be written as $F = \overline{\bar{X} + \bar{Y}}$
By de-Morgan's $F = X + Y$
$= \bar{A}.B + \bar{B}.A$

Truth table.

B	A	$\bar{A}$	$\bar{B}$	$\overline{\bar{A}.B}$	$\overline{A.\bar{B}}$	$(\overline{\bar{A}.B}).(\overline{A.\bar{B}})$	$\overline{(\overline{\bar{A}.B}).(\overline{A.\bar{B}})}$	
0	0	1	1	1	1	1	0	EX OR
0	1	0	1	1	0	0	1	unction
1	0	1	0	0	1	0	1	
1	1	0	0	1	1	1	0	

3) a) $F = \overline{\overline{A.B} + \overline{C.D}}$ Which can be written as $F = \overline{\bar{X} + \bar{Y}}$
By de-Morgan's $F = X.Y$
Therefore $F = A.B.C.D$

3) b) $F = \overline{\overline{A+B} \;.\; \overline{C+D}}$ which can be written as $F = \overline{\bar{X}.\bar{Y}}$
By de-Morgan's $F = X + Y$
Therefore $F = A+B+C+D$

Karnaugh Mapping

SAQ 1

a) $F = \bar{A}.C$ b) $F = A.\bar{C}$ c) $F = \bar{A}.\bar{B}.\bar{C}$ d) $F = \bar{B}.\bar{C}$
e) $F = B.C$ f) $F = B + A.\bar{C}$ g) $F = A.B + \bar{A}.\bar{B}$ h) $F = A.C + \bar{A}.\bar{C}$

SAQ 2

X		Y		F	
A	**B**	**C**	**D**		
0	0	0	0	0	
0	0	0	1	0	
0	0	1	0	0	
0	0	1	1	0	
0	1	0	0	1	$\bar{A}.B.\bar{C}.\bar{D}$
0	1	0	1	0	
0	1	1	0	0	
0	1	1	1	0	
1	0	0	0	1	$A.\bar{B}.\bar{C}.\bar{D}$
1	0	0	1	1	$A.\bar{B}.\bar{C}.D$
1	0	1	0	0	
1	0	1	1	0	
1	1	0	0	1	$A.B.\bar{C}.\bar{D}$
1	1	0	1	1	$A.B.\bar{C}.D$
1	1	1	0	1	$A.B.C.\bar{D}$
1	1	1	1	0	

Mapping.

AB \ CD	00	01	11	10
00	0	0	0	0
01	1	0	0	0
11	1	1	0	1
10	1	1	0	0

C = 1 (columns 11, 10); D = 1 (columns 01, 11); B = 1 (rows 01, 11); A = 1 (rows 11, 10).

Loops: $B\bar{C}\bar{D}$, $A\bar{C}$, $A.B.\bar{D}$

The Map can be split into three loops.

i) $A.\bar{C}$ ii) $B.\bar{C}.\bar{D}$ iii) $A.B.\bar{D}$

Hence the Boolean expression becomes:

$$F = A.\bar{C} + B.\bar{C}.\bar{D} + A.B.\bar{D}$$

To obtain the NAND equivalent form the complement:

$$\bar{F} = \overline{A.\bar{C} + B.\bar{C}.\bar{D} + A.B.\bar{D}}$$

$$= \overline{(A.\bar{C}).(B.\bar{C}.\bar{D}).(A.B.\bar{D})} \text{ by de-Morgan's}$$

Finally re-complementing gives:

$$F = \overline{\overline{(A.\bar{C})}.\overline{(B.\bar{C}.\bar{D})}.\overline{(A.B.\bar{D})}}$$

Which is now in the NAND form.

NAND Gate Diagram

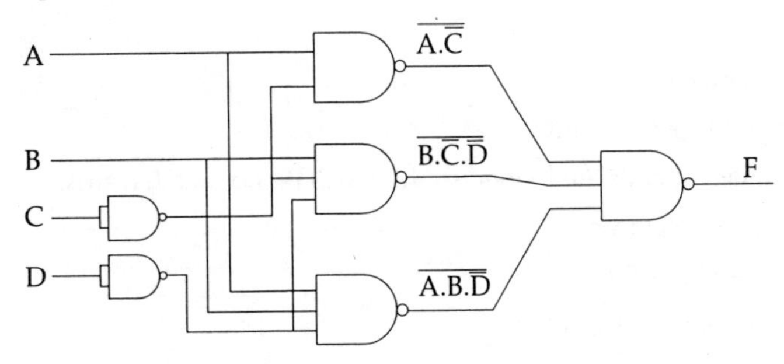

Self Assessment Exercise

Step 1. Construct the truth table

Denary	Binary inputs D	C	B	A	Output
0	0	0	0	0	0
1	0	0	0	1	0
2	0	0	1	0	1
3	0	0	1	1	0
4	0	1	0	0	0
5	0	1	0	1	1
6	0	1	1	0	1
7	0	1	1	1	1
8	1	0	0	0	0
9	1	0	0	1	0

Mapping gives:

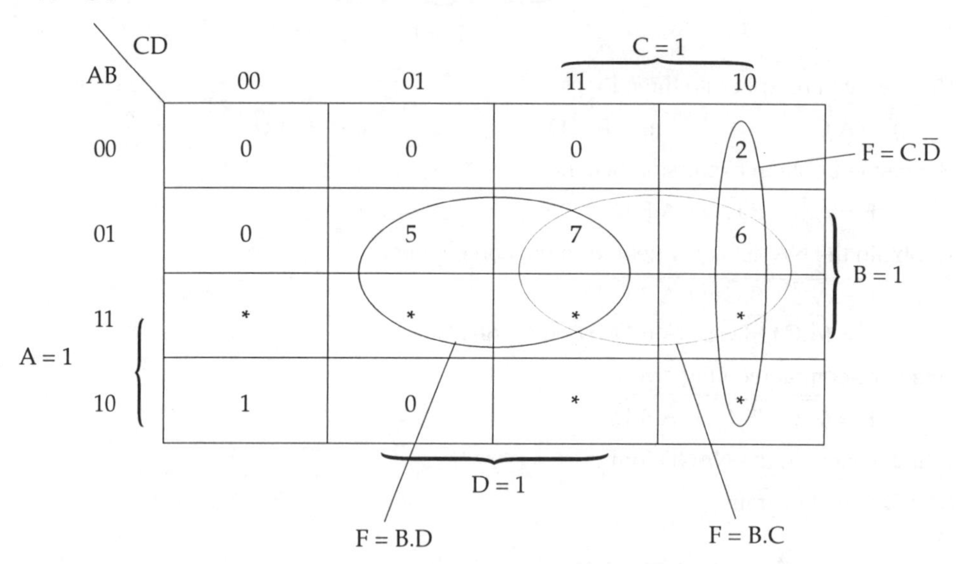

Note * = don't care terms.

From which $F = B.D + C.\overline{D} + B.C$

Note: All the adjacent cells are looped so there will be no race hazards.

Converting to all NAND form.

$$F = B.D + C.\overline{D} + B.C$$

Complementing both sides gives:

$$\overline{F} = \overline{B.D + C.\overline{D} + B.C}$$

$$= (\overline{B.D}).(\overline{C.\overline{D}}).(\overline{B.C}) \text{ by De-Morgan's}$$

Re-complementing gives:

$$F = \overline{(\overline{B.D}).(\overline{C.\overline{D}}).(\overline{B.C})}$$

Logic gate diagram.

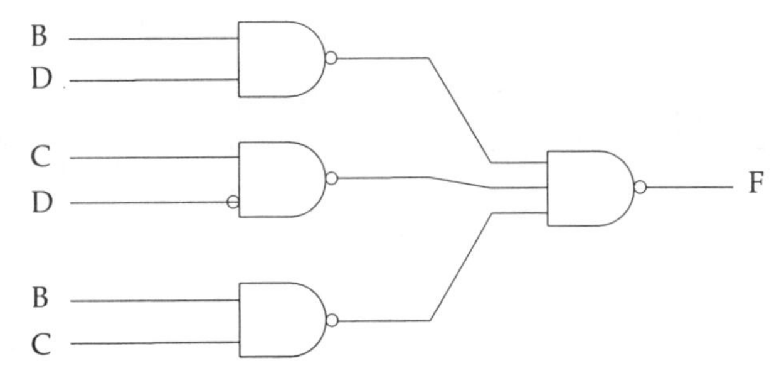

The bi-stable

Self Assessment Question

a) The current flowing in the lamp will be 3/12 amps = 250mA

The hfe of the transistor is 50 so if the transistor is to be saturated (switched fully ON) the base current will have to be greater than;

$$Ib = \frac{Ic}{hfe} = \frac{250}{50} = 5mA$$

The Vbe of the transistor will be about 0.7V therefore the voltage across any base resistor (R1,R2 or R3) if we assume a supply voltage of 12V will be 12V – 0.7V = 11.3V.

Therefore the value of R1, R2 and R3 will be:

$$\frac{11.3}{5} \text{ k ohms} = 2.26k$$

Nearest preferred value 2k2.

b) A Vce greater than 12V

Ic_{max} greater than 250mA

Pc_{max} greater than $Vce_{sat} \times Ic_{max}$ = 0.2V x 250mA = 50mW

If however the transistor, due to a fault, were only to be switched half ON then the power dissipation of the transistor would be: 6V x 125mA = 750mW

i.e. only half the current would be flowing but half the supply voltage would now be across TR1.

c) If any switch is closed then TR1 is ON (output LOW)

All switches open the TR1 is OFF (output HIGH)

C	B	A	output
0	0	0	1
0	0	1	0
0	1	0	0
0	1	1	0
1	0	0	0
1	0	1	0
1	1	0	0
1	1	1	0

The circuit behaves as a NOR gate.

R.S. & J.K. Bi-stables

1)

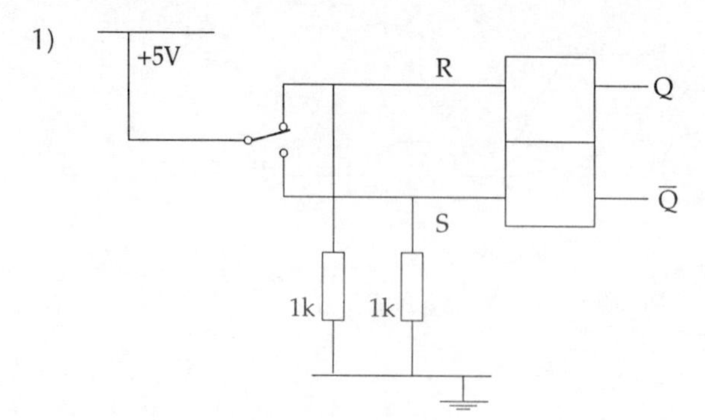

2)

Inputs		Output before C.P		Output after C.P	
S	R	Q	Q	Q	Q
0	0	0	1	0	1
0	0	1	0	1	0
0	1	0	1	0	1
0	1	1	0	0	1
1	0	0	1	1	0
1	0	1	0	1	0
1	1	0	1	?	?
1	1	1	0	?	?

? = indeterminate

Truth table for clocked R.S. bi-stable.

3) a) E to reset and then buttons A,B,C or D

b) Button F c) 15 d) Button E e) Parallel

4) a) A,B, and then C

b) When Q3 output is HIGH, current will flow via R1 and the diode into the base of the transistor. The transistor conducts and energises the relay.

When the output at Q3 is LOW, current will flow through the other diode to ground (Q3 output LOW) diverting current away from the base of the transistor. The transistor will switch OFF and the relay coil de-energises.

c) The SET or RESET inputs require an ACTIVE HIGH (logic 1) and this is the case with a cross coupled NOR bi-stable.

5)

Inputs		Output before C.P		Output after C.P	
S	R	Q	Q	Q	Q
0	0	0	1	0	1
0	0	1	0	1	0
0	1	0	1	0	1
0	1	1	0	0	1
1	0	0	1	1	0
1	0	1	0	1	0
1	1	0	1	1	0
1	1	1	0	0	1

Truth table for clocked J.K. bi-stable.

6) a) Whenever the CLEAR input is made active then the output Q will go to logic 0 overriding the J and K and clock pulse signals.

b) Whenever the PRESET input is made active then the output Q will go to logic 1 overriding the J and K and clock pulse signals.

c) Both J and K must be at logic 1 (+5V)

d)

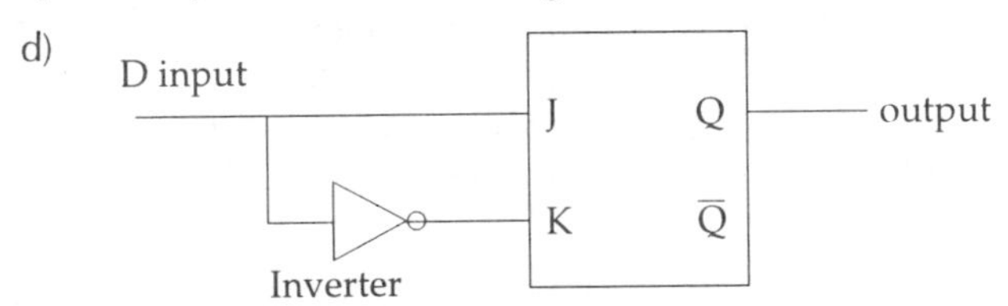

Binary counters

SAQ 1

1) A bi-stable that can toggle.

2) A ripple counter is a serial counter. The output of the first bi-stable toggles the second bi-stable and the second toggles the third and so on.

3) With a 3-stage counter the time taken for all the bi-stables to change state, when changing from the count of 3 to 4 for example, will be 3 times the propagation delay of one bi-stable. This limits the maximum frequency of operation.

4) Four. 3 bi-stables will only allow a maximum count of 7.

5) In combinational logic circuits the outputs do not depend upon previous input or output conditions, only upon the present inputs, that is, they have no memory.

   Sequential logic circuits have memory and their outputs can depend upon previous inputs or outputs in addition to the present inputs. They allow logical operations to be performed in sequence.

6) A modulo 7 counter has a maximum count of 6. Therefore when the count of 7 occurs the J.K bi-stables must be reset. i.e. the state of 7 is only transitory.

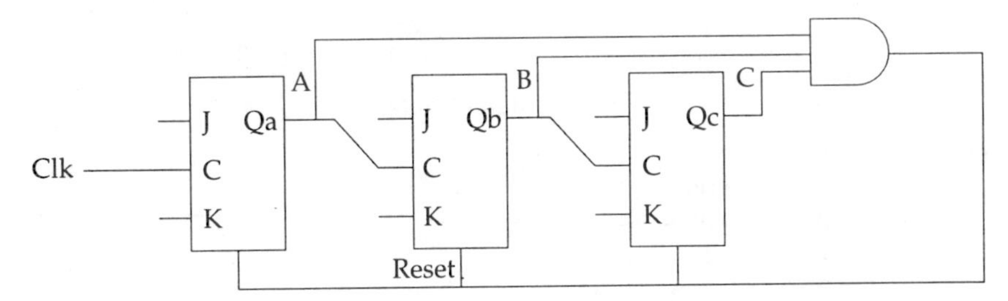

Modulo 7 counter

7) Assume that the D-type bi-stables are constructed from negative edge triggered J.K bi-stables, with an inverter between the J and K inputs.

   With 3 bi-stables and no feedback to modify the count the counter must be a modulo 8 (max count 7) counter.

The operation of a D type bi-stable is that; Output Q follows input D after a clock pulse.

The input D in each bi-stable is fed from the NOT Q output which is always be opposite to the Q output. The bi-stable will only change state when its clock input goes from a logic 1 to a logic 0 (negative edge).

Therefore on the first input pulse the NOT Q output of bi-stable A will clock bi-stable B. Bi-stable B changing will clock bi-stable C. Therefore after the first input pulse all the outputs will go to logic 1 (maximum count) Further input pulses will decrease the count. The counter is therefore a modulo 8 down counter.

Truth Table.

C	B	A	Input Pulse
0	0	0	0
1	1	1	1
1	1	0	2
1	0	1	3
1	0	0	4
0	1	1	5
0	1	0	6
0	0	1	7
0	0	0	8

Alternative. If the D-types are positive edge triggered then the operation is opposite to that above. i.e. the counter becomes a modulo 8 UP counter.

8) The circuit will have to divide by 5. A counter may be used with a modulo of 5 then for every 5 input pulses to bi-stable A there will be 1 complete pulse at the output of bi-stable C.

The counter circuit is the same as that used in question 6 except that it is modified to give a modulo of 5. i.e. it must reset immediately the count of 5 is detected.

SAQ 2 *It is assumed that negative edge triggered J.K bi-stables are employed.*

1) Modulo 3 UP counter.

B	A	Input	
0	0	0	
0	1	1	
1	0	2	
0	0	3	only 3 different states

2) Modulo 4 UP counter

B	A	Input	
0	0	0	
0	1	1	
1	0	2	
1	1	3	4 different states
0	0	4	

3) Modulo 5 UP counter

C	B	A	Input	
0	0	0	0	
0	0	1	1	
0	1	0	2	
0	1	1	3	
1	0	0	4	
1	0	1	0	5 different states

Shift Registers

1) P

2) N

3) Input to M, Parallel outputs from B,D,F,H

4) Input to M; output from H

5) Inputs to A,C,E,G: outputs from B,D,F,H

6) Inputs to A,C,E,G: output from H

7) Apply the data word 1011 to input M , apply four clock pulses to input P to shift the data in.

8) Apply an input to N to clear the register. Activate the preset inputs A,E and G. (the right hand bi-stable will contain the lsb). Now apply four clock pulses to shift the data word out serially from terminal H.

9) This inverter converts the first bi-stable into a D-type. The output logic level at Q will always follow the logic level at the D input after a clock pulse.

10) Because of the connections from Q and NOT Q the remaining bi-stables also behave as D-types because the J and K inputs will always be complementary.

Index

Mathematics for Engineering

An Active-Learning Approach

D Clarke

The book is up-to-date and relates to current engineering practice. Mathematical techniques make full use of readily available information technology. It has been written specifically for engineering students by an engineer, and the activity-based approach gives a realistic context in which to follow the problem-solving routine of an engineer.

It provides a full course of study in mathematics for students following BTEC National Engineering and covers all the NII and NIII objectives.

The book is known to be used on the following courses: BTEC National Engineering, Access to Maths and Engineering, HND first year Engineering.

Review comments

'Excellent text for self-directed learning of mathematics.'

'Good clear examples and layout, very readable.' Lecturers

1st edition • 304 pp • 245 x 176 mm • 1993
ISBN 1 85805 043 X